PROGRESS IN

Molecular Biology and Translational Science

Volume 101

PROGRESS IN

Molecular Biology and Translational Science

Modifications of Nuclear DNA and its Regulatory Proteins

edited by

Xiaodong Cheng

Department of Biochemistry Emory
University School of Medicine, Atlanta, Georgia, USA

Robert M. Blumenthal

Department of Medical Microbiology & Immunology
and Program in Bioinformatics & Proteomics/Genomics
University of Toledo Health Science Campus, Toledo, Ohio, USA

Volume 101

ELSEVIER

AMSTERDAM • BOSTON • HEIDELBERG • LONDON
NEW YORK • OXFORD • PARIS • SAN DIEGO
SAN FRANCISCO • SINGAPORE • SYDNEY • TOKYO
Academic Press is an imprint of Elsevier

Academic Press is an imprint of Elsevier
32 Jamestown Road, London, NW1 7BY, UK
Radarweg 29, PO Box 211, 1000 AE Amsterdam, The Netherlands
225 Wyman Street, Waltham, MA 02451, USA
525 B Street, Suite 1900, San Diego, CA 92101-4495, USA

This book is printed on acid-free paper. ♾

Library of Congress Cataloging-in-Publication Data
A catalog record for this book is available from the Library of Congress

British Library Cataloguing in Publication Data
A catalogue record for this book is available from the British Library

ISBN: 978-0-12-387685-0
ISSN: 1877-1173

For information on all Academic Press publications
visit our website at elsevierdirect.com

Printed and bound by CPI Group (UK) Ltd, Croydon, CR0 4YY

Transferred to Digital Print 2011

Contents

Section III DNA Methylation and Demethylation. 287

DNA Methylation Changes in Cancer 447

Minoru Toyota and Eiichiro Yamamoto

Genome-Wide Distribution of DNA Methylation at Single-Nucleotide Resolution 459

Eleanor Wong and Chia-Lin Wei

Contributors

Numbers in parentheses indicate the pages on which the authors' contributions begin.

Saraswathi Abhiman, National Center for Biotechnology Information, National Library of Medicine, National Institutes of Health, Bethesda, Maryland, USA (25, 105)

L. Aravind, National Center for Biotechnology Information, National Library of Medicine, National Institutes of Health, Bethesda, Maryland, USA (25, 105)

Robert M. Blumenthal, Department of Medical Microbiology & Immunology, and Program in Bioinformatics & Proteomics/Genomics, University of Toledo Health Science Campus, Toledo, Ohio, USA (1)

Frédéric Chédin, Department of Molecular and Cellular Biology, University of California, Davis, California, USA (255)

Taiping Chen, Developmental and Molecular Pathways, Novartis Institutes for Biomedical Research, Cambridge, Massachusetts, USA (335)

Xiaodong Cheng, Department of Biochemistry, Emory University School of Medicine, Atlanta, Georgia, USA (1)

Pierre-Antoine Defossez, CNRS UMR7216, Université Paris 7, Paris Cedex 13, France (377)

Mary G. Goll, Developmental Biology Program, Sloan-Kettering Institute, New York, USA (193)

Marnie E. Halpern, Carnegie Institution for Science, Department of Embryology, Baltimore, Maryland, USA (193)

Lakshminarayan M. Iyer, National Center for Biotechnology Information, National Library of Medicine, National Institutes of Health, Bethesda, Maryland, USA (25, 105)

Shannon R. Morey Kinney, New England Biolabs, Ipswich, Massachusetts, USA (311)

Haruhiko Koseki, RIKEN Research Center for Allergy & Immunology (RCAI), Tsurumi-ku, Yokohama, Kanagawa, Japan (289)

Veiko Krauss, Developmental Genetics, Institute of Biology, Martin Luther University Halle-Wittenberg, Weinbergweg, Halle/S, Germany (177)

Sriharsa Pradhan, New England Biolabs, Ipswich, Massachusetts, USA (311)

Gunter Reuter, Developmental Genetics, Institute of Biology, Martin Luther University Halle-Wittenberg, Weinbergweg, Halle/S, Germany (177)

Jafar Sharif, RIKEN Research Center for Allergy & Immunology (RCAI), Tsurumi-ku, Yokohama, Kanagawa, Japan (289)

Irina Stancheva, Wellcome Trust Centre for Cell Biology, University of Edinburgh, Edinburgh, United Kingdom (377)

Željko M. Svedružić, Medical Biochemistry, PB Rab, Faculty of Medicine, University of Rijeka, Rab, Croatia (221)

Minoru Toyota, Department of Biochemistry, Sapporo Medical University, Chuo-ku, Sapporo, Japan (447)

Francisco Úbeda, Department of Ecology and Evolutionary Biology, University of Tennessee, Knoxville, Tennessee, USA (401)

Chia-Lin Wei, Genome Technology and Biology, Genome Institute of Singapore, and Department of Biological Sciences, National University of Singapore, Singapore; and Department of Genome Technology, Joint Genome Institute, California, USA (459)

Jon F. Wilkins, Santa Fe Institute, Santa Fe, New Mexico, USA (401)

Eleanor Wong, Genome Technology and Biology, Genome Institute of Singapore; and Department of Biological Sciences, National University of Singapore, Singapore (459)

Anton Wutz, Wellcome Trust Centre for Stem Cell Research, University of Cambridge, Cambridge, United Kingdom (351)

Eiichiro Yamamoto, Department of Biochemistry, Sapporo Medical University, Chuo-ku, Sapporo, Japan (447)

Preface

It hardly seems possible that we are approaching two decades since the discovery that DNA methyltransferases act by everting the target base out of the double helix Klimasauskas et al., HhaI methyltransferase flips its target base out of the DNA helix. Cell 1994 76:357–69. Since 1994, a very substantial body of work has emerged from the attempts to understand both the molecular mechanisms and biological roles of methylation, not only of DNA but also of the associated histone proteins.

As is typical of important biological questions, there is no good stopping point at which we can say that the basic answers are known. Nevertheless, we feel the time is right for a summing up that allows establishment of context. For that reason, we are delighted that colleagues who study chromatin methylation from a variety of perspectives have agreed to contribute to this volume.

This book is divided into five sections, following an introduction by us that is intended to consolidate the most basic background material into one chapter. The next five paragraphs should have appropriate section numbers.

Section I explores the evolution of the DNA- and histone-methylating (and demethylating) enzymes and of the biological roles of this methylation. The former is presented in two chapters by Iyer et al. and Aravind et al., while the latter involves chapters on chromatin methylation in *Drosophila* (Krauss and Reuter) and zebrafish (Goll and Halpern).

Section II is focused on the major mammalian DNA methyltransferases, Dnmt1 (Svedruzic) and Dnmt3/3L (Chédin). One of the take-home messages is that the division of labor between *de novo* and maintenance methylation is much less strict than had been thought earlier.

Section III examines the processes and control of DNA methylation and demethylation. There are chapters on recruitment of Dnmt1 (Sharif and Koseki) and modulation of its stability (Morey-Kinney and Pradhan), along with the functional linkages between demethylation of DNA and histones (Chen).

Section IV covers the association between DNA methylation and silencing of gene expression. This can occur via an RNA-directed process (Wutz) or the involvement of proteins bearing a methyl-binding domain (Defossez and Stancheva).

Finally, Section V describes biomedical and biotechnological aspects of these epigenetic processes. There are chapters on chromatin methylation and cancer (Toyota and Yamamoto) and diseases associated with imprinting

(Wilkins and Úbeda). The section, and the volume, is brought to a close by a description of the methods to determine, and the implications of knowing, the distribution of DNA methylation at the single-nucleotide level (Wong and Wei).

We are happy that these chapters will be available online in addition to the print format, so it will be possible to update them periodically.

We would like to thank those who made this volume possible. First, of course, the authors of the various chapters, from whom we have learned so much. Second, Delsy Retchagar and the team at Elsevier, for laboring mightily to keep us on schedule. Third, our respective institutions and departments for encouragement and staff support (especially Suzanne Payne of the Department of Medical Microbiology and Immunology at the University of Toledo). Fourth, grant support that allows our ongoing research in this field—the U.S. National Institutes of Health (Grants GM-049245-17 and GM-068680-06 to X. C.) and the U.S. National Science Foundation (Grant MCB-0964728 to R. M. B.). Last, but by no means least, we thank our families and look forward to being able to spend a bit more time with them.

February, 2011

Xiaodong Cheng and Robert M. Blumenthal

Introduction—Epiphanies in Epigenetics

Xiaodong Cheng* and Robert M. Blumenthal[†,‡]

*Department of Biochemistry, Emory University School of Medicine, Atlanta, Georgia, USA

[†]Department of Medical Microbiology & Immunology, University of Toledo Health Science Campus, Toledo, Ohio, USA

[‡]Program in Bioinformatics & Proteomics/Genomics, University of Toledo Health Science Campus, Toledo, Ohio, USA

The combinatorial pattern of DNA and histone modifications and their associated histone variants constitute an epigenetic code that shapes gene expression patterns by increasing or decreasing the transcriptional potential of genomic domains. The epigenetic coding status, at any given chromosomal location, is subject to modulation by noncoding RNAs and remodeling complexes. DNA methylation is associated with histone modifications, particularly the absence of histone H3 lysine 4 methylation (H3K4me0) and the presence of histone H3 lysine 9 methylation (H3K9m). We briefly discuss four protein domains (ADD, CXXC, MBD, and SRA), and the functional implications of their architecture in linking histone methylation to that of DNA in mammalian cells. We also consider the domain structure of the DNA methyltransferase DNMT1, its accessory protein UHRF1, and their associated proteins. Finally,

Progress in Molecular Biology
and Translational Science, Vol. 101
DOI: 10.1016/B978-0-12-387685-0.00001-9

1877-1173/11 $35.00

we discuss a mechanism by which methylation of DNA and of histones may be coordinately maintained during mitotic cell division, allowing for the transmission of parental methylation patterns to newly replicated chromatin.

I. What is an Epigenetic Code?

All cells face the problem of controlling the amounts and timing of expression of their various genes. In some cases, this control involves nonpermanent but relatively long-term and heritable modifications to the chromatin. Such modifications, that do not change the DNA sequence, are referred to as "epigenetic." The resulting epigenetic effects maintain the various patterns of gene expression in different cell types.[1] Epigenetic modifications include DNA methylation and posttranslational histone modifications. Nucleosomes consist of approximately 146 bp of DNA wrapped approximately 1.8 times around a histone octamer that is evolutionarily conserved.[2] The combinatorial pattern of DNA and histone modifications constitutes an epigenetic code that shapes transcriptional patterns (Fig. 1A). The coding modification is "written" by sequence- and site-specific enzymes, and "interpreted" by effector molecules that mediate the assembly of higher-order chromatin structures involving remodeling complexes, histone variants, and noncoding RNAs (ncRNAs).[12]

A. DNA Methylation

In mammals and other vertebrates, DNA methylation occurs at the C5 position of cytosine, resulting in 5-methylcytosine (5mC), mostly within CpG dinucleotides (Fig. 1B). The Dnmt enzymes use a conserved mechanism[13] that has been studied best in the bacterial 5mC MTase M.HhaI.[14–18] Briefly, this mechanism involves MTase binding to the DNA, eversion of the target nucleotide so that it projects out of the double helix (base flipping), covalent attack of a conserved Cys nucleophile on cytosine C6, transfer of the methyl group from S-adenosyl-L-methionine (AdoMet) to the activated cytosine C5, and the various release steps.

B. DNA Hydroxymethylation

5-Hydroxymethylcytosine (5hmC) has long been noted in bacterial phage DNA.[19–23] Its presence in mammalian cells[24] was originally believed to be a byproduct of oxidative DNA damage.[25] Recently, Kriaucionis and Heintz found that significant fractions (~40%) of cytosine nucleotides correspond to 5hmC, the amount of which inversely correlates with 5mC and nuclear heterochromatin, in isolated and relatively homogenous populations of Purkinje and granule neuronal nuclei of adult mouse brains.[26] Even more fascinating, a conserved mammalian-specific family of TET (ten–eleven translocation) proteins was

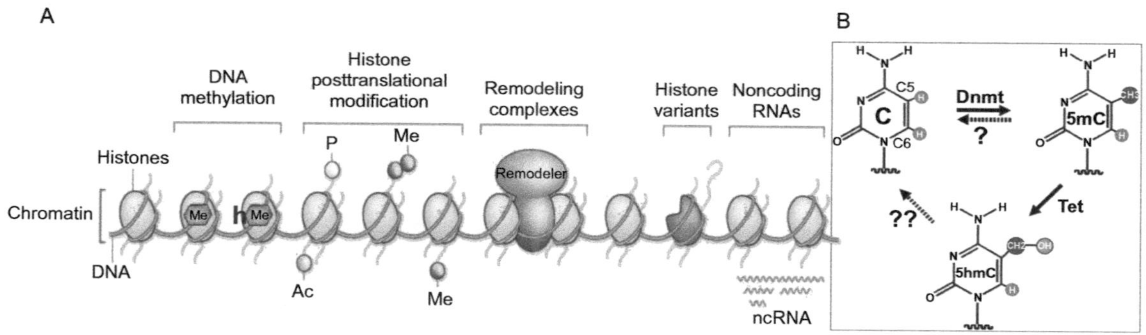

FIG. 1. Possible components of an epigenetic "code." (A) Five broad and interrelated phenomena affect chromatin structure: DNA methylation, histone modification, chromatin-remodeling, histone replacement by variants, and the effects of noncoding RNAs. All five have been shown to be essential contributors to the epigenetics, though DNA methylation and histone modifications have, so far, been much more extensively investigated. Adopted and modified from Ref. 3. (B) DNA cytosine methylation, hydroxylation, and demethylation. The question mark indicates possible activity of DNA demethylases.[4–9] "Tet" indicates conversion of 5mC to 5hmC in mammalian DNA, by the MLL fusion partner TET1.[10] It is currently unknown whether 5hmC is an end product or an intermediate in active DNA demethylation. The double question mark indicates a possible MTase-assisted removal of the C5-bound hydroxymethyl group.[11]

identified that converts 5mC to 5hmC (Fig. 1B).[10] One of these proteins, TET1, is fused to the MLL protein in a subset of acute myeloid leukemia patients. Overproduction of TET1 in human cells led to the appearance of 5hmC. A concomitant reduction in DNA 5mC suggests that 5hmC results from enzymatic oxidation of 5mC. 5hmC was also detected in ES cells, and its amount decreased with RNAi knockdown of TET1. The surprising finding of a 5mC oxidation pathway raises numerous questions, such as whether oxidation of 5mC is an important epigenetic modification. 5hmC could be an end product, or an intermediate in active DNA demethylation, as supported by the existence of a 5hmC DNA excision repair glycosylase.[27] It is intriguing that the bacterial 5mC MTase M.HhaI can promote the reverse reaction *in vitro*—the removal of formaldehyde from 5hmC to yield the unmodified cytosine[11] (Fig. 1C). New lines of research will likely be catalyzed by the presence of 5hmC in mammalian DNA.

C. Histone Posttranslational Modifications

Histones are subject to a multitude of posttranslational modifications, including acetylation, methylation, ubiquitination and sumoylation of lysines, and methylation of arginines.[28] There is extensive crosstalk among histone modifications.[29] As a result, different histone markings often have nonadditive effects. Structural and biochemical characterization of more of the players in histone modification are beginning to clarify the underlying mechanisms of crosstalk, and the interplay with other pathways, such as chromatin remodeling and DNA methylation and repair.[30–33] Mechanistically, crosstalk occurs when one or more binding modules and catalytic domains reside in the same complex or polypeptide, allowing coordination of different activities. Crosstalk can occur prior to catalysis, in which case the recognition of one mark (or its absence) can serve to recruit an enzyme to its substrate in the generation or removal of a second mark. For example, phosphorylation of H3 serine 10 by Ipl1/aurora kinase interferes with methylation of H3 lysine 9 by SUV39H1, whereas acetylation of lysine 9 or lysine 14 enhances serine 10 phosphorylation by Ipl1/aurora kinase.[34] Ubiquitination of lysine 123 of histone H2B by Rad6 regulates methylation of H3 lysines 4[35] and 79 in *Saccharomyces cerevisiae*.[36,37] PHF8 and KIAA1718 harbor two domains in their respective N-terminal halves: a PHD domain that binds trimethylated histone H3 lysine 4 (H3K4me3; a modification associated with transcriptional activation), and their linked Jumonji domains that remove methyl marks associated with transcriptional repression (di- and monomethylated histone H3 lysine 9, H3K9me2/1) via PHF8,[38–41] H3K27me2/1 via KIAA1718,[38,42,43] or H4K20me1 via PHF8.[44,45]

D. Remodeling Complexes

In eukarya, the packaging of DNA into chromatin provides a barrier that limits access to the genome. Condensed chromatin is refractory to processes acting on DNA, including transcription, replication, and repair.[46,47] ATP-dependent chromatin-remodeling machinery can overcome this barrier—or strengthen it in silencing processes—by sliding or displacing nucleosomes and by altering their histone content.[48]

The modifications that constitute the histone code show significant interplay with the ATP-dependent chromatin-remodeling machinery. The PHD finger of a component of the nucleosome remodeling factor (NURF) complex binds H3K4me3.[49,50] INO80 and SWR1 chromatin-remodeling complexes are recruited through H2A phosphoryation at serine129,[51] and SWI/SNF associates with acetylated histones via bromodomains.[52–54] It is unclear to what extent histone code modifications and their binding modules serve as adaptors for the chromatin-remodeling machinery, and to what extent they regulate chromatin structure directly.

E. Histone Variants

The H3 variants (H3.1, 3.2, and 3.3) differ at just five positions.[55,56] Importantly, the N-terminal 31 residues are identical, so there is no difference in the immediate contexts of K4 and K9; but other residues in the core histones affect H3 methylation—at least of K4.[57] Even before incorporation into nucleosomes, some methylation at H3K9 has been reported[58]; and this methylation is substantially more abundant on H3.1 than on H3.3 (which may only play a role in gametogenesis[59]).

F. Noncoding RNAs (Including lincRNAs Like HOTAIR)

ncRNAs are transcribed from DNA but are not translated into proteins. Many are functional, and are involved in the processing and regulation of other RNAs. Small ncRNAs, such as microRNAs (miRNAs) and short-interfering RNAs (siRNAs that cause RNA interference RNAi), are involved in the regulation of target mRNAs and chromatin. The ncRNAs can increase or decrease transcription, inhibit translation, or guide methylation.[60–63] Longer ncRNAs (typically > 200 nt) have also been implicated in gene regulation[64,65] and have roles in epigenetic processes.[66] Some of these roles have been known for some time (e.g., X-chromosome inactivation[67–70] and gene imprinting[71–74]). Further, long ncRNAs are components of histone-modifying complexes. The male-specific lethal (MSL) complex, responsible for dosage compensation in *Drosophila*, is an example of a class of complexes with five protein subunits (including a histone acetyltransferase, MOF) and one of the two ncRNAs, RNA on the X1 and 2 (rox1 and rox2).[75] HOTAIR, a long ncRNA, serves as a scaffold for at least two distinct histone modification complexes.[76] A 5′ domain of HOTAIR binds polycomb repressive complex 2 (PRC2) for histone H3 lysine 27 methylation, whereas a 3′ domain of HOTAIR binds the LSD1/CoREST/REST complex for coupled lysine 4 demethylation.

II. Inheritance

It is well accepted that DNA methylation patterns are replicated in a semiconservative fashion during cell division, via mechanisms discussed below (and in subsequent chapters). However, one of the fundamental unresolved questions is how, and indeed whether, histone modifications are similarly "inherited." Recent work has shed light on this question. Through a combination of kinetic isotope labeling and mass spectroscopy studies, Xu et al. showed that histone H3.1–H4 tetramers remain intact and are partitioned evenly during DNA replication-dependent chromatin assembly.[77] Considering that the well-studied lysine methylation events reside on histone H3 (K4, K9, K27, K36, K79) or H4 (K20), this result suggests a model in which histone methylation patterns may be copied onto newly deposited tetramers from neighboring parental nucleosomes. Indeed, many of the SET domain histone methyltransferases contain intrinsic or associated "reader" domains that recognize the same mark that they generate, theoretically allowing the copying of these marks from old to new nucleosomes. For example, G9a/GLP catalyzes formation of H3K9me1/2 and contains an ankyrin repeat domain that binds H3K9me1/2.[78] Likewise, SUV39H1/2, the H3K9me3 writer interacts with HP1, the H3K9me3 reader.[79] Similarly, yeast Clr4 methylates H3K9 and contains a chromodomain that binds H3K9me3.[80] These observations suggest coordinated mechanisms of deposition and replicative transmission of repressive chromatin marks.

The identification of UHRF1 and its potential role in modulating the specificity of Dnmt1 for hemimethylated CpG sites (see Chapter by Jafar Sharif and Haruhiko Koseki) adds another layer to the mechanisms ensuring the faithful transmission of epigenetic information during DNA replication. Hemimethylated CpG sites, where only one DNA strand is methylated, are transiently generated during semiconservative DNA replication. Methylation patterns are faithfully preserved with somatic cell division by a "maintenance" methylation reaction. UHRF1 has the potential to interact with both hemimethylated CpGs (via the SRA domain) and H3K9me2/3 (via the Tudor domain), and is known to interact with a wide variety of epigenetic regulators including Dnmt1,[81,82] the H3K9 di- and monomethyltransferase G9a,[83] and a histone acetyltransferase Tip60.[84] Thus it is possible that UHRF1 and the proteins in this larger complex play a more central role in coupling the transmission of DNA and histone (H3K9 in particular) methylation during mitotic cell division.

III. Modularity of Epigenetic Modifiers

One broad emerging theme is that a web of interactions tightly coordinates the modification of a segment of DNA and its associated histones, particularly histone H3. This section focuses on four protein domains (ADD, CXXC, MBD, and SRA) in characterizing the functional links between histone and DNA modification in mammalian cells. In particular, we consider here the relationship between DNA CpG methylation and histone H3 methylation on lysines 4 and 9.[85–91] DNA methylation and histone lysine methylation are intimately connected with one another.[85,87–89] In fact, genome-scale DNA methylation profiles suggest that DNA methylation is correlated with histone methylation patterns.[86] Specifically, DNA methylation is associated with the absence of H3K4 methylation (H3K4me0) and the presence of H3K9 methylation; by comparison, it has little correlation with methylation of H3K27.[90] *In vivo* studies support a molecular link between the mechanisms that maintain DNA methylation and H3K9 methylation. Studies in *Neuropora* and *Arabidopsis* have shown a strict dependence of DNA methylation on the H3K9 methyltransferases Dim-5 and KRYPTONITE.[92–94] Functional linkage of the mechanisms defining genomic patterns of DNA methylation, H3K4 methylation, and H3K9 methylation is underscored by the finding that treatment with a DNA methyltransferase inhibitor (5-aza-2′-deoxycytidine or 5azadC) leads to depletion of DNA methylation, with concomitant loss of H3K9 methylation and increase in H3K4 methylation.[95]

A. ADD Domains of Dnmt3L/Dnmt3a/Dnmt3b Link DNA Methylation to Unmethylated H3K4

DNA methylation in mammals is coordinately established and maintained by two DNA methyltransferase families, the so-called *de novo* enzymes of the Dnmt3 family and the "maintenance" methyltransferase Dnmt1. The Dnmt3 family includes two active *de novo* Dnmts, Dnmt3a and Dnmt3b, and one regulatory factor, Dnmt3-like protein (Dnmt3L; Fig. 2A). Dnmt3a and Dnmt3b have similar domain arrangements: both contain a variable region at the N-terminus, followed by a PWWP domain, an ADD (ATRX-Dnmt3-Dnmt3L) domain, and a C-terminal catalytic domain. The amino acid sequence of the Dnmt3L ADD domain is very similar to those of Dnmt3a and Dnmt3b (Fig. 2A). Dnmt3L associates *in vivo* not only with Dnmt3b and Dnmt3a2 (a shorter isoform of Dnmt3a[99]) but also with the four core histones.[100] Peptide interaction assays and cocrystallization of Dnmt3L with the amino tail of H3 showed that Dnmt3L specifically interacts with the amino terminus of histone H3, only when H3K4 is not modified (H3K4me0).[100] These

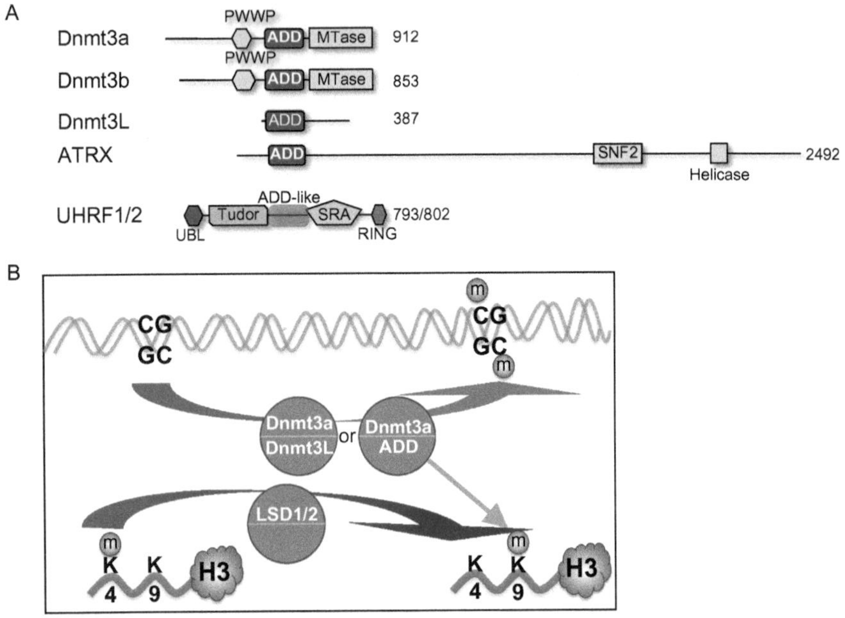

Fig. 2. H3K4me0 interacting proteins. (A) Domain architecture of H3K4me0-interacting proteins containing ADD domains. ATRX interacts with histone H3.3,[96,97] and mutating the K4 residue of H3.3 significantly diminishes ATRX–H3.3 interaction.[98] (B) Model of the reactions that regulate DNA methylation by Dnmt3a/3L. Recognition of H3K4me0 by the ADD domain of DNMT3 directs the DNA methylation reaction. The "m" in a red circle indicates one or more methyl groups in DNA (5mC) or histone lysines (Km). (For interpretation of the references to color in this figure legend, the reader is referred to the Web version of this chapter.)

observations suggest that Dnmt3L acts as a sensor for H3K4 methylation, such that H3K4m0 is bound by Dnmt3L, that consequently promotes *de novo* DNA methylation by docking Dnmt3a to the nucleosome (Fig. 2B).

Histone–Dnmt3L–Dnmt3a–DNA interactions have recently been studied in the budding yeast *S. cerevisiae*,[101] which has no detectible native DNA methylation[102] and lacks Dnmt orthologs. Introduction of the murine methyltransferase Dnmt1 or Dnmt3a into yeast leads to detectible, but extremely low levels of DNA methylation.[103] In contrast, a substantially higher level of *de novo* methylation could be achieved in yeast by coexpressing murine Dnmt3a and Dnmt3L.[101] This induced DNA methylation was found preferentially in heterochromatic regions where H3K4 methylation is rare. When genes for components of the H3K4-methylating complex COMPASS/Set1 were disrupted in the context of Dnmt3a/Dnmt3L overexpression, a greater level of genomic DNA methylation was observed. Deletions or targeted mutations in the ADD of Dnmt3L inhibited both global DNA methylation and the ability of Dnmt3L to associate with an H3K4me0 peptide. These same Dnmt3L mutants failed to restore normal DNA methylation to a specific promoter when introduced into ES cells from Dnmt3L$^{-/-}$ mice.[101] That H3K4 methylation status plays an important role in establishing DNA methylation is further supported by studies with mammalian LSD1 and LSD2, two related lysine-specific demethylases whose substrates include mono- and dimethylated H3K4 (H3K4me1/2; Fig. 2B). LSD1 and LSD2 are essential for maintaining global DNA methylation,[104] or establishing maternal DNA genomic imprints,[105] respectively. Indeed, disruption of *LSD1* results in earlier embryonic lethality and a more severe hypomethylation defect than disruption of the *Dnmt* genes themselves.[104]

The above data suggest that perhaps Dnmt3L binds to H3K4me0 (via its ADD domain) and recruits Dnmt3a to regions of chromatin where H3K4 is unmethylated. Such a model could explain part of the puzzle of how DNA methylation patterns are established *de novo* during embryonic and germ cell development, when both Dnmt3a and Dnmt3L are expressed.[106] However, while Dnmt3a and Dnmt3b expression occurs in somatic cells, Dnmt3L is expressed poorly if at all in differentiated cells. This raises the questions of how *de novo* DNA methylation is restricted in somatic cells, whether Dnmt3a and Dnmt3b alone are capable of discriminating H3K4 methylation status, and (if so) the structural basis for that discrimination. To this end, recent work by Jeong *et al.* showed that in nuclei from HCT116 human colon cancer cells (which do not express DNMT3L), almost all of the cellular DNMT3a and DNMT3b (but not DNMT1) were associated with nucleosomes.[107] Chromatin binding of DNMT3a and DNMT3b required an intact nucleosomal structure, though no other chromatin factors (such as the HP1 or MBD proteins), suggesting that DNMT3a and DNMT3b themselves can interact directly with chromatin components in addition to DNA. Further, recent *in vitro* studies indicate that the ADD domains of Dnmt3a or Dnmt3b possess the same H3 tail binding specificity as that of Dnmt3L.[108] Indeed, a structure of the Dnmt3a ADD domain in complex with an amino-terminal tail peptide from histone H3 indicates that the ADD domain is sufficient to recognize H3K4me0.[109]

B. The CXXC Domain Links H3K4 Methylation to Unmethylated CpGs

In humans, there are at least eight histone lysine methyltransferases that act on H3K4. These include products of the mixed lineage leukemia (*MLL*) genes, MLL1 through 5, hSET1a, hSET1b, and ASH1. MLL1/SET1-associated methyltransferase activity appears to be functional only in the context of multiprotein complexes; characterization of these reveals distinct multiprotein complexes for each of the eight, though with several shared components.[110] The MLL family plays an important role in embryonic development and is necessary for methylation of H3K4 at a subset of genes in the human and mouse genomes, particularly the HOX gene clusters.[111] Translocations involving *MLL* genes are involved in the etiology of myeloid and lymphoid leukemias. Considering the aforementioned inverse relationship between H3K4 methylation and DNA methylation, it is interesting that disruption of the *MLL1* gene in mice results in loss of both H3K4 methylation and *de novo* DNA methylation at some Hox gene promoters,[112,113] suggesting that MLLs directly or indirectly (through H3K4 methylation) prevent DNA methylation or perhaps stabilize unmethylated DNA. In fact, MLL proteins contain CXXC, an evolutionarily conserved domain that mediates selective binding to unmethylated CpGs (Fig. 3A).[114–116] This interaction has now been confirmed by a solution structure of an MLL1-CXXC domain complexed with unmethylated DNA.[117]

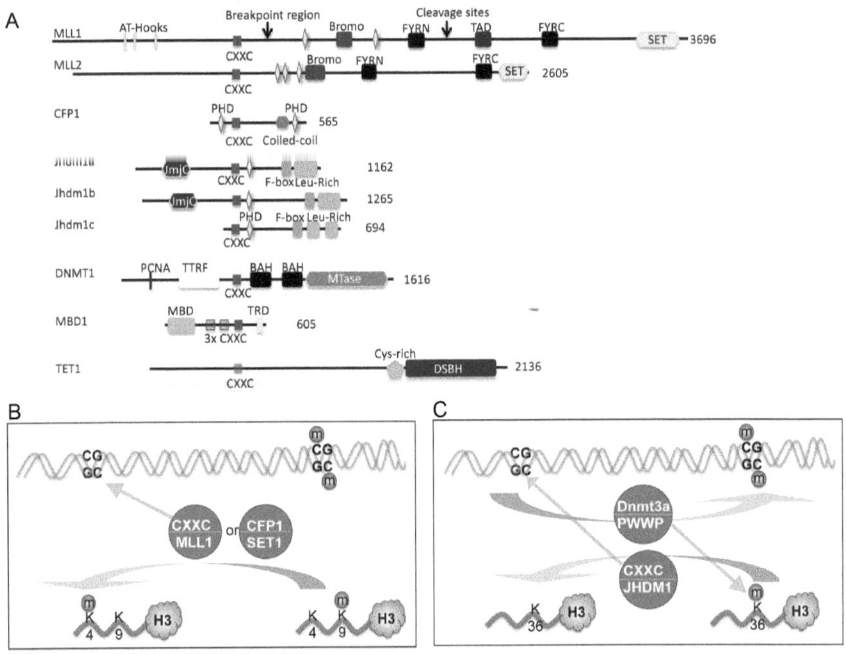

FIG. 3. CpG interacting proteins (A) Domain architecture of proteins containing a CXXC domain (two copies of CxxCxxC motif). (B) Model of the H3K4 methylation reaction by MLL/SET1 proteins. CXXC domain-mediated binding to unmethylated CpGs directs the methylation of H3K4 by MLL/SET1 proteins.

The Set1 H3K4 methyltransferases also appear to interact with unmethylated DNA, although in this case it is via an accessory protein, as with Dnmt3a/Dnmt3L (discussed above). Set1 lacks a CXXC domain, but interacts directly with a protein that contains one—the CXXC finger protein 1 (Cfp1)[118–120] (Fig. 3A). High-throughput sequencing of Cfp1-bound chromatin revealed a notable concordance between H3K4me3 and Cfp1 at unmethylated CpG islands, in the mouse brain.[121] At loci that exhibit allele-specific DNA methylation (e.g., imprinted loci, *Xist* gene), Cfp1 bound specifically to the unmethylated allele. Depletion of Cfp1 resulted in a marked reduction in H3K4me3 genome wide. The targeting of Cfp1 to CpG islands was independent of promoter activity as the insertion of an unmethylated CpG-dense construct into the genome of ES cells was sufficient to nucleate Cfp1 binding and H3K4me3. This suggests that unmethylated CpGs recruit Cfp1, and the associated methyltransferase Set1 then creates new H3K4me3 marks on the local chromatin.

Like the histone H3K4 methyltransferases of the MLL/SET1 family, the Jumonji domain-containing histone demethylase JHDM1a (known as KDM2A) also has a CXXC domain[122] (Fig. 3A). Recent work indicates that, like Cfp1, JHDM1 is recruited to unmethylated CpG islands on a genome-wide scale, and that this is dependent on the CXXC domain.[123] The localization to CpG islands was independent of promoter activity and gene expression levels, and correlated with the selective depletion of H3K36me1/2 within the CpG island but not surrounding regions or the bodies of genes; knockdown of KDM2A resulted in the selective accumulation of H3K36me2 in these regions. Consistent with a role for DNA methylation in determining the localization of CXXC proteins in Dnmt1−/− mice, KDM2A was mislocalized to pericentric heterochromatin. Further, the lack of methylation alone does not appear to be sufficient for KDM2A recruitment *in vivo*, as KDM2A does not localize to unmethylated non-CpG island promoters. Although *in vitro* studies suggest that CXXC domains can bind a single CpG site with micromolar affinity, both the Cfp1 and KDM2A studies suggest that the targeting of CXXC proteins *in vivo* is dependent on CpG density as well as methylation status. It is possible that these proteins oligomerize and form nucleoprotein filaments on CpG-dense DNA, in a manner similar to that described for Dnmt3a and Dnmt3L.[124]

CXXC domains are also found in DNMT1, the enzyme responsible for postreplication maintenance of 5-methyl-Cyt,[125] MBD1, the methyl-CpG-binding protein,[126] and Tet1, a Jumonji-like 2-oxoglutarate- and Fe(II)-dependent enzyme that catalyzes conversion of 5mC to 5hmC[10] (Fig. 3A). Interestingly, a recurrent t[19,20](q22;q23) translocation has been described in acute myelogeous leukemias, and results in a fusion transcript that juxtaposes the first six exons of MLL (containing AT hook and CXXC) to the C-terminal one-third of TET1 thus “replacing” the TET1 CXXC with the MLL CXXC.[127,128] Whether this leads to altered targeting of methyl hydroxylation remains to be determined.

C. A Role for MBDs in the Linkage Between Histone Modification and DNA Methylation

The methyl-CpG-binding domain (MBD) is present in a family of proteins conserved throughout the eukaryotic lineage. This domain, in some but not all cases, confers the ability to bind methylated CpGs. Mammals have five well-characterized members of this family, each with unique biological characteristics (reviewed in Ref. 129; Fig. 4A). Of particular interest are recent reports indicating that in zebrafish, MBD4 (containing both a N-terminal MBD and a

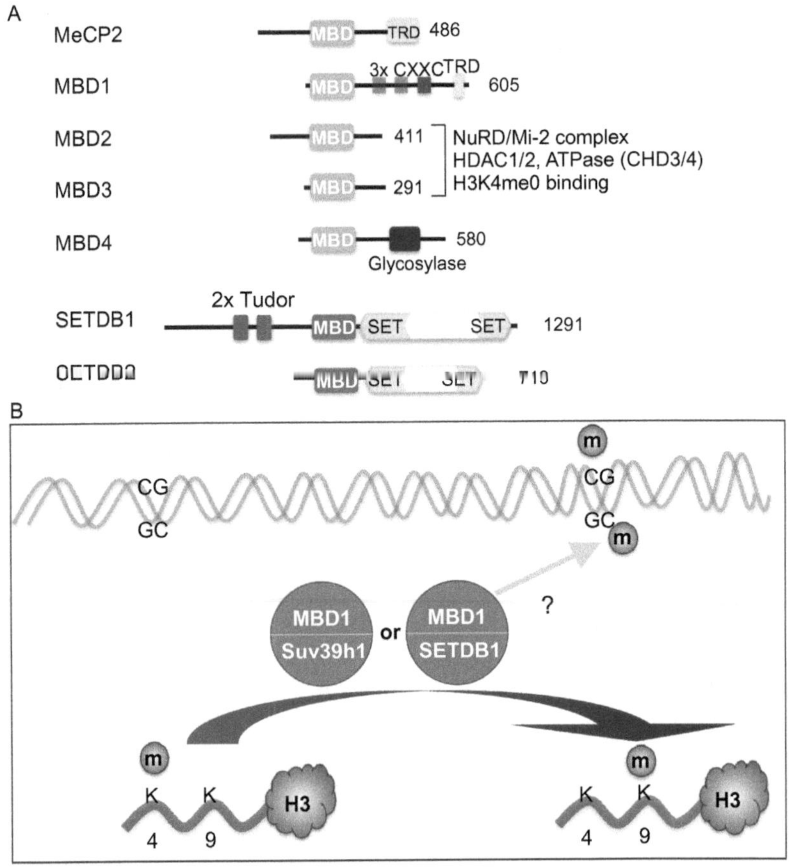

FIG. 4. MBD containing proteins. (A) Domain architecture of known and putative proteins containing a MBD domain. (B) Model of the reactions that regulate H3K9 methylation by SUV39H1 or SETDB1. Recognition of methylated CpGs by the methyl-CpG domain of MBD1 directs methylation of H3K9me3. The "?" refers to the interactions need to be experimentally verified.

C-terminal thymine glycosylase domain,[130]) and the cytidine deaminase AID cooperate to demethylate DNA[4] (see also Chapter by Mary G. Goll and Marnie E. Halpern). Consistent with a role in DNA demethylation, erasure of DNA methylation in primordial germ cells is less efficient in AID-deficient animals,[131] and AID is required to demethylate pluripotency genes during reprogramming of the somatic genome in ES cell fusions.[132] It is noteworthy that phosphorylation of MBD4 enhances DNA demethylation.[133]

A particularly interesting observation, for the purposes of this review, is the fact that MBD1 forms a stable complex with SETDB1,[134,135] an H3K9me3 methyltransferase responsible for silencing of endogenous retroviruses[136–138] as well as the Suv39h1/HP1 complex,[139] a heterochromatin-specific H3K9me3 writer and reader. SETDB1 also contains an intrinsic putative MBD domain with two conserved DNA-interacting arginine residues (Fig. 4B) that make direct contact with DNA in the structures of the MBD domain from MBD1[140] and MeCP2.[141] It remains to be seen whether the putative MBD domain of SETDB1 is similarly able to selectively bind methylated DNA. The intrinsic or associated coupling of a DNA methylation "reader" with H3K9me3 "writers" implies an interdependent mechanism for the propagation or maintenance of these marks.

As MBD1 contains two DNA-binding domains—an MBD domain (Fig. 4), which recognizes methylated CpG, and a CXXC domain (Fig. 3), which binds unmethylated CpG—it is unclear whether these two domains function independently to facilitate the recruitment of MBD1 to repressive complexes or active complexes, or if they cooperate in some way. Recently, Clouaire et al. found that the MBD domain of MBD1 binds more efficiently to methylated DNA within a specific sequence context, and that a functional MBD domain is necessary and sufficient for recruitment of MBD1 to these loci, while DNA binding by the CXXC domain is largely dispensable.[142]

D. UHRF1 Links Hemimethylated CpGs to Histone Modifications During Replication

UHRF1 harbors five recognizable functional domains (Fig. 2A): a ubiquitin-like domain (UBL) at the N-terminus, followed by a tandem Tudor domain that binds H3K9me2/3,[143,144] an ADD-like domain that binds the histone H3 tail,[145,146] a SET- and RING-associated (SRA) domain that binds hemimethylated CpG-containing DNA,[81,82,147–149] and a really interesting new gene (*RING*) domain at the C-terminus that may endow UHRF1 with E3 ubiquitin ligase activity for histones.[145] As discussed above, Dnmt1 contains a CXXC domain (Fig. 3A) that binds unmethylated CpGs.[125] The interaction with UHRF1 provides an SRA domain that recognizes hemimethylated CpGs, the substrate of Dnmt1. It is tempting to speculate that these two DNA-binding domains function synergically to maintain the fidelity of DNA methylation during replication, by preventing Dnmt1 from modifying unmethylated CpGs (through the masking of such sites by the CXXC domain) and directing methylation to newly synthesized hemimethylated CpG sites (via the SRA domain). Indeed, molecular modeling suggests that the Dnmt1 catalytic domain and UHRF1 SRA domain are very unlikely to bind simultaneously to the same hemimethylated CpG site.[148] Somehow, the SRA domain must be displaced from the site to allow methylation by Dnmt1.[150]

We therefore suggest that the SRA–DNA interaction (through recognition of and flipping of the parental strand 5mC out of the helix[147–149]) serves as an anchor to keep UHRF1 at the hemimethylated CpG site where it then recruits Dnmt1 to methylate the opposing CpG in the daughter strand, which maintains the fidelity of DNA methylation.

Acknowledgments

The work in the authors' laboratories is currently supported by the U.S. National Institutes of Health (GM049245-17, GM068680-06, and GM092035-01 to X. C.) and the National Science Foundation (MCB-0964728 to R. M. B.). X. C. is a Georgia Research Alliance Eminent Scholar. This chapter is modified and updated from Ref. 91, with permission from American Chemical Society Publications.

References

1. Turner BM. Defining an epigenetic code. *Nat Cell Biol* 2007;**9**:2–6.
2. Clapier CR, Chakravarthy S, Petosa C, Fernandez-Tornero C, Luger K, Muller CW. Structure of the Drosophila nucleosome core particle highlights evolutionary constraints on the H2A-H2B histone dimer. *Proteins* 2008;**71**:1–7.
3. Dulac C. Brain function and chromatin plasticity. *Nature* 2010;**465**:728–35.
4. Rai K, Huggins IJ, James SR, Karpf AR, Jones DA, Cairns BR. DNA demethylation in zebrafish involves the coupling of a deaminase, a glycosylase, and gadd45. *Cell* 2008;**135**:1201–12.
5. Kress C, Thomassin H, Grange T. Active cytosine demethylation triggered by a nuclear receptor involves DNA strand breaks. *Proc Natl Acad Sci USA* 2006;**103**:11112–7.
6. Vairapandi M. Characterization of DNA demethylation in normal and cancerous cell lines and the regulatory role of cell cycle proteins in human DNA demethylase activity. *J Cell Biochem* 2004;**91**:572–83.
7. Kangaspeska S, Stride B, Metivier R, Polycarpou-Schwarz M, Ibberson D, Carmouche RP, et al. Transient cyclical methylation of promoter DNA. *Nature* 2008;**452**:112–5.
8. Metivier R, Gallais R, Tiffoche C, Le Peron C, Jurkowska RZ, Carmouche RP, et al. Cyclical DNA methylation of a transcriptionally active promoter. *Nature* 2008;**452**:45–50.
9. Ooi SK, Bestor TH. The colorful history of active DNA demethylation. *Cell* 2008;**133**:1145–8.

10. Tahiliani M, Koh KP, Shen Y, Pastor WA, Bandukwala H, Brudno Y, et al. Conversion of 5-methylcytosine to 5-hydroxymethylcytosine in mammalian DNA by the MLL fusion partner TET1. *Science* 2009;**324**:930–5.
11. Liutkeviciute Z, Lukinavicius G, Masevicius V, Daujotyte D, Klimasauskas S. Cytosine-5-methyltransferases add aldehydes to DNA. *Nat Chem Biol* 2009;**5**:400–2.
12. Kouzarides T. Chromatin modifications and their function. *Cell* 2007;**128**:693–705.
13. Cheng X, Roberts RJ. AdoMet-dependent methylation, DNA methyltransferases and base flipping. *Nucleic Acids Res* 2001;**29**:3784–95.
14. Klimasauskas S, Kumar S, Roberts RJ, Cheng X. HhaI methyltransferase flips its target base out of the DNA helix. *Cell* 1994;**76**:357–69.
15. Sheikhnejad G, Brank A, Christman JK, Goddard A, Alvarez E, Ford Jr. H, et al. Mechanism of inhibition of DNA (cytosine C5)-methyltransferases by oligodeoxyribonucleotides containing 5,6-dihydro-5-azacytosine. *J Mol Biol* 1999;**285**:2021–34.
16. Wu JC, Santi DV. Kinetic and catalytic mechanism of HhaI methyltransferase. *J Biol Chem* 1987;**262**:4778–86.
17. Youngblood B, Shieh FK, Buller F, Bullock T, Reich NO. S-adenosyl-L-methionine-dependent methyl transfer: observable precatalytic intermediates during DNA cytosine methylation. *Biochemistry* 2007;**46**:8766–75.
18. Zhang X, Bruice TC. The mechanism of M.HhaI DNA C5 cytosine methyltransferase enzyme: a quantum mechanics/molecular mechanics approach. *Proc Natl Acad Sci USA* 2006;**103**:6148–53.
19. Wyatt GR, Cohen SS. The bases of the nucleic acids of some bacterial and animal viruses: the occurrence of 5-hydroxymethylcytosine. *Biochem J* 1953;**55**:774–82.
20. Wiberg JS. Amber mutants of bacteriophage T4 defective in deoxycytidine diphosphatase and deoxycytidine triphosphatase. On the role of 5-hydroxymethylcytosine in bacteriophage deoxyribonucleic acid. *J Biol Chem* 1967;**242**:5824–9.
21. Anisymova NI, Gabrilovich IM, Soshina NV, Cherenkevich SN. 5-Hydroxymethylcytosine-containing Klebsiella bacteriophage. *Biochim Biophys Acta* 1969;**190**:225–7.
22. Fleischman RA, Cambell JL, Richardson CC. Modification and restriction of T-even bacteriophages. In vitro degradation of deoxyribonucleic acid containing 5-hydroxymethylctosine. *J Biol Chem* 1976;**251**:1561–70.
23. Warren RA. Modified bases in bacteriophage DNAs. *Annu Rev Microbiol* 1980;**34**:137–58.
24. Penn NW, Suwalski R, O'Riley C, Bojanowski K, Yura R. The presence of 5-hydroxymethylcytosine in animal deoxyribonucleic acid. *Biochem J* 1972;**126**:781–90.
25. Castro GD, Diaz Gomez MI, Castro JA. 5-Methylcytosine attack by hydroxyl free radicals and during carbon tetrachloride promoted liver microsomal lipid peroxidation: structure of reaction products. *Chem Biol Interact* 1996;**99**:289–99.
26. Kriaucionis S, Heintz N. The nuclear DNA base 5-hydroxymethylcytosine is present in Purkinje neurons and the brain. *Science* 2009;**324**:929–30.
27. Cannon SV, Cummings A, Teebor GW. 5-Hydroxymethylcytosine DNA glycosylase activity in mammalian tissue. *Biochem Biophys Res Commun* 1988;**151**:1173–9.
28. Strahl BD, Allis CD. The language of covalent histone modifications. *Nature* 2000;**403**:41–5.
29. Suganuma T, Workman JL. Crosstalk among histone modifications. *Cell* 2008;**135**:604–7.
30. Hung T, Binda O, Champagne KS, Kuo AJ, Johnson K, Chang HY, et al. ING4 mediates crosstalk between histone H3 K4 trimethylation and H3 acetylation to attenuate cellular transformation. *Mol Cell* 2009;**33**:248–56.
31. Lee JS, Shukla A, Schneider J, Swanson SK, Washburn MP, Florens L, et al. Histone crosstalk between H2B monoubiquitination and H3 methylation mediated by COMPASS. *Cell* 2007;**131**:1084–96.

32. Adhvaryu KK, Selker EU. Protein phosphatase PP1 is required for normal DNA methylation in Neurospora. *Genes Dev* 2008;**22**:3391–6.
33. van Attikum H, Gasser SM. Crosstalk between histone modifications during the DNA damage response. *Trends Cell Biol* 2009;**19**:207–17.
34. Rea S, Eisenhaber F, O'Carroll D, Strahl BD, Sun ZW, Schmid M, et al. Regulation of chromatin structure by site-specific histone H3 methyltransferases. *Nature* 2000;**406**:593–9.
35. Sun ZW, Allis CD. Ubiquitination of histone H2B regulates H3 methylation and gene silencing in yeast. *Nature* 2002;**418**:104–8.
36. Ng HH, Xu RM, Zhang Y, Struhl K. Ubiquitination of histone H2B by Rad6 is required for efficient Dot1-mediated methylation of histone H3 lysine 79. *J Biol Chem* 2002;**277**:34655–7.
37. Briggs SD, Xiao T, Sun ZW, Caldwell JA, Shabanowitz J, Hunt DF, et al. Gene silencing: trans-histone regulatory pathway in chromatin. *Nature* 2002;**418**:498.
38. Horton JR, Upadhyay AK, Qi HH, Zhang X, Shi Y, Cheng X. Enzymatic and structural insights for substrate specificity of a family of jumonji histone lysine demethylases. *Nat Struct Mol Biol* 2010;**17**:38–43.
39. Fortschegger K, de Graaf P, Outchkourov NS, van Schaik FM, Timmers HT, Shiekhattar R. PHF8 targets histone methylation and RNA polymerase II to activate transcription. *Mol Cell Biol* 2010;**30**:3286–98.
40. Kleine-Kohlbrecher D, Christensen J, Vandamme J, Abarrategui I, Bak M, Tommerup N, et al. A functional link between the histone demethylase PHF8 and the transcription factor ZNF711 in X-linked mental retardation. *Mol Cell* 2010;**38**:165–78.
41. Feng W, Yonezawa M, Ye J, Jenuwein T, Grummt I. PHF8 activates transcription of rRNA genes through H3K4me3 binding and H3K9me1/2 demethylation. *Nat Struct Mol Biol* 2010;**17**:445–50.
42. Yokoyama A, Okuno Y, Chikanishi T, Hashiba W, Sekine H, Fujiki R, et al. KIAA1718 is a histone demethylase that erases repressive histone methyl marks. *Genes Cells* 2010;**15**:867–73.
43. Huang C, Xiang Y, Wang Y, Li X, Xu L, Zhu Z, et al. Dual-specificity histone demethylase KIAA1718 (KDM7A) regulates neural differentiation through FGF4. *Cell Res* 2010;**20**:154–65.
44. Liu W, Tanasa B, Tyurina OV, Zhou TY, Gassmann R, Liu WT, et al. PHF8 mediates histone H4 lysine 20 demethylation events involved in cell cycle progression. *Nature* 2010;**466**:508–12.
45. Qi HH, Sarkissian M, Hu GQ, Wang Z, Bhattacharjee A, Gordon DB, et al. Histone H4K20/H3K9 demethylase PHF8 regulates zebrafish brain and craniofacial development. *Nature* 2010;**466**:503–7.
46. Li B, Carey M, Workman JL. The role of chromatin during transcription. *Cell* 2007;**128**:707–19.
47. Groth A, Corpet A, Cook AJ, Roche D, Bartek J, Lukas J, et al. Regulation of replication fork progression through histone supply and demand. *Science* 2007;**318**:1928–31.
48. Cairns BR. Chromatin remodeling: insights and intrigue from single-molecule studies. *Nat Struct Mol Biol* 2007;**14**:989–96.
49. Li H, Ilin S, Wang W, Duncan EM, Wysocka J, Allis CD, et al. Molecular basis for site-specific read-out of histone H3K4me3 by the BPTF PHD finger of NURF. *Nature* 2006;**442**:91–5.
50. Wysocka J, Swigut T, Xiao H, Milne TA, Kwon SY, Landry J, et al. A PHD finger of NURF couples histone H3 lysine 4 trimethylation with chromatin remodelling. *Nature* 2006;**442**:86–90.
51. Downs JA, Allard S, Jobin-Robitaille O, Javaheri A, Auger A, Bouchard N, et al. Binding of chromatin-modifying activities to phosphorylated histone H2A at DNA damage sites. *Mol Cell* 2004;**16**:979–90.

52. Hassan AH, Neely KE, Workman JL. Histone acetyltransferase complexes stabilize swi/snf binding to promoter nucleosomes. *Cell* 2001;**104**:817–27.
53. Hassan AH, Prochasson P, Neely KE, Galasinski SC, Chandy M, Carrozza MJ, et al. Function and selectivity of bromodomains in anchoring chromatin-modifying complexes to promoter nucleosomes. *Cell* 2002;**111**:369–79.
54. Syntichaki P, Topalidou I, Thireos G. The Gcn5 bromodomain co-ordinates nucleosome remodelling. *Nature* 2000;**404**:414–7.
55. Hake SB, Allis CD. Histone H3 variants and their potential role in indexing mammalian genomes: the "H3 barcode hypothesis" *Proc Natl Acad Sci USA* 2006;**103**:6428–35.
56. Hake SB, Garcia BA, Duncan EM, Kauer M, Dellaire G, Shabanowitz J, et al. Expression patterns and post-translational modifications associated with mammalian histone H3 variants. *J Biol Chem* 2006;**281**:559–68.
57. Nakanishi S, Sanderson BW, Delventhal KM, Bradford WD, Staehling-Hampton K, Shilatifard A. A comprehensive library of histone mutants identifies nucleosomal residues required for H3K4 methylation. *Nat Struct Mol Biol* 2008;**15**:881–8.
58. Loyola A, Bonaldi T, Roche D, Imhof A, Almouzni G. PTMs on H3 variants before chromatin assembly potentiate their final epigenetic state. *Mol Cell* 2006;**24**:309–16.
59. Hodl M, Basler K. Transcription in the absence of histone H3.3. *Curr Biol* 2009;**19**:1221–6.
60. Collins LJ, Chen XS. Ancestral RNA: the RNA biology of the eukaryotic ancestor. *RNA Biol* 2009;**6**:495–502.
61. Collins LJ, Penny D. The RNA infrastructure: dark matter of the eukaryotic cell? *Trends Genet* 2009;**25**:120–8.
62. Collins RE, Cheng X. Structural and biochemical advances in mammalian RNAi. *J Cell Biochem* 2006;**99**:1251–66.
63. Collins RE, Cheng X. Structural domains in RNAi. *FEBS Lett* 2005;**579**:5841–9.
64. Mercer TR, Dinger ME, Mattick JS. Long non-coding RNAs: insights into functions. *Nat Rev Genet* 2009;**10**:155–9.
65. Mattick JS. The genetic signatures of noncoding RNAs. *PLoS Genet* 2009;**5**:e1000459.
66. Kurokawa R, Rosenfeld MG, Glass CK. Transcriptional regulation through noncoding RNAs and epigenetic modifications. *RNA Biol* 2009;**6**:233–6.
67. Chow J, Heard E. X inactivation and the complexities of silencing a sex chromosome. *Curr Opin Cell Biol* 2009;**21**:359–66.
68. Lee JT. Lessons from X-chromosome inactivation: long ncRNA as guides and tethers to the epigenome. *Genes Dev* 2009;**23**:1831–42.
69. Tian D, Sun S, Lee JT. The long noncoding RNA, jpx, is a molecular switch for x chromosome inactivation. *Cell* 2010;**143**:390–403.
70. Lee JT. The X as model for RNA's niche in epigenomic regulation. *Cold Spring Harb Perspect Biol* 2010;**2**:a003749.
71. Royo H, Cavaille J. Non-coding RNAs in imprinted gene clusters. *Biol Cell* 2008;**100**:149–66.
72. Latos PA, Barlow DP. Regulation of imprinted expression by macro non-coding RNAs. *RNA Biol* 2009;**6**:100–6.
73. Koerner MV, Pauler FM, Huang R, Barlow DP. The function of non-coding RNAs in genomic imprinting. *Development* 2009;**136**:1771–83.
74. Koerner MV, Barlow DP. Genomic imprinting—an epigenetic gene-regulatory model. *Curr Opin Genet Dev* 2010;**20**:164–70.
75. Lucchesi JC. The structure-function link of compensated chromatin in Drosophila. *Curr Opin Genet Dev* 2009;**19**:550–6.
76. Tsai MC, Manor O, Wan Y, Mosammaparast N, Wang JK, Lan F, et al. Long noncoding RNA as modular scaffold of histone modification complexes. *Science* 2010;**329**:689–93.

77. Xu M, Long C, Chen X, Huang C, Chen S, Zhu B. Partitioning of histone H3-H4 tetramers during DNA replication-dependent chromatin assembly. *Science* 2010;**328**:94–8.
78. Collins RE, Northrop JP, Horton JR, Lee DY, Zhang X, Stallcup MR, et al. The ankyrin repeats of G9a and GLP histone methyltransferases are mono- and dimethyllysine binding modules. *Nat Struct Mol Biol* 2008;**15**:245–50.
79. Grewal SI, Jia S. Heterochromatin revisited. *Nat Rev Genet* 2007;**8**:35–46.
80. Zhang K, Mosch K, Fischle W, Grewal SI. Roles of the Clr4 methyltransferase complex in nucleation, spreading and maintenance of heterochromatin. *Nat Struct Mol Biol* 2008;**15**:381–8.
81. Bostick M, Kim JK, Esteve PO, Clark A, Pradhan S, Jacobsen SE. UHRF1 plays a role in maintaining DNA methylation in mammalian cells. *Science* 2007;**317**:1760–4.
82. Sharif J, Muto M, Takebayashi S, Suetake I, Iwamatsu A, Endo TA, et al. The SRA protein Np95 mediates epigenetic inheritance by recruiting Dnmt1 to methylated DNA. *Nature* 2007;**450**:908–12.
83. Kim JK, Esteve PO, Jacobsen SE, Pradhan S. UHRF1 binds G9a and participates in p21 transcriptional regulation in mammalian cells. *Nucleic Acids Res* 2009;**37**:493–505.
84. Achour M, Fuhrmann G, Alhosin M, Ronde P, Chataigneau T, Mousli M, et al. UHRF1 recruits the histone acetyltransferase Tip60 and controls its expression and activity. *Biochem Biophys Res Commun* 2009;**390**:523–8.
85. Bernstein BE, Meissner A, Lander ES. The mammalian epigenome. *Cell* 2007;**128**:669–81.
86. Meissner A, Mikkelsen TS, Gu H, Wernig M, Hanna J, Sivachenko A, et al. Genome-scale DNA methylation maps of pluripotent and differentiated cells. *Nature* 2008;**454**:766–70.
87. Fan S, Zhang MQ, Zhang X. Histone methylation marks play important roles in predicting the methylation status of CpG islands. *Biochem Biophys Res Commun* 2008;**374**:559–64.
88. Cedar H, Bergman Y. Linking DNA methylation and histone modification: patterns and paradigms. *Nat Rev Genet* 2009;**10**:295–304.
89. Dindot SV, Person R, Strivens M, Garcia R, Beaudet AL. Epigenetic profiling at mouse imprinted gene clusters reveals novel epigenetic and genetic features at differentially methylated regions. *Genome Res* 2009;**19**:1374–83.
90. Laurent L, Wong E, Li G, Huynh T, Tsirigos A, Ong CT, et al. Dynamic changes in the human methylome during differentiation. *Genome Res* 2010;**20**:320–31.
91. Cheng X, Blumenthal RM. Coordinated chromatin control: structural and functional linkage of DNA and histone methylation. *Biochemistry* 2010;**49**:2999–3008.
92. Tamaru H, Zhang X, McMillen D, Singh PB, Nakayama J, Grewal SI, et al. Trimethylated lysine 9 of histone H3 is a mark for DNA methylation in Neurospora crassa. *Nat Genet* 2003;**34**:75–9.
93. Tamaru H, Selker EU. A histone H3 methyltransferase controls DNA methylation in Neurospora crassa. *Nature* 2001;**414**:277–83.
94. Jackson JP, Lindroth AM, Cao X, Jacobsen SE. Control of CpNpG DNA methylation by the KRYPTONITE histone H3 methyltransferase. *Nature* 2002;**416**:556–60.
95. Nguyen CT, Weisenberger DJ, Velicescu M, Gonzales FA, Lin JC, Liang G, et al. Histone H3-lysine 9 methylation is associated with aberrant gene silencing in cancer cells and is rapidly reversed by 5-aza-2′-deoxycytidine. *Cancer Res* 2002;**62**:6456–61.
96. Elsaesser SJ, Goldberg AD, Allis CD. New functions for an old variant: no substitute for histone H3.3. *Curr Opin Genet Dev* 2010;**20**:110–7.
97. Goldberg AD, Banaszynski LA, Noh KM, Lewis PW, Elsaesser SJ, Stadler S, et al. Distinct factors control histone variant H3.3 localization at specific genomic regions. *Cell* 2010;**140**:678–91.

98. Wong LH, McGhie JD, Sim M, Anderson MA, Ahn S, Hannan RD, et al. ATRX interacts with H3.3 in maintaining telomere structural integrity in pluripotent embryonic stem cells. *Genome Res* 2010;**20**:351–60.
99. Chen T, Ueda Y, Xie S, Li E. A novel Dnmt3a isoform produced from an alternative promoter localizes to euchromatin and its expression correlates with active de novo methylation. *J Biol Chem* 2002;**277**:38746–54.
100. Ooi SKT, Qiu C, Bernstein E, Li K, Jia D, Yang Z, et al. DNMT3L connects unmethylated lysine 4 of histone H3 to de novo methylation of DNA. *Nature* 2007;**448**:714–7.
101. Hu JL, Zhou BO, Zhang RR, Zhang KL, Zhou JQ, Xu GL. The N-terminus of histone H3 is required for de novo DNA methylation in chromatin. *Proc Natl Acad Sci USA* 2009;**106**:22187–92.
102. Proffitt JH, Davie JR, Swinton D, Hattman S. 5-Methylcytosine is not detectable in Saccharomyces cerevisiae DNA. *Mol Cell Biol* 1984;**4**:985–8.
103. Bulkowska U, Ishikawa T, Kurlandzka A, Trzcinska-Danielewicz J, Derlacz R, Fronk J. Expression of murine DNA methyltransferases Dnmt1 and Dnmt3a in the yeast Saccharomyces cerevisiae. *Yeast* 2007;**24**:871–82.
104. Wang J, Hevi S, Kurash JK, Lei H, Gay F, Bajko J, et al. The lysine demethylase LSD1 (KDM1) is required for maintenance of global DNA methylation. *Nat Genet* 2009;**41**:125–9.
105. Ciccone DN, Su H, Hevi S, Gay F, Lei H, Bajko J, et al. KDM1B is a histone H3K4 demethylase required to establish maternal genomic imprints. *Nature* 2009;**461**:359–60.
106. Kato Y, Kaneda M, Hata K, Kumaki K, Hisano M, Kohara Y, et al. Role of the Dnmt3 family in de novo methylation of imprinted and repetitive sequences during male germ cell development in the mouse. *Hum Mol Genet* 2007;**16**:2272–80.
107. Jeong S, Liang G, Sharma S, Lin JC, Choi SH, Han H, et al. Selective anchoring of DNA methyltransferases 3A/3B to nucleosomes containing methylated DNA. *Mol Cell Biol* 2009; **29**:5366–76.
108. Zhang Y, Jurkowska R, Soeroes S, Rajavelu A, Dhayalan A, Bock I, et al. Chromatin methylation activity of Dnmt3a and Dnmt3a/3L is guided by interaction of the ADD domain with the histone H3 tail. *Nucleic Acids Res* 2010;**38**:4246–53.
109. Otani J, Nankumo T, Arita K, Inamoto S, Ariyoshi M, Shirakawa M. Structural basis for recognition of H3K4 methylation status by the DNA methyltransferase 3A ATRX-DNMT3-DNMT3L domain. *EMBO Rep* 2009;**10**:1235–41.
110. Cosgrove MS, Patel A. Mixed lineage leukemia: a structure-function perspective of the MLL1 protein. *FEBS J* 2010;**277**:1832–42.
111. Ansari KI, Mandal SS. Mixed lineage leukemia: roles in gene expression, hormone signaling and mRNA processing. *FEBS J* 2010;**277**:1790–804.
112. Terranova R, Agherbi H, Boned A, Meresse S, Djabali M. Histone and DNA methylation defects at Hox genes in mice expressing a SET domain-truncated form of Mll. *Proc Natl Acad Sci USA* 2006;**103**:6629–34.
113. Milne TA, Briggs SD, Brock HW, Martin ME, Gibbs D, Allis CD, et al. MLL targets SET domain methyltransferase activity to Hox gene promoters. *Mol Cell* 2002;**10**:1107–17.
114. Birke M, Schreiner S, Garcia-Cuellar MP, Mahr K, Titgemeyer F, Slany RK. The MT domain of the proto-oncoprotein MLL binds to CpG-containing DNA and discriminates against methylation. *Nucleic Acids Res* 2002;**30**:958–65.
115. Ayton PM, Chen EH, Cleary ML. Binding to nonmethylated CpG DNA is essential for target recognition, transactivation, and myeloid transformation by an MLL oncoprotein. *Mol Cell Biol* 2004;**24**:10470–8.
116. Allen MD, Grummitt CG, Hilcenko C, Min SY, Tonkin LM, Johnson CM, et al. Solution structure of the nonmethyl-CpG-binding CXXC domain of the leukaemia-associated MLL histone methyltransferase. *EMBO J* 2006;**25**:4503–12.

117. Cierpicki T, Risner LE, Grembecka J, Lukasik SM, Popovic R, Omonkowska M, et al. Structure of the MLL CXXC domain-DNA complex and its functional role in MLL-AF9 leukemia. *Nat Struct Mol Biol* 2010;**17**:62–8.
118. Lee JH, Voo KS, Skalnik DG. Identification and characterization of the DNA binding domain of CpG-binding protein. *J Biol Chem* 2001;**276**:44669–76.
119. Lee JH, Skalnik DG. CpG-binding protein (CXXC finger protein 1) is a component of the mammalian Set1 histone H3-Lys4 methyltransferase complex, the analogue of the yeast Set1/COMPASS complex. *J Biol Chem* 2005;**280**:41725–31.
120. Lee JH, Tate CM, You JS, Skalnik DG. Identification and characterization of the human Set1B histone H3-Lys4 methyltransferase complex. *J Biol Chem* 2007;**282**:13419–28.
121. Thomson JP, Skene PJ, Selfridge J, Clouaire T, Guy J, Webb S, et al. CpG islands influence chromatin structure via the CpG-binding protein Cfp1. *Nature* 2010;**464**:1082–6.
122. Tsukada Y, Fang J, Erdjument-Bromage H, Warren ME, Borchers CH, Tempst P, et al. Histone demethylation by a family of JmjC domain-containing proteins. *Nature* 2006;**439**:811–6.
123. Blackledge NP, Zhou JC, Tolstorukov MY, Farcas AM, Park PJ, Klose RJ. CpG islands recruit a histone H3 lysine 36 demethylase. *Mol Cell* 2010;**38**:179–90.
124. Jurkowska RZ, Anspach N, Urbanke C, Jia D, Reinhardt R, Nellen W, et al. Formation of nucleoprotein filaments by mammalian DNA methyltransferase Dnmt3a in complex with regulator Dnmt3L. *Nucleic Acids Res* 2008;**36**:6656–63.
125. Pradhan M, Esteve PO, Chin HG, Samaranayke M, Kim GD, Pradhan S. CXXC domain of human DNMT1 is essential for enzymatic activity. *Biochemistry* 2008;**47**:10000–9.
126. Jorgensen HF, Ben-Porath I, Bird AP. Mbd1 is recruited to both methylated and nonmethylated CpGs via distinct DNA binding domains. *Mol Cell Biol* 2004;**24**:3387–95.
127. Ono R, Taki T, Taketani T, Taniwaki M, Kobayashi H, Hayashi Y. LCX, leukemia-associated protein with a CXXC domain, is fused to MLL in acute myeloid leukemia with trilineage dysplasia having t(10;11)(q22;q23). *Cancer Res* 2002;**62**:4075–80.
128. Lorsbach RB, Moore J, Mathew S, Raimondi SC, Mukatira ST, Downing JR. TET1, a member of a novel protein family, is fused to MLL in acute myeloid leukemia containing the t(10;11)(q22;q23). *Leukemia* 2003;**17**:637–41.
129. Dhasarathy A, Wade PA. The MBD protein family-reading an epigenetic mark? *Mutat Res* 2008;**647**:39–43.
130. Hendrich B, Hardeland U, Ng HH, Jiricny J, Bird A. The thymine glycosylase MBD4 can bind to the product of deamination at methylated CpG sites. *Nature* 1999;**401**:301–4.
131. Popp C, Dean W, Feng S, Cokus SJ, Andrews S, Pellegrini M, et al. Genome-wide erasure of DNA methylation in mouse primordial germ cells is affected by AID deficiency. *Nature* 2010;**463**:1101–5.
132. Bhutani N, Brady JJ, Damian M, Sacco A, Corbel SY, Blau HM. Reprogramming towards pluripotency requires AID-dependent DNA demethylation. *Nature* 2010;**463**:1042–7.
133. Kim MS, Kondo T, Takada I, Youn MY, Yamamoto Y, Takahashi S, et al. DNA demethylation in hormone-induced transcriptional derepression. *Nature* 2009;**461**:1007–12.
134. Sarraf SA, Stancheva I. Methyl-CpG binding protein MBD1 couples histone H3 methylation at lysine 9 by SETDB1 to DNA replication and chromatin assembly. *Mol Cell* 2004;**15**:595–605.
135. Lyst MJ, Nan X, Stancheva I. Regulation of MBD1-mediated transcriptional repression by SUMO and PIAS proteins. *EMBO J* 2006;**25**:5317–28.
136. Schultz DC, Ayyanathan K, Negorev D, Maul GG, Rauscher 3rd FJ. SETDB1: a novel KAP-1-associated histone H3, lysine 9-specific methyltransferase that contributes to HP1-mediated silencing of euchromatic genes by KRAB zinc-finger proteins. *Genes Dev* 2002;**16**:919–32.

137. Rowe HM, Jakobsson J, Mesnard D, Rougemont J, Reynard S, Aktas T, et al. KAP1 controls endogenous retroviruses in embryonic stem cells. *Nature* 2010;**463**:237–40.
138. Matsui T, Leung D, Miyashita H, Maksakova IA, Miyachi H, Kimura H, et al. Proviral silencing in embryonic stem cells requires the histone methyltransferase ESET. *Nature* 2010;**464**:927–31.
139. Fujita N, Watanabe S, Ichimura T, Tsuruzoe S, Shinkai Y, Tachibana M, et al. Methyl-CpG binding domain 1 (MBD1) interacts with the Suv39h1-HP1 heterochromatic complex for DNA methylation-based transcriptional repression. *J Biol Chem* 2003;**278**:24132–8.
140. Ohki I, Shimotake N, Fujita N, Jee J, Ikegami T, Nakao M, et al. Solution structure of the methyl-CpG binding domain of human MBD1 in complex with methylated DNA. *Cell* 2001;**105**:487–97.
141. Ho KL, McNae IW, Schmiedeberg L, Klose RJ, Bird AP, Walkinshaw MD. MeCP2 binding to DNA depends upon hydration at methyl-CpG. *Mol Cell* 2008;**29**:525–31.
142. Clouaire T, de Las Heras JI, Merusi C, Stancheva I. Recruitment of MBD1 to target genes requires sequence-specific interaction of the MBD domain with methylated DNA. *Nucleic Acids Res* 2010;**38**:4620–34.
143. Walker JR, Avvakumov GV, Xue S, Li Y, Bountra C, Weigelt J, et al. Structure of the replication foci-targeting sequence of human DNA cytosine methyltransferase DNMT1. 2008, http://www.rcsb.org/pdb/explore.do?structureid=3epz
144. Rottach A, Frauer C, Pichler G, Bonapace IM, Spada F, Leonhardt H. The multi-domain protein Np95 connects DNA methylation and histone modification. *Nucleic Acids Res* 2010;**38**:1796–804.
145. Citterio E, Papait R, Nicassio F, Vecchi M, Gomiero P, Mantovani R, et al. Np95 is a histone-binding protein endowed with ubiquitin ligase activity. *Mol Cell Biol* 2004;**24**:2526–35.
146. Karagianni P, Amazit L, Qin J, Wong J. ICBP90, a novel methyl K9 H3 binding protein linking protein ubiquitination with heterochromatin formation. *Mol Cell Biol* 2008;**28**:705–17.
147. Hashimoto H, Horton JR, Zhang X, Bostick M, Jacobsen SE, Cheng X. The SRA domain of UHRF1 flips 5-methylcytosine out of the DNA helix. *Nature* 2008;**455**:826–9.
148. Arita K, Ariyoshi M, Tochio H, Nakamura Y, Shirakawa M. Recognition of hemi-methylated DNA by the SRA protein UHRF1 by a base-flipping mechanism. *Nature* 2008;**455**:818–21.
149. Avvakumov GV, Walker JR, Xue S, Li Y, Duan S, Bronner C, et al. Structural basis for recognition of hemi-methylated DNA by the SRA domain of human UHRF1. *Nature* 2008;**455**:822–5.
150. Hashimoto H, Vertino PM, Cheng X. Molecular coupling of DNA methylation and histone methylation. *Epigenomics* 2010;**2**:657–69.

Section I

Evolution of DNA Methylation

Natural History of Eukaryotic DNA Methylation Systems

Lakshminarayan M. Iyer,
Saraswathi Abhiman, and
L. Aravind

National Center for Biotechnology Information, National Library of Medicine, National Institutes of Health, Bethesda, Maryland, USA

Progress in Molecular Biology
and Translational Science, Vol. 101
DOI: 10.1016/B978-0-12-387685-0.00002-0

1877-1173/11 $35.00

Methylation of cytosines and adenines in DNA is a widespread epigenetic mark in both prokaryotes and eukaryotes. In eukaryotes, it has a profound influence on chromatin structure and dynamics. Recent advances in genomics and biochemistry have considerably elucidated the functions and provenance of these DNA modifications. DNA methylases appear to have emerged first in bacterial restriction–modification (R–M) systems from ancient RNA-modifying enzymes, in transitions that involved acquisition of novel catalytic residues and DNA-recognition features. DNA adenine methylases appear to have been acquired by ciliates, heterolobosean amoeboflagellates, and certain chlorophyte algae. Six distinct clades of cytosine methylases, including the DNMT1, DNMT2, and DNMT3 clades, were acquired by eukaryotes through independent lateral transfer of their precursors from bacteria or bacteriophages. In addition to these, multiple adenine and cytosine methylases were acquired by several families of eukaryotic transposons. In eukaryotes, the DNA-methylase module was often combined with distinct modified and unmodified peptide recognition domains and other modules mediating specialized interactions, for example, the RFD module of DNMT1 which contains a permuted Sm domain linked to a helix-turn-helix domain. In eukaryotes, the evolution of DNA methylases appears to have proceeded in parallel to the elaboration of histone-modifying enzymes and the RNAi system, with functions related to counter-viral and counter-transposon defense, and regulation of DNA repair and differential gene expression being their primary ancestral functions. Diverse DNA demethylation systems that utilize base-excision repair via DNA glycosylases and cytosine deaminases appear to have emerged in multiple eukaryotic lineages. Comparative genomics suggests that the link between cytosine methylation and DNA glycosylases probably emerged first in a novel R–M system in bacteria. Recent studies suggest that the 5mC is not a terminal DNA modification, with enzymes of the Tet/JBP family of 2-oxoglutarate- and iron-dependent dioxygenases further hydroxylating it to form 5-hydroxymethylcytosine (5hmC). These enzymes emerged first in bacteriophages and appear to have been transferred to eukaryotes on one or more occasions. Eukaryotes appear to have recruited three major types of DNA-binding domains (SRA/SAD, TAM/MBD, and CXXC) in discriminating DNA with methylated or unmethylated cytosines. Analysis of the domain architectures of these domains and the DNA methylases suggests that early in eukaryotic evolution they developed a close functional link with SET-domain methylases and Jumonji-related demethylases that operate on peptides in chromatin proteins. In several eukaryotes, other functional connections were elaborated in the form of various combinations between domains related to DNA methylation and those involved in ATP-dependent chromatin remodeling and RNAi. In certain eukaryotes, such as mammals and angiosperms, novel dependencies on the DNA methylation system emerged, which resulted in it affecting unexpected aspects of the biology of these organisms such as parent–offspring interactions. In genomic terms, this was reflected in the emergence of new proteins related to methylation, such as Stella. The well-developed methylation systems of certain heteroloboseans, stramenopiles, chlorophytes, and haptophyte indicate that these might be new model systems to explore the relevance of DNA modifications in eukaryotes.

I. Introduction

A. Methylation and Other Modifications of Bases in Nucleic Acids

Catalytic modification of bases in DNA and RNA occurs universally across the three primary superkingdoms of life (bacteria, archaea, and eukaryotes) and also in several viruses.[1–3] Some of these modifications, such as methylation, thiouridylation, and pseudouridylation of bases in rRNAs and tRNAs, are traceable to the last universal common ancestor (LUCA) of all life and are absolutely required for survival.[1,2,4] Other RNA base modifications are more limited in their distribution. For example, wybutosine is found only in eukaryotic tRNAs, whereas related modifications like 4-demethylwyosine and its derivatives are restricted to the archaeal tRNAs.[1,5] Certain forms of methylation and thiouridylation of different RNAs might show even more restricted phyletic profiles.[1,2] As a rule, modifications of bases in DNA are apparently less diverse and more sporadic in their distribution.[3,6–8] The enzymes catalyzing these modifications are often not essential for basic survival in several lineages of life.[2,9–12] The lower diversity and relatively restricted distributions of DNA modifications appear to be a consequence of the selective constraints imposed by the need to maintain double-helical pairing in DNA, and protecting the genetic material from the potentially mutagenic effects of base modifications. Hence, it is conceivable that the emergence of DNA as the primary genetic material allowed RNAs to retain biochemical diversity essential for their function through a panoply of modifications while safeguarding the genetic material in a relatively unmodified state. Nevertheless, modifications of DNA represent a layer of information beyond that offered by the four typical bases (epigenetic information). As a result, a relatively small set of DNA modifications have emerged in course of evolution, and have been widely used to specify several distinct biological functions.

The most frequent DNA modification in all the three superkingdoms of life is the methylation of cytosine at the 5th position of the pyrimidine ring (5mC).[7,13] The next most frequent DNA methylation is that of adenine on the NH_2 group attached to the 6th position of the purine ring (N6mA), which is fairly common in prokaryotes and certain eukaryotic lineages.[7,13] Prokaryotes also possess a related methylation of the NH_2 group attached to the 4th position of the cytosine ring (N4mC).[7,13] DNA modifications other than methylation are primarily known from caudate bacteriophages and include a spectacular array of modified bases such as 5-hydroxymethylpyrimidines and their mono- or diglycosylated derivatives, α-putrescinylated or α-glutamylated thymines, sugar-substituted 5-hydroxypentyl uracil, and N6-carbamoylmethyl adenines (called Momylation after the Mom enzyme of phage Mu that catalyzes this modification).[3,7] Other DNA base modifications have more recently become apparent in eukaryotes, the simplest of which is the catalytic deamination of cytosine that has thus far only been confirmed in vertebrates.[14–16] Another well-studied eukaryotic modification is the formation of β-D-glucosyl-hydroxymethyluracil (base J) from thymine in euglenozoans, including the parasites *Trypanosoma* and *Leishmania*.[6] A related modification namely 5hmC was first observed in the DNA of caudate phages.[3,7] It has more recently been shown to occur in animals and is predicted to occur more widely across eukaryotes.[8,17] In this chapter, we primarily focus on DNA methylation, with an emphasis on cytosine methylation and its further modification in eukaryotes and their viruses.

The biological consequences of DNA modification are rather diverse across the three superkingdoms of life. The 5C, N6A, and N4C methylation in prokaryotes is primarily catalyzed by methylases from restriction–modification (R–M) systems.[18–20] These systems are widely mobile between diverse bacterial and archaeal genomes. Some can be considered selfish elements that ensure their retention by acting as "addiction" elements, by launching a restriction endonucleolytic attack on the genomes that have lost or disrupted the methylase gene.[21,22] However, they also potentially enhance host fitness by selectively targeting invading DNA such as those of phages, plasmids, and conjugative transposons for endonucleolytic cleavage, while simultaneously protecting the host DNA.[23,24] This self versus nonself recognition is primarily achieved by the action of the methylases encoded by these systems, which provide an epigenetic mark to distinguish one type of DNA molecule from another. The above-mentioned diverse, atypical hypermodified bases observed in the DNA of diverse phages are adaptations, mainly to counter the action of restriction enzymes from the host genome.[25] Some derivatives of the R–M systems, especially the methylase genes, have been co-opted by the prokaryotic hosts as potential defensive elements against restriction attacks by the selfish R–M systems.[21] Further, in several prokaryotes, the epigenetic mark provided by DNA methylation has been reused to distinguish the DNA strands for directing DNA repair. For example, the *vsr–dcm* gene pair in *Escherichia coli* represents a "domesticated" R–M system that is utilized for very short patch repair to correct C-to-T mutations, as well as a defense against selfish R–Ms.[21,26] Several distinct DNA cytosine methylases related to the bacterial R–M methylases are also found in eukaryotes where they primarily function in regulating chromatin organization. Of the other modifications in eukaryotic DNA, cytosine deamination has been shown to play a role in the diversification of immunity molecules in vertebrates.[14–16] In trypanosomes, base J has been shown to be an epigenetic mark that is localized to subtelomeric repetitive DNA and might help in the assembly of transcriptionally silent chromatin associated with the expression of surface antigens in these organisms.[6] The more recently discovered 5hmC has also been shown in vertebrates and predicted in fungi and other eukaryotes to have a key role in organization of chromatin in several cell types.[17,27–29]

B. Enzymes Catalyzing Base-Modifications in DNA, and Domains which Recognize Modifications

A combination of computational analysis of protein sequences, X-ray crystallography, and biochemical studies have helped in identifying and elucidating several aspects of the functions of DNA-modifying enzymes.[5,8,30–38] Some of the enzymes generating modified bases in bacteriophage DNA act prior to DNA replication, synthesizing premodified bases that are then incorporated into DNA during viral synthesis. The best studied of these are the 5hmC and 5-hydroxymethyluracil synthases of several DNA viruses (e.g., T-even phages), which have evolved from the classical thymidylate synthases.[34] In contrast, most other enzymes modify DNA bases *in situ*. The catalytic domains of these DNA-modifying enzymes belong to a relatively small set of structurally distinct folds. Of these, the phage DNA base glycosyltransferases, that further modify the 5-hydroxymethylpyrimidines through the transfer of sugar moieties, belong to two structurally unrelated folds: (1) The glycogen synthase/glycogen phosphorylase fold, which contains enzymes such as the α-glucosyltransferase and β-glucosyltransferase. (2) The Fringe-like glucosyltransferase fold that includes the β-glucosyl-hmC-α-glucosyltransferase.[32,33,39] The phage Mu Mom enzyme and its relatives from diverse organisms, which catalyze the momylation reaction (i.e., addition of carbamoylmethyl or a related adduct to adenines), belong to the GCN5-like acetyltransferase fold.[8] Enzymes catalyzing *in situ* base hydroxylations in DNA, such as those in the first step of base J biosynthesis, and in 5hmC biosynthesis are iron- and 2-oxoglutarate-dependent members of the vast double-stranded β-helix fold, which includes the DNA repair protein AlkB (which oxidatively removes alkyl adducts on adenine), protein hydroxylases, and histone demethylases.[5,8] All currently known deaminases belong to the deaminase-JAB fold of metal-dependent enzymes and include the deaminases that act on bases in RNA (e.g., ADAR and TAD1), DNA (AID), and also free nucleotides.[16] S-adenosine methionine (AdoMet)-dependent methyltransferases belong to five major folds, namely the Rossmann fold, the β-clip fold (i.e., SET-domain methylases), the SPOUT fold,[40–42], and two others not known to methylate DNA or protein.[43] Of these, RNA methylases are known from both the Rossmann and SPOUT folds, whereas all confirmed DNA methylases only belong to the Rossmann fold. Of the protein methylases, those methylating the ε-NH_2 group of lysines contain either a SET domain or a Rossman-fold catalytic domain, whereas all studied protein arginine methylases belong to the Rossmann fold.

Modified bases in DNA are recognized by a set of conserved protein domains, which play a major role as the primary "discriminators" of the epigenetic code.[44–50] While these domains are found in both prokaryotes and eukaryotes, they are particularly diverse and abundant in the latter clade. This is because, unlike in prokaryotes, most of the eukaryotic DNA modifications have a regulatory function—they help in targeting the assembly of specialized chromatin–protein complexes. These complexes establish structurally and functionally distinct chromatin in regions associated with the DNA modification. In this article, we first systematically survey the structure and evolution of enzymes catalyzing DNA methylation, demethylation, and further modifications of methyl groups. We then consider the domains which recognize methylated DNA and the significance of their domain architectures. We present this information as a synthetic overview of the natural history and functional implications of these protein domains.

II. DNA Methyltransferases

A. The Basic Morphology of Rossmann-Fold Methyltransferases

The Rossmannoid folds are a vast assemblage of catalytic domains, typical of diverse enzymes that utilize nucleotide substrates.[13,42,51–53] These folds are characterized by a three-layered sandwich structure made up of multiple β–α units, with a largely parallel central β-sheet sandwiched between two layers of α-helices (Fig. 1). All active members of this fold have a substrate-binding site in the loop bounded by the first β–α unit. Among these, the catalytic domains of methyltransferases, FAD/NAD-dependent dehydrogenases, E1-like adenylating/thiolating enzymes, and the Sir2-like enzymes are closer to each other and form a distinct monophyletic clade of Rossmannoid folds.[51] They are all unified

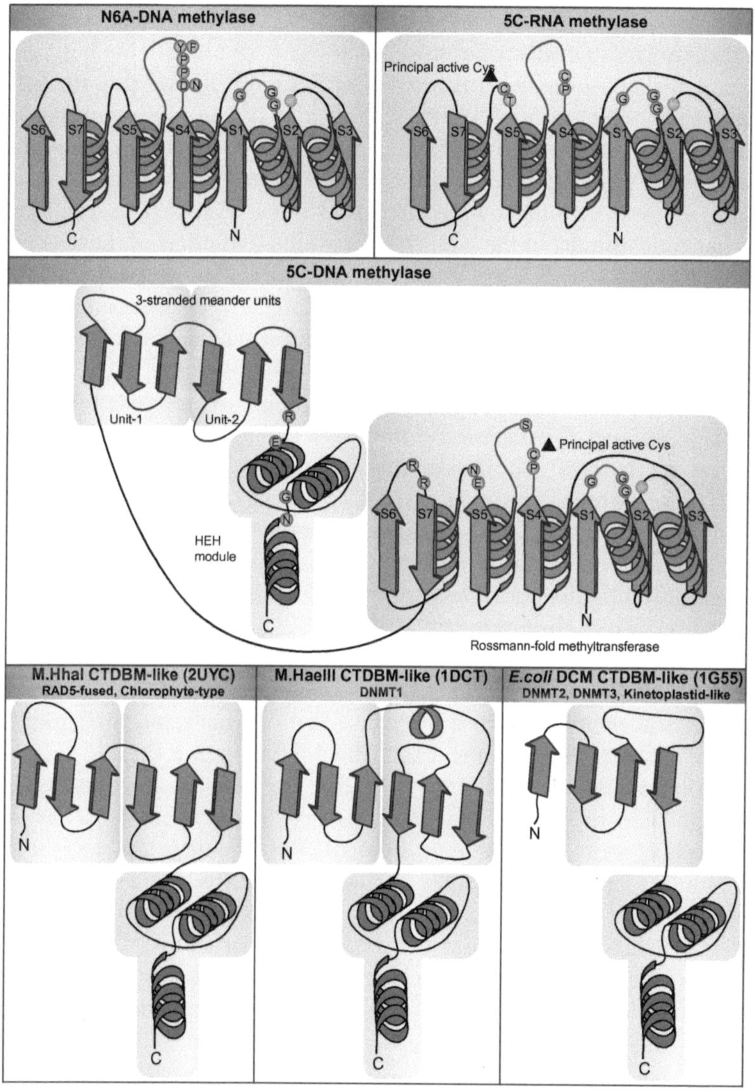

FIG. 1. Structure and sequence features of DNA and RNA methylases. The methylases, and distinct variants of the DNA 5C-MTase CTDBM, are depicted as cartoon topology diagrams. Strands and helices of the Rossmannoid fold core of the methylases are colored green and orange, whereas those of the CTDBM are colored blue and red, respectively. Strands of the core Rossmannoid fold are labeled S1–S6. Key sequence features described in the text, including those involved in AdoMet binding, catalysis, lineage-specific residues, and residues that are frequently mutated in human DNMT3A in acute myeloid leukemia,[84] are shown in gray circles with the residue abbreviation at the corresponding structural element. The blue circle corresponds to the highly conserved polar position in methylases at the end of strand 2, that H-bonds the AdoMet ribose. (See Color Insert.)

by the presence of a glycine-rich loop bracketed by the first β–α unit which binds their nucleotide substrate and a "cross-over" (topological switch point) in their core β-sheet after the 3rd conserved β-strand placing the 4th β-strand adjacent to the 1st strand. The Rossmann fold of the methyltransferases is differentiated from the other domains, in the above-mentioned monophyletic assemblage, by virtue of its specificity for AdoMet and the presence of a unique β-hairpin at the C-terminal end of the core β-sheet (Fig. 1). The second strand of this hairpin (strand 7 of the core) is antiparallel to the rest of the sheet and is inserted between strand 5 and strand 6 of the core. The AdoMet specificity is achieved in large part by the several contacts made by the binding loop in the first β–α unit with the cofactor and also a conserved polar residue (usually acidic) at the end of strand 2 of the core, which H-bonds the sugar of the AdoMet. While some variations to this basic template are encountered in the AdoMet-dependent Rossmann-fold methylases, majority of nucleic acid base-modifying methylases conform to it. The methyl transfer reaction usually depends on one or more residues at the C-terminus of strand 4. In this respect, the methylases follow the ancestral Rossmannoid condition, wherein a catalytic residue is often found at the end of strand 4, as is also observed in several other Rossmannoid folds that catalyze various unrelated reactions.[51] In the case of DNA methylases, these residues play a key role in initiating the attack on the substrate atom to facilitate acceptance of the methyl group from AdoMet. However, because the target atoms of the 5C and N6A/N4C methylases are very distinct in their properties, the conserved residue/s and their role in the respective catalytic mechanisms drastically differ between them.

In evolutionary terms, all DNA methylases belong to a large monophyletic assemblage, which is unified by the presence of a characteristic large loop immediately C-terminal to the core strand 4 (Fig. 1), and is distinguished from other families of Rossmann-fold methylases such as the neurotransmitter biosynthesis methylases and the RNA methylases GCD10 and GCD14 (which methylate adenine-58 at the 1st position in $tRNA^{Met}$) that lack this loop.[4] Most members of this assemblage methylate bases in nucleic acids, or amino acid side chains of nucleoproteins.[4] The characteristic post-β4 loop shared by them plays a major role in binding their nucleic acid substrates, typically in conjunction with lineage-specific and unrelated globular domains fused to the N- or C-terminus of the core Rossmann-fold domain. Within this assemblage, the N6A/N4C and 5C methylases show specific relationships to distinct sets of RNA or nucleoprotein methylases.[4] Typically, these RNA/nucleoprotein methylase families have a much wider phyletic distribution, suggesting that many of them had emerged in the LUCA or at the base of the bacterial or archaeo-eukaryotic lineages.[4] In contrast, the DNA methylases are sporadically distributed and presumably derived within the prokaryotic lineages from the more ancient RNA methylases. In discussing the evolution of the DNA methylases, we first consider the origin of the DNA N6A methylases (including the related N4C methylases) and their hitherto underappreciated presence in eukaryotes. We then consider the origin and diversification of the various families of 5C methylases in detail.

B. DNA Adenine Methyltransferases

The N6A methylases and the related N4C methylases contain a characteristic signature at the C-terminus of strand 4, that is typically of the form [NDS] PP[YFW][4,42,53–55] (Fig. 1). They share this signature with several highly conserved RNA methylases, such as RsmC/RsmD/YcbY(RlmL), which methylate the N2 position in various Gs in rRNAs; TrmA, which methylates U54 at the 5th position in most tRNAs; and the nucleoprotein methylases like HemK and YfcB, which methylate the glutamine side chain in the ribosomal protein L3 and peptide release factors.[4] Of these, the classical DNA N6A methylases appear to be most closely related to the HemK–RsmC–RsmD clade, which is consistent with the similarity in their substrates: an $-NH_2$ group.[4] Studies on the bacterial N6A methylases from R–M systems such as M.TaqI indicate that the aromatic [YFW] residue from the above signature stacks against the flipped-out base via π–π interactions.[56] The conserved polar [NDS] residue, and the proline after it in this motif, interacts via hydrogen bonds with the target—NH_2 group on adenine.[57] It is believed that these residues either decouple the lone electron pair of the target nitrogen from the aromatic ring, or increase its charge density for a nucleophilic attack to facilitate the transfer of the methyl group from AdoMet. Most prokaryotic N6A DNA methylases are found in R–M systems, which have been widely disseminated via lateral transfer across distantly related lineages.[20] However, on multiple occasions, in several bacterial lineages, N6A methylases derived from R–M operons, such as Dam in γ-proteobacteria and CcrM in α-proteobacteria, have been exapted for cellular roles.[58,59] The Dam methylases provide an epigenetic mark to distinguish the two strands of the duplex during DNA repair by the MutHLS system (MutH was derived from the endonuclease component of an ancestral R–M system). Methyl marks produced by the above enzymes are also implicated in the assembly of the replication initiation complex at the methylated *oriC* and the regulation of transcription by modification of promoters and other transcription factor target sites on DNA.[58,59] Thus, bacterial and phage Dam/CcrM methylases represent some of the earliest instances of the recruitment of originally selfish R–M-derived methylases for purely cellular regulatory functions. Such methylases (e.g., the phage T4 Dam) have also been acquired by certain phages where they appear to have a comparable regulatory role.[60,61] Among the N6A methylases there is a distinctive group of circularly permuted forms, typified by M.MunI and the *Caulobacter cresentus* CcrM. These versions may have one or more additional N-terminal strands which might insert into the core sheet of the methylase domain.[62,63] N4C methylases related to both the typical and permuted forms of N6A methylases have been uncovered. This suggests that N4C methylases have evolved independently on multiple occasions from both types of the N6A methylases within R–M systems.

N6mA is relatively uncommon in most eukaryotes, but has been positively identified in several lineages of ciliates, chlorophyte algae, and dinoflagellates, where it constitutes 0.5–10% of the adenines in genome.[7,64] To date none of the enzymes involved in these DNA methylation events have been identified. Making use of the currently available genome sequences from several of these organisms, we were able to confidently identify numerous potential N6A methylases related to Dam across the eukaryotic superkingdom (Fig. 2; Supplementary Material: ftp://ftp.ncbi.nlm.nih.gov/pub/aravind/chromatin/methylase/supplementary.html). Of these, several distinct versions appear to be specified by different types of mobile elements. *Trichomonas* possesses several paralogous N6A methylases that are often fused to a domain found in phage structural proteins (e.g., gi: 121901620, TVAG_056220). These appear to have been ultimately derived from a phage version and are encoded by a virus-like transposable element that is highly expanded in the genome of this organism (Supplementary Material). A second subset of eukaryotic N6A methylase domains are encoded by a distinct family of retroposons, whose archetypal member is the *Dictyostelium* DIRS-1 element,[65,66] that has widely disseminated across eukaryotes and expanded in several distantly related organisms (Fig. 2; e.g., gi: 167739, *Dictyostelium* DIRS1 ORF3; Supplementary Material). The main protein specified by the complete versions of these retroposons contains N-terminal reverse transcriptase (RT) and RNaseH domains fused to a C-terminal Dam-like methylase domain. The methylase domain appears to be inactive due to disruption of the AdoMet-binding loop and the key motif at the end of strand 4 in most DIRS-1-like retroposons from animals (e.g., the fishes *Tetraodon* and *Danio rerio*, the frog *Xenopus*, and the nematodes *Caenorhabditis briggsae*, *C. remanei*, and *Nematostella*) and *Dictyostelium.* Thus, it is more likely that the inactive methylase domain of the animal versions of these retroposons functions as a DNA-binding regulatory protein rather than a DNA-modifying enzyme. However, in some chlorophyte algae (e.g., *Volvox* retroposon ORF-B, gi: 22415757) at least one of the copies of the retroposon codes for an active methylase domain, which might generate a part of the N6mA detected in the genomes of chlorophyte algae. A version of eukaryotic Dam-like methylases is encoded by the CrRem1-like LTR-containing retroposons,[67] currently only found in chlorophytes (e.g., *Volvox* and *Chlamydomonas*). The complete versions of this element encode a polyprotein with the Dam-like methylase fused to C-terminal aspartyl protease and RT domains. Additionally, these elements also specify a protein with a chromodomain and PHD finger that might regulate the methylation catalyzed by the Dam-like

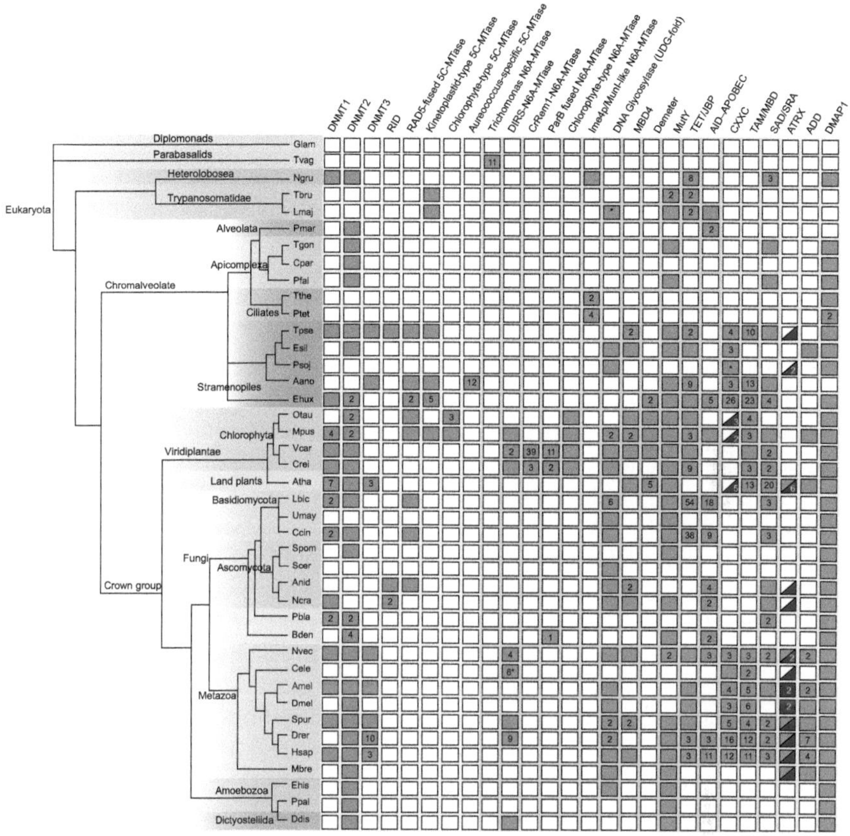

FIG. 2. Phyletic patterns of DNA methylases and functionally related enzymes and proteins. These are shown to the right of the eukaryotic tree. A filled box with numbers depicts the presence, and number of representatives, of a protein or domain family shown in the column for a given species. These numbers represent an approximate count, for they might include pseudogenes in some organisms whose genomes are poorly studied. Numbers are not shown in the filled boxes for species with a single representative. An asterisk is used in a box if a protein or domain family, though absent in the given species, was present in a closely related species. Divided boxes were used for the CXXC domains and ATRX proteins to distinguish the mono- and bi-CXXC units, and the ATRX proteins with and without the ADD module, respectively. The DRD1-like proteins of plants have been included in the ATRX column. In these instances, the darker half of the box is used to depict the presence and numbers of the mono-CXXC domains and the ATRX proteins without the ADD module. Species abbreviations in the eukaryotic tree are as follows: Aano, *Aureococcus anophagefferens*; Amel, *Apis mellifera*; Anid, *Aspergillus nidulans*; Atha, *Arabidopsis thaliana*; Bden, *Batrachochytrium dendrobatidis*; Ccin, *Coprinopsis cinerea*; Cele, *Caenorhabditis elegans*; Cpar, *Cryptosporidium parvum*; Crei, *Chlamydomonas reinhardtii*; Ddis, *Dictyostelium discoideum*; Dmel, *Drosophila melanogaster*; Drer, *Danio rerio*; Ehis, *Entamoeba histolytica*; Ehux, *Emiliania huxleyi*; Esil, *Ectocarpus siliculosus*; Glam, *Giardia lamblia*; Hsap, *Homo sapiens*; Lbic, *Laccaria bicolor*; Lmaj, *Leishmania major*; Mbre, *Monosiga brevicollis*; Mpus, *Micromonas pusilla*; Ncra,

enzyme. A fourth subset of Dam-like methylases is currently found only in chlorophytes such as *Chlamydomonas* and *Volvox*, and chythrid fungi, and are specified by a previously uncharacterized type of transposon (Fig. 2; Supplementary Material). These Dam-like proteins are fused to an N-terminal bacterial chromosome partition protein ParB-type HTH domain[68] (e.g., *Volvox* gi: 302854263, VOLCADRAFT_108225). The pervasive presence of Dam-like methylases associated with distinct groups of transposons suggests that they might act in *cis* to control their own gene expression and mobility through methylation of specific adenines within themselves or in their vicinity. This is reminiscent of the regulation of the movement of certain bacterial transposable elements by DNA methylation.[69]

In addition to these transposon-coded enzymes, there are other potential eukaryotic N6A methylases, which appear to be cellular enzymes with a role in chromatin organization. One of these, found across chlorophyte algae, but not land plants, is a multidomain protein with the N6A methylase domain fused to one or more N-terminal BMB/PWWP and C-terminal PHD-X/ZF-CW domains (e.g., *Volvox* VOLCADRAFT_89771, gi: 302835622). Additionally, they often contain PHD finger domains N-terminal to the methylase domain (Figs. 2 and 3). These fusions, to multiple domains implicated in binding trimethylated lysines on histones, suggest that these enzymes localize to specific regions of chromatin which bear such marks to catalyze localized N6A or N4C methylation. Thus, these enzymes could possibly represent the first dedicated eukaryotic methylases generating modifications other than 5mC in chromatin organization. A Dam-like methylase, typified by the human PCIF1, was also acquired by eukaryotes from bacteria prior to their radiation from the last eukaryotic common ancestor (LECA), and is fused to an N-terminal WW domain.[4,70] This version interacts with the phosphorylated CTD of the RNA polymerase-II via the WW domain[70] and is conserved throughout eukaryotes, even among organisms in which there is no evidence for N6A DNA methylation. This phyletic pattern is typical of RNA methylases and, given its role in coupling pre-mRNA processing to transcription, it is likely to function as an RNA N6A methylase rather than a DNA methylase. A similar transfer of an N6A methylase from bacteria to eukaryotes prior to their radiation from the LECA occurred in the form of the IME4-like (also called MT-A70) family which is also widely conserved in eukaryotes.[4,71] These are related to the MunI-like

---

Neurospora crassa; Ngru, *Naegleria gruberi*; Nvec, *Nematostella vectensis*; Otau, *Ostreococcus tauri*; Pbla, *Phycomyces blakesleeanus*; Pfal, *Plasmodium falciparum*; Pmar, *Perkinsus marinus*; Ppal, *Polysphondylium pallidum*; Psoj, *Phytophthora sojae*; Ptet, *Paramecium tetraurelia*; Scer, *Saccharomyces cerevisiae*; Spom, *Schizosaccharomyces pombe*; Spur, *Strongylocentrotus purpuratus*; Tbru, *Trypanosoma brucei*; Tgon, *Toxoplasma gondii*; Tpse, *Thalassiosira pseudonana*; Tthe, *Tetrahymena thermophila*; Tvag, *Trichomonas vaginalis*; Umay, *Ustilago maydis*; Vcar, *Volvox carteri*.

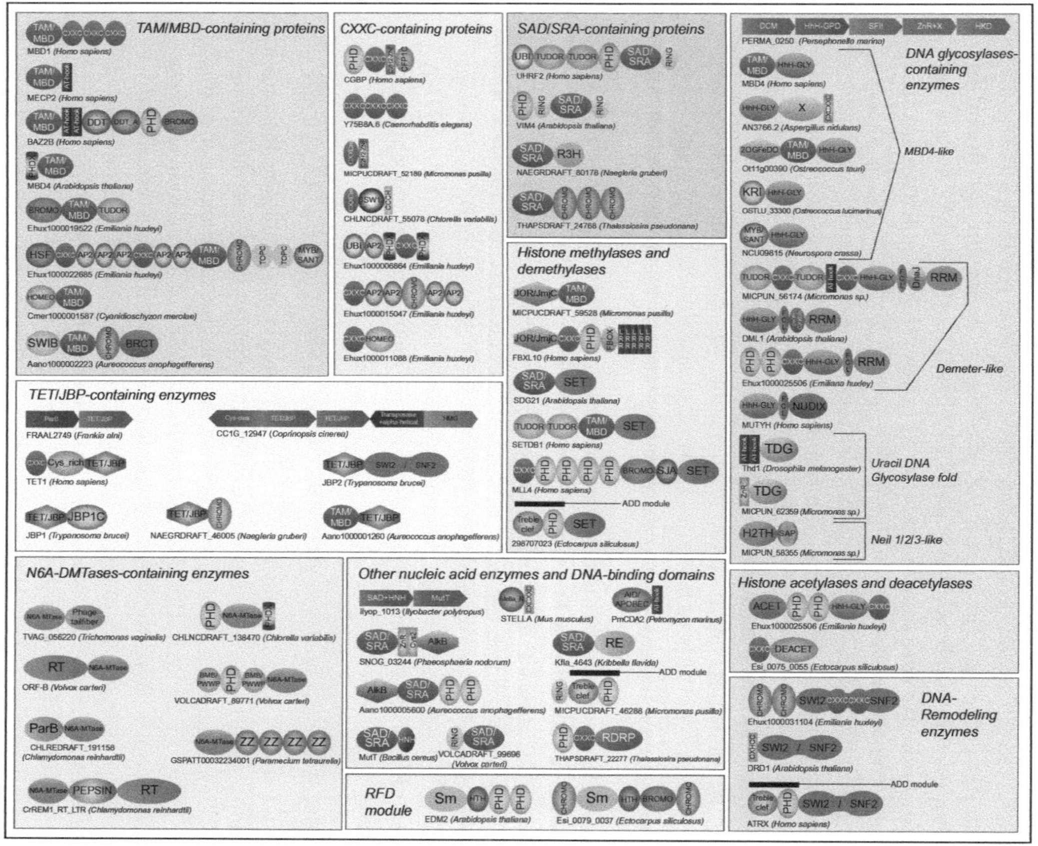

FIG. 3. Domain architectures and gene neighborhoods of various proteins related to DNA methylation. These are arranged based on various groups of enzymatic and DNA-binding domains. Proteins are labeled with their gene id and source species name. Standard abbreviations are used for most domains; "X" refers to unknown globular domains. A comprehensive list of nonstandard domain names can be found in the legend to Fig. 6. Refer to the Supplementary Material for a comprehensive list of architectures and gene neighborhoods. Temporary gene names are used for proteins from the unpublished sequences of *Emiliania*, *Aureococcus anophagefferens*, and *Micromonas pusilla*. To access these protein sequences, refer to the Supplementary Material in the FTP site.

circularly permuted methylases of bacterial R–M systems, rather than to the classical Dam methylases. Representatives of this family (like IME4) methylate mRNA rather than DNA, suggesting an early substrate shift after the transfer to eukaryotes. Certain members of this family (like *Saccharomyces* KAR4) are inactive and have been exapted to function as a transcription factor rather than a methylase.[72] In ciliates, we found a distinctive version of the IME4-like family, which is fused to four N-terminal ZZ Zn-fingers, a domain also found in chromatin proteins such as ADA2 and CBP/p300 (Figs. 2 and 3). Given that all ciliates studied to date show substantial N6mA in DNA[7,64] and have no other candidate methylases to catalyze this reaction, we suggest that these ZZ-domain containing methylases indeed perform this function. Additionally, orthologous methylases of this ciliate version are found in the heterolobosean amoebo-flagellate *Naegleria* and the rhodophyte alga *Cyanidioschyzon*, suggesting a wide distribution for this form of adenine methylation across eukaryotes (Fig. 2). Beyond these more conserved versions, we also found evidence for sporadic lateral transfers of bacterial R–M or phage-derived N6A methylases in *Naegleria* and the stramenopile alga *Emiliania* (Fig. 2; Supplementary Material).

There has been a report that a plant protein of the TRM11 family of RNA methylases functions as a DNA adenine methylase in plant mitochondria.[73] However, this appears dubious, since these proteins belong to a class of conserved RNA methylases with the RNA-binding THUMP domain that have been demonstrated to methylate tRNA at the G10 nucleotide to generate an m2G.[74,75] Further, the plant proteins appear to lack a mitochondrial DNA targeting peptide that would be needed for it to methylate the mitochondrial genome.

C. Origin of 5C DNA Cytosine Methylases

Unlike the 5C DNA methylases, 5C RNA methylases (typified by the Sun/Fmu-Nop2 family) have a universal distribution across the three superkingdoms of life, suggesting an origin in the LUCA.[4] The bacterial member of this family (Fmu) methylates 16S rRNA to generate 5mC at nucleotide 967 in a conserved loop.[76] Given the sporadic distribution of 5C DNA cytosine methylases across prokaryotic genomes,[4,20] it is likely that they emerged from an RNA methylase of the Sun/Fmu-Nop2 family in bacterial R–M systems. The 5C DNA methylases share with the 5C RNA cytosine methylases a conserved PC motif found at the C-terminus of strand 4.[77,78] Studies on the 5C DNA methylases suggest that this cysteine in the above motif is central to the catalytic mechanism by forming a covalent adduct with the C6 carbon to facilitate methylation of the C5 carbon.[79,80] Interestingly, while this cysteine plays a certain role in optimal catalysis by RNA methylases, it does not appear to have a primary catalytic role in these enzymes.[81] Instead, the equivalent catalytic role is performed by a second cysteine found at the C-terminus of strand 5 in a TC motif that is only conserved among the RNA methylases. Thus, the emergence of the DNA cytosine methylases from a Sun/Fmu-like precursor appears to have been accompanied by the loss of the TC motif and the complete shift of the catalytic activity to the cysteine associated with strand 4.

Additionally, emergence of the DNA methylases from their RNA-modifying counterparts involved acquisition of several features that allowed for flipping out of cytosine and specific interaction with the base after its eversion from double stranded DNA.[79,82] The most prominent of these features was acquisition of the C-terminal DNA-binding module (CTDBM). The CTDBM is a composite module that emerged through the fusion of two distinct closely interacting domains (Fig. 1). First is a C-terminal trihelical unit that is a derived version of the helix-extension-helix (HEH) domain[83] (Fig. 1). It lacks the small N-terminal helix of the HEH but has an additional C-terminal helix. However the core HEH structure, comprising the first and second helices and the extended connector between them, contacts DNA in a manner comparable to the classical HEH-fold domains such as SAP.[83] Specifically, a highly conserved GN motif at the end of the second helix of this unit contacts the flipped-out cytosine.[79,82] Recurrent mutations compromising the conformation of this derived HEH domain in DNMT3A, which are likely to affect its affinity or specificity, are observed in patients with acute myeloid leukemia.[84] Second is an N-terminal element comprising two copies of a 3-stranded β-meander unit, which is typified by large loops assuming the "hammer-head" configuration and connecting the successive strands of each unit (Fig. 1). A salt bridge, between an arginine in the last strand of this element and a glutamate in the first helix of the derived HEH unit, tightly links the two domains of the CTDBM. Each 3-stranded unit of the N-terminal element might contain large inserts in the "hammer-head" loops and show extreme sequence divergence. The two copies of the 3-stranded unit might also show considerable differences in the spatial arrangement with respect to each other. The hammer-head loops from one or both the units play an important role in recognition of the target sequence, and insert deeply into the DNA duplex to facilitate flipping out of the target base.[79,82] Comparison of different 5C DNA methylase CTDBM structures suggests that having two tandem copies of the 3-stranded unit placed in immediate succession after each other is probably the ancestral condition of this element in the DNA methylases (e.g., in M.HhaI).[85] Further development is seen in versions (typified by M.HaeIII) wherein a long insert separates the two 3-stranded units of the N-terminal element of the CTDBM.[35] Finally, there are versions (such as *E. coli* Dcm) in which only the C-terminal 3-stranded unit is intact, whereas the N-terminal unit has lost the first two strands (PDB: 3LX6). We discuss the exact condition of the CTDBM in the eukaryotic 5C DNA methylases further as we consider each of them individually.

Concomitant with the acquisition of the CTDBM, the core catalytic domain of the 5C DNA methylases also acquired several distinctive features to interact with and capture the flipped-out base in a suitable conformation for catalysis.[85–87] The chief of these features are shown in Fig. 1. First is a highly conserved glutamate at the C-terminus of strand 5 that makes a salt-bridge with the 4-NH_2 and 3N positions of the cytosine to hold the flipped-out base in place. Second is a conserved arginine (part of a highly conserved RxR motif) at the beginning of strand 7, that makes a polar interaction with the cytosine 2-oxo, and also helps in positioning the flipped-out base. It is possible that this arginine also acts as the general base to complete the methylation reaction by restoring the aromaticity of the pyrimidine that is broken by the covalent interaction with the catalytic cysteine. Third is a highly conserved serine, four residues downstream to the PC motif C-terminal of strand 4, that makes a polar interaction with the phosphate backbone of DNA, stabilizing the phosphoester bond torsion that accompanies the base flipping. These three features, together with those in the CTDBM, form an intricate mechanism to present the cytosine to the catalytic cysteine and the bound AdoMet substrate. The complete absence of all these elements in the Sun/Fmu-Nop2 family strongly supports a single origin for all 5C DNA methylases from the RNA-modifying precursor, with subsequent elaborations as a part of the diversification of R–M systems across prokaryotes. In addition to the "in-built" DNA-binding domain in the form of the CTDBM, several methylases in R–M systems acquired additional DNA-binding domains, which might have a role in refining the target specificity or aiding in more complex contacts with DNA. One notable example of this is the fusion of the methylase domain (e.g., *Frankia* gi: 288919493, FrEUN1f-DRAFT_3521) to the iron–sulfur cluster-coordinating, redox-senstive FCL DNA-binding domain (also found in MutY-like DNA glycosylases and certain nucleases with RecB-type nuclease domains).[88,89] This domain might help these methylases to modify DNA in a redox-sensitive manner. Further, there are multiple independent fusions to diverse types of helix-turn-helix domains in methylases from various R–M operons. Interestingly, certain cyanobacterial Dcm-like 5C DNA methylases display a fusion to a similar ParB-like HTH, similar to the one fused to the Dam-like methylase domain from the above-described eukaryotic transposons (e.g., *Nostoc* Npun_F2574, gi: 186682875). Beyond fusions to distinct DNA-binding domains, the methylases also developed fusions to their cognate restriction endonucleases (REases) in several R–M systems of prokaryotes.[20]

D. Diversity of 5C DNA Methylases in Eukaryotes and Their Viruses

5mC has been observed in the genomes of a wide range of eukaryotes, albeit with patchy phyletic patterns (Fig. 2). Even members of a given lineage might differ widely in their 5C methylation status. For example, while most animal lineages have 5mC and the enzymes catalyzing this modification, it appears to have been entirely lost in nematodes such as *Caenorhabditis elegans* (Fig. 2). Likewise, within arthropods, most dipterans like *Drosophila* and *Aedes* have at best very limited methylation (along with loss of most of their 5C methylases; Fig. 2)[90] (see Chapter by Veiko Krauss and Gunter Reuter), whereas hymenopterans like honeybees, ants, and wasps have extensive 5mC having considerable significance to their biology.[91–94] Since the cloning and characterization of the first eukaryotic 5C DNA methylases, their relationship to the cognates found in bacterial R–M operons has been recognized.[13,95,96] Despite this, there is considerable confusion in the literature regarding the actual interrelationships between the eukaryotic members and the bacterial representatives to which they are most closely related.[97–99] We used the currently available wealth of data from bacterial and eukaryotic genomes and structures to elucidate this issue, and also present several examples of novel DNA 5C methylases beyond those that have been well characterized in the model organisms. Accordingly, we first discuss the evolution and domain architectures of the well-studied DNMT methylases and their close relatives, and then discuss the other novel groups of poorly studied 5C DNA methylases. The best-studied 5C methylases of eukaryotes, namely the DNMT methylases,[100,101] can be classified on the basis of sequence conservation patterns and phylogenetic analysis into three major monophyletic groups that have very distinct evolutionary histories (Figs. 3 and 4). The first of these is the DNMT1-chromomethylase-RIP methylase group, the second is the DNMT3 group, and the third is the DNMT2 group (see also Chapters by Željko M. Svedružić; Frédéric Chédin).

1. The DNMT1-Chromomethylase-RID Methylase Group

One of the first eukaryotic methylases to be extensively characterized was the DNMT1 enzyme from mammals,[95] which is thought to function as the primary maintenance methylase that reestablishes the methylation marks at CpG sites on both strands of the duplex after replication[102–105] (though see Chapter by Željko M. Svedružić). In vertebrates, it appears to be an essential gene, with DNMT1 knockout mice showing embryonic lethality.[106] It is also critical for egg cell reprogramming, and controlling gene silencing in both transposons and euchromatic regions. In plants, disruption of DNMT1 orthologs results in partial sterility and homeotic transformations during floral development.[107–109] Thus, in both animals and plants, the disruption of normal methylation by this enzyme results in loss of integrity of the germline.[98,110]

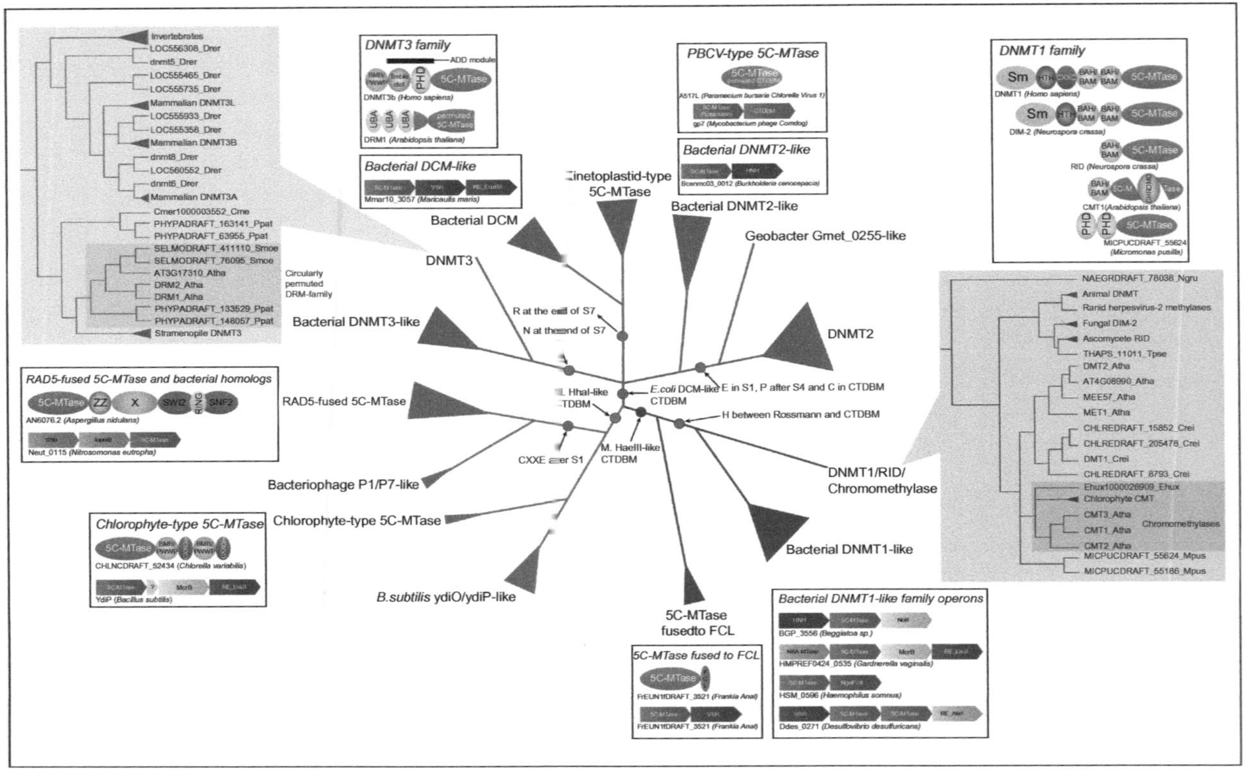

FIG. 4. Evolution of 5C-MTases. The maximum-likelihood (ML) tree of the 5C MTases was derived from a comprehensive multiple alignment (Supplementary Material) of different 5C MTases using the FastTree and Mega programs.[278,279] The higher order relationships were constrained using structural information based on the three distinct CTDBMs shown in Fig. 1. The links of each of the eukaryotic clades to their respective bacterial representatives was supported by $>85\%$ Bootstrap support in the ML trees. The central tree shows the overall relationships of the different 5C MTase

The cognate of this methylase in fungi is typified by the *Neurospora* DIM-2 protein, which is required for both *de novo* and maintenance methylation.[11] While all stable methylation observed in this organism depends on DIM-2, unlike in animals and plants, its deletion does not result in developmental defects.[11] Plants possess a second group of 5C DNA methylases related to DNMT1, the chromomethylases (CMTs), which are characterized by the distinctive insertion of a chromodomain into the methylase domain (Fig. 4).[50,111] In the multicellular plant *Arabidopsis*, one of the CMTs (CMT3) is involved in the methylation of CpNpG rather than CpG and is a critical player in the RNA-directed DNA methylation process observed in plants.[112–114] Ascomycete fungi possess a second distinct methylase related to DNMT1, exemplified by RID (*R*epeat-*I*nduced point mutation *D*efective) from *Neurospora* and Masc1 from *Ascobolus*.[10,115] These methylases are implicated in a related set of phenomena: repeat-induced point mutation (RIP) in *Neurospora* and probably *Uncinocarpus reesii* and methylation-induced premeiotically (MIP) in *Ascobolus*.[10,99,115] In RIP, pairwise linked or unlinked DNA repeats are methylated densely in ascogenous tissue followed by point mutation of the methylated copy through 5mC deamination.[10] In MIP, short sequences are methylated on CpG while longer sequences are methylated throughout and targeted for gene silencing. Both *Ascobolus* Masc1 and the *Aspergillus* ortholog are required for proper sexual development, suggesting that methylation by these enzymes might be required for the integrity of the germline as observed for DNMT1 in animals and plants.[115,116] The plant and animal DNMT1, fungal DIM-2, the CMTs, and the RID-like methylases are unified and differentiated from all other DNMTs of eukaryotes by the presence of two 3-stranded units in the N-terminal element of their CTDBM (see above). Moreover, the two 3-stranded units of the CTDBM of this clade are separated by an insert comparable to that seen in the CTDBM of M.HaeIII[35] (Fig. 1).

---

families described in the text. The branches of the DNMT1 and DNMT3 clades are shown in greater detail to the right and left, respectively, to illustrate the presence of multiple lineage-specific duplications described in the text. The phycodnaviral and iridoviral methylases are not shown in the tree, due to their extreme divergence and architectural reorganization. A comprehensive overall tree and trees of individual families can be accessed from the Supplementary Material. Sequence motifs and structural features that further support various relationships are shown next to filled circles. Relevant domain architectures and operons are arranged around the tree. Operons are shown as boxed arrows with the arrowhead pointing from the 5′ gene to the 3′ gene. Domain architectures and operons are labeled with the gene and species name of a given protein. For operons, the gene name corresponds to the 5C DMTase in the operon. Species abbreviations of organism depicted in the trees are as follows: Atha, *Arabidopsis thaliana*; Cmer, *Cyanidioschyzon merolae*; Crei, *Chlamydomonas reinhardtii*; Drer, *Danio rerio*; Mpus, *Micromonas pusilla*; Ngru, *Naegleria gruberi*; Ppat, *Physcomitrella patens*; Smoe, *Selaginella moellendorffii*; Tpse, *Thalassiosira pseudonana*. Standard gene names are not available for proteins from genomes whose translations are currently not accessible from Genbank: *Emiliania*, *Aureococcus anophagefferens*, and *Cyanidioschyzon merolae* (protein sequences available in Supplementary Material).

Further, this clade of methylases is also unified by the presence of a conserved histidine present immediately downstream of the last (7th) strand of the core Rossmann domain in the extended linker that connects the former domain to the CTDBM (Supplementary Material).

A combination of phylogenetic trees and analysis of phyletic pattern suggests that these methylases diverged from a single precursor within eukaryotes (Fig. 4). The core of this clade of methylases is the DNMT1 methylase from which the CMTs and RID-like methylases arose as lineage-specific branches. A representative of the classical DNMT1 methylase is found in animals, fungi (DIM-2), land plants, their basal chlorophyte relatives, and the early-branching eukaryote *Naegleria*. This suggests that DNMT1 was acquired early in eukaryotic evolution, prior to the divergence of the heteroloboseans, followed by multiple losses in lineages such as kinetoplastids, alveolates, stramenopiles, and amoebozoans (Fig. 2). Recent work proposed that the fungal DIM-2 represents a distinct paralog, closer to the plant CMTs that was lost in animals rather than being the DNMT1 ortholog in fungi.[98,99] However, this view conflicts with multiple lines of evidence. First, the parsimony principle and the basal position of the *Naegleria* DNMT1 with respect to the other eukaryotic versions both suggest that the fungal version is merely a divergent ortholog of DNMT1 (the above proposal posits a greater number of duplications and losses than necessary to explain the observed phyletic patterns; Fig. 4). Second, they are the only methylases produced by fungi that retain the ancestral domain architecture of the eukaryotic DNMT1. Hence, this suggested relationship between DIM-2 and the CMTs is likely to be an artifact of not including basal versions (e.g., from *Naegleria*, the RID-like methylases and the actual bacterial cognates of this group of methylases) in a phylogenetic analysis.

The ancestral architecture of DNMT1 can be reconstructed as comprising a methylase module (including the catalytic domain and the CTDMB) fused to the N-terminal RFD module and two BAM(BAH) domains (Fig. 4). Structural analysis of the RFD module reveals two distinct globular domains,[117] an N-terminal circularly permuted version of the Sm domain, and a C-terminal HTH domain of the four-helical variety.[118,119] Sequence analysis shows that this RFD module occurs independently of DNA methylation across a wide range of eukaryotes, either as a stand-alone protein or fused to PHD (e.g., *Arabidopsis* EDM2, gi: 9758171) or chromo- and bromodomains (e.g., *Ectocarpus* Esi_0079_0037, gi: 298714686, Fig. 3). In *Schizosaccharomyces pombe*, the RAF2 protein with a solo RFD module is implicated in establishing heterochromatin at centromeres.[120] In vertebrates, the RFD module of DNMT1 recruits the histone deacetylase HDAC2 and DMAP1 (a SANT domain protein) to replication foci during S-phase, to maintain repressive chromatin through replication.[117] Thus, emergence of DNMT1 appears to have proceeded via fusion of the RFD module and the BAM(BAH) domains to an ancestral DNA methylase derived from a bacterial R–M system. These fusions provide a means of recruiting it to repressive chromatin and also potentially maintaining repressive chromatin, not just by the action of the methylase domain but also via recruitment of repressors through the RFD module. Given that the Sm domain binds RNA in other contexts,[119] it would be useful to know whether the RFD–Sm domain has a role in RNA-mediated regulation of DNA methylation that has been observed in several eukaryotes[114,121,122] (see also Chapter by Anton Wutz). In metazoans, the architectural complexity of DNMT1 increased further via the insertion of a DNA-binding CXXC domain, between the RFD–HTH domain and the first BAM(BAH) domain.[123] Additionally, the metazoan RFD module has gained a neomorphic Zn-chelating site, characterized by a CXXC motif N-terminal to the RFD–Sm domain and an HxC motif within the RFD–Sm domain itself. The metazoan DNMT1s are also characterized by the emergence of a low-complexity sequence in the form of KG dipeptide repeats just N-terminal of the methylase module.[96] It is possible that these lysines are targets for methylation by SET-domain proteins to regulate the activity of the DNA methyltransferase. While most eukaryotes possess only a single DNMT1, some plants exhibit independent lineage-specific expansions of DNMT1.[109] For example, both the basal chlorophyte *Chlamydomonas* and the land plants like *Arabidopsis* have independently acquired four distinct paralogs of DNMT1 through lineage-specific duplications.

The CMTs appear to have emerged in the plant lineage through duplication and divergence from DNMT1. This proposal is supported by the presence of a synapomorphic HP sequence signature present within helix 2 of the methylase catalytic domain that is uniquely shared with the plant DNMT1s (Supplementary Material). Their presence in the chlorophyte algae such *Chlamydomonas* and *Chlorella* indicates that the precursor of CMT diverged from DNMT1 prior to the radiation of land plants and chlorophyte algae from their common ancestor. This divergence was accompanied by the loss of the N-terminal RFD module and insertion of the chromodomain just downstream of strand 3 of the catalytic domain (Figs. 2 and 4), suggesting a clear functional differentiation with respect to the ancestral DNMT1, perhaps in relation to RNA-directed methylation. CMT appears to have been transferred from the plant lineage to a haptophyte alga *Emiliania*, which shares a common environment with several chlorophyte algae. Within the plant lineage, multiple independent duplications of CMTs have occurred in both certain chlorophytes and angiosperms such as *Arabidopsis* (three CMT paralogs).[109] In addition to the CMTs, in certain chlorophyte algae like *Micromonas* (e.g., gi: 303273542, MICPUCDRAFT_55624), a distinct type of methylase arose via duplication and divergence from the DNMT1s, characterized by two N-terminal copies of PHD finger domains (Fig. 4). Since these algae lack CMTs, it remains to be seen if this group of PHD-containing methylases have taken up their role in recognizing methylated lysines. Profile–profile comparisons show that the RID-like methylases are closest to the fungal DNMT1 orthologs, that is, the DIM-2 methylases. Within fungi, they are limited to the lineage of filamentous ascomycetes known as the leotiomyctes (Fig. 2); hence, they appear to have emerged relatively late in fungal evolution through loss of the N-terminal RFD domain and one of the BAM(BAH) domains, and rapid divergence of the other copy seen in the ancestral DNMT1s. Interestingly, outside fungi, RID-like methylases are found in the diatom *Thalassiosira* (Fig. 2). Given the clear affinities of the RID-like methylases to the fungal DNMT1 orthologs, and the sporadic presence in this single stramenopile lineage, it is likely that the RID-like methylase was horizontally transferred to the diatoms. In addition to cellular eukaryotes, multiple paralogs of DNMT1 are present in certain herpesviruses, such as the Ranid herpesvirus-2 that infects frogs (e.g., RHV-2 gp86 and gp120 proteins). Phylogenetic trees and domain architecture analysis suggest that these viral versions were derived from the metazoan DNMT1 through the loss of the N-terminal RFD and CXXC domains, while retaining the BAM/BAH domain. The genome of this virus is highly methylated[124]; hence, these enzymes could be deployed to methylate the viral genome, perhaps as a mechanism to evade host DNA sensors.[125]

Outside eukaryotes, the closest relatives of DNMT1 and allied methylases are a distinct group of methylases found in bacterial R–M systems typified by M.NgoF-VII. They share with the eukaryotic members of the DNMT1 clade a CTDBM with two 3-stranded units in its N-terminal element, and also the conserved histidine in the extended linker between the Rossmann fold and the CTDBM. These methylases in turn belong to a large group of methylases including M.HaeIII, the FCL domain-containing versions, and some phage Dcms (e.g., *Thermus* phage P23p14 gi: 157265308), which have a similarly structured N-terminal element of the CTDBM along with a conserved histidine in the second strand of the first 3-stranded unit (Fig. 4; Supplementary Material). Gene neighborhood analysis suggests that they are nearly always associated with REases, including those of the HNH, AlwI subfamily, NotI, Vsr-like, and NgoFVII-like families that have widely disseminated across bacteria. This picture indicates that the origin of the eukaryote DNMT1-like clade is nested deep within the bacterial radiation of methylases of R–M systems with a single transfer seeding the eukaryotes.

2. The DNMT3 Methylase Group

The DNMT3 methylase clade is prototyped by the mammalian DNMT3 methylases, which were first characterized as the *de novo* methylase required for the reestablishment of the methylation patterns after they have been erased by demethylation[37,80,104] (see Chapter by Frédéric Chédin). One member of this clade DNMT3B is disrupted in the human ICF (immunodeficiency, centromere instability, and facial anomalies) syndrome and has been specifically implicated in the methylation of minor satellite repeats.[126,127] Multiple independent mutations in human DNMT3A have been reported in individuals suffering from *de novo* acute myeloid leukemia and they are correlated with poor disease outcome.[84] DNMT3A knockout in mouse results in impaired fetal growth and postnatal mortality.[127] In female placental mammals, a member of this clade, DNMT3L, is necessary for methylation imprinting at maternally imprinted loci in oocytes, whereas in males it protects the germline by methylating retrotransposons in the nondividing prospermatogonia.[128,129] These phenomena are described further in Section 6 of this volume. The plant member of this clade DRM is involved in *de novo* methylation of transgenes and inverted repeats and also in RNA-directed DNA methylation.[109,114] Thus, it appears that this clade ancestrally possessed *de novo* methylase activity, though this activity probably existed alongside the ancestral *de novo* methylase activity of DNMT1 orthologs when DNMT3 was acquired by eukaryotes. This clade is characterized by the presence of single intact 3-stranded unit in the N-terminal element of the CTDBM similar to the condition typified by the *E. coli* Dcm (Fig. 1). More specifically the DNMT3 clade is defined by the presence of a synapomorphic asparagine at the end of strand 7.

The analysis of phyletic patterns suggests that DNMT3 is found primarily in the animal, plant, and stramenopile lineages, indicating that it has been entirely lost in the fungal and amoebozoan lineages (Fig. 2). In the plant lineage, it is found in the rhodophyte alga *Cyanidioschyzon* but has been lost in several chlorophyte algae (Fig. 2). In land plants, one of the copies underwent a circular permutation within the methylase module, which resulted in strand 5 of the Rossmann-fold domain moving to the N-terminus of the entire methylase module (including the CTDBM; Fig. 4). While mosses possess both a permuted and a regular version, the latter has been lost in the angiosperms. In both plants and animals, the evolutionary history of the DNMT3 clade is marked by a propensity for lineage-specific expansions (Figs. 2 and 4). In plants such as *Arabidopsis*, there are three members of this clade. In metazoans, independent lineage-specific duplications resulting in 2–10 paralogs of DNMT3 are observed in urochordates like *Ciona* and vertebrates.

At the base of the vertebrate lineage, a single ancestral DNMT3 ortholog duplicated to yield two lineages defined by the mammalian DNMT3A and DNMT3B proteins. In fishes, these two lineages further proliferated resulting in at least 10 distinct paralogs in the zebrafish (see Chapter by Mary G. Goll and Marnie E. Halpern). In the common ancestor of the therian mammals (marsupials and placentals), there was a further duplication resulting in the DNMT3L paralog. In this paralog the catalytic domain has been disrupted by mutations and it functions as an inactive partner for both DNMT3A and DNMT3B in aiding their localization to regions with unmethylated H3K4 for *de novo* methylation.[37] In a comparable situation, the plant DNMT3 paralog DRM3 is catalytically inactive.[130] Given the role of DNMT3B in heterochromatinization of a specific set of repeats and DNMT3L in silencing retroposons, it appears likely that the lineage-specific expansion observed in different lineages, especially in fishes, is probably related to the specialization of different DNMT3 paralogs for targeting specific repeat and selfish elements. Unlike DNMT1, DNMT3 shows dramatic differences in domain architectures between the animal and plant lineages. In metazoans the methylase module is fused at the N-terminus to the BMB/PWWP domain, which has been shown to bind H3K36 trimethyllysine by DNMT3A,[131] followed by a multinuclear Zn-chelating module shared with the SWI2/SNF2 ATPase ATRX1, referred to as the "ADD" module.[132,133] The ADD module comprises an N-terminal mono-nuclear treble-clef domain and a C-terminal PHD finger domain, which is a binuclear version of the treble clef. The latter domain binds unmethylated H3K4,[132,134] while the N-terminal treble-clef domain has been proposed to be a DNA-binding domain by comparison to the GATA-type Zn-finger.[133] While both the GATA-type Zn-finger and this N-terminal domain of the ADD module share the treble-clef fold, we found no evidence for a specific relationship between them in structure similarity searches. Hence, in the absence of direct evidence for DNA binding by this domain, this proposal should be viewed with circumspection. In DNMT3L the BMB/PWWP domain has been lost, consistent with its specific role in binding unmethylated lysines.[37] Interestingly, in land plants the methylase module is fused to three N-terminal UBA domains which are known to bind ubiquitin (Fig. 4).[135] This suggests that, unlike the trimethyllysine recognized by the animal versions, the localization of the plant versions is likely to depend on ubiquitinated histones or other chromatin proteins. In light of this, and given the extensive deployment of treble-clef domains in Ub-recognition,[136] it is worth exploring whether the N-terminal treble-clef domain of the animal ADD modules might have a role in Ub-recognition.

Outside eukaryotes, the DNMT3 clade includes a specific group of bacterial methylases which are united with the eukaryotic versions by the synapomorphic asparagine after strand 7. These bacterial versions are well conserved in firmicutes (low-GC Gram-positive bacteria) and also found in *Bacteroidetes* (e.g., *Bacteroides* BSFG_03198, gi: 254883949; Fig. 4). The exact role of these bacterial versions of DNMT3 is rather unclear. Given their conservation in firmicutes, independently of R–M systems, it is possible that they have been recruited for a distinct cellular role; such as, perhaps, providing an epigenetic mark for DNA repair.

3. The DNMT2 Group

The DNMT2 methylases have been at the center of controversy over whether they function as DNA or RNA methylases or both.[90,137,138] Studies in various eukaryotic models convincingly demonstrate that DNMT2 specifically methylates $tRNA^{Asp}$ on cytosine 38.[139] However, studies in *Dictyostelium* clearly demonstrate a role for its DNMT2 ortholog DnmA in the developmentally regulated methylation of the Skipper retroposons, perhaps, also the DIRS-1 retroposon.[140] Given the increased mobility of the Skipper element upon deletion of DnmA, it appears possible that in *Dictyostelium* this methylase is also involved in DNA methylation for transposon repression. In contrast, there is currently no evidence for methylation of the $tRNA^{Asp}$ by DnmA in *Dictyostelium.*[141] Likewise, the evidence from *Entamoeba* supports a role for its DNMT2 ortholog in DNA methylation.[142] It is conceivable that even methylation of $tRNA^{Asp}$ could affect the mobility of certain retroposons because they use this tRNA as a primer for their RT.[143] *Drosophila* (which lacks both DNMT1 and DNMT3) shows early embryonic DNA methylation, primarily at non-CpG sites that is ascribed to DNMT2 activity[138,144] (see Chapter by Veiko Krauss and Gunter Reuter). This DNA methylase activity has also been linked to the silencing of Invader4 retroposons.[138] Despite counterclaims regarding the genuineness of this DNA methylase activity,[137] we suspect *Drosophila* DNMT2 is a *bona fide* DNA methylase based on the indirect evidence for the presence of a catalytically active Tet enzyme that uses 5mC as a substrate (see below). Like DNMT3, the CTDBM of DNMT2 contains a single intact 3-stranded unit in the N-terminal as seen in the structural prototype presented by the *E. coli* Dcm (Fig. 1). However, the methylase module of the DNMT2 clade is distinguished by several distinctive features from all other 5C DNA methylases, namely a glutamate in strand 1 of the Rossmann-fold domain, a proline two positions downstream of the catalytic cysteine associated with strand 4, and a highly conserved cysteine in the "hammer-head" loop of the 3-stranded unit of the CTDBM. This latter cysteine is spatially close to the active site cysteine and is required for optimal activity.[141] Thus, like the Sun/Fmu RNA cytosine methylases, DNMT2 has convergently evolved two distinct cysteines that appear to be required for optimal activity. This observation suggests that unlike pure DNA methylases, the RNA methylases might require cooperation between two cysteines at the active site for their catalysis. While the exact basis for this remains unclear, it is possible that the methylation of RNA occurs in a loop rather than a flipped-out base in a duplex; thus, presenting a different local environment to the active site of the methylase.

DNMT2 is the most widely distributed DNMT clade in eukaryotes, being present in the animal lineage, fungi, amoebozoans, the plant lineage, stramenopiles, apicomplexans, and the heterolobosean *Naegleria* (Fig. 2). Thus, it appears to have been acquired early in eukaryotic evolution and has been vertically inherited ever since. Nevertheless, it has been entirely lost in several eukaryotic lineages such as ciliates, and sporadically within others like in the animal lineage (e.g., *C. elegans*) and fungi (e.g., *Saccharomyces cerevisiae*). This suggests that the modification of the $tRNA^{Asp}$ is not an essential feature for all eukaryotes. As noted in earlier phylogenetic studies, outside eukaryotes the

DNMT2 clade is found in the bacterium *Geobacter*[97]; the function of this bacterial version remains unclear. While it was proposed that it might methylate tRNA in light of a similar sequence of the $tRNA^{Asp}$ in *Geobacter*,[139] this proposal is not entirely supported because of conservation of comparable tRNA sequences even in organisms lacking a DNMT2 representative.[139] While *Geobacter* is the only currently known bacterium with a classical representative of the DNMT2 clade, in phylogenetic trees, they appear to be nested within a larger group of bacterial R–M system methylases with a single intact 3-stranded element in the CTDBM (Fig. 4). This indicates that DNMT2 first emerged within this radiation in bacteria and was transferred to eukaryotes early in their evolution. Unlike DNMT1 and DNMT3, DNMT2 shows a simple domain architecture with no fusions to other chromatin protein domains in eukaryotes. This observation, together with their more widespread phyletic pattern and presence in organisms with no detectable genomic 5mC, suggests that they were primarily recruited as an RNA methylase upon acquisition by the eukaryotes.[90] Only in certain lineages, where the other 5C DNA methylases were lost, there appears to have been an atavistic resumption of their DNA methylation role. In this respect, they appear to mirror the evolutionary history of the IME4 (MT-A70) clade of methylases.

4. Other 5C DNA Methylases of Eukaryotes

In addition to the three DNMT clades, there are several other 5C DNA methylases in eukaryotes that have been poorly characterized or are entirely unstudied (Figs. 2 and 4). Their domain architectures are suggestive of key roles in chromatin dynamics in the organisms in which they are present.

5. The Methylases Fused to Rad5-like SWI2/SNF2 ATPases

These methylases are found in both ascomycete and basidiomycete fungi, chlorophyte algae, and stramenopiles.[50] While they are likely to have been present in the common ancestor of most of the above groups, they have been frequently lost in several members. However, their overall distribution in eukaryotes is best interpreted as a consequence of lateral gene transfers occurring early in the evolution of these groups. They differ from most other methylases, in that the methylase module is part of a large multidomain architecture with other enzymatic domains in the same polypeptide. The methylase module is fused at the C-terminus to a distinctive domain with a treble-clef fold related to the ZZ domain,[145] followed by an uncharacterized globular domain, which in turn is followed by a C-terminal SWI2/SNF2 ATPase module (Fig. 4). This SWI2/SNF2 ATPase module specifically belongs to the RAD5-clade of SWI2/SNF2 ATPase, which is characterized by the insertion of a RING finger domain within their ATPase module.[50] The RING finger domain could act as an ubiquitin E3 ligase that operates on chromatin proteins. The domain architecture suggests that the methylation catalyzed by these enzymes is likely to function in close coordination with the ATP-dependent chromatin remodeling and ubiquitination of chromatin proteins. In this respect, they are similar to the kinetoplastid JBP2 proteins that combine the DNA-modifying dioxygenase domain with a C-terminal SWI2/SNF2 module.[146] This domain combination is also consistent with the functional collaboration between chromatin remodeling catalyzed by the SWI2/SNF2 ATPases and DNA methylation that is evidenced by DRD1, which assists RNA-directed methylation in plants[147–149] and ATRX in vertebrates.[133,150,151] The occurrence of this type of DNA methylase in organisms such as *Aspergillus*, in which there is little detectable DNA methylation through most of their lifecycle, suggests that the methylation catalyzed by these enzymes might occur only under specific circumstances, such as during DNA repair. The methylase module of these proteins is characterized by a CxxxE signature in the AdoMet-binding loop of the Rossmann-fold domain, which it shares uniquely with a group of methylases encoded by bacteriophages like P1 and P7 (the *Dmt* gene of P1 where the 5C DNA methylase module is fused to a Dam methylase domain).[152] Here, they occur in operons closely linked to the origin of these viruses along with the single-strand-binding protein and chromosome partitioning topoisomerases. This suggests that the bacterial versions might methylate the origins of the virus to regulate DNA replication and partitioning of the chromosomes. In structural terms, the CTDBM of this clade of methylases (both the bacterial and eukaryotic versions) are similar to the structural prototype offered by M.HhaI, wherein the two 3-stranded units of the N-terminal element are closely placed, without any intervening insert.[85]

6. The Kinetoplastid-Type 5C DNA Methylases

The kinetoplastids encode a conserved 5C DNA methylase typified by *Leishmania* LmjF25.1200,[50] whose cognate in *Trypanosoma brucei* has been termed TbDMT. Additionally, representatives of this methylase family are found in several stramenopiles and the chlorophyte alga *Micromonas* (Figs. 1 and 4). Recently, it was demonstrated that TbDMT methylates cytosine at retroposon insertion hotspots and variable surface antigen gene (*VSG*) loci in the *T. brucei* genome.[153] This is consistent with a potential function for these methylases in repression of retroposon and regulation of the expression of the multigene VSG loci. It remains to be seen if this methylation of VSG loci might have a mutagenic role similar to *Neurospora* RIP in generating antigenic diversity in the VSG products.[154] While these proteins show fairly long extensions N-terminal to the methylase domain, they do not bear detectable similarity to previously characterized domains. These eukaryotic methylases are united into a clade with the bacterial Dcms (e.g., *E. coli* Dcm) and related methylases from R–M systems with which they share a highly conserved arginine at the end of strand 7 of the Rossmann-fold domain. The bacterial versions are commonly associated in operons with Vsr-like or *Eco*RII-like nucleases (Fig. 4).

7. The Chlorophyte-Type 5C DNA Methylases

This group of methyltransferases is exclusively found in chlorophyte algae such as *Ostreococcus*, *Micromonas*, and *Chlorella* (Fig. 2).[50] Their methylase module is fused to two C-terminal BMB/PWWP domains that sandwich a distinct divergent CXXC domain (see below). Certain chlorophyte versions additionally have a second CXXC domain C-terminal to the methylase module (Fig. 4). This architecture bears some resemblance to both the animal DNMT3s, which are instead fused to N-terminal BMB/PWWP domains and DNMT1s, which have a CXXC domain. Hence, it is likely that these chlorophyte-type methylases localize to particular trimethyllysine marks on histones and modify DNA in their vicinity. Given the absence of DNMT3 orthologs in the chlorophyte lineages that contain these chlorophyte-type methylases, we propose that the latter have displaced the ancestral DNMT3s and perform an equivalent role. However, they are not closely related to DNMT3 and are instead close to a group of methylases of bacterial and phage R–M systems typified by *Bacillus subtilis* YdiO/YdiP protein (gi: 16077674), whose gene is linked to a *LlaJI*-like REase and an McrB-like AAA+ GTPase (Fig. 4). In structural terms, they follow the M.HhaI type of methylases with two closely placed 3-stranded units in the N-terminus of the CTDBM[85] (Fig. 1).

8. Other Miscellaneous 5C DNA Methylases of Eukaryotes and Their Viruses

There are some other sporadic 5C DNA methylases specified by selfish elements in eukaryotic genomes and viruses that infect eukaryotes. One of these is carried by a novel retroposon that has proliferated in the genome of the stramenopile alga *Aureococcus*, where the methylase is combined to a C-terminal RT domain (Figs. 2 and 3). Many of the copies of this element appear to be inactive with disruption of both the RT and methylase domains. In terms of general organization, that is, combination of a methylase domain with a RT domain, they resemble the DIRS1-like elements, which instead specify Dam methylases (see above).[65] This suggests that both adenine and cytosine methylases might have a role in DNA modification-dependent autoregulation of transposons. Phycodnaviruses of the chlorella virus group, which infect chlorophyte algae, code for multiple R–M systems with both DNA cytosine methylases and adenine methylases.[155–157] For example, the *Paramecium bursaria Chlorella* virus-1 possess three DNA cytosine methylases and two adenine methylases. These represent rare examples of R–Ms present in eukaryotic systems, and protect viral DNA via methylation while launching a restriction attack on the host DNA.[155] The chlorophyte alga *Micromonas* specifies a genomic version (*Micromonas* MICPUN_59797, gi: 255079758) that appears to have been acquired from a phycodnavirus and might provide defense against the viral restriction attack (Supplementary Material). Likewise, a sporadic 5C methylase found only in the stramenopile alga *Emiliania* might also provide protection in this organism against viral attacks (Supplementary Material). These viral 5C methylases are unified in a clade with bacteriophage-coded versions that infect actinobacteria like *Mycobacterium* (Fig. 4). In these viruses, they might be involved in methylation of the origin site as they are associated in operons with chromosome partitioning proteins. An examination of the alignment reveals that the CTDBM of PBCV is circularly permuted, such that the last helix of the HEH unit and the helix that follows it are moved to the N-terminus of the CTDBM. Interestingly, some of the related methylases from the above bacteriophages lack a CTDBM, but it occurs as a separate adjacent gene in the same operon (Fig. 4). Hence, it is likely that the original permutation happened in the stand-alone CTDBM of such a system, followed by a fusion with the Rossmann-fold catalytic domain, prior to acquisition by the chlorella viruses. Iridoviruses, such as the Lymphocystis disease virus, which infect aquatic vertebrates, specify a distinct cytosine DNA methylase of unclear function that is related to certain bacterial 5C DNA methylases.[158] These methylases are defined by a characteristic small CTDBM, that contains three conserved cysteines and a histidine which might stabilize the domain through chelation of a cation (Supplementary Material). An early study had shown that a significant fraction of the cytosines in the iridoviral genome are methylated in a pattern distinct from the host genomes.[159] This methylation could be mediated by the virally coded cytosine methylase, and could aid in both evasion of host foreign-DNA surveillance systems and perhaps even epigenetic regulation of viral chromatin.

Beyond these methylases, the shotgun genomic sequences of various eukaryotes (like the moss *Physcomitrella*, the frog *Xenopus*, and *Trichoplax*) show some sporadic 5C DNA methylases. Currently it remains unclear if these are novel DNA methylases actually produced by these organisms, or if they are bacterial sequence contaminants of the genomic sequences (Supplementary Material).

III. 5mC Demethylation and Potential DNA Demethylases

A. Evidence for Active Demethylation and Different Proposed Demethylase Mechanisms

In eukaryotes, demethylation of 5mC has consequences in the maintenance of epigenetic information (see Chapter by Taiping Chen). This phenomenon has been best characterized in mammalian and plant genomes. In mammalian genomes, several distinct demethylation events have been reported. The most drastic of these occurs in the fertilized egg, where the paternal genome is first demethylated about 6–8 h postfertilization before the first round of zygotic DNA replication.[160–162] This is accompanied by a large-scale remodeling of the sperm chromatin and establishment of parent genome-specific gene-expression patterns. However, several imprinted loci and the maternal genome escape demethylation at this stage.[161] Subsequently, after cleavage has divided the zygote to 4–32 cells, the maternal genome undergoes large-scale demethylation and chromatin reorganization. However, the complete demethylation of all imprinted loci occurs only after the primordial germ cells are specified and the epigenetic marks are erased to reprogram the genome for totipotency.[163] This reprogramming occurs independently of DNA replication during the G2 phase of the cell cycle. In addition to these global demethylation events during embryonic development in vertebrates, localized demethylation has also been observed at certain regulatory DNA regions in adult cells. One well-studied example is that of the interleukin-2 promoter in T cells, which is induced in response to stimulation of the T cell receptor with an antigen.[164] Prior to induction, the promoter is methylated at CpG sites but is rapidly demethylated during T cell activation. It has been reported that the pS2/TFF1 gene promoter undergoes periodic and strand-specific methylation and demethylation as a part of the transcriptional cycling process that depends on estrogen.[165] A comparable phenomenon is observed in the activation of the cytochrome p450 27B1 gene by the parathyroid hormone, where active demethylation releases the repressive state established by vitamin D.[166] Demethylation of the promoter in this system is central to the activation of the gene by the estrogen signal. Much less in known regarding demethylation outside vertebrates, but the distribution of methylation marks in *Drosophila* suggests that, unlike in the former organisms, major demethylation might occur relatively late in development, probably after the completion of the larval stage[144] (see Chapter by Veiko Krauss and Gunter Reuter). In plants, demethylation has been studied in the context of endosperm development and transgene expression. In course of endosperm development, the uniparental expression of certain genes, like the maternal Medea allele, is achieved via allele-specific demethylation.[167] Other demethylation events in plants appear to function as an editing mechanism to alleviate certain genes from the methylation-repression mechanism that are laid down by *de novo* methylation or by RNAi-dependent mechanisms.[168]

DNA demethylation at 5mC is thus a critical process across eukaryotes. Nevertheless, the phenomenon is not well understood in terms of biochemistry or possible mechanisms. While a number of distinct enzymes and mechanisms have been proposed for the catalysis of demethylation, several of these appear either unlikely or dubious.[169] We briefly survey the major proposed enzymes and their mechanisms, and then focus only on the more likely and better-confirmed candidates for further discussion of their phylogenetic spread and natural history. We also present evolutionary arguments that favor these candidates as potential demethylases.

The most unlikely of all the proposed demethylases is mammalian MBD2, which was claimed to remove the methyl group by generating formaldehyde.[170] This protein contains a TAM/MBD domain, which specifically binds methylated DNA (see Chapter by Pierre-Antoine Defossez and Irina Stancheva), but does not possess any conserved residues or structural features that could support the kind of reaction mechanism proposed for this protein.[171] Other than this domain, the rest of the protein does not contain any globular domains, which strongly suggests that it is unlikely to be able to support demethylase activity by itself. Consistent with this, the demethylase activity of MBD2 has not been successfully reproduced by other experimental groups.[162,171] Another potentially uncertain demethylase candidate is the transcription elongation complex protein ELP3.[172] This is a highly conserved protein found in both archaea and eukaryotes and comprises two distinct globular domains, an N-terminal radical SAM domain and a C-terminal acetyltransferase domain.[173,174] This protein is clearly a bifunctional protein, the acetyltransferase domain of which is required for its role in transcription elongation.[175] While the intact radical SAM domain is needed for its role in transcription elongation, there is no evidence that its catalytic activity is required for transcription elongation.[174] The ELP3 protein is also required for the synthesis of two modified uracils, namely, 5-methoxycarbonylmethyl and 5-carbamoylmethyl uracil in the wobble position of tRNAs.[176] These modifications are likely to require the radical SAM domain for their catalysis. RNAi knockdown of ELP3 and other elongation complex proteins such as ELP1 and ELP4 were shown to impair paternal genome demethylation in mammals.[172] Introduction of mutant ELP3 mRNA with a disrupted metal-binding cluster in the radical SAM domain impaired demethylation. Based on this, it was proposed that the ELP3 protein might directly function as a demethylase. However, this proposal is dubious on multiple grounds. First, the intact radical SAM domain is required for both the structural integrity and effective functioning of the elongation complex, even though ELP3 catalytic activity is not involved. Second, the basic reaction catalyzed by the radical SAM domain is cleavage of AdoMet to generate a deoxyadenosyl radical that is then used as a free radical to abstract protons from other molecules. The deoxyadenosyl radical generated by these enzymes has been implicated in several nucleic acid and protein modifications, but none of these involve removal of a methyl group.[173,174] Finally, the ELP3 protein is highly conserved throughout eukaryotes and archaea, whether or not their DNA contains 5mC, and ELP3 shows no specific differences between these two groups.[50] In light of this it is, at best, possible that the transcription elongation complex (i.e., ELP1-6) has a secondary role in the demethylation process; for example, in recruiting the actual demethylation machinery. We also discuss below the possibility of an indirect role for the radical SAM catalytic domain in demethylation, though there is currently no evidence that this is indeed the case.

The remaining proposed demethylation mechanisms involve different types of DNA repair processes. These may act either directly or indirectly and typically invoke base excision repair (BER) involving a DNA glycosylase. DNA glycosylases are classified as monofunctional or bifunctional, depending on the reaction they catalyze.[177] The former enzymes simply break the glycosidic linkage between the base and the sugar and leave behind an abasic lesion in the DNA. This lesion is then acted on by an AP-endonuclease, which cleaves the backbone at the abasic site. In contrast, the bifunctional enzymes not only remove the base but also exhibit lyase activity; that is, they cleave the DNA backbone to leave a free 5′ phosphate. These lesions are then processed by the BER system to digest a patch of DNA, followed by refilling by a repair DNA polymerase and ligation. The direct action of DNA glycosylases has been demonstrated in plants and is catalyzed by the Demeter-like family of glycosylases.[167,168,178–180] These glycosylases show specificity for 5mC and catalyze both removal of the base and subsequent cleavage of the backbone through lyase activity.[180] Similarly, multiple studies in vertebrates (e.g., demethylation of the cytochrome p450 27B1 promoter) have demonstrated MBD4 to be a bifunctional DNA glycosylase that removes 5mC in addition to G/T mismatches, generating a strand break.[166,181–186] The unrelated thymine DNA glycosylase Tdg may also possess this activity,[187,188] though this has not been reproduced *in vitro* by other groups.[165] However, support for its potential role in DNA demethylation has been obtained in a screen for demethylation regulators.[189] This study suggests that regulation of Tdg by sumoylation might be critical for its demethylase activity. Indirect DNA repair mechanisms for DNA demethylation through BER typically posit a deamination step prior to the action of the DNA glycosylases. An example, proposed in the zebrafish system, implicates the action of the deaminases AID or APOBEC2a/b in deamination of 5mC to thymine.[181] This deamination is believed to be followed by the action of MBD4 in removing the T:G mismatch through its glycosylase/endonuclease action. The nonenzymatic pelota domain protein Gadd45a/b was also implicated in this system,[181] though other researchers have questioned the role of this protein in demethylation.[190] Biochemical studies have demonstrated that AID and APOBECs prefer C or 5mC and that MBD4 prefers U (the deamination product of C) over T (the deamination product of 5mC).[182] In light of these observations, it is rather unclear if the highly mutagenic deamination step is indeed a prerequisite for DNA demethylation by MBD4. Another route for deamination of 5mC is suggested by studies on the estrogen-dependent activation of the pS2/TFF1 gene promoter. Here DNMT3A and

DNMT3B are implicated as 5mC deaminases.[165] This unusual activity of the *de novo* DNA methylases is supported by experiments on DNA methylases from bacterial R–M systems, which show that under low AdoMet concentrations or presence of AdoMet competitors, the methylase domain can function as a deaminase.[191,192] Subsequent to the deamination of 5mC to T by the DNMT3s, it is believed to become the substrate for DNA repair by the glycosylase Tdg.[165] In light of this, it is worth determining whether depletion of AdoMet, through cleavage by the radical SAM domain of a protein such as ELP3, might indirectly regulate demethylation via a deamination pathway.

Multiple studies also demonstrate involvement of other BER components in different demethylation events. For example, the erasure of imprinting in primordial germ cells involves the appearance of single DNA breaks associated with BER.[163] Specifically, inhibition of the AP endonuclease APE1 disrupts the demethylation process. However, it remains unclear as to how BER is initiated in primordial germ cells because there appears to be no concomitant expression of the DNA glycosylases, deaminases, or DNMT3s previously implicated in demethylation.[163] Studies on MBD4 have also shown that its DNA glycosylase activity is strongly inhibited by RNA.[184] Interestingly, the other DNA glycosylase Tdg has been shown to form a complex with the RNA helicase p68.[188] These observations suggest that DNA demethylation could additionally be regulated by RNA-dependent mechanisms. The weight of the currently available evidence points in the direction of DNA glycosylases as the best candidates for DNA demethylases in eukaryotes. We discuss below their structure and evolution.

B. The Structural Features and Classes of DNA Glycosylases Related to DNA Demethylation

The catalytic domains of all currently known DNA glycosylases belong to four structurally unrelated folds, two of which contain members that have currently been implicated in DNA demethylation.[193–195] The first of these, the uracil DNA glycosylase (UDG) superfamily, typified by human Tdg and *E. coli* Mug and Ung, contains an α/β domain with a central β-sheet formed by four conserved strands.[193,194] These enzymes are strictly monofunctional and only catalyze the removal of the base from the nucleotide. They contain three conserved motifs, which constitute their active site, and are, respectively, associated with the C-termini of strand 1, strand 2, and strand 4. The motif associated with the C-terminus of strand 2 usually contains an asparagine or aspartate and interacts with the mismatched base.[194,196] The motif associated with strand 3 is involved in stabilization of the enzyme-coupled reaction intermediate.

The second superfamily of DNA glycosylases implicated in demethylation (HhH-glycosylase) is typified by the catalytic domains of MBD4, Demeter, and their bacterial counterparts such as *E. coli* MutY and Endonuclease

III (Nth).[194,197,198] This catalytic domain comprises four copies of the helix-hairpin-helix (HhH) motif, which also occurs independently as a DNA-binding domain in diverse DNA repair proteins and the bacterial RNA polymerase α-subunit.[194,199] In practically all the latter cases, the HhH motif is a non-catalytic DNA-binding element[199]; however, in these DNA glycosylases, they do not just bind DNA but also contribute residues involved in catalyzing DNA glycosylase/lyase activity. The four HhH motifs of this domain are deeply inserted into the duplex around the mismatch site and make extensive contacts with the DNA via the hairpin loops between the two helical segments of the HhH. As a consequence, they hold the DNA in a "pincer grip," and the conformational change in DNA structure induced by this interaction appears to be critical for catalysis of the glycosylase reaction. Except for the clade defined by eukaryotic MBD4 and prokaryotic AlkA and Ogg1, other members of the HhH-DNA glycosylase superfamily have an FCL domain C-terminal to the catalytic domain.[88] This domain contains four conserved cysteines that bind an iron–sulfur cluster, supporting a flap-like structure in the protein that makes a deep minor groove contact with DNA.[88] Certain members of the HhH-DNA glycosylase superfamily, such as *E. coli* MutY and human MYH, contain a further C-terminal extension in the form of a catalytically inactive version of the Nudix domain.[198] This domain binds DNA and allows these versions to form a complete ring around DNA in conjunction with the HhH-glycosylase domain that is positioned opposite to the Nudix domain. Different members of the HhH-DNA glycosylase superfamily have been shown to function as either monofunctional or bifunctional, enzymes with both simple glycosylase and lyase activity. However, both activities have been proposed to proceed via a reaction intermediate that involves formation of a Schiff's base between a basic residue on the enzyme and the sugar.[196]

The third distinct fold of DNA glycosylases, typified by *E. coli* Endonuclease VIII and vertebrate Neil1/2/3, has currently not been implicated in DNA demethylation.[195] Nevertheless, versions of this superfamily from chlorophyte algae show fusions to the SAP domain (Fig. 3), which specifically functions in tethering various DNA modification and repair activities to regions of chromatin such as SARs/MARs.[200] In light of this, a role in DNA demethylation or related epigenetic DNA modifications cannot be ruled out for this class of DNA glycosylases in certain eukaryotes.

The fourth distinct class of DNA glycosylases is typified by the *B. subtilis* AlkD protein, which is implicated in alkylated purine repair.[193] This enzyme is unusual in that its catalytic domain is almost entirely comprised of HEAT repeats, which are normally typical of structural rather than enzymatic domains; its α-helical catalytic domain convergently mimics that of the HhH-DNA glycosylase superfamily. Though certain eukaryotes with 5mC in their genomes specify orthologous enzymes, currently there is no evidence for their participation in a demethylation process.

C. Evolution of the Tdg-Like Enzymes of the Uracil DNA Glycosylase Superfamily

The UDG domain can be traced back to the last common ancestor of all life forms and appears to have functioned primarily as a DNA repair enzyme, which removes uracil produced as a cytosine deamination product or due to misincorporation by the polymerase.[194] This superfamily comprises one family that first emerged in the archaea and five families that first radiated in the bacteria.[194] Interestingly, eukaryotes did not inherit the archaeal version; instead they independently acquired at least three of the five bacterial families through lateral gene transfer at different points in their evolution. Two of these, namely the cognates of *E. coli* Ung and Smug1/ssUDG, are highly specific for uracil and appear to function primarily in DNA repair, removing uracil from dsDNA and ssDNA.[194] The third, which is the cognate of the *E. coli* Mug, has given rise to the eukaryotic Tdg that operates on T:G mismatches and thus plays a role in removal of deaminated 5mC. Tdg is currently known from animals, fungi, chlorophyte algae, and stramenopiles, suggesting that it was transferred from bacteria to the eukaryotes prior to the radiation of the eukaryotic crown group that encompasses these lineages (Fig. 2). Following transfer to the eukaryotes, Tdg has often acquired additional extensions, usually in the form of low-complexity sequences on either side of the globular UDG catalytic domain. The N-terminal extension is often positively charged and resembles the tails of histones. In vertebrates, these extensions contain target sites for sumoylation by the E3 Sumo-ligase Rnf4, a process that appears critical for DNA demethylation through BER.[189] In insects, the Tdg ortholog is characterized by an N-terminal extension with two AT-hook motifs that are known to bind the minor groove of DNA.[194] It is conceivable that these AT-hooks help target the Tdg ortholog to specific chromatin regions, such as matrix attachment or scaffold attachments regions, and initiate BER at such chromosomal locations.[49] The versions from certain chlorophyte algae and stramenopiles contain a Zn-ribbon domain just N-terminal to the UDG catalytic domain (Fig. 3). The Tdg family is frequently lost in lineages that entirely lack DNA methylation, such as *S. cerevisiae* among the fungi and *C. elegans* among the animals. While Tdg has also been lost in land plants, which show abundant DNA methylation, these plants show a proliferation of other, unrelated DNA glycosylases (see below). This phyletic pattern, together with the acquisition of additional domains in eukaryotes, suggests that Tdg was probably acquired as a defense against the mutagenic effects of extensive genomic methylation, and also perhaps for resetting some of these methyl marks through BER.

D. Evolution of Demeter, MBD4, and Other HhH-DNA Glycosylases Related to DNA Methylation

Like the UDG superfamily, the HhH-glycosylase superfamily is found in organisms across the three superkingdoms of life. However, the versions from both eukaryotes and archaea are nested within the bacterial radiation of this superfamily. Therefore, they probably emerged in bacteria originally and were dispersed by lateral transfer to the two other superkingdoms.[89] In bacteria, the HhH-glycosylase superfamily radiated into three major clades: the Ogg1–AlkA clade, whose catalytic domain comprises just the four HhH modules and which further diverged into the Ogg1-like and AlkA-like clades; the Endonuclease III (Nth)-like clade, in which the FCL domain was added to the C-terminus of the core catalytic domain; and the MutY-like clade, which has acquired an inactive C-terminal Nudix domain. In bacteria, these distinct clades appear to have diversified to perform distinct roles in BER.[196] The AlkA clade appears to have specialized in removing alkylated DNA bases such as methyladenine. However, the related Ogg1-clade, in bacterial lineages such as firmicutes, appears to have specialized in acting on the highly mutagenic 7,8-dihydro-8-oxoguanine that can cause G→T transversions. Likewise, the MutY clade acquired a role in excision of oxoguanine lesions in other bacterial lineages like the proteobacteria. The Nth clade appears to have specialized in removal of pyrimidines damaged by oxidation, dihydrothymine, and also strand cleavage at abasic sites. The direct connection between 5mC and Nth-like HhH-glycosylase appears to have emerged first in the prokaryotes. We uncovered a novel R–M system, which is distributed across phylogenetically distant archaea and bacteria such as *Persephonella*, *Chloroflexus*, and *Methanosarcina*, whose core consists of four tightly linked genes: A 5C DNA methylase, an Nth-like HhH-glycosylase, a SFII helicase, and a large protein with an N-terminal Zn-ribbon domain (Fig. 3). Some versions of this system might additionally specify an HKD phosphoesterase/nuclease protein found in several R–M systems. This organization indicates that the 5C DNA methylases are the modification component, while the Nth-like HhH-glycosylase is the endonuclease, which most probably recognizes the site modified by the former enzyme and cleaves the DNA in the manner of Type IV restriction systems.[19,20]

HhH-glycosylases, of the different clades that had diversified in bacteria, were independently transferred laterally to eukaryotes on several occasions. The most ancient transfer was that of the Nth clade that occurred prior to radiation of the eukaryotes from the LECA, as evidenced by its presence in the early-branching eukaryotic lineages such as *Giardia* and *Trichomonas* and also those with reduced genomes such as the microsporidians. The classical Nth homologs, like the mammalian Nthl1,[201] are primarily implicated in BER rather than DNA demethylation, consistent with both their phyletic patterns (i.e., presence in species lacking DNA methylation) and absence of fusions to domains suggestive of a role in chromatin (Fig. 2). Another independent transfer of the Nth group appears to have happened later in eukaryotic evolution, giving rise to a group of Nth-like paralogs that are found in plants and fungi. Members of the Ogg1–AlkA clade appear to have been introduced to the eukaryotes on multiple occasions. A member of the classical Ogg1 subgroup archetyped by human Ogg1 that was probably acquired, after the divergence of lineages such as *Giardia* and *Trichomonas* (in which it is absent), is widely conserved in most eukaryotic lineages and appears to function as a DNA repair enzyme like its bacterial cognates. A member of the AlkA subgroup, typified by *S. cerevisiae* MAG1, is found in plants and fungi and appears to have been derived from a late transfer from bacteria into one of these two phylogenetically distant eukaryotic lineages, followed by further transfer between them. This enzyme appears to function primarily in protecting DNA against alkylation damage.[202] Transfer of an MutY-like glycosylase from bacteria, relatively early in eukaryotic evolution, appears to have given rise to yet another group of DNA glycosylases in eukaryotes, whose archetype is the human MYH. While this clade has not yet been implicated in DNA demethylation, its absence in several eukaryotic clades lacking DNA methylation makes it a candidate that could be considered in future investigations for BER dependent demethylation (Fig. 2). The origins of the two groups of enzymes of the HhH-glycosylase superfamily that are currently implicated in DNA demethylation appear to have distinct histories from the above families. The first of these, the Mbd4-like clade, lacks any close bacterial cognates; however, it is clear that it was derived from the Ogg1–AlkA clade as it shares with them the core HhH-based catalytic scaffold without a C-terminal FCL domain. Hence, this clade probably diverged rapidly from an ancestral Ogg1-like version within the eukaryotes. However, the Demeter-like clade has clear cognates within the vast bacterial Nth-like radiation, from which it appears to have been derived. Given that these bacterial cognates are found in the cyanobacteria, and that the Demeter-like clade is restricted to plants and stramenopiles, it is possible that its ancestor was first acquired during the cyanobacterial endosymbiosis that gave rise to the plant lineage (Fig. 2).

The Mbd4-like clade is the most widely distributed of the HhH-glycosylase clades implicated in DNA demethylation. MBD4 orthologs are known from animals, fungi, plants, and certain stramenopiles (Fig. 2). The phyletic pattern of MBD4 in eukaryotes usually shows a strong correlation with notable levels of genomic 5mC and has been repeatedly lost in many of the lineages with low levels of, or no genomic methylation. In animals, basal members of the plant lineage (chlorophyte algae) and diatoms, MBD4 is fused to a TAM/MBD domain. This fusion suggests that the ancestral version of the MBD4 family probably directly translocated to sites enriched in methylated CpG by means of its TAM/MBD domain. However, this domain has been lost in the land plants and fungi (Fig. 2). In land plants the MBD4 ortholog contains a long N-terminal extension with one to six copies of a short peptide repeat with a consensus motif [VI]SPxh (where x is any amino acid and h a hydrophobic residue). Though the function of these repeats is currently unclear, it is possible that these repeats are the sites of posttranscriptional modification that regulates these enzymes. Chlorophyte algae possess a second paralog of MBD4 which contains, in place of the TAM/MBD domain, a distinct module known as the KRI motif which is found in diverse eukaryotic chromatin proteins (Fig. 3).[203] Based on analysis of KRI motif architectures, we predict that it is likely to have a role in recognizing epigenetic modification of histones, in particular, histone methylation. Thus, this paralog of MBD4 might localize to regions in chromatin that have specific histone modifications and locally catalyze demethylation. Fungal MBD4s display several distinct architectures where the HhH-glycosylase is fused to different N- or C-terminal domains (Fig. 3). For example, the *Neurospora* MBD4 ortholog includes a fusion to a divergent version of the Myb domain that could potentially help it recognize specific DNA sequences. In *Aspergillus*, one of the MBD4 paralogs (e.g., AN3766.2; gi: 67526617) is fused to a distinct C-terminal globular domain that contains a conserved CxCxxC motif, which is also found in the mammalian Stella proteins that protect imprinted sites from demethylation (see below). A second *Aspergillus* MBD4 paralog (ANIA_10443; gi: 259481685) is instead fused to an N-terminal conserved globular domain whose provenance is unclear. It is possible that these distinct fungal specific domains help binding and recognizing specific DNA or chromatin based signals that are distinct from those recognized by animal MBD4s.

Eukaryotic representatives of the Demeter-like clade are characterized by a distinct C-terminal region, which is a divergent version of the RNA-recognition motif (RRM) domain (Fig. 3, Supplementary Material). Versions of this domain have been implicated in binding single-stranded nucleic acids,[204] and it may thus facilitate interaction of the catalytic domain with ssDNA or perhaps even regulatory RNAs. The Demeter orthologs of chlorophyte algae and stramenopiles show a diverse range of architectures, including fusions to diverse domains that bind methylated histone peptides, such as multiple PHD fingers and tudor domains (Fig. 3). Several of the Demeter orthologs of these algae display one or more CXXC DNA-binding domains, either to the N-terminus of the HhH-glycosylase module or to the C-terminus of the RRM domain (Fig. 3). Further, some of these algal versions also show an insertion of the DNAJ domain between the catalytic and RRM domains. In the Demeter-like proteins of land plants, (e.g., the *Arabidopsis* Demeter), a divergent permuted CXXC domain appears to have been inserted between the HhH-glycosylase module and the RRM domain. In general these architectures suggest that, even in basal plant lineages (Fig. 2), Demeter-like proteins have acquired a role in modifying DNA in conjunction with recognizing epigenetic modifications on chromatin proteins, thereby strongly implicating these versions in DNA demethylation. Fusion to the DNAJ domain, which interacts specifically with the chaperone Hsp70,[205] suggests that the algal Demeter-like proteins are probably regulated via the recruitment of this chaperone. In light of this, it would be worth exploring whether DNA demethylation in these organisms might occur in response to protein misfolding stresses.

IV. Further Modifications of 5mC in Eukaryotic DNA

A. 5-Hydroxymethyl Cytosine in Eukaryotic DNA

Until recently it was thought that 5mC is a terminal DNA modification whose only further fate is removal by demethylation during the erasure of epigenetic marks. Studies in euglenozoans, such as the human parasites *Trypanosoma* and *Leishmania*, revealed the presence of two enzymes, JBP1 and JBP2, which catalyzed the hydroxylation of the methyl group in thymine forming hydroxymethyl thymine.[6,206] This base is further modified by glycosylation of the hydroxyl group resulting in the base "J." Sequence analysis of the JBP hydroxylase domains revealed that they were members of a distinctive family of 2-oxoglutarate and Fe^{2+}-dependent dioxygenases (2OGFeDOs), whose previously undetected representatives were found in several organisms.[50,206] In particular, these studies showed that the metazoan Tet proteins (Tet1, Tet2, and Tet3 oncogenes in humans) are members of this family of 2OGFeDOs.[8] Given that their domain architecture closely parallels that of the metazoan DNMT1, with an N-terminal DNA-binding CXXC domain combined to a C-terminal catalytic domain, it was proposed that they would act on 5mC and hydroxylate it to form 5hmC.[8,50] Follow-up experimental studies showed that indeed the Tet proteins were 2OGFeDOs that generated 5hmC *in situ* from the 5mC in DNA.[17] Though the presence of 5hmC had been noted earlier in mammalian DNA, there was some debate over whether it was an artifact of nonbiological oxidation or a genuine modified base.[29] With the discovery of the catalytic activity of Tet proteins, it became clear that this further modification of 5mC is indeed a biologically relevant modification with possible significance as a novel epigenetic mark. Studies are only just beginning to reveal the regulatory potential of this modification. 5hmC generated by Tet1 was detected in embryonic stem cells (ESCs) and was found to be required for their maintenance by affecting the methylation status of critical ESC maintenance genes such as Nanog.[17,27] Additionally, 5hmC generated by Tet1 has been shown to be required for maintenance of the trophoectoderm-inner cell mass balance in mammalian embryos, with loss of 5hmC favoring the former cellular state.[27] Further, Tet2-generated 5hmC was shown to be required for maintenance of proper balance in the differentiated progeny of hematopoietic precursors: knockdown of Tet2 skewed their differentiation toward monocyte/macrophage lineages.[28] Consistent with this, Tet2 disruption and consequent reduction in genomic 5hmC is associated with several myeloid malignancies. Higher levels of 5hmC were also detected in the Purkinje neurons of the mammalian cerebellum, which have large and euchromatic nuclei, as compared to associated cells such as the granule cells which have small nuclei with typical heterochromatin distribution.[29] Interestingly, overexpression of Tet1 in cell culture also resulted in nuclei with increased size.[17] In biochemical terms, it was found that conversion of 5mC to 5hmC resulted in loss of binding for certain TAM/MBD proteins such as the mammalian MeCP2 and also impairs the recognition of CpG sites by DNMT1.[17,207] These observations suggest that 5hmC could interfere with the recognition of methylated DNA and maintenance of methyl marks, thereby favoring retention of certain differentiation states that are probably characterized by more open chromatin.

Another problem for which a definitive solution remains to be found is the connection between 5hmC and DNA demethylation. Overexpression of Tet1 resulted in a significant decrease of 5mC in cell lines, whereas knockdown of Tet1 resulted in methylation at certain promoters in ESCs.[17,27] Further, those patients with myeloid neoplasms undergoing treatment with methylation inhibitors (such as 5-azacytidine and decitabine) show significantly poorer prognosis if they have a mutant *Tet2* gene than patients with intact *Tet2* genes.[208] This result could be interpreted as a case for weakened demethylation in the Tet2 patients, reducing the effectiveness of the methylation inhibitor treatment. Under high pH conditions, 5hmC spontaneously reverts to C with the release of formaldehyde.[209] Hence, it is technically possible that 5hmC serves as an intermediate in a direct demethylation pathway. However, other lines of evidence point to a more indirect role for 5hmC in demethylation. First, there appears to be strong expression of Tet1 in mammalian primordial germ cells around the time the complete erasure of methyl marks and BER occurs.[163] Second, an uncharacterized DNA glycosylase activity has been identified in bovine thymus extracts that is specific to 5hmC.[209] This observation, together with the poor recognition of 5hmC by DNMT1, suggests that the 5hmC could not only favor a form of BER that replaces it with C but also attenuate maintenance methylation. Other recent results suggest that the relationship between the two modifications might be more complicated. In patients with Tet2 mutations, there is a clear hypomethylation, relative to controls, at the majority of differentially methylated CpG sites.[17] This is in apparent contradiction to the expected situation if Tet2 were to directly function in demethylation. However, it is possible that this phenomenon is not a direct consequence of loss of Tet2 catalytic activity but the preferential proliferation of hypomethylated cells in the neoplasms.

B. Structure and Evolution of the Tet/JBP Family of Enzymes

The catalytic domain of the Tet/JBP family displays a double stranded β-helix fold (DSBH). This is characteristic of a vast class of 2OGFeDOs that catalyze dioxygenase reactions on a wide range of substrates, including peptides, nucleic acids, and small molecules.[5] The conserved core of the DSBH contains eight strands: the second strand bears a conserved HxD motif while the seventh strand bears a conserved His; together these residues chelate an Fe^{2+} ion. The eighth strand bears a conserved Arg that binds the 2-oxoacid cofactor via a salt bridge. In the dioxygenase reaction catalyzed by these enzymes one of the oxygen atoms from molecular oxygen is used to oxidize the 2-oxoglutarate cofactor to form succinate, whereas the second one is inserted into the substrate. This allows these enzymes to catalyze a variety of hydroxylations or hydroxylation-dependent removal of alkyl groups as their aldehydes. The Tet/JBP family of enzymes is widely, albeit sporadically, distributed across the tree of life.[8] The minimal versions of these domains are found in bacteriophages, where the relevant gene is positioned close to the replication origin of the viral genome, in an operon with a gene for a chromosome partitioning protein with a ParB-type HTH domain.[8] This association suggests that these bacteriophage Tet/JBP-like enzymes probably generate 5hmC from the 5mC found at the origins of these viruses and regulate their replication. All eukaryotic versions appear to have been derived via lateral transfers of the bacteriophage versions on more than one occasion. In eukaryotes the Tet/JBP proteins have diversified into five distinct subfamilies. The first of these, archetyped by the Tet proteins, is restricted to Metazoa and is strictly correlated with presence of DNA cytosine methylation. This subfamily is distinguished by the remarkable insertion of a cysteine-rich domain into the N-terminal region of the catalytic 2OGFeDO domains just upstream of the HxD motif.[5,8] Additionally, all members of the Tet subfamily contain a giant low-complexity insert right in the middle of the core DSBH domain, just after strand 4. This insert is likely to undergo regulatory posttranslational modifications such as sumoylation.[8] Most animals have just a single Tet ortholog, which is characterized by an N-terminal DNA-binding CXXC domain and a C-terminal catalytic domain. In gnathostome vertebrates, after the divergence of the cyclostomes like the lamprey, there was a triplication of the *Tet* genes resulting in three paralogous versions, of which Tet1 and Tet3 retain their CXXC domains. In the case of Tet2, the CXXC domain has broken away from the catalytic domain due to a chromosomal inversion and is encoded by an adjacent gene (*CXXC4*) in the opposite direction.[8] The *CXXC4* gene is regulated by the Wnt pathway and could possibly physically associate with the Tet2 protein to reconstitute a functional protein similar to the other two paralogs.[210] It is possible that the function of Tet2 is hence controlled via the Wnt pathway.

The next major Tet/JBP subfamily, the transposon-associated subfamily, is currently known from chlorophyte algae like *Chlamydomonas* and *Volvox*, and mushrooms.[8] It is particularly expanded in the mushrooms with at least 40–60 copies in the genomes of *Coprinopsis* and *Laccaria*. The minimal complete versions of these transposons are characterized by at least three genes, which specify the Tet/JBP-type 2OGFeDO, a transposase with a derived RNAse H-fold catalytic domain and a protein with a specialized version of the HMG domain. The genes for the 2OGFeDO and the HMG-domain protein are codirectional, whereas that for the transposase is nearly always in the opposite direction. Thus, these transposons present a parallel to the above-described transposons that carry their own DNA-modifying adenine and cytosine methylases. These transposons appear to be located predominantly in the subtelomeric regions, which is often heterochromatic across most eukaryotes and might also show enrichment in methylation in the mushrooms.[8,211] This suggests that the Tet/JBP-like enzymes encoded by these transposons might generate 5hmC, which could have an important role in regulating their gene expression and mobility. Given the organization of genes in these transposons, it is conceivable that the action of the 2OGFeDO is influenced by the protein with the specialized HMG-domain-binding specific DNA sequences. Further, given that several copies of these transposons encode their own 2OGFeDO, it is plausible that each 2OGFeDO acts largely in *cis* to regulate the element that produces it. Of the remaining subfamilies, the JBP family is currently only known from euglenozoans. These versions occur either fused to a Swi2/Snf2 ATPase module (JBP2) or fused to a poorly characterized JBP1C domain that also occurs in a stand-alone form in the trypanosomes.[8] While they are currently only implicated in hydroxylation of thymine, it remains to be seen if they might also act, in a manner similar to the Tets, on the 5mC that has been detected in the trypanosome genomes.[153] The 4th subfamily is currently only known from the heterolobosean amoeboflagellate *Naegleria*, at least one of which is fused to a C-terminal chromodomain.[8] Given the inference of the presence of 5mC (see above) in *Naegleria*, it is possible that these proteins generate 5hmC like their homologs in other eukaryotes. The 5th subfamily is currently known from chlorophyte algae and stramenopiles. One version of this family is fused to an N-terminal TAM/MBD domain, suggesting that it is likely to recognize

DNA with 5mC and modify the base to 5hmC (Fig. 3). However, the domain architectures of the remaining members of this subfamily are characterized by fusions to various RNA-binding or RNA-modifying enzymatic domains.[8] It is likely that they generate a range of lineage-specific hmC or hT modifications in tRNAs and other small RNAs in these lineages.

C. The AID–APOBEC Family of Deaminases and the Deamination of 5mC

As noted above, another modification of 5mC that has been implicated in demethylation is the deamination of 5mC to T resulting in a G:T mismatch that can then be corrected by BER to restore a C at that position.[181] Though there is still uncertainty about the role of this modification in demethylation,[182] deamination appears to be a potentially important fate of C and 5mC in DNA as organisms with genomic 5mC typically carry genes for more than one G:T mismatch-specific DNA glycosylase.[194] Currently, the only enzymes that have been demonstrated to catalyze this reaction are the vertebrate AID and Apobec2a/b. AID was originally identified as the enzyme involved in a variety of mutagenic processes related to maturation of antibodies in gnathostome vertebrates.[14,15] Across gnathostomes, breaks in DNA induced by AID mutagenesis have been implicated in antibody class-switching and gene conversion, which play major roles in generation of antibody diversity. In certain mammals, the direct action of AID also plays an important part in the antibody diversification through hypermutation.[15] More recently two AID homologs were identified in the cyclostome vertebrates, and available evidence suggests that they are involved in generating diversity in their variable lymphocyte receptors that are structurally unrelated to gnathostome antibodies.[16] Given the greater efficiency of AID-catalyzed deamination on C rather than 5mC, it appears likely that its role in the diversification of immunity receptors is the primary one.[185] However, demonstration of 5mC deamination activity in Apobec2a/b on single-stranded DNA substrates raises the question if this enzyme might have a function, distinct from AID, which is directed toward the methylated base.[181] Most of the remaining members of the Apobec–Aid family of deaminases mediate RNA-editing through deamination of C to U.[212,213] Apobec1 is required for generating the intestinal isoform of apolipoprotein B by editing its mRNA to generate a premature stop codon.[213] The Apobec3 group comprising multiple closely related paralogs has been shown to be involved in defense against various retroviruses and hepadnaviruses by hypermutation of their template RNAs to disrupt their coding capacity.[212] Indeed, viruses, such as HIV, have evolved counter-Apobec3 defenses, such as the VIF protein that helps them replicate in the presence of this deaminase by targeting it for ubiquitination.[214] The targets of Apobec4 remain unclear to date. All these deaminases share a common catalytic domain with a core sheet formed with five strands. The active site comprises two motifs, HxE and $CX_{2-6}C$, respectively, associated with the C-termini of strand 2 and strand 3 of the core, which chelate a Zn^{2+} ion essential for the deamination reaction.[16]

Classical members of the Aid–Apobec family are currently known only from vertebrates.[16] The primary split appears to have separated the Aid-like group from the Apobec4 clade, both of which were present in the common ancestor of all extant vertebrates. In gnathostomes, the Aid-like lineage appears to have diversified further resulting in distinct Apobec2 and Aid versions. Within mammals, these appear to have spawned Apobec3 and Apobec1 through rapid sequence divergence. Thus, the DNA- and RNA-modifying activities are not strongly separated in phylogenetic terms within the Aid–Apobec family, consistent with the *in vitro* DNA modification capabilities of many of these proteins.[214] The Aid–Apobec family shares a set of distinct structural features (strands 4 and 5 are parallel to each other and two C-terminal helices), and some sequence motifs, with the Tad2–TadA family that is widely conserved across eukaryotes and bacteria.[16] These latter enzymes deaminate adenosine to form inosine at the wobble position in several tRNAs. This observation indicated that the Aid–Apobec family was ultimately derived from the more widespread Tad2–TadA family, suggesting that the ancestral Aid–Apobec-like proteins also probably modified RNAs like the latter family.[16] However, it remained unclear if the Aid–Apobec family was derived from the Tad2–TadA family in the common ancestor of vertebrates, or whether they entered the animal lineage through lateral transfers. Analysis of the genomic data indicates that the Aid–Apobec family was most probably derived within a large radiation of divergent deaminases in bacteria that were in turn derived from the Tad2–TadA family (L.M.I. and L.A., manuscript in preparation). These bacterial deaminases are secreted by several bacteria, including pathogenic and symbiotic bacteria such as *Listeria*, *Wolbachia*, and *Bacillus anthracis*, and are likely to function as toxins that target host nucleic acids for mutation. Interestingly, these deaminases appear to have been transferred on multiple occasions from bacteria to different eukaryotic lineages such as animals, plants, and fungi. The Aid–Apobec family appears to be one such group, whereas there are other groups which were independently transferred from bacteria to fungi and basal animals such as *Trichoplax* (Fig. 2). Hence, the likely origin of the Aid–Apobec family was via lateral transfer from an intracellular bacterial symbiont or parasite of the animal lineage. Presence of multiple such deaminases in other eukaryotic lineages raises the possibility that Aid–Apobec-like deamination of C or 5mC could be more widespread in eukaryotes.

V. Domains Involved in Discrimination of Methylated Versus Nonmethylated Cytosines in DNA

A. Discriminating Epigenetic Marks in DNA

Epigenetic information stored in modified DNA is interpreted via dedicated DNA-binding domains that are able to discriminate between modified and nonmodified bases and target different chromatin-remodeling and -modifying activities to sites with or without the modification (see Chapter by Pierre-Antoine Defossez and Irina Stancheva). The best-known modified DNA-recognition domains are those that recognize methylated cytosine. These DNA-binding domains are often fused to other domains, which might catalyze distinct modifications of chromatin proteins, for example, methylation, demethylation, or ubiquitination, or they might nucleate the assembly of protein complexes such as the repressive histone deacetylase complex.[50] However, DNA-binding domains that specifically recognize nonmethylated cytosine could protect these sites from the action of methylases by setting up particular chromatin states or recruiting catalytic domains (including DNA methylases) to unmethylated target sequences. Currently two major 5mC-recognizing DNA-binding domains (TAM/MBD and SAD/SRA) and one DNA-binding domain primarily recognizing unmodified C (CXXC) have been characterized. In addition, the conserved motif present in the mammalian Stella protein could define another 5mC-recognition module.

B. The TAM/MBD Domain

The so-called methylated DNA-binding domain (MBD) is a conserved domain first observed in the avian SAR-binding protein ARBP, its mammalian ortholog being the methylated CpG-binding protein MeCP2 and another methylated DNA-binding protein MeCP1/PCM1/MBD1.[215,216] While this conserved domain was recovered in other *bona fide* methylated CpG-binding proteins such as MBD2, MBD3, and MBD4, sequence profile analysis showed that a related domain was also found in a number of other proteins in which it was not originally recognized such as the mammalian BAZ2A/B (TTF-IIP5) and SETDB2; several *C. elegans* proteins, such as Flt-1; and *Drosophila* Toutatis.[49] These versions of the domain, while clearly related to the 5mCpG-binding MBDs, did not contain all the conserved residues required for 5mCpG binding.[49] Further, they were found in one or more copies in species with no detectable CpG methylation (such as *C. elegans*) and those with very limited or no CpG methylation at the time of action of these proteins (e.g., Toutatis in adult *Drosophila*). Hence, it became clear that not all versions of this domain are likely to bind 5mCpG-containing DNA and the more inclusive superfamily of these domains was accordingly named TAM (after

*T*TF-IIP5, ARBP, and *M*eCP2).[49] Despite this suggestion, more accurately reflecting the natural history of this domain, the term MBD has unfortunately been used indiscriminately in the literature. We caution against this as it does not accurately reflect the biochemical role of the entire superfamily, and suggest that the domain more appropriately be designated as TAM/MBD or just TAM. Consistent with this suggestion, some of the more divergent mammalian members within the extended TAM/MBD superfamily, which were later named MBD5 and MBD6, have been shown not to bind methylated CpG-containing DNA.[217]

In structural terms, the TAM/MBD is a simple domain of three strands forming a β-sheet followed by a single α-helix, and a C-terminal, less-structured polar extension, which packs against the rest of the fold due to two conserved aromatic residues[218] (Fig. 5). The main determinants for the recognition of the symmetrically methylated CpG dinucleotide come from elements within the three strands that are inserted deeply within the major groove of DNA bearing this dinucleotide.[207,218,219] The C-terminus of the first strand contains an arginine, whose guanido group shows π–π stacking interaction with the pyrimidine ring of the methylated C. An aspartate (which forms a salt-bridge with the above arginine) and a tyrosine from the middle of strand 2 form a complementary pocket to accommodate the methyl group on the first C of the dinucleotide. The alkyl stem of the side chain of an arginine at the C-terminus of strand 3 forms a pocket to accommodate the methyl group of the second C from the complementary dinucleotide, while its guanido group forms a π–π stacking interaction with the pyrimidine ring. The guanido group of this arginine also contacts the $-NH_2$ group of the first C, indicating that it is the key constraint for strict recognition of CpG rather than 5mC occurring in other contexts (Fig. 5). The two conserved aromatic residues from the C-terminal extension appear to be critical for stabilizing the conformation of this arginine at the end of strand 3, while a polar residue immediately downstream of them makes a nonspecific DNA contact (Fig. 5). Additional DNA contacts with the minor groove appear to arise from C-terminal AT-hook domains in some TAM/MBD proteins like MeCP2 and the vertebrate BAZ2A/B (Fig. 3).[49] The TAM/MBD–DNA complex cocrystal structures reveal that the hydroxymethylation of the CpG sequences by the Tet/JBP family proteins would result in bulkier exocyclic adducts to the pyrimidine that would result in steric hindrance. This is consistent with the observed loss of DNA binding of MeCP2 upon hydroxymethylation of the CpG dinucleotide.[207,219] Of the above-mentioned residues, which are central to recognition of 5mCpG, most of them are substituted in *C. elegans* by residues ill suited for such interactions (Supplementary Material). This suggests that, with the loss of CpG methylation in the nematodes, there was a concomitant divergence of the binding sites of TAM/MBD superfamily members, without loss of the DNA-binding domain

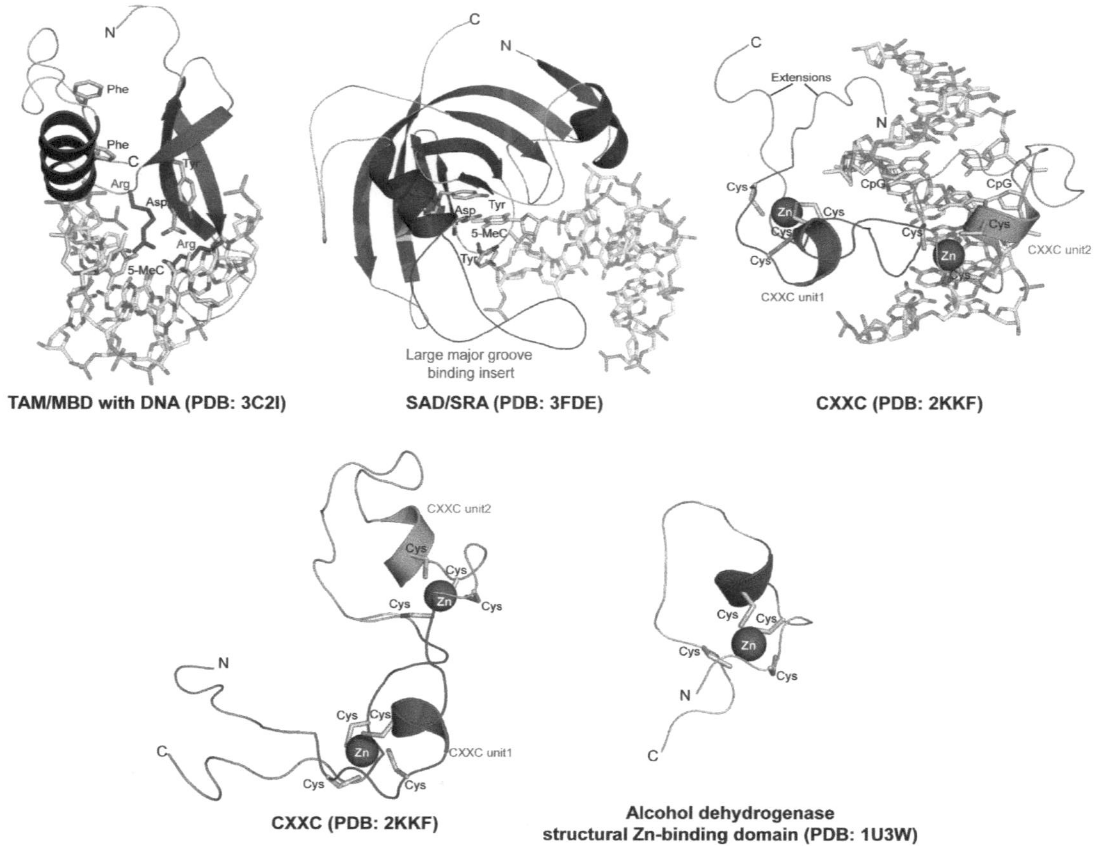

FIG. 5. DNA methylation-discriminating domains. The top panel illustrates the DNA-recognition mode of the TAM/MBD, SAD/SRA, and CXXC domains. β-Strands are colored green and α-helices brick-red. The two repeat units of the bi-CXXC domain are each shown in magenta and blue, respectively. DNA is shown as a semitransparent stick model with the interacting bases in yellow. Key interacting and zinc-binding residues of the domains are marked. The bottom panel illustrates the duplication in the bi-CXXC domain and its similarity of each unit to the structural zinc-binding domain of medium-chain alcohol dehydrogenases. (See Color Insert.)

itself. The TAM/MBD found in the SETDB2 and the BAZ2A/B homologs from across animals show unfavorable substitutions of one or more of the 5mCpG-recognizing residues in strands 1 and 2 of the domain (Supplementary Material). Hence, it is possible that they lost their 5mCpG specificity rather early in animal evolution. However, the retention of the conserved arginine from the third strand in most of them suggests that they may retain the means of at least recognizing unmethylated CpG dinucleotides. The mammalian MBD5 and MBD6 also show substitutions of most of these residues, consistent with their lack of 5mCpG-binding capabilities.[217] MBD5 additionally appears to have gained a potential metal-chelating insert in the C-terminal extension (Supplementary Material). However, given these substitutions it remains to be seen if their binding sites might have been adapted for hemimethylated CpG binding. Based on the conservation patterns, it can also be predicted that the *Arabidopsis* MBD10 might have lost 5mCpG-binding capabilities.

The TAM/MBD domain shows a rather distinctive phyletic pattern, being found in animals, plants, and stramenopiles (Fig. 2). As noted above, within animals it is retained even in the lineages that have secondarily lost cytosine methylation, such as in nematodes. Its phyletic pattern suggests that it emerged in the common ancestor of animals, plants, and fungi followed by a lateral transfer to stramenopiles from plants with which they show an intimate endosymbiosis.[220] The complete loss of this domain in fungi is intriguing, because several fungi display noticeable amounts of CpG methylation.[98,99] Methylation patterns in fungi suggest that the ancestral fungus is likely to have possessed transposon and repeat element methylation, but not the gene body methylation observed in both animals and plants.[98,99] Hence, we speculate that the loss of gene body methylation in the common ancestor of most extant fungi might be correlated with the loss of the TAM/MBD. Therefore, the ancestral role of the TAM/MBD domain might have primarily been in the context of gene body methylation and control of gene expression via methylation. This regulatory function in gene expression might have resulted in the retention of this domain in certain animal lineages even after the loss of DNA methylation—here the TAM/MBD probably helps in nucleating a particular chromatin state even in the absence of 5mC. The TAM/MBD domain has to date been found only among eukaryotes; however, given their rapid divergence, they could have originated in bacterial R–M systems and have currently diverged beyond recognition.

C. The SAD/SRA Domain

This domain was first identified in Np65, certain plant SET-domain histone methylases, and a *Deinococcus* McrA-like REase, and was accordingly named the SET-*a*ssociated *Deinococcus* endonuclease domain (SAD).[221,222] The same domain was subsequently given names such as YDG after an eponymous motif found in a subset of these domains and SRA (for *S*ET and *R*ing finger

Associated) by other workers.[223,224] A number of studies on the eukaryotic SAD/SRA domains have shown that they bind hemimethylated CpG dinucleotides and also other 5mC containing dinucleotides.[225–227] Functional studies have shown that the mammalian SAD/SRA domain protein UHRF1/NP95/ICBP90 plays an important role in maintenance of methylation at CpG dinucleotides by recruiting the maintenance methylase DNMT1 to hemimethylated sites associated with replication forks.[225,226] In plants, genetic evidence suggests that the SAD/SRA domain found in the SET-domain protein KRYPTONITE might play a similar role.[227] Further evidence from different eukaryotic SAD/SRA domains suggests that they might have evolved different sequence specificities, with some being specific to hemimethylated CpGs while others target 5mC in other sequence contexts.[227]

The SAD/SRA domain adopts the β-barrel-like PUA fold, with a core of eight strands (Fig. 5). The prototypical members of the PUA-like fold are the PUA and ASCH domains which bind different types of RNA.[228] For example, the PUA domain in the archaeo-eukaryotic pseudouridine synthases binds the box H/ACA guide RNAs to direct pseudouridylation of target sequences in the maturing rRNAs.[229] The version of the PUA fold found in the SAD/SRA domain is somewhat modified by additional decoration in the form of large inserts, one of which plays a major role in inserting into the major groove of DNA (Fig. 5). Other residues involved in DNA binding by the SAD/SRA domain are located in a position similar to the RNA-binding residues of the PUA and ASCH domains; however, the interface of the SAD/SRA domain with the DNA is located opposite to the RNA-binding face of the PUA-like domains.[45–48,228] The SAD/SRA domain is rather distinctive in recognizing methylated cytosine by flipping the base out of the double helix.[45–48] Deep insertion of the long loop of the SAD/SRA domain into the major groove results in destabilization of the double helix preparing the base for flipping out. The flipped-out base is sandwiched between the two highly conserved tyrosines in the domain, which form aromatic stacking interactions with the pyrimidine ring on either side of it (Fig. 5). Further, a conserved aspartate, three positions downstream of the first conserved tyrosine, forms hydrogen bonds with the 5mC, thereby mimicking the base-pairing interactions in DNA. Thus, the flipped-out base is held firmly in place by the SAD/SRA domain. The recognition of the methyl group in 5mC is achieved via a specific recognition pocket formed primarily by the backbone of a glycine-rich patch immediately downstream of the second conserved tyrosine. This asymmetric mode of binding the flipped-out 5mC is radically different from what is observed in the TAM/MBD (Fig. 5) and provides the structural explanation for the recognition of hemimethylated CpG and non-CpG sites by this domain. In this respect, it is closer to enzymatic domains that operate on single bases, such as the DNA methylases, AlkB-like dioxygenases, Udg- and HhH-superfamily DNA glycosylases, and certain endonucleases like *Hin*P1I REase (the nontarget base in this case).[45,230] In contrast, this mode of binding, with few exceptions like the DNA-clamps of the polIIIβ-PCNA superfamily, is rarely observed in nonenzymatic DNA-binding domains.[231] This raises the possibility that at least certain versions of the SAD/SRA domain might possess some cryptic enzymatic activity that operates on 5mCs. Further, its binding to flipped-out bases suggests that it could remain stationed on DNA and act as a "size amplifier" of the mark, demarcating the differentially methylated strands, and play a role during repair or in postreplication chromatin deposition. The rare versions of the SAD/SRA domains that lack the above features for 5mC recognition include those found in apicomplexans and the highly derived versions fused to the AlkB-type 2OGFeDO domains in fungi.[8] Given that AlkB operates on methylated adenines rather than cytosines, it is conceivable that these fungal SAD/SRA domains have diverged to recognize alkylated adenines.[8] Unlike the fungal versions, the apicomplexan versions are closely related to the typical SAD/SRA domains except for the lack of the key 5mC-recognition features. Given the apparent lack of DNA methylation in apicomplexa, it is possible that they have lost 5mC binding while retaining unmodified cytosine-binding capability.

In bacteria, the SAD/SRA domain is usually fused to or found in an operon with either of two distinct REases of the EndoVII/HNH-fold or a domain of the classical restriction endonuclease fold.[221] Additionally, some of these restriction systems also encode an MutT-like nudix nucleotidase (Fig. 3). One of the EndoVII/HNH-fold ENases of this system is closely related to the MrcA enzyme, which targets DNA sequences containing 5mC and 5hmC.[19,20] This suggests that these restriction systems are likely to specialize in cutting methylated target sites (analogous to REases such as DpnI) and that the SAD/SRA domain helps in the recognition of methylated DNA sequences. We speculate that the MutT-like nucleotidases specified by some of these systems perhaps hydrolyze 5hmC-triphosphate, providing an additional line of defense against phages using a 5hmC-based counter-restriction mechanism. Eukaryotes appear to have acquired the SAD/SRA domain through a single lateral transfer from such a restriction system. In eukaryotes, the domain is found in animals, fungi, plants, stramenopiles, apicomplexans, and heteroloboseans like *Naegleria* (Fig. 2). While certain versions, as noted above, might have evolved to recognize targets other than 5mC, the vast majority of eukaryotic SAD/SRA domains appear to contain the necessary determinants to bind 5mC (Supplementary Material). Indeed, in many lineages, such as fungi and *Naegleria,* this is currently the primary 5mC-recognizing domain. Given its wider phyletic spread in eukaryotes than TAM/MBD and its clear bacterial antecedents, SAD/SRA appears to have been the first dedicated 5mC recognizing domain to have been acquired and recruited by the eukaryotes rather early in their evolution (Fig. 2). This role suggests that asymmetric and hemimethylated

CpG binding might have been the primary mode of recognition of the methyl mark, with the symmetric CpG recognition emerging only later with the origin of the TAM/MBD domain.

D. The CXXC Domain

This domain was originally identified in the vertebrate MeCP2, in the N-terminal region of the vertebrate SET-domain histone methyltransferase MLL1, and in the animal DNMT1.[104,215,216] These architectures indicated that this domain played an important role in connection to DNA methylation (Fig. 3). Subsequent studies have showed that, unlike versions of the TAM/MBD and SAD/SRA domains, it primarily recognizes unmethylated CpG nucleotides and thus plays a role complementary to theirs in discriminating epigenetic marks.[44,123,232–234] However, it is possible that some versions of this domain are more promiscuous in their DNA-binding properties (see below). A mammalian CXXC domain protein, CXXC1/CFP1, is required for recruitment of the histone H3K4-trimethylating enzyme SETD1A/B and also for maintaining proper levels of cytosine methylation by DNMT1.[235,236] This result, together with the presence of this domain in MLL1, suggests that it is important in the recruitment of both DNA and protein methylating activities to CpG-containing DNA, and in mediating the cross-talk between these two systems in regulation of genes.[232,235,236] The CXXC domain is characterized by eight conserved cysteines, whose arrangement includes multiple CXXC motifs that give the domain its name.[123] Analysis of its sequence and structure showed that the classical CXXC domain comprises a peculiar internal duplication, in which the second unit is inserted into the first one.[50] Each of these units, the mono-CXXC domain, is characterized by four conserved cysteines displaying a signature of the form CXXCXXCX(n)C, that together chelate a Zn^{2+} ion (Fig. 5). This proposal for the origin of the classical CXXC domain, that is, "the bi-CXXC domain," as a duplication of two modules is strongly supported by the observation that, in the plant lineage, the only version of this domain is the type comprising a single unit; that is, a "mono-CXXC domain" (Figs. 2 and 3). The second and third cysteines of each individual mono-CXXC domain are situated on a single turn of the helix, while the third and fourth cysteines border a flap-like loop inserted into the double helix (Fig. 5). Outside the core metal-chelating part, the N- and the C-terminal extensions of both the mono- and bi-CXXC domains are typically enriched in basic residues. The NMR structures of the bi-CXXC domain–DNA structure complex reveal that the two CXXC units form a crescent-shaped clasp around both grooves of the DNA bearing the target CpG dinucleotide[44] (Fig. 5). The second unit (i.e., the one nested in the first one) makes the key contacts within the major groove by recognizing the CpG. The protein backbone of the flap-like loop between the third and fourth cysteine of this unit come very close to the 5th position of the pyrimidine rings of the cytosines. As a result, methylation at this position would result in a potential steric hindrance, thereby providing a structural basis for the specific recognition of unmethylated cytosines in DNA. The first CXXC unit predominantly makes DNA backbone contacts via conserved basic residues. The basic N-terminal extension adopts an extended conformation and is inserted into the minor groove of the DNA, while the C-terminal extension makes DNA backbone contacts with both the strands of the DNA simultaneously. Based on this structure, the mono-CXXC domains are inferred to make less extensive contacts and primarily preserve the major groove contacts with the CpG. In bi-CXXC domains, the less-specific DNA contacts made by the strongly charged N-terminal extension and the first unit could result in DNA binding, irrespective of the CpG methylation status or even the presence of this dinucleotide. However, such promiscuous contacts could be modulated by accompanying domains or associated proteins.

Sequence and structure comparisons show that the mono-CXXC domain is homologous to the structural Zn-binding domain of the medium-chain dehydrogenases/reductases (MDRs), which is inserted into the β-barrel GroES-like domain of the latter enzymes.[237] Both the mono-CXXC and the structural Zn-binding domain share a characteristic CXXCXXCX(n)C signature and the geometry of the Zn-chelating site (Fig. 5). However, the latter domain does not bind DNA; instead it appears to be critical for homodimerization of the MDRs.[237] As the version of the domain found in MDRs is present across the three superkingdoms of life, it is likely to represent the ancestral form. The DNA-binding properties of the CXXC domain appear to be a later innovation on the core scaffold offered by the MDR Zn-binding domain. In eukaryotes, the CXXC domain is found only in stramenopiles, plants, and metazoans (Fig. 2). In land plants, the only version appears to be a highly derived, permuted mono-CXXC version seen in the C-terminus of the Demeter-like proteins. In contrast, all currently identified animal and stramenopile versions appear to have the bi-CXXC version (Fig. 2). This unusual phyletic pattern, combined with the state of the duplication of the domain, poses an evolutionary conundrum in terms of their point of origin and dissemination across eukaryotes. The bi-CXXC version is considerably expanded in animals and stramenopiles, whereas the mono-CXXC version is expanded in chlorophyte algae. In large part the phyletic patterns of the CXXC domain mirror that of the TAM/MBD domain, with a comparable absence in the fungi (Fig. 2). This suggests that CXXC might be used as a discriminator between methylated and nonmethylated cytosines in conjunction with the TAM/MBD domain, in the lineages in which they co-occur. In land plants there are no other detectable copies of the CXXC domain beside the derived version in Demeter-like proteins, suggesting that its role might have been taken up by other DNA-binding domains (Fig. 3). One possible candidate is the AP2 domain, which is considerably expanded in plants and specifically recognizes targets with GpC sequences.[50] Consistent with this, representatives of the AP2 domain have been shown to display impaired DNA binding in the presence of methylated cytosines in their target sequences.[238] Also in line with this proposal is the frequent combination of the TAM/MBD, CXXC, and AP2 domains in the same polypeptide in multiple proteins from stramenopiles (Figs. 3 and 6).

E. Stella and H2AZ: Other Miscellaneous Proteins Involved in Affecting Accessibility of Cytosine for Methylation

Other than these domains, which recognize methylated or unmethylated CpG directly, there are a few other proteins that might detect the methylation status of cytosine in the genome. One of these is the mammalian protein, PGC7/Stella/Dppa3, which localizes to the nucleus and maintains methylation of the maternal genome at imprinted loci, thereby perpetuating the imprinting asymmetry between the parental genomes during early development.[161] Given its role in protecting imprinted regions from demethylation during postfertilization, it may bind methylated sequences directly and alter the chromatin state to protect it from demethylation. Stella belongs to a fast-evolving family of small proteins that are currently known only from placental mammals. The conserved core shared by these proteins includes a positively charged helical segment, followed by a C-terminal CXCXXC motif that could potentially chelate a metal ion (Supplementary Material). The conservation of the Stella family only within placental mammals, coupled with its rapid evolution, suggests that it may help to deploy DNA methylation-based imprints in the intersexual conflict posited to play out during early mammalian development. According to the sexual-conflict hypothesis paternal alleles would demand greater resources from the maternal environment than the maternal alleles, which in contrast would try to reduce the demand on maternal resources[239–241] (see Chapter by Jon F. Wilkins and Francisco Úbeda). In placental mammals, the origin of the placenta provided new opportunities for channelizing maternal resources to the developing fetus. This conflict appears to have resulted in differential methylation of several loci including those pertinent to placental, fetal, and neonatal growth.[239–241] Thus, we speculate that the sudden origin of Stella in the placental mammals was perhaps an evolutionary response to this conflict as a mechanism to protect maternal methylation when paternal methylation is being erased. Most placental mammals contain 3–6 paralogs of the Stella family; the greatest number of paralogs (six) is currently seen in *Rattus norvegicus* (Supplementary Material). At least two of these, respectively, typified by Stella and FAM156A, are inferred to have been present in the common ancestor of most extant placental mammals, with independent lineage-specific

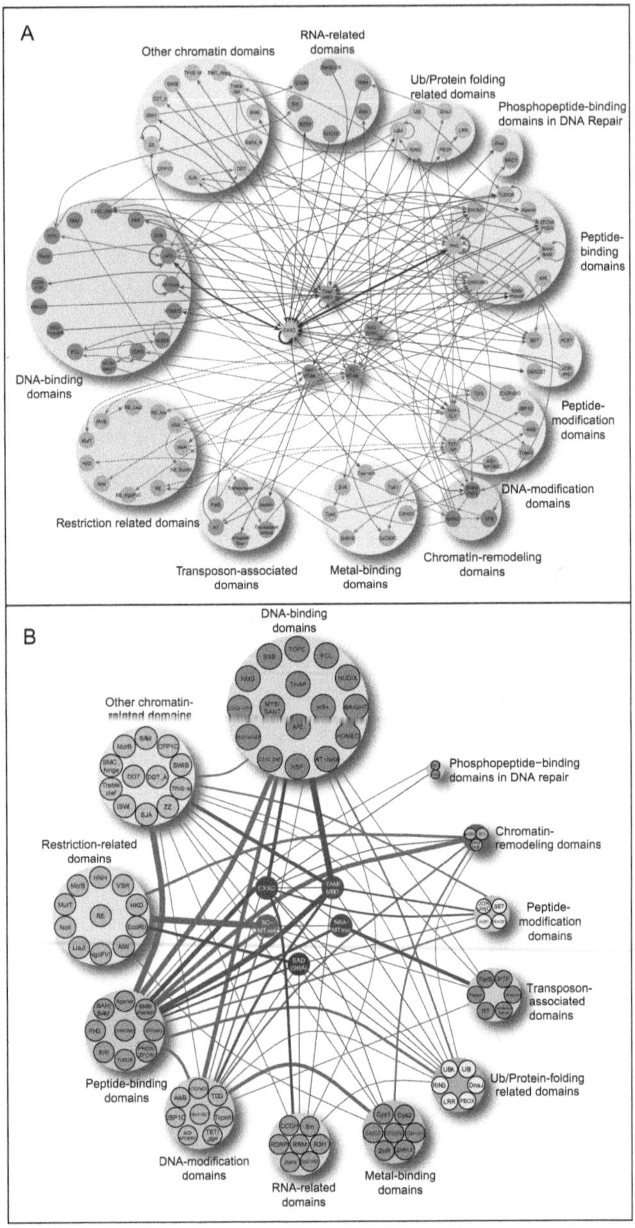

FIG. 6. Domain architecture and gene neighborhood network. These are shown as a network graph with nodes representing domains related to DNA methylation, and edges their physical connectivity in a polypeptide or gene neighborhood. The metanetwork is used to highlight the

duplications among both these paralogous groups. It is worth investigating whether these paralogs play similar roles in protecting other chromosomal regions, distinct from the regions targeted by Stella, from demethylation. It is conceivable that the rapid divergence between orthologs and paralogs in the Stella family might be linked to positive selection for recognizing changing landscapes of the imprinted genes. The CXCXXC motif of Stella is, interestingly, also conserved in a subset of fungal MBD4-like proteins (Fig. 3), though its role in interacting with methylated sequences remains unclear.

In mammalian systems, the histone variant H2A.Z and di- or trimethylated histone H3K4 is strongly anticorrelated with DNA methylation, whereas trimethylated histone H3K9 and the histone variant macroH2A show an overlap and synergistic functional interaction with DNA methylation.[242–249] The H2A.Z anticorrelation with DNA methylation is highly conserved across eukaryotes.[99] More generally, H2A.Z deposition and H3K4 di/trimethylation is correlated with active chromatin and the prevention of the spread of repressive heterochromatin into euchromatic regions, even in eukaryotes with no DNA methylation such as *S. cerevisiae*.[250] Hence, it might be argued that H2A.Z deposition potentially prevents the spread of various distinct mechanisms promoting the heterochromatic state, irrespective of whether it is via interaction with the DNA methylation system or through independent histone modifications. Nevertheless, the conservation of the striking anticorrelation between H2A.Z deposition and DNA methylation across a wide phylogenetic range raises the possibility that H2A.Z binding to DNA might directly shield cytosine (CpG sites in particular) from the DNA methylases. However, another explanation, albeit not mutually exclusive, is also possible. In mammals, DNMT1 interacts

---

overall trends of associations between different functional types of domains involved in DNA methylation. The arrow heads depict directionality; for domain architectures they point from the N-terminal to the C-terminal domain and for gene neighborhoods from the 5′ gene to the 3′ gene. Gene neighborhood associations are shown as dashed lines. Domains with similar functional roles are in the same color and further grouped into metanodes in the metanetwork. Edges are colored based on the principal domain of an association; 5C MTases: orange, N6A MTase: green, CXXC: blue, TAM/MBD: magenta, and SAD/SRA: purple. Edges not involving these principal domains are colored gray. The edge thickness is proportional to the relative frequency with which linkages between two domains or metanodes reoccur in distinct polypeptides and gene neighborhoods. Conventional abbreviations are used for domain nomenclature. Other domains with nonstandard abbreviations include CFP1C; CFP1 C-terminal domain; ACET, GCN5-like acetyltransferase; AuxRF, a novel version of the chromo-fold predicted to bind methylated histones; Cys1, domain with conserved cysteines associated with fungal TET/JBP-containing transposons; Cys2, a domain with conserved cysteines associated with the AlkB and SAD family of proteins in fungi; Cys-rich, a domain with conserved cysteines inserted in the 2OGFeDO domain of the metazoan TET family; DEACET, RPD3/HDAC-like histone deacetylase; DDT_A, DDT associated domain; RT, reverse transcriptase; RE, restriction endonuclease; and ZnR, zinc ribbon. (See Color Insert.)

with and is activated by the highly conserved SANT domain protein DMAP1.[117,251] DMAP1 is, interestingly, also in other chromatin-modifying complexes such as the repressive histone deacetylase HDAC2 complex, the NuA4 histone acetylase complex, and the SWR1 SWI2/SNF2 ATPase-dependent complex required for deposition of H2A.Z.[252–254] This link between DMAP1 and the complex involved in H2A.Z deposition raises the possibility that SWR1 and DNMT1 compete for DMAP1. H2A.Z could draw DMAP1 away from DNMT1, as a part of the SWR1 complex, and thereby depress DNA methylation in regions of the genome where it is present. In evolutionary terms, SWR1, DMAP1, and H2A.Z are ancient proteins, which are present in all eukaryotic lineages with an ancestral DNMT1 ortholog (Fig. 2), though they are also present in eukaryotic lineages that have secondarily lost 5C DNA methylation (consistent with their more extensive roles). However, they are absent from the basal-most eukaryotes such as *Trichomonas* and *Giardia*, that appear to lack DNMT1 orthologs and also apparently do not have 5C DNA methylation. Thus, the point of origin of the DNMT1 clade in eukaryotes appears to be coeval with the point of origin of SWR1, DMAP1, and H2A.Z suggesting that they could have developed functional interactions from an early period in their evolutionary history (Fig. 2).

The mammalian ATRX protein has been characterized as the SWI2/SNF2 ATPase subunit of a complex required for proper 5C DNA methylation.[150] As noted above, it shares conserved PHD and treble-clef Zn-chelating domains (so-called ADD module) with the metazoan DNMT3 clade proteins.[132,133,151] ATRX proteins from both the plant and animal lineages contain an ADD module, while among the DNMT3 orthologs the module is only present in the metazoan representatives (Figs. 3 and 4). This suggests that the ADD module first emerged in the context of the ATRX proteins and was then acquired via N-terminal domain accretion by the DNMT3 clade only in the metazoan lineage. The ADD module has also independently fused to a SUMO-ligase and a SET methylase domain in chlorophytes and the haptophyte alga *Emiliania* (Fig. 3). In addition to histone tail recognition, in DNMT3 the ADD module is required for interaction with MBD3 and the SWI2/SNF2 ATPase BRG1,[151] while in ATRX it mediates interaction with MeCP2.[255] This suggests that the ADD module might facilitate indirect discrimination of 5mC via interactions with TAM/MBD proteins. In support of this observation, the ADD module is only present in ATRX orthologs of organisms with multiple 5mCpG-recognizing TAM/MBD proteins; it has been lost from the fungal ATRX orthologs (e.g., *Neurospora*), concomitant with the loss of TAM/MBD and CXXC proteins in fungi (Fig. 2). The ATRX subgroup arose within the older RAD54 clade of SWI2/SNF2 ATPases that are universally conserved across eukaryotes.[50] The point of origin of the ATRX subgroup appears to have corresponded to the point of origin of DNMT1, SWR1, DMAP1, and

H2A.Z, and its phyletic pattern correlates well with the presence of 5mC in the genome (Fig. 2). ATRX versions with the ADD module appear to have first emerged within the "crown group" of eukaryotes; that is, the common ancestor of the plants and animals (Fig. 2). Within plants there appears to have been a further duplication of ATRX resulting in a paralogous group typified by the *Arabidopsis* proteins CHR31, CHR34, CHR38, CHR40, CHR42, and DRD1 proteins, of which DRD1 is required for RNA-directed 5C DNA methylation.[147–149] These proteins lost the N-terminal ADD module and instead acquired a distinct Zn-finger with the Zn-chelating residues showing a CHCC pattern (Fig. 3; Supplementary Material). This feature might be critical for RNA-dependent recruitment of methylases in plants.

VI. Domain Architectural Logic of Proteins Related to DNA Methylation

A. Visualizing Domain Architectures as Networks

The functional properties of the domains related to DNA methylation are reflected in their domain architectures—that is, linkages between various catalytic domains, modified-histone discriminator domains, DNA-binding domains, and chromatin–protein interaction domains. Despite the dramatic diversity of these domains and domain architectures seen across eukaryotes, natural selection for relevant interactions appears to have channelized architectures into certain themes, which often have considerable predictive value for functional inferences.[50] A useful representation to discern these functional themes is the domain architecture network: all domain architectures of a given functional system are displayed as an ordered graph, in which the domains are the nodes and the edges connecting them stand for two domains occurring adjacent to each other within the same polypeptide.[50] Further, the edges can be weighted using the number of times a pair of domains independently co-occur as adjacent neighbors in different proteins. This graph can further be supplemented with co-occurrence in operons in the case of prokaryotes and physical domain–domain interactions if a detailed protein interaction map is available. Within this network, different set of domains can then be grouped depending on their function to give information regarding the interactions between whole groups of domains with similar function. Fig. 6 shows such a domain architecture network encompassing all proteins with domains relevant to DNA methylation, demethylation, further modifications or discrimination of methylation status of DNA. It primarily uses information from domain architectures and gene neighborhoods, as detailed domain–domain interaction maps for these domains are currently unavailable.

B. 5mC and Unmethylated-C Recognition Domains, and Their Interplay with Histone Methylation and Other Modifications

Examination of this network and domain architectures reveals several key themes related to the linkages of domains related to DNA methylation (Fig. 6). Firstly, though the CXXC and TAM/MBD domains co-occur in the same polypeptide, neither of them co-occurs with the SAD/SRA domain in any protein (Fig. 6). This strong exclusion is correlated with the symmetric recognition of methylated or unmethylated sites by the former, and the recognition of primarily asymmetric methylated sites by the latter.[46,48,227] Thus, there appears to be complete functional compartmentalization of TAM/MBD and CXXC on the one hand, and SAD/SRA on the other, based on their DNA-binding mode. The independent co-occurrence of CXXC and TAM/MBD in proteins from multiple, distantly related eukaryotes, suggests that these two domains might often cooperate within a polypeptide to form a regulatory switch by, respectively, sensing methylated or unmethylated CpG dinucleotides.[234] The CXXC domain is found in the same polypeptide as the 5C DNA methylase module on at least three independent occasions (Figs. 3 and 6), but neither TAM/MBD nor SAD/SRA is ever found in the same polypeptide with any DNA methylase domain. However, both TAM/MBD and SAD/SRA protein interact physically with different 5C DNA methylases.[151,225,226] This observation points to a direct role for the CXXC domain in assisting 5C methylase sensing of unmethylated targets,[233] whereas the two methylated DNA-sensing domains appear to only regulate methylase activity (after an initial methyl mark is established) as independent, diffusible, accessory factors. The CXXC domain is also linked in the same polypeptide to other methylated DNA-modifying enzymatic domains, such as Demeter-like DNA glycosylases and Tet/JBP 5mC hydroxylases (Figs. 3 and 6). Though the TAM/MBD is never linked to methylases in the same polypeptide, like the CXXC domain, it is combined with the DNA glycosylase and Tet/JBP domains in different proteins (Figs. 3 and 6). Hence, both the TAM/MBD and the CXXC domain might be utilized as internal switches in these proteins, perhaps acting oppositely, in helping them distinguish methylated substrates from unmethylated DNA with CpG sequences. The CXXC and TAM/MBD domains also appear to be found in the same polypeptide with distinct chromatin-remodeling ATPases modules such as SWI2/SNF2 and MORC (Figs. 3 and 6).[256] These 5mC-discriminating domains may recruit these ATPase modules to mediate local or large-scale chromatin remodeling. However, they may also help in furthering methylation marks, as suggested by the recovery of an *Arabidopsis* Smc-hinge domain protein, similar to the version fused to the MORC ATPase module in other eukaryotes, as a factor required for 5C DNA methylation.[256–258] Interestingly, these ATPase modules are also seen in bacterial R–M systems and appear to play an analogous role in mediating long-distance interactions between the REase-recognition site and DNA cleavage site.[50,256] Another notable linkage is the fusion of the CXXC domain to the RNA-dependent-RNA polymerase of the RNAi system in stramenopiles, suggesting that it might play a role in recruiting this enzyme involved in posttranscriptional gene silencing to particular regions of chromatin (Figs. 3 and 6).

TAM/MBD, SAD/SRA, and CXXC are each found frequently in the same polypeptide as the peptide-methylating SET domains and demethylating Jumonji-related (JOR/JmjC) domains (Figs. 3 and 6).[5,50] However, these 5mC-discriminating domains are only rarely, if ever, found associated with peptide acetylase and deacetylase domains. Thus, sensing of DNA methylation status mainly appears to directly regulate histone methylating and demethylating enzymes rather than the acetylases. These architectural trends might also have bearing on the observed anticorrelation between 5mC and certain histone methylation marks such as H3K4 di/trimethylation and the positive correlation with other histone methylation marks such as H3K9.[44,108,134,148,227] In particular, CXXC and versions of the TAM/MBD domain, which do not bind methylated CpGs, could target SET-domain proteins to unmethylated CpG sites[44,236] and help establish histone methylation patterns that are inversely related to DNA methylation status. The primary domain that directly links SET domains to methylated regions of DNA is the SAD/SRA domains and could play an important role in directing repressive histone methylation marks.[224,227] The cognate apicomplexan version, predicted to bind C rather than 5mC, which appears to have been acquired through lateral transfer from the plant lineage, might still recruit the histone methylases to establish repressive chromatin by binding unmethylated C-rich regions associated with genes and promoters in these organisms.[259,260] At least in some organisms both TAM/MBD and CXXC domains might recruit the JOR/JmjC protein to remove certain histone methyl marks, probably with distinct consequences in each case (Fig. 6).[232] In stramenopiles, the CXXC domain is also linked to the histone deacetylase domain, suggesting that it might also be used to establish repressive chromatin by removing acetyl marks in these organisms (Figs. 3 and 6). The SAD/SRA domain is the only known domain that directly links recognition of DNA methylation to chromatin–protein ubiquitination.[45,223] Accordingly, it has been combined with the ubiquitin E3 ligase RING domain, independently on more than one occasion, and also other Ub-binding domains, such as the Ub-like β-grasp and UBA domains (Fig. 6). Just as the domains discriminating the cytosine methylation status of DNA are fused to the histone methylase catalytic domains, a number of modified peptide-binding domains have been combined with DNA methylase domains on several independent occasions (Figs. 4 and 6). The BMB/PWWP domains have been fused independently to both 5C and N6A DNA methylases in different lineages. Additionally, multiple

Chromo/Tudor-like SH3-fold domains, namely the BAM/BAH and chromodomains, and the PHD finger and its derivatives are combined in the same polypeptide with 5C DNA methylases. Parallel to the situation between the 5mC-discriminating DNA-binding domains and the histone acetylase catalytic domains, there is not a single case of combination of the methylase domain with bromodomains. Hence, though there is strong tendency for the DNA methylases to recognize lysine di/trimethylation patterns in histones, they appear to be rather strictly decoupled from recognition of comparable acetyl marks, consistent with the typically repressive role of DNA methylation. There are also a number of links of the 5C methylase modules to ubiquitination-related domains (Fig. 6). First, the UBA domains are fused to the plant 5C DNA methylases[130]; second, the SAD/SRA domain protein UHRF1, which is a separate partner of DNMT1, also contains Ub-like and RING domains[45,225,226]; and third, the DCMs fused to the Rad5-like SWI2/SNF2 are linked to a RING domain that is inserted within the SWI2/SNF2 domain. These connections suggest that, in addition to histone methylation, ubiquitination of chromatin proteins might also be an important signal recognized by different 5C DNA methylases.[45,130,223,225,226]

There are also numerous combinations in the same polypeptide between the above-discussed methylated-DNA-discrimination domains and diverse methylated peptide-binding domains such as those belonging to the Chromo/Tudor-like SH3 fold and the PHD finger and its derivatives.[50,261] Such linkages more often involve the TAM/MBD and CXXC domains than the SAD/SRA domain, suggesting that recognition of histone modifications is linked to a greater degree to the CpG dinucleotide either in a completely modified or unmodified state rather than the recognition of hemimethylated CpGs or other 5mCs. Unlike the case of catalytic domains modifying DNA and histones, the 5mC-discriminating DNA-binding domains are often linked to a bromodomain that recognizes acetylated peptides of chromatin proteins.[261,262] The high frequency of the combinations between 5mC discrimination DNA-binding domains and different types of modified-histone peptide-binding domains, which have often independently emerged in the major lineages, strongly point to an important role for simultaneous recognition of methylation status of both DNA and different epigenetic marks on histones across eukaryotes (Figs. 3, 4, and 6). The ADD module appears to have been combined in different proteins to the DNA-methylase module, the SWI2/SNF2 ATPase module, the SET methyltransferase domain, and the SUMO E3-ligase-type RING domain. Further, in insects there are stand-alone versions of the ADD module. Thus, the ADD module appears to represent a distinct theme; that is, an adaptor that senses the status of methyl marks on histones and (indirectly) on DNA, and connects them to other chromatin remodeling or modifying activities. Finally, in bacteria the domains related to the biochemistry of DNA methylation are found primarily as part of R–M systems. Indeed, the loss of the operon organization in eukaryotes appears to have in large part disfavored the retention of linked-gene systems, such as the R–Ms, in cellular genomes. The cellular genomes of eukaryotes do not encode combinations of REase domains and methylases in the same protein[263,264] (Fig. 6). It is possible that the development of the link between methylation and heterochromatin in large part precluded the elaboration of such systems in eukaryotes because methylated DNA tended to be associated with condensed chromatin and was segregated from transcriptionally active open chromatin.

VII. Evolutionary Considerations

While there have been previous phylogenetic analyses of the eukaryotic DNA methylases, these have been hampered by lack of proper identification of the bacterial cognates of each group, the imprecise analysis of domain architectures, and lack of consideration of the structural features distinguishing the CTDBM of each group.[97–99] In the current work, we have remedied these issues through systematic analysis of these features and also used a much greater phyletic spread of eukaryotes to clarify the global evolutionary picture of the eukaryotic DNA methylases (Fig. 4). The emerging picture points to multiple independent acquisitions of different DNA methylases by eukaryotes, through lateral transfer from bacteria at different points in their evolution. Beyond those N6A methylases and 5C methylases that were incorporated into the core genomes of eukaryotes, there are the mobile versions of both types borne by transposons and viruses. The core genomes appear to have acquired N6A methylases on at least three independent occasions, with two of these transfers occurring prior to the LECA. The phylogenetic tree of the 5C methylases shows that there were six notable independent transfers of these methylases from bacteria to core genomes of eukaryotes. These, in addition to the DNMT1-RID, DNMT2, and DNMT3 clades, also spawned the kinetoplastid-type methylases, Rad5-like SWI2/SNF2 fused methylases, and chlorophyte-type methylases (Fig. 4). None of these major 5C methylase families are currently known from two basal eukaryotic lineages, the parabasalids (e.g., *Trichomonas*), and diplomonads (e.g., *Giardia*; Fig. 2).[265] However, both DNMT1 and DNMT2 are seen in *Naegleria*, which belongs to another ancient eukaryotic lineage (the heteroloboseans) that are a sister group of the kinetoplastids (e.g., *Trypanosoma*).[266] This suggests that the first 5C DNA methylases were probably not acquired in the LECA, but after the divergence of the diplomonads and parabasalids from the rest of the eukaryotes and before the divergence of the kinetoplastid–heterolobosean clade. Multiple chromatin-related adaptations appear to have emerged around the same time just prior to the divergence of the kinetoplastid–heterolobosean clade from other eukaryotes (Fig. 2), such as histone acetylases and deacetylases, histone methylases and demethylases, polyADP ribosyl transferases, SWI2/SNF2 ATPases (e.g., ATRX), and diverse adaptor proteins in chromatin (e.g., DMAP1).[50] This suggests that after the early eukaryotic lineages (diplomonads and parabasalids) diverged, there was a second phase of innovation among chromatin proteins which included for the first time recruitment of 5C DNA methylases as generators of epigenetic marks. However, there is some uncertainty regarding the actual relationships between the basal eukaryotes,[265,266] and also extensive lateral transfer and gene loss between different unicellular eukaryotes.[50,220] Hence, the details of this reconstruction might change with increasing availability of genomic data from basal eukaryotes.

As noted above, most bacterial cognates of each of the major eukaryotic cellular 5C and N6A methylases have primarily radiated as a part of the R–M systems of bacteria. Thus, the selective pressures, which favor diversification of R–M systems, appear to have driven evolution of a great variety of DNA methylases that were then repeatedly acquired by eukaryotes. However, the same epigenetic codes utilized by the R–M systems appear to have been deployed in the very distinct context of chromatin dynamics in eukaryotes. Indeed, several other components of R–M systems and other selfish elements have been acquired in parallel to the methylases and utilized in different facets of eukaryotic chromatin dynamics. The most prominent of these include chromatin-remodeling enzymes like SWI2/SNF2 ATPases and MORCs, the Tet/JBP-like DNA base hydroxylases, DNA-binding domains such as SAD/SRA and HIRAN, and DNA repair enzymes like the VRR-NUCs.[8,50,96,256,263] The DNMT2 clade and two of the clades of N6A methylases appear to have been recruited to a role primarily in RNA methylation. Likewise, at least one clade of Tet/JBP hydroxylases appears to have undergone a substrate shift to function as RNA-modifying enzymes in eukaryotes. Thus, it can be said that the bacterial mobile selfish systems have served as the "development labs" for not just the DNA methylases but also other key players in eukaryotic chromatin and RNA-related functions.

In eukaryotes, DNA methylation-dependent epigenetic marks have been combined with two other forms of regulatory information, namely peptide modifications of chromatin proteins and the RNAi systems of posttranscription gene regulation.[99,110,114,122,147,267,268] Usually, DNA methylation-based systems act in concert with RNAi systems to negatively regulate gene expression, and to establish heterochromatic states in specific chromosomal regions.[268,269] In contrast, peptide modification of histones and other proteins can function either agonistically or antagonistically with respect to DNA methylation-dependent regulatory mechanisms.[269] In evolutionary terms, the RNAi and peptide-modification systems such as histone acetylation/deacetylation and histone methylation can be traced back to the LECA[5,50]; hence, they are likely to have preceded the emergence of DNA methylation-based regulation in eukaryotes. While eukaryotes maintained the histones and their nucleosome organization from their common ancestor with archaea, they showed a simple but notable evolutionary innovation in the form of positively charged tails linked to the globular domains of the nucleosomal histones.[270] This appears to have provided a niche for the early expansion of peptide-modification systems: at least six potential methylases, four acetylases, and two deacetylases modifying chromatin proteins, along with several adaptor proteins that recognized peptides modified by these enzymes, can be traced back to the LECA.[50] These ancient histone-modifying enzymes are also strongly retained across eukaryotes, and appear to be essential for the very existence of a functional eukaryotic cell.[271] In contrast, both the DNA methylation and RNAi systems are retained to a much lower degree across eukaryotes (Fig. 2).[50,267] Either or both of these systems have been completely or partially lost in several eukaryotic lineages (e.g., the yeast *S. cerevisiae* or the chordate *Oikopleura*).[98,99,267] Organisms lacking these systems do not necessarily show drastic differences in terms of body-plan or organization relative to their sister groups that have them intact (e.g., *Oikopleura* vs. *Ciona*). Therefore, both DNA methylation and RNAi appear to be potentially dispensable back-ups (i.e., partially redundant) for the core peptide-modification-dependent regulatory systems. Evidence from fungi, plants, and animals strongly suggest that 5C DNA methylation is directed to specific chromosomal sites by RNA[99,110,114,122,147,267] (see Chapter by Anton Wutz). In vertebrates, there is evidence for piRNAs generated by the RNAi system playing a role in the methylation of transposons.[122] Thus the 5C DNA methylation and RNAi systems are likely to have developed a close functional connection early in eukaryotic evolution.

Both the DNA methylation and the RNAi systems appear to have been deployed as a defense against transposons in several eukaryotic lineages.[98,122,267] Indeed, this could be one of the ancestral functions of both these systems. As a corollary to this idea, it has been proposed that 5C DNA methylation might serve as a mechanism to control spread of transposons from a genome bearing them to one lacking them during sexual reproduction.[98] It was suggested that this might be an important reason for vertebrates and land plants displaying strong methylation patterns. It was also stated that, because unicellular eukaryotes are primarily asexual, they might have lower costs for the loss of DNA methylases.[98] While there is evidence in favor of DNA methylation preventing sexual transmission of transposons,[92,104,110,122] the latter claim regarding unicellular eukaryotes is largely unjustified, both in terms of the observed propensity for sexual reproduction in unicellular forms[272] and also the presence of DNA methylase genes in them (Fig. 2). Conversely, in several animal lineages, such as insects and nematodes, there is little or no methylation of transposons, suggesting that, even when present, this system is not universally used in antitransposon defense.[98] DNA methylation might have other defensive roles. For example, in algae, it could protect against the restriction systems of phycodnaviruses, whereas in vertebrates, it helps in distinguishing highly methylated "self" DNA from poorly methylated nonself DNA.[98,125,155]

In addition to defensive roles, recent studies also point to conservation of gene body methylation patterns, suggesting that regulation of gene expression might also be an evolutionarily early function of the DNA methylation systems.[92,110,246] This might be compared to the miRNA-dependent branch of the RNAi system that is directed primarily at regulating genes posttranscriptionally.[267,269] Another somewhat neglected role of DNA methylation is suggested by the finding that, upon homology-directed repair or gene conversion (using an undamaged sister of a dsDNA break in a damaged duplex), the two recombinant DNA molecules are differentially methylated.[273] This differential methylation of the two duplexes results in divergent gene-expression patterns between them. As homologous recombination repair could alter the genetic information in the repaired region, selection could subsequently favor either the copy with the gene silenced due to methylation (if the postrepair version were deleterious) or the copy which is unmethylated (if the expression of the repaired gene were advantageous).[273] Hence, DNA methylation could serve as a protective mechanism against the consequence of DNA repair errors and also provide "evolvability" to the organism. Taken together, these observations suggest that both the DNA methylation and RNAi systems might provide multiple, functionally overlapping layers of defense against distinct genetic threats impacting the genome. Therefore, the retention or loss of these systems in particular eukaryotic lineages might be dependent on the benefits and costs they offer to an organism with respect to the unique combination of life history factors that it faces.[274] Accordingly, once retention of these systems is favored in a given lineage, new functional dependencies on these systems could develop among certain representatives of that lineage. Phenomena such as imprinting, which is observed in mammalian lineages and angiosperms, appear to be new dependencies on the DNA methylation and RNAi system that appear to have developed from their older role in counter-transposon defense.[275] The emergence of mammalian behaviors such as suckling could have favored the emergence of imprinting at loci such as the Gnasxl and Peg3. They, respectively, code for a G-protein and a Zn-finger transcription factor—the uniparental expression of the alleles of these are required for fetal growth and/or proper suckling and maternal behavior in placental mammals.[276,277]

VIII. General Conclusions

A combination of ancient functions and newly emergent dependencies has resulted in 5C DNA methylation profoundly influencing numerous aspects of mammalian and angiosperm biology. Despite the recent advances in uncovering the many ramifications of DNA methylation in these systems, there remain aspects of its function that are yet poorly understood. Even among the better-studied aspects, we lack a clear understanding of their relative importance and the biochemical foundations of connections of various aspects to other regulatory systems, such as RNAi. The discovery of the hydroxylation of 5mC catalyzed by the Tet/JBP family adds a further wrinkle to our understanding—even the preliminary results relating to its functions point to ramifications comparable to DNA methylation.[8,17] For example, the role of 5hmC levels in defining the balance between the trophoectoderm and inner cell mass and different hemal cell lineages in placental mammals suggests that these further modifications of 5mC could also be recruited to the regulation of parent–kin interactions that emerged in mammals or development of the immune system across gnathostomes.[27,28] Genomic data suggest that, just as in mammals, there might be interesting lineage-specific dependencies of DNA methylation in other organisms. For instance, expansion of the DNMT3 clade in fishes suggests a distinctive role for specific methylation events in these organisms (see Chapter by Mary G. Goll and Marnie E. Halpern), even as imprinting emerged in therian mammals. Importantly, the genomic data shows that the chlorophytes, haptophytes, stramenopiles, and heterolobosean amoeboflagellates possess well-developed DNA modifications systems that are of comparable complexity to those seen in vertebrates and plants. In some of these organisms, 5C DNA methylation appears to be combined with other modifications like N6A methylation and equivalents of modifications such as Momylation catalyzed by the bacteriophage Mom protein.[8] Ciliates and heteroloboseans, however, appear to possess a unique N6A methylation system. These offer a virtually unexplored area for understanding better the spectrum of biological process that might be controlled by DNA modifications. Studies on these microbial eukaryotes have the potential for informing studies in mammals and other vertebrate models. In this regard, it should be noted that the discovery of the Tet/JBP family was sparked by the studies on the microbial eukaryotes such as trypanosomes.[6] The above-presented analysis of domain architectures shows that the linkages from microbial eukaryotes point to interesting possibilities regarding unexplored functional connections. Examples include the possible role for the MORC ATPases in regulating methylation and the recruitment of the RNA-dependent RNA polymerase of the RNAi system to regions of chromatin. In particular, studies on microbial eukaryotes could help in teasing out the common denominator from lineage-specific roles of the DNA methylation system and thereby clarify the hierarchical links between the different consequences of DNA modifications. Hence, we hope that the systematic survey of the comparative genomics of DNA methylation systems presented in this chapter might help in these endeavors.

Acknowledgments

Work by the authors is supported by the intramural funds of the National Library of Medicine, National Institutes of Health, USA. We would like to acknowledge the numerous contributions of various researchers in the DNA methylation and chromatin field which we were regrettably unable to cite due to sheer enormity of the literature under review.

Appendix. Supplementary Material

A systematic collection of the different DNA methylases and functionally related enzymes, chromatin-associated and DNA-binding proteins, and multiple alignments of particular protein families discussed in the text can be found at the following FTP site:

ftp://ftp.ncbi.nih.gov/pub/aravind/chromatin/methylase/supplementary.html

Note Added in Proof

While this article was being prepared for publication there was a publication demonstrating the role for 5hmC in mammalian paternal genome reprogramming immediately after fertilization. This is possibly catalyzed by Tet3 which is expressed in this time window. This supports the possibility of 5hmC serving as an intermediate for demethylation (Iqbal K, Jin SG, Pfeifer GP, Szabó PE.; Proc Natl Acad Sci USA 2011 vol. 108 no. 9 3642–3647).

References

1. Czerwoniec A, Dunin-Horkawicz S, Purta E, Kaminska KH, Kasprzak JM, Bujnicki JM, et al. MODOMICS: a database of RNA modification pathways. 2008 update. *Nucleic Acids Res* 2008;**37**:D118–21.
2. Grosjean H. *DNA and RNA modification enzymes: structure, mechanism, function, and evolution*. Austin, Texas: Landes Bioscience; 2009.
3. Warren RA. Modified bases in bacteriophage DNAs. *Annu Rev Microbiol* 1980;**34**:137–58.
4. Anantharaman V, Koonin EV, Aravind L. Comparative genomics and evolution of proteins involved in RNA metabolism. *Nucleic Acids Res* 2002;**30**:1427–64.
5. Iyer LM, Abhiman S, de Souza RF, Aravind L. Origin and evolution of peptide-modifying dioxygenases and identification of the wybutosine hydroxylase/hydroperoxidase. *Nucleic Acids Res* 2010;**38**:5261–79.
6. Borst P, Sabatini R. Base J: discovery, biosynthesis, and possible functions. *Annu Rev Microbiol* 2008;**62**:235–51.
7. Gommers-Ampt JH, Borst P. Hypermodified bases in DNA. *FASEB J* 1995;**9**:1034–42.
8. Iyer LM, Tahiliani M, Rao A, Aravind L. Prediction of novel families of enzymes involved in oxidative and other complex modifications of bases in nucleic acids. *Cell Cycle* 2009;**8**:1698–710.

9. Cao X, Jacobsen SE. Locus-specific control of asymmetric and CpNpG methylation by the DRM and CMT3 methyltransferase genes. *Proc Natl Acad Sci USA* 2002;**99**(Suppl. 4):16491–8.
10. Freitag M, Williams RL, Kothe GO, Selker EU. A cytosine methyltransferase homologue is essential for repeat-induced point mutation in *Neurospora crassa*. *Proc Natl Acad Sci USA* 2002;**99**:8802–7.
11. Kouzminova E, Selker EU. dim-2 encodes a DNA methyltransferase responsible for all known cytosine methylation in Neurospora. *EMBO J* 2001;**20**:4309–23.
12. Malagnac F, Gregoire A, Goyon C, Rossignol JL, Faugeron G. Masc2, a gene from Ascobolus encoding a protein with a DNA-methyltransferase activity in vitro, is dispensable for in vivo methylation. *Mol Microbiol* 1999;**31**:331–8.
13. Fauman EB, Blumenthal RM, Cheng X. Structure and evolution of AdoMet-dependent methyltransferases. In: Cheng X, Blumenthal RM, editors. *S-adenosylmethionine-dependent methyltransferases: structures and functions*. River Edge: World Scientific; 1999. p. 1–54.
14. Arakawa H, Hauschild J, Buerstedde JM. Requirement of the activation-induced deaminase (AID) gene for immunoglobulin gene conversion. *Science* 2002;**295**:1301–6.
15. Muramatsu M, Kinoshita K, Fagarasan S, Yamada S, Shinkai Y, Honjo T. Class switch recombination and hypermutation require activation-induced cytidine deaminase (AID), a potential RNA editing enzyme. *Cell* 2000;**102**:553–63.
16. Rogozin IB, Iyer LM, Liang L, Glazko GV, Liston VG, Pavlov YI, et al. Evolution and diversification of lamprey antigen receptors: evidence for involvement of an AID-APOBEC family cytosine deaminase. *Nat Immunol* 2007;**8**:647–56.
17. Tahiliani M, Koh KP, Shen Y, Pastor WA, Bandukwala H, Brudno Y, et al. Conversion of 5-methylcytosine to 5-hydroxymethylcytosine in mammalian DNA by MLL partner TET1. *Science* 2009;**324**:930–5.
18. Roberts RJ. Restriction and modification enzymes and their recognition sequences. *Gene* 1980;**8**:329–43.
19. Roberts RJ, Belfort M, Bestor T, Bhagwat AS, Bickle TA, Bitinaite J, et al. A nomenclature for restriction enzymes, DNA methyltransferases, homing endonucleases and their genes. *Nucleic Acids Res* 2003;**31**:1805–12.
20. Roberts RJ, Vincze T, Posfai J, Macelis D. REBASE—a database for DNA restriction and modification: enzymes, genes and genomes. *Nucleic Acids Res* 2010;**38**:D234–6.
21. Takahashi N, Naito Y, Handa N, Kobayashi I. A DNA methyltransferase can protect the genome from postdisturbance attack by a restriction-modification gene complex. *J Bacteriol* 2002;**184**:6100–8.
22. Kobayashi I. Behavior of restriction-modification systems as selfish mobile elements and their impact on genome evolution. *Nucleic Acids Res* 2001;**29**:3742–56.
23. Sadykov M, Asami Y, Niki H, Handa N, Itaya M, Tanokura M, et al. Multiplication of a restriction-modification gene complex. *Mol Microbiol* 2003;**48**:417–27.
24. Bickle TA. Neidhardt H, editor. *E. coli* and *S. typhimurium*. In cellular and molecular biology. Washington, DC: ASM Press; 1987. p. 692–6.
25. Rocha EP, Danchin A, Viari A. Evolutionary role of restriction/modification systems as revealed by comparative genome analysis. *Genome Res* 2001;**11**:946–58.
26. Bhagwat AS, Lieb M. Cooperation and competition in mismatch repair: very short-patch repair and methyl-directed mismatch repair in *Escherichia coli*. *Mol Microbiol* 2002;**44**:1421–8.
27. Ito S, D'Alessio AC, Taranova OV, Hong K, Sowers LC, Zhang Y. Role of Tet proteins in 5mC to 5hmC conversion, ES-cell self-renewal and inner cell mass specification. *Nature* 2010;**466**:1129–33.

28. Ko M, Huang Y, Jankowska AM, Pape UJ, Tahiliani M, Bandukwala HS, et al. Impaired hydroxylation of 5-methylcytosine in myeloid cancers with mutant TET2. *Nature* 2010;**468**:839–43.
29. Kriaucionis S, Heintz N. The nuclear DNA base 5-hydroxymethylcytosine is present in Purkinje neurons and the brain. *Science* 2009;**324**:929–30.
30. Prochnow C, Bransteitter R, Klein MG, Goodman MF, Chen XS. The APOBEC-2 crystal structure and functional implications for the deaminase AID. *Nature* 2007;**445**:447–51.
31. Kaminska KH, Bujnicki JM. Bacteriophage Mu Mom protein responsible for DNA modification is a new member of the acyltransferase superfamily. *Cell Cycle* 2008;**7**:120–1.
32. Morera S, Lariviere L, Kurzeck J, Aschke-Sonnenborn U, Freemont PS, Janin J, et al. High resolution crystal structures of T4 phage beta-glucosyltransferase: induced fit and effect of substrate and metal binding. *J Mol Biol* 2001;**311**:569–77.
33. Morera S, Imberty A, Aschke-Sonnenborn U, Ruger W, Freemont PS. T4 phage beta-glucosyltransferase: substrate binding and proposed catalytic mechanism. *J Mol Biol* 1999;**292**:717–30.
34. Song HK, Sohn SH, Suh SW. Crystal structure of deoxycytidylate hydroxymethylase from bacteriophage T4, a component of the deoxyribonucleoside triphosphate-synthesizing complex. *EMBO J* 1999;**18**:1104–13.
35. Reinisch KM, Chen L, Verdine GL, Lipscomb WN. The crystal structure of HaeIII methyltransferase convalently complexed to DNA: an extrahelical cytosine and rearranged base pairing. *Cell* 1995;**82**:143–53.
36. Tran PH, Korszun ZR, Cerritelli S, Springhorn SS, Lacks SA. Crystal structure of the DpnM DNA adenine methyltransferase from the DpnII restriction system of streptococcus pneumoniae bound to S-adenosylmethionine. *Structure* 1998;**6**:1563–75.
37. Jia D, Jurkowska RZ, Zhang X, Jeltsch A, Cheng X. Structure of Dnmt3a bound to Dnmt3L suggests a model for de novo DNA methylation. *Nature* 2007;**449**:248–51.
38. Horton JR, Liebert K, Bekes M, Jeltsch A, Cheng X. Structure and substrate recognition of the *Escherichia coli* DNA adenine methyltransferase. *J Mol Biol* 2006;**358**:559–70.
39. Holm L, Sander C. Evolutionary link between glycogen phosphorylase and a DNA modifying enzyme. *EMBO J* 1995;**14**:1287–93.
40. Iyer LM, Aravind L. The emergence of catalytic and structural diversity within the beta-clip fold. *Proteins* 2004;**55**:977–91.
41. Anantharaman V, Koonin EV, Aravind L. SPOUT: a class of methyltransferases that includes spoU and trmD RNA methylase superfamilies, and novel superfamilies of predicted prokaryotic RNA methylases. *J Mol Microbiol Biotechnol* 2002;**4**:71–5.
42. Bujnicki JM. Comparison of protein structures reveals monophyletic origin of the AdoMet-dependent methyltransferase family and mechanistic convergence rather than recent differentiation of N4-cytosine and N6-adenine DNA methylation. *In Silico Biol* 1999;**1**:175–82.
43. Schubert HL, Blumenthal RM, Cheng X. Many paths to methyltransfer: a chronicle of convergence. *Trends Biochem Sci* 2003;**28**:329–35.
44. Cierpicki T, Risner LE, Grembecka J, Lukasik SM, Popovic R, Omonkowska M, et al. Structure of the MLL CXXC domain-DNA complex and its functional role in MLL-AF9 leukemia. *Nat Struct Mol Biol* 2010;**17**:62–8.
45. Hashimoto H, Horton JR, Zhang X, Cheng X. UHRF1, a modular multi-domain protein, regulates replication-coupled crosstalk between DNA methylation and histone modifications. *Epigenetics* 2009;**4**:8–14.
46. Arita K, Ariyoshi M, Tochio H, Nakamura Y, Shirakawa M. Recognition of hemi-methylated DNA by the SRA protein UHRF1 by a base-flipping mechanism. *Nature* 2008;**455**:818–21.

47. Avvakumov GV, Walker JR, Xue S, Li Y, Duan S, Bronner C, et al. Structural basis for recognition of hemi-methylated DNA by the SRA domain of human UHRF1. *Nature* 2008;**455**:822–5.
48. Hashimoto H, Horton JR, Zhang X, Bostick M, Jacobsen SE, Cheng X. The SRA domain of UHRF1 flips 5-methylcytosine out of the DNA helix. *Nature* 2008;**455**:826–9.
49. Aravind L, Landsman D. AT-hook motifs identified in a wide variety of DNA-binding proteins. *Nucleic Acids Res* 1998;**26**:4413–21.
50. Iyer LM, Anantharaman V, Wolf MY, Aravind L. Comparative genomics of transcription factors and chromatin proteins in parasitic protists and other eukaryotes. *Int J Parasitol* 2008;**38**:1–31.
51. Burroughs AM, Iyer LM, Aravind L. Natural history of the E1-like superfamily: implication for adenylation, sulfur transfer, and ubiquitin conjugation. *Proteins* 2009;**75**:895–910.
52. Aravind L, Mazumder R, Vasudevan S, Koonin EV. Trends in protein evolution inferred from sequence and structure analysis. *Curr Opin Struct Biol* 2002;**12**:392–9.
53. Cheng X. Structure and function of DNA methyltransferases. *Annu Rev Biophys Biomol Struct* 1995;**24**:293–318.
54. Malone T, Blumenthal RM, Cheng X. Structure-guided analysis reveals nine sequence motifs conserved among DNA amino-methyltransferases, and suggests a catalytic mechanism for these enzymes. *J Mol Biol* 1995;**253**:618–32.
55. Willcock DF, Dryden DT, Murray NE. A mutational analysis of the two motifs common to adenine methyltransferases. *EMBO J* 1994;**13**:3902–8.
56. Schluckebier G, Labahn J, Granzin J, Saenger W. M.TaqI: possible catalysis via cation-pi interactions in N-specific DNA methyltransferases. *Biol Chem* 1998;**379**:389–400.
57. Goedecke K, Pignot M, Goody RS, Scheidig AJ, Weinhold E. Structure of the N6-adenine DNA methyltransferase M.TaqI in complex with DNA and a cofactor analog. *Nat Struct Biol* 2001;**8**:121–5.
58. Collier J. Epigenetic regulation of the bacterial cell cycle. *Curr Opin Microbiol* 2009;**12**:722–9.
59. Kahng LS, Shapiro L. The CcrM DNA methyltransferase of Agrobacterium tumefaciens is essential, and its activity is cell cycle regulated. *J Bacteriol* 2001;**183**:3065–75.
60. Horton JR, Liebert K, Hattman S, Jeltsch A, Cheng X. Transition from nonspecific to specific DNA interactions along the substrate-recognition pathway of dam methyltransferase. *Cell* 2005;**121**:349–61.
61. Urig S, Gowher H, Hermann A, Beck C, Fatemi M, Humeny A, et al. The *Escherichia coli* dam DNA methyltransferase modifies DNA in a highly processive reaction. *J Mol Biol* 2002;**319**:1085–96.
62. Bujnicki JM. Sequence permutations in the molecular evolution of DNA methyltransferases. *BMC Evol Biol* 2002;**2**:3.
63. Gong W, O'Gara M, Blumenthal RM, Cheng X. Structure of pvu II DNA-(cytosine N4) methyltransferase, an example of domain permutation and protein fold assignment. *Nucleic Acids Res* 1997;**25**:2702–15.
64. Hattman S, Kenny C, Berger L, Pratt K. Comparative study of DNA methylation in three unicellular eucaryotes. *J Bacteriol* 1978;**135**:1156–7.
65. Poulter RT, Goodwin TJ. DIRS-1 and the other tyrosine recombinase retrotransposons. *Cytogenet Genome Res* 2005;**110**:575–88.
66. Goodwin TJ, Poulter RT. A new group of tyrosine recombinase-encoding retrotransposons. *Mol Biol Evol* 2004;**21**:746–59.
67. Perez-Alegre M, Dubus A, Fernandez E. REM1, a new type of long terminal repeat retro-transposon in *Chlamydomonas reinhardtii*. *Mol Cell Biol* 2005;**25**:10628–38.

68. Leonard TA, Butler PJ, Lowe J. Structural analysis of the chromosome segregation protein Spo0J from *Thermus thermophilus*. *Mol Microbiol* 2004;**53**:419–32.
69. Roberts D, Hoopes BC, McClure WR, Kleckner N. IS10 transposition is regulated by DNA adenine methylation. *Cell* 1985;**43**:117–30.
70. Fan H, Sakuraba K, Komuro A, Kato S, Harada F, Hirose Y. PCIF1, a novel human WW domain-containing protein, interacts with the phosphorylated RNA polymerase II. *Biochem Biophys Res Commun* 2003;**301**:378–85.
71. Bujnicki JM, Feder M, Radlinska M, Blumenthal RM. Structure prediction and phylogenetic analysis of a functionally diverse family of proteins homologous to the MT-A70 subunit of the human mRNA:m(6)A methyltransferase. *J Mol Evol* 2002;**55**:431–44.
72. Lahav R, Gammie A, Tavazoie S, Rose MD. Role of transcription factor Kar4 in regulating downstream events in the *Saccharomyces cerevisiae* pheromone response pathway. *Mol Cell Biol* 2007;**27**:818–29.
73. Fedoreyeva LI, Vanyushin BF. N(6)-Adenine DNA-methyltransferase in wheat seedlings. *FEBS Lett* 2002;**514**:305–8.
74. Aravind L, Koonin EV. THUMP—a predicted RNA-binding domain shared by 4-thiouridine, pseudouridine synthases and RNA methylases. *Trends Biochem Sci* 2001;**26**:215–7.
75. Purushothaman SK, Bujnicki JM, Grosjean H, Lapeyre B. Trm11p and Trm112p are both required for the formation of 2-methylguanosine at position 10 in yeast tRNA. *Mol Cell Biol* 2005;**25**:4359–70.
76. Foster PG, Nunes CR, Greene P, Moustakas D, Stroud RM. The first structure of an RNA m5C methyltransferase, Fmu, provides insight into catalytic mechanism and specific binding of RNA substrate. *Structure* 2003;**11**:1609–20.
77. Kumar S, Cheng X, Klimasauskas S, Mi S, Posfai J, Roberts RJ, et al. The DNA (cytosine-5) methyltransferases. *Nucleic Acids Res* 1994;**22**:1–10.
78. Posfai J, Bhagwat AS, Roberts RJ. Sequence motifs specific for cytosine methyltransferases. *Gene* 1988;**74**:261–5.
79. O'Gara M, Klimasauskas S, Roberts RJ, Cheng X. Enzymatic C5-cytosine methylation of DNA: mechanistic implications of new crystal structures for HhaI methyltransferase-DNA-AdoHcy complexes. *J Mol Biol* 1996;**261**:634–45.
80. Jeltsch A. Molecular enzymology of mammalian DNA methyltransferases. *Curr Top Microbiol Immunol* 2006;**301**:203–25.
81. Liu Y, Santi DV. m5C RNA and m5C DNA methyl transferases use different cysteine residues as catalysts. *Proc Natl Acad Sci USA* 2000;**97**:8263–5.
82. Klimasauskas S, Nelson JL, Roberts RJ. The sequence specificity domain of cytosine-C5 methylases. *Nucleic Acids Res* 1991;**19**:6183–90.
83. Aravind L, Koonin EV. Prokaryotic homologs of the eukaryotic DNA-end-binding protein Ku, novel domains in the Ku protein and prediction of a prokaryotic double-strand break repair system. *Genome Res* 2001;**11**:1365–74.
84. Ley TJ, Ding L, Walter MJ, McLellan MD, Lamprecht T, Larson DE, Welch J, et al. DNMT3A mutations in acute myeloid leukemia. *N Engl J Med* 2010;**363**(25):2424–33.
85. O'Gara M, Zhang X, Roberts RJ, Cheng X. Structure of a binary complex of HhaI methyltransferase with S-adenosyl-L-methionine formed in the presence of a short non-specific DNA oligonucleotide. *J Mol Biol* 1999;**287**:201–9.
86. Cheng X, Roberts RJ. AdoMet-dependent methylation, DNA methyltransferases and base flipping. *Nucleic Acids Res* 2001;**29**:3784–95.
87. Shieh FK, Youngblood B, Reich NO. The role of Arg165 towards base flipping, base stabilization and catalysis in M.HhaI. *J Mol Biol* 2006;**362**:516–27.
88. Lukianova OA, David SS. A role for iron-sulfur clusters in DNA repair. *Curr Opin Chem Biol* 2005;**9**:145–51.

89. Aravind L, Walker DR, Koonin EV. Conserved domains in DNA repair proteins and evolution of repair systems. *Nucleic Acids Res* 1999;**27**:1223–42.
90. Schaefer M, Lyko F. Solving the Dnmt2 enigma. *Chromosoma* 2010;**119**:35–40.
91. Lyko F, Foret S, Kucharski R, Wolf S, Falckenhayn C, Maleszka R. The honey bee epigenomes: differential methylation of brain DNA in queens and workers. *PLoS Biol* 2010;**8**: e1000506.
92. Feng S, Cokus SJ, Zhang X, Chen PY, Bostick M, Goll MG, et al. Conservation and divergence of methylation patterning in plants and animals. *Proc Natl Acad Sci USA* 2010;**107**:8689–94.
93. Wang Y, Jorda M, Jones PL, Maleszka R, Ling X, Robertson HM, et al. Functional CpG methylation system in a social insect. *Science* 2006;**314**:645–7.
94. Bonasio R, Zhang G, Ye C, Mutti NS, Fang X, Qin N, et al. Genomic comparison of the ants *Camponotus floridanus* and *Harpegnathos saltator. Science* 2010;**329**:1068–71.
95. Bestor T, Laudano A, Mattaliano R, Ingram V. Cloning and sequencing of a cDNA encoding DNA methyltransferase of mouse cells. The carboxyl-terminal domain of the mammalian enzymes is related to bacterial restriction methyltransferases. *J Mol Biol* 1988;**203**:971–83.
96. Bestor TH. DNA methylation: evolution of a bacterial immune function into a regulator of gene expression and genome structure in higher eukaryotes. *Philos Trans R Soc Lond B Biol Sci* 1990;**326**:179–87.
97. Ponger L, Li WH. Evolutionary diversification of DNA methyltransferases in eukaryotic genomes. *Mol Biol Evol* 2005;**22**:1119–28.
98. Zemach A, Zilberman D. Evolution of eukaryotic DNA methylation and the pursuit of safer sex. *Curr Biol* 2010;**20**:R780–5.
99. Zemach A, McDaniel IE, Silva P, Zilberman D. Genome-wide evolutionary analysis of eukaryotic DNA methylation. *Science* 2010;**328**:916–9.
100. Cheng X, Blumenthal RM. Mammalian DNA methyltransferases: a structural perspective. *Structure* 2008;**16**:341–50.
101. Goll MG, Bestor TH. Eukaryotic cytosine methyltransferases. *Annu Rev Biochem* 2005;**74**:481–514.
102. Ooi SK, Bestor TH. Cytosine methylation: remaining faithful. *Curr Biol* 2008;**18**:R174–6.
103. Svedruzic ZM. Mammalian cytosine DNA methyltransferase Dnmt1: enzymatic mechanism, novel mechanism-based inhibitors, and RNA-directed DNA methylation. *Curr Med Chem* 2008;**15**:92–106.
104. Bestor TH. The DNA methyltransferases of mammals. *Hum Mol Genet* 2000;**9**:2395–402.
105. Grandjean V, Yaman R, Cuzin F, Rassoulzadegan M. Inheritance of an epigenetic mark: the CpG DNA methyltransferase 1 is required for de novo establishment of a complex pattern of non-CpG methylation. *PLoS ONE* 2007;**2**:e1136.
106. Li E, Bestor TH, Jaenisch R. Targeted mutation of the DNA methyltransferase gene results in embryonic lethality. *Cell* 1992;**69**:915–26.
107. Chan SW, Henderson IR, Jacobsen SE. Gardening the genome: DNA methylation in *Arabidopsis thaliana*. *Nat Rev Genet* 2005;**6**:351–60.
108. Tariq M, Paszkowski J. DNA and histone methylation in plants. *Trends Genet* 2004;**20**:244–51.
109. Finnegan EJ, Kovac KA. Plant DNA methyltransferases. *Plant Mol Biol* 2000;**43**:189–201.
110. Zilberman D, Henikoff S. Silencing of transposons in plant genomes: kick them when they're down. *Genome Biol* 2004;**5**:249.
111. Henikoff S, Comai L. A DNA methyltransferase homolog with a chromodomain exists in multiple polymorphic forms in *Arabidopsis*. *Genetics* 1998;**149**:307–18.
112. Papa CM, Springer NM, Muszynski MG, Meeley R, Kaeppler SM. Maize chromomethylase Zea methyltransferase2 is required for CpNpG methylation. *Plant Cell* 2001;**13**:1919–28.

113. Bartee L, Malagnac F, Bender J. *Arabidopsis* cmt3 chromomethylase mutations block non-CG methylation and silencing of an endogenous gene. *Genes Dev* 2001;**15**:1753–8.
114. Cao X, Aufsatz W, Zilberman D, Mette MF, Huang MS, Matzke M, et al. Role of the DRM and CMT3 methyltransferases in RNA-directed DNA methylation. *Curr Biol* 2003;**13**:2212–7.
115. Malagnac F, Wendel B, Goyon C, Faugeron G, Zickler D, Rossignol JL, et al. A gene essential for de novo methylation and development in Ascobolus reveals a novel type of eukaryotic DNA methyltransferase structure. *Cell* 1997;**91**:281–90.
116. Lee DW, Freitag M, Selker EU, Aramayo R. A cytosine methyltransferase homologue is essential for sexual development in *Aspergillus nidulans*. *PLoS ONE* 2008;**3**:e2531.
117. Rountree MR, Bachman KE, Baylin SB. DNMT1 binds HDAC2 and a new co-repressor, DMAP1, to form a complex at replication foci. *Nat Genet* 2000;**25**:269–77.
118. Aravind L, Anantharaman V, Balaji S, Babu MM, Iyer LM. The many faces of the helix-turn-helix domain: transcription regulation and beyond. *FEMS Microbiol Rev* 2005;**29**:231–62.
119. Anantharaman V, Aravind L. Novel conserved domains in proteins with predicted roles in eukaryotic cell-cycle regulation, decapping and RNA stability. *BMC Genomics* 2004;**5**:45.
120. Horn PJ, Bastie JN, Peterson CL. A Rik1-associated, cullin-dependent E3 ubiquitin ligase is essential for heterochromatin formation. *Genes Dev* 2005;**19**:1705–14.
121. Mohammad F, Mondal T, Guseva N, Pandey GK, Kanduri C. Kcnq1ot1 noncoding RNA mediates transcriptional gene silencing by interacting with Dnmt1. *Development* 2010;**137**:2493–9.
122. Aravin AA, Sachidanandam R, Bourc'his D, Schaefer C, Pezic D, Toth KF, et al. A piRNA pathway primed by individual transposons is linked to de novo DNA methylation in mice. *Mol Cell* 2008;**31**:785–99.
123. Allen MD, Grummitt CG, Hilcenko C, Min SY, Tonkin LM, Johnson CM, et al. Solution structure of the nonmethyl-CpG-binding CXXC domain of the leukaemia-associated MLL histone methyltransferase. *EMBO J* 2006;**25**:4503–12.
124. Davison AJ, Cunningham C, Sauerbier W, McKinnell RG. Genome sequences of two frog herpesviruses. *J Gen Virol* 2006;**87**:3509–14.
125. de Souza RF, Iyer LM, Aravind L. Diversity and evolution of chromatin proteins encoded by DNA viruses. *Biochim Biophys Acta* 2010;**1799**:302–18.
126. Hansen RS, Wijmenga C, Luo P, Stanek AM, Canfield TK, Weemaes CM, et al. The DNMT3B DNA methyltransferase gene is mutated in the ICF immunodeficiency syndrome. *Proc Natl Acad Sci USA* 1999;**96**:14412–7.
127. Okano M, Bell DW, Haber DA, Li E. DNA methyltransferases Dnmt3a and Dnmt3b are essential for de novo methylation and mammalian development. *Cell* 1999;**99**:247–57.
128. Kato Y, Kaneda M, Hata K, Kumaki K, Hisano M, Kohara Y, et al. Role of the Dnmt3 family in de novo methylation of imprinted and repetitive sequences during male germ cell development in the mouse. *Hum Mol Genet* 2007;**16**:2272–80.
129. Kaneda M, Okano M, Hata K, Sado T, Tsujimoto N, Li E, et al. Essential role for de novo DNA methyltransferase Dnmt3a in paternal and maternal imprinting. *Nature* 2004;**429**:900–3.
130. Henderson IR, Deleris A, Wong W, Zhong X, Chin HG, Horwitz GA, et al. The de novo cytosine methyltransferase DRM2 requires intact UBA domains and a catalytically mutated paralog DRM3 during RNA-directed DNA methylation in *Arabidopsis thaliana*. *PLoS Genet* 2010;**6**:e1001182.
131. Dhayalan A, Rajavelu A, Rathert P, Tamas R, Jurkowska RZ, Ragozin S, et al. The Dnmt3a PWWP domain reads histone 3 lysine 36 trimethylation and guides DNA methylation. *J Biol Chem* 2010;**285**:26114–20.

132. Otani J, Nankumo T, Arita K, Inamoto S, Ariyoshi M, Shirakawa M. Structural basis for recognition of H3K4 methylation status by the DNA methyltransferase 3A ATRX-DNMT3-DNMT3L domain. *EMBO Rep* 2009;**10**:1235–41.
133. Argentaro A, Yang JC, Chapman L, Kowalczyk MS, Gibbons RJ, Higgs DR, et al. Structural consequences of disease-causing mutations in the ATRX-DNMT3-DNMT3L (ADD) domain of the chromatin-associated protein ATRX. *Proc Natl Acad Sci USA* 2007;**104**:11939–44.
134. Zhang Y, Jurkowska R, Soeroes S, Rajavelu A, Dhayalan A, Bock I, et al. Chromatin methylation activity of Dnmt3a and Dnmt3a/3L is guided by interaction of the ADD domain with the histone H3 tail. *Nucleic Acids Res* 2010;**38**:4246–53.
135. Hofmann K, Bucher P. The UBA domain: a sequence motif present in multiple enzyme classes of the ubiquitination pathway. *Trends Biochem Sci* 1996;**21**:172–3.
136. Aravind L, Iyer LM, Koonin EV. Scores of RINGS but no PHDs in ubiquitin signaling. *Cell Cycle* 2003;**2**:123–6.
137. Schaefer M, Lyko F. Lack of evidence for DNA methylation of Invader4 retroelements in *Drosophila* and implications for Dnmt2-mediated epigenetic regulation. *Nat Genet* 2010;**42**:920–1 (author reply 921).
138. Phalke S, Nickel O, Walluscheck D, Hortig F, Onorati MC, Reuter G. Retrotransposon silencing and telomere integrity in somatic cells of *Drosophila* depends on the cytosine-5 methyltransferase DNMT2. *Nat Genet* 2009;**41**:696–702.
139. Goll MG, Kirpekar F, Maggert KA, Yoder JA, Hsieh CL, Zhang X, et al. Methylation of tRNAAsp by the DNA methyltransferase homolog Dnmt2. *Science* 2006;**311**:395–8.
140. Kuhlmann M, Borisova BE, Kaller M, Larsson P, Stach D, Na J, et al. Silencing of retrotransposons in Dictyostelium by DNA methylation and RNAi. *Nucleic Acids Res* 2005;**33**:6405–17.
141. Jurkowski TP, Meusburger M, Phalke S, Helm M, Nellen W, Reuter G, et al. Human DNMT2 methylates tRNA(Asp) molecules using a DNA methyltransferase-like catalytic mechanism. *RNA* 2008;**14**:1663–70.
142. Fisher O, Siman-Tov R, Ankri S. Characterization of cytosine methylated regions and 5-cytosine DNA methyltransferase (Ehmeth) in the protozoan parasite *Entamoeba histolytica*. *Nucleic Acids Res* 2004;**32**:287–97.
143. Neumann P, Pozarkova D, Koblizkova A, Macas J. PIGY, a new plant envelope-class LTR retrotransposon. *Mol Genet Genomics* 2005;**273**:43–53.
144. Kunert N, Marhold J, Stanke J, Stach D, Lyko F. A Dnmt2-like protein mediates DNA methylation in *Drosophila*. *Development* 2003;**130**:5083–90.
145. Ponting CP, Blake DJ, Davies KE, Kendrick-Jones J, Winder SJ. ZZ and TAZ: new putative zinc fingers in dystrophin and other proteins. *Trends Biochem Sci* 1996;**21**:11–3.
146. DiPaolo C, Kieft R, Cross M, Sabatini R. Regulation of trypanosome DNA glycosylation by a SWI2/SNF2-like protein. *Mol Cell* 2005;**17**:441–51.
147. Kanno T, Huettel B, Mette MF, Aufsatz W, Jaligot E, Daxinger L, et al. Atypical RNA polymerase subunits required for RNA-directed DNA methylation. *Nat Genet* 2005;**37**:761–5.
148. Chan SW, Henderson IR, Zhang X, Shah G, Chien JS, Jacobsen SE. RNAi, DRD1, and histone methylation actively target developmentally important non-CG DNA methylation in *Arabidopsis*. *PLoS Genet* 2006;**2**:e83.
149. Kanno T, Mette MF, Kreil DP, Aufsatz W, Matzke M, Matzke AJ. Involvement of putative SNF2 chromatin remodeling protein DRD1 in RNA-directed DNA methylation. *Curr Biol* 2004;**14**:801–5.
150. Gibbons RJ, McDowell TL, Raman S, O'Rourke DM, Garrick D, Ayyub H, et al. Mutations in ATRX, encoding a SWI/SNF-like protein, cause diverse changes in the pattern of DNA methylation. *Nat Genet* 2000;**24**:368–71.

151. Datta J, Majumder S, Bai S, Ghoshal K, Kutay H, Smith DS, et al. Physical and functional interaction of DNA methyltransferase 3A with Mbd3 and Brg1 in mouse lymphosarcoma cells. *Cancer Res* 2005;**65**:10891–900.
152. Lobocka MB, Rose DJ, Plunkett 3rd G, Rusin M, Samojedny A, Lehnherr H, et al. Genome of bacteriophage P1. *J Bacteriol* 2004;**186**:7032–68.
153. Militello KT, Wang P, Jayakar SK, Pietrasik RL, Dupont CD, Dodd K, et al. African trypanosomes contain 5-methylcytosine in nuclear DNA. *Eukaryot Cell* 2008;**7**:2012–6.
154. Barry JD, McCulloch R. Antigenic variation in trypanosomes: enhanced phenotypic variation in a eukaryotic parasite. *Adv Parasitol* 2001;**49**:1–70.
155. Agarkova IV, Dunigan DD, Van Etten JL. Virion-associated restriction endonucleases of chloroviruses. *J Virol* 2006;**80**:8114–23.
156. Nelson M, Burbank DE, Van Etten JL. Chlorella viruses encode multiple DNA methyltransferases. *Biol Chem* 1998;**379**:423–8.
157. Que Q, Zhang Y, Nelson M, Ropp S, Burbank DE, Van Etten JL. Chlorella virus SC-1A encodes at least five functional and one nonfunctional DNA methyltransferases. *Gene* 1997;**190**:237–44.
158. Tidona CA, Schnitzler P, Kehm R, Darai G. Identification of the gene encoding the DNA (cytosine-5) methyltransferase of lymphocystis disease virus. *Virus Genes* 1996;**12**:219–29.
159. Doerfler W. In pursuit of the first recognized epigenetic signal–DNA methylation: a 1976 to 2008 synopsis. *Epigenetics* 2008;**3**:125–33.
160. Mayer W, Niveleau A, Walter J, Fundele R, Haaf T. Demethylation of the zygotic paternal genome. *Nature* 2000;**403**:501–2.
161. Nakamura T, Arai Y, Umehara H, Masuhara M, Kimura T, Taniguchi H, et al. PGC7/Stella protects against DNA demethylation in early embryogenesis. *Nat Cell Biol* 2007;**9**:64–71.
162. Santos F, Hendrich B, Reik W, Dean W. Dynamic reprogramming of DNA methylation in the early mouse embryo. *Dev Biol* 2002;**241**:172–82.
163. Hajkova P, Jeffries SJ, Lee C, Miller N, Jackson SP, Surani MA. Genome-wide reprogramming in the mouse germ line entails the base excision repair pathway. *Science* 2010;**329**:78–82.
164. Bruniquel D, Schwartz RH. Selective, stable demethylation of the interleukin-2 gene enhances transcription by an active process. *Nat Immunol* 2003;**4**:235–40.
165. Metivier R, Gallais R, Tiffoche C, Le Peron C, Jurkowska RZ, Carmouche RP, et al. Cyclical DNA methylation of a transcriptionally active promoter. *Nature* 2008;**452**:45–50.
166. Kim MS, Kondo T, Takada I, Youn MY, Yamamoto Y, Takahashi S, et al. DNA demethylation in hormone-induced transcriptional derepression. *Nature* 2009;**461**:1007–12.
167. Gehring M, Huh JH, Hsieh TF, Penterman J, Choi Y, Harada JJ, et al. DEMETER DNA glycosylase establishes MEDEA polycomb gene self-imprinting by allele-specific demethylation. *Cell* 2006;**124**:495–506.
168. Penterman J, Uzawa R, Fischer RL. Genetic interactions between DNA demethylation and methylation in *Arabidopsis*. *Plant Physiol* 2007;**145**:1549–57.
169. Ooi SK, Bestor TH. The colorful history of active DNA demethylation. *Cell* 2008;**133**:1145–8.
170. Bhattacharya SK, Ramchandani S, Cervoni N, Szyf M. A mammalian protein with specific demethylase activity for mCpG DNA. *Nature* 1999;**397**:579–83.
171. Ng HH, Zhang Y, Hendrich B, Johnson CA, Turner BM, Erdjument-Bromage H, et al. MBD2 is a transcriptional repressor belonging to the MeCP1 histone deacetylase complex. *Nat Genet* 1999;**23**:58–61.
172. Okada Y, Yamagata K, Hong K, Wakayama T, Zhang Y. A role for the elongator complex in zygotic paternal genome demethylation. *Nature* 2010;**463**:554–8.
173. Anantharaman V, Koonin EV, Aravind L. TRAM, a predicted RNA-binding domain, common to tRNA uracil methylation and adenine thiolation enzymes. *FEMS Microbiol Lett* 2001;**197**:215–21.

174. Greenwood C, Selth LA, Dirac-Svejstrup AB, Svejstrup JQ. An iron-sulfur cluster domain in Elp3 important for the structural integrity of elongator. *J Biol Chem* 2009;**284**:141–9.
175. Wittschieben BO, Otero G, de Bizemont T, Fellows J, Erdjument-Bromage H, Ohba R, et al. A novel histone acetyltransferase is an integral subunit of elongating RNA polymerase II holoenzyme. *Mol Cell* 1999;**4**:123–8.
176. Huang B, Johansson MJ, Bystrom AS. An early step in wobble uridine tRNA modification requires the Elongator complex. *RNA* 2005;**11**:424–36.
177. Krokan HE, Standal R, Slupphaug G. DNA glycosylases in the base excision repair of DNA. *Biochem J* 1997;**325**:1–16.
178. Morales-Ruiz T, Ortega-Galisteo AP, Ponferrada-Marin MI, Martinez-Macias MI, Ariza RR, Roldan-Arjona T. DEMETER and REPRESSOR OF SILENCING 1 encode 5-methylcytosine DNA glycosylases. *Proc Natl Acad Sci USA* 2006;**103**:6853–8.
179. Gong Z, Morales-Ruiz T, Ariza RR, Roldan-Arjona T, David L, Zhu JK. ROS1, a repressor of transcriptional gene silencing in *Arabidopsis*, encodes a DNA glycosylase/lyase. *Cell* 2002;**111**:803–14.
180. Agius F, Kapoor A, Zhu JK. Role of the *Arabidopsis* DNA glycosylase/lyase ROS1 in active DNA demethylation. *Proc Natl Acad Sci USA* 2006;**103**:11796–801.
181. Rai K, Huggins IJ, James SR, Karpf AR, Jones DA, Cairns BR. DNA demethylation in zebrafish involves the coupling of a deaminase, a glycosylase, and gadd45. *Cell* 2008;**135**:1201–12.
182. Jiricny J, Menigatti M. DNA Cytosine demethylation: are we getting close? *Cell* 2008;**135**:1167–9.
183. Yoon JH, Iwai S, O'Connor TR, Pfeifer GP. Human thymine DNA glycosylase (TDG) and methyl-CpG-binding protein 4 (MBD4) excise thymine glycol (Tg) from a Tg:G mispair. *Nucleic Acids Res* 2003;**31**:5399–404.
184. Zhu B, Zheng Y, Angliker H, Schwarz S, Thiry S, Siegmann M, et al. 5-Methylcytosine DNA glycosylase activity is also present in the human MBD4 (G/T mismatch glycosylase) and in a related avian sequence. *Nucleic Acids Res* 2000;**28**:4157–65.
185. Hendrich B, Hardeland U, Ng HH, Jiricny J, Bird A. The thymine glycosylase MBD4 can bind to the product of deamination at methylated CpG sites. *Nature* 1999;**401**:301–4.
186. Bellacosa A, Cicchillitti L, Schepis F, Riccio A, Yeung AT, Matsumoto Y, et al. MED1, a novel human methyl-CpG-binding endonuclease, interacts with DNA mismatch repair protein MLH1. *Proc Natl Acad Sci USA* 1999;**96**:3969–74.
187. Zhu B, Benjamin D, Zheng Y, Angliker H, Thiry S, Siegmann M, et al. Overexpression of 5-methylcytosine DNA glycosylase in human embryonic kidney cells EcR293 demethylates the promoter of a hormone-regulated reporter gene. *Proc Natl Acad Sci USA* 2001;**98**:5031–6.
188. Jost JP, Schwarz S, Hess D, Angliker H, Fuller-Pace FV, Stahl H, et al. A chicken embryo protein related to the mammalian DEAD box protein p68 is tightly associated with the highly purified protein-RNA complex of 5-MeC-DNA glycosylase. *Nucleic Acids Res* 1999;**27**:3245–52.
189. Hu XV, Rodrigues TM, Tao H, Baker RK, Miraglia L, Orth AP, et al. Identification of RING finger protein 4 (RNF4) as a modulator of DNA demethylation through a functional genomics screen. *Proc Natl Acad Sci USA* 2010;**107**:15087–92.
190. Jin SG, Guo C, Pfeifer GP. GADD45A does not promote DNA demethylation. *PLoS Genet* 2008;**4**:e1000013.
191. Sharath AN, Weinhold E, Bhagwat AS. Reviving a dead enzyme: cytosine deaminations promoted by an inactive DNA methyltransferase and an S-adenosylmethionine analogue. *Biochemistry* 2000;**39**:14611–6.

192. Zingg JM, Shen JC, Yang AS, Rapoport H, Jones PA. Methylation inhibitors can increase the rate of cytosine deamination by (cytosine-5)-DNA methyltransferase. *Nucleic Acids Res* 1996;**24**:3267–75.
193. Rubinson EH, Metz AH, O'Quin J, Eichman BF. A new protein architecture for processing alkylation damaged DNA: the crystal structure of DNA glycosylase AlkD. *J Mol Biol* 2008;**381**:13–23.
194. Aravind L, Koonin EV. The alpha/beta fold uracil DNA glycosylases: a common origin with diverse fates. *Genome Biol* 2000;**1**: RESEARCH0007.
195. Qi Y, Spong MC, Nam K, Banerjee A, Jiralerspong S, Karplus M, et al. Encounter and extrusion of an intrahelical lesion by a DNA repair enzyme. *Nature* 2009;**462**:762–6.
196. Slupphaug G, Mol CD, Kavli B, Arvai AS, Krokan HE, Tainer JA. A nucleotide-flipping mechanism from the structure of human uracil-DNA glycosylase bound to DNA. *Nature* 1996;**384**:87–92.
197. Zhang QM, Yonekura S, Takao M, Yasui A, Sugiyama H, Yonei S. DNA glycosylase activities for thymine residues oxidized in the methyl group are functions of the hNEIL1 and hNTH1 enzymes in human cells. *DNA Repair (Amst)* 2005;**4**:71–9.
198. Fromme JC, Banerjee A, Huang SJ, Verdine GL. Structural basis for removal of adenine mispaired with 8-oxoguanine by MutY adenine DNA glycosylase. *Nature* 2004;**427**:652–6.
199. Doherty AJ, Serpell LC, Ponting CP. The helix-hairpin-helix DNA-binding motif: a structural basis for non-sequence-specific recognition of DNA. *Nucleic Acids Res* 1996;**24**:2488–97.
200. Aravind L, Koonin EV. SAP—a putative DNA-binding motif involved in chromosomal organization. *Trends Biochem Sci* 2000;**25**:112–4.
201. Dizdaroglu M, Karahalil B, Senturker S, Buckley TJ, Roldan-Arjona T. Excision of products of oxidative DNA base damage by human NTH1 protein. *Biochemistry* 1999;**38**:243–6.
202. Alseth I, Osman F, Korvald H, Tsaneva I, Whitby MC, Seeberg E, et al. Biochemical characterization and DNA repair pathway interactions of Mag1-mediated base excision repair in *Schizosaccharomyces pombe*. *Nucleic Acids Res* 2005;**33**:1123–31.
203. Birtle Z, Ponting CP. Meisetz and the birth of the KRAB motif. *Bioinformatics* 2006;**22**:2841–5.
204. Clery A, Blatter M, Allain FH. RNA recognition motifs: boring? Not quite. *Curr Opin Struct Biol* 2008;**18**:290–8.
205. Walsh P, Bursac D, Law YC, Cyr D, Lithgow T. The J-protein family: modulating protein assembly, disassembly and translocation. *EMBO Rep* 2004;**5**:567–71.
206. Cliffe LJ, Kieft R, Southern T, Birkeland SR, Marshall M, Sweeney K, et al. JBP1 and JBP2 are two distinct thymidine hydroxylases involved in J biosynthesis in genomic DNA of African trypanosomes. *Nucleic Acids Res* 2009;**37**:1452–62.
207. Valinluck V, Liu P, Kang Jr. JI, Burdzy A, Sowers LC. 5-halogenated pyrimidine lesions within a CpG sequence context mimic 5-methylcytosine by enhancing the binding of the methyl-CpG-binding domain of methyl-CpG-binding protein 2 (MeCP2). *Nucleic Acids Res* 2005;**33**:3057–64.
208. Pollyea DA, Raval A, Kusler B, Gotlib JR, Alizadeh AA, Mitchell BS. Impact of TET2 mutations on mRNA expression and clinical outcomes in MDS patients treated with DNA methyltransferase inhibitors. *Hematol Oncol.* 2010. DOI: 10.1002/hon.976.
209. Privat E, Sowers LC. Photochemical deamination and demethylation of 5-methylcytosine. *Chem Res Toxicol* 1996;**9**:745–50.
210. Hino S, Kishida S, Michiue T, Fukui A, Sakamoto I, Takada S, et al. Inhibition of the Wnt signaling pathway by Idax, a novel Dvl-binding protein. *Mol Cell Biol* 2001;**21**:330–42.
211. Freedman T, Pukkila PJ. De novo methylation of repeated sequences in Coprinus cinereus. *Genetics* 1993;**135**:357–66.
212. Conticello SG. The AID/APOBEC family of nucleic acid mutators. *Genome Biol* 2008;**9**:229.

213. Blanc V, Davidson NO. APOBEC-1-mediated RNA editing. *Wiley Interdiscip Rev Syst Biol Med* 2010;**2**:594–602.
214. Hamilton CE, Papavasiliou FN, Rosenberg BR. Diverse functions for DNA and RNA editing in the immune system. *RNA Biol* 2010;**7**:220–8.
215. Weitzel JM, Buhrmester H, Stratling WH. Chicken MAR-binding protein ARBP is homologous to rat methyl-CpG-binding protein MeCP2. *Mol Cell Biol* 1997;**17**:5656–66.
216. Cross SH, Meehan RR, Nan X, Bird A. A component of the transcriptional repressor MeCP1 shares a motif with DNA methyltransferase and HRX proteins. *Nat Genet* 1997;**16**:256–9.
217. Laget S, Joulie M, Le Masson F, Sasai N, Christians E, Pradhan S, et al. The human proteins MBD5 and MBD6 associate with heterochromatin but they do not bind methylated DNA. *PLoS ONE* 2010;**5**:e11982.
218. Ho KL, McNae IW, Schmiedeberg L, Klose RJ, Bird AP, Walkinshaw MD. MeCP2 binding to DNA depends upon hydration at methyl-CpG. *Mol Cell* 2008;**29**:525–31.
219. Lao VV, Darwanto A, Sowers LC. Impact of base analogues within a CpG dinucleotide on the binding of DNA by the methyl-binding domain of MeCP2 and methylation by DNMT1. *Biochemistry* 2010;**49**:10228–36.
220. Baurain D, Brinkmann H, Petersen J, Rodriguez-Ezpeleta N, Stechmann A, Demoulin V, et al. Phylogenomic evidence for separate acquisition of plastids in cryptophytes, haptophytes, and stramenopiles. *Mol Biol Evol* 2010;**27**:1698–709.
221. Makarova KS, Aravind L, Wolf YI, Tatusov RL, Minton KW, Koonin EV, et al. Genome of the extremely radiation-resistant bacterium *Deinococcus radiodurans* viewed from the perspective of comparative genomics. *Microbiol Mol Biol Rev* 2001;**65**:44–79.
222. Makarova KS, Aravind L, Daly MJ, Koonin EV. Specific expansion of protein families in the radioresistant bacterium *Deinococcus radiodurans*. *Genetica* 2000;**108**:25–34.
223. Citterio E, Papait R, Nicassio F, Vecchi M, Gomiero P, Mantovani R, et al. Np95 is a histone-binding protein endowed with ubiquitin ligase activity. *Mol Cell Biol* 2004;**24**:2526–35.
224. Baumbusch LO, Thorstensen T, Krauss V, Fischer A, Naumann K, Assalkhou R, et al. The *Arabidopsis thaliana* genome contains at least 29 active genes encoding SET domain proteins that can be assigned to four evolutionarily conserved classes. *Nucleic Acids Res* 2001;**29**:4319–33.
225. Sharif J, Muto M, Takebayashi S, Suetake I, Iwamatsu A, Endo TA, et al. The SRA protein Np95 mediates epigenetic inheritance by recruiting Dnmt1 to methylated DNA. *Nature* 2007;**450**:908–12.
226. Bostick M, Kim JK, Esteve PO, Clark A, Pradhan S, Jacobsen SE. UHRF1 plays a role in maintaining DNA methylation in mammalian cells. *Science* 2007;**317**:1760–4.
227. Johnson LM, Bostick M, Zhang X, Kraft E, Henderson I, Callis J, et al. The SRA methyl-cytosine-binding domain links DNA and histone methylation. *Curr Biol* 2007;**17**:379–84.
228. Iyer LM, Burroughs AM, Aravind L. The ASCH superfamily: novel domains with a fold related to the PUA domain and a potential role in RNA metabolism. *Bioinformatics* 2006;**22**:257–63.
229. Normand C, Capeyrou R, Quevillon-Cheruel S, Mougin A, Henry Y, Caizergues-Ferrer M. Analysis of the binding of the N-terminal conserved domain of yeast Cbf5p to a box H/ACA snoRNA. *RNA* 2006;**12**:1868–82.
230. Cheng X, Blumenthal RM. Finding a basis for flipping bases. *Structure* 1996;**4**:639–45.
231. Georgescu RE, Kim SS, Yurieva O, Kuriyan J, Kong XP, O'Donnell M. Structure of a sliding clamp on DNA. *Cell* 2008;**132**:43–54.
232. Blackledge NP, Zhou JC, Tolstorukov MY, Farcas AM, Park PJ, Klose RJ. CpG islands recruit a histone H3 lysine 36 demethylase. *Mol Cell* 2010;**38**:179–90.
233. Pradhan M, Esteve PO, Chin HG, Samaranayke M, Kim GD, Pradhan S. CXXC domain of human DNMT1 is essential for enzymatic activity. *Biochemistry* 2008;**47**:10000–9.

234. Jorgensen HF, Ben-Porath I, Bird AP. Mbd1 is recruited to both methylated and nonmethylated CpGs via distinct DNA binding domains. *Mol Cell Biol* 2004;**24**:3387–95.
235. Tate CM, Lee JH, Skalnik DG. CXXC finger protein 1 contains redundant functional domains that support embryonic stem cell cytosine methylation, histone methylation, and differentiation. *Mol Cell Biol* 2009;**29**:3817–31.
236. Tate CM, Lee JH, Skalnik DG. CXXC finger protein 1 restricts the Setd1A histone H3K4 methyltransferase complex to euchromatin. *FEBS J* 2010;**277**:210–23.
237. Auld DS, Bergman T. Medium- and short-chain dehydrogenase/reductase gene and protein families: The role of zinc for alcohol dehydrogenase structure and function. *Cell Mol Life Sci* 2008;**65**:3961–70.
238. Nole-Wilson S, Krizek BA. DNA binding properties of the *Arabidopsis* floral development protein AINTEGUMENTA. *Nucleic Acids Res* 2000;**28**:4076–82.
239. Branco MR, Oda M, Reik W. Safeguarding parental identity: Dnmt1 maintains imprints during epigenetic reprogramming in early embryogenesis. *Genes Dev* 2008;**22**:1567–71.
240. Wilkins JF, Haig D. Parental modifiers, antisense transcripts and loss of imprinting. *Proc Biol Sci* 2002;**269**:1841–6.
241. Reik W, Walter J. Genomic imprinting: parental influence on the genome. *Nat Rev Genet* 2001;**2**:21–32.
242. Barzily-Rokni M, Friedman N, Ron-Bigger S, Isaac S, Michlin D, Eden A. Synergism between DNA methylation and macroH2A1 occupancy in epigenetic silencing of the tumor suppressor gene p16(CDKN2A). *Nucleic Acids Res* 2010; **39** (4): 1326–1335.
243. Conerly ML, Teves SS, Diolaiti D, Ulrich M, Eisenman RN, Henikoff S. Changes in H2A.Z occupancy and DNA methylation during B-cell lymphomagenesis. *Genome Res* 2010;**20**:1383–90.
244. Edwards JR, O'Donnell AH, Rollins RA, Peckham HE, Lee C, Milekic MH, et al. Chromatin and sequence features that define the fine and gross structure of genomic methylation patterns. *Genome Res* 2010;**20**:972–80.
245. Kobor MS, Lorincz MC. H2A.Z and DNA methylation: irreconcilable differences. *Trends Biochem Sci* 2009;**34**:158–61.
246. Zilberman D, Coleman-Derr D, Ballinger T, Henikoff S. Histone H2A.Z and DNA methylation are mutually antagonistic chromatin marks. *Nature* 2008;**456**:125–9.
247. Tamaru H, Zhang X, McMillen D, Singh PB, Nakayama J, Grewal SI, et al. Trimethylated lysine 9 of histone H3 is a mark for DNA methylation in Neurospora crassa. *Nat Genet* 2003;**34**:75–9.
248. Johnson L, Cao X, Jacobsen S. Interplay between two epigenetic marks. DNA methylation and histone H3 lysine 9 methylation. *Curr Biol* 2002;**12**:1360–7.
249. Jackson JP, Lindroth AM, Cao X, Jacobsen SE. Control of CpNpG DNA methylation by the KRYPTONITE histone H3 methyltransferase. *Nature* 2002;**416**:556–60.
250. Venkatasubrahmanyam S, Hwang WW, Meneghini MD, Tong AH, Madhani HD. Genome-wide, as opposed to local, antisilencing is mediated redundantly by the euchromatic factors Set1 and H2A.Z. *Proc Natl Acad Sci USA* 2007;**104**:16609–14.
251. Lee GE, Kim JH, Taylor M, Muller MT. DNA methyltransferase 1 associated protein (DMAP1) is a co-repressor that stimulates DNA methylation globally and locally at sites of double strand break repair. *J Biol Chem* 2010;**285**:37630–40.
252. Doyon Y, Selleck W, Lane WS, Tan S, Cote J. Structural and functional conservation of the NuA4 histone acetyltransferase complex from yeast to humans. *Mol Cell Biol* 2004;**24**:1884–96.
253. Krogan NJ, Keogh MC, Datta N, Sawa C, Ryan OW, Ding H, et al. A Snf2 family ATPase complex required for recruitment of the histone H2A variant Htz1. *Mol Cell* 2003;**12**:1565–76.

254. Mizuguchi G, Shen X, Landry J, Wu WH, Sen S, Wu C. ATP-driven exchange of histone H2AZ variant catalyzed by SWR1 chromatin remodeling complex. *Science* 2004;**303**:343–8.
255. Nan X, Hou J, Maclean A, Nasir J, Lafuente MJ, Shu X, et al. Interaction between chromatin proteins MECP2 and ATRX is disrupted by mutations that cause inherited mental retardation. *Proc Natl Acad Sci USA* 2007;**104**:2709–14.
256. Iyer LM, Abhiman S, Aravind L. MutL homologs in restriction-modification systems and the origin of eukaryotic MORC ATPases. *Biol Direct* 2008;**3**:8.
257. Law JA, Ausin I, Johnson LM, Vashisht AA, Zhu JK, Wohlschlegel JA, et al. A protein complex required for polymerase V transcripts and RNA-directed DNA methylation in *Arabidopsis*. *Curr Biol* 2010;**20**:951–6.
258. Kanno T, Bucher E, Daxinger L, Huettel B, Bohmdorfer G, Gregor W, et al. A structural-maintenance-of-chromosomes hinge domain-containing protein is required for RNA-directed DNA methylation. *Nat Genet* 2008;**40**:670–5.
259. Abrahamsen MS, Templeton TJ, Enomoto S, Abrahante JE, Zhu G, Lancto CA, et al. Complete genome sequence of the apicomplexan, *Cryptosporidium parvum*. *Science* 2004;**304**:441–5.
260. Gardner MJ, Tettelin H, Carucci DJ, Cummings LM, Aravind L, Koonin EV, et al. Chromosome 2 sequence of the human malaria parasite *Plasmodium falciparum*. *Science* 1998;**282**:1126–32.
261. Yap KL, Zhou MM. Keeping it in the family: diverse histone recognition by conserved structural folds. *Crit Rev Biochem Mol Biol* 2010;**45**:488–505.
262. Dhalluin C, Carlson JE, Zeng L, He C, Aggarwal AK, Zhou MM. Structure and ligand of a histone acetyltransferase bromodomain. *Nature* 1999;**399**:491–6.
263. Iyer LM, Babu MM, Aravind L. The HIRAN domain and recruitment of chromatin remodeling and repair activities to damaged DNA. *Cell Cycle* 2006;**5**:775–82.
264. Aravind L, Makarova KS, Koonin EV. Holliday junction resolvases and related nucleases: identification of new families, phyletic distribution and evolutionary trajectories. *Nucleic Acids Res* 2000;**28**:3417–32.
265. Arisue N, Hasegawa M, Hashimoto T. Root of the Eukaryota tree as inferred from combined maximum likelihood analyses of multiple molecular sequence data. *Mol Biol Evol* 2005;**22**:409–20.
266. Simpson AG, Inagaki Y, Roger AJ. Comprehensive multigene phylogenies of excavate protists reveal the evolutionary positions of "primitive" eukaryotes. *Mol Biol Evol* 2006;**23**:615–25.
267. Muljo SA, Kanellopoulou C, Aravind L. MicroRNA targeting in mammalian genomes: genes and mechanisms. *Wiley Interdiscip Rev Syst Biol Med* 2010;**2**:148–61.
268. Grewal SI. RNAi-dependent formation of heterochromatin and its diverse functions. *Curr Opin Genet Dev* 2010;**20**:134–41.
269. Allis CD, Jenuwein T, Reinberg D. *Epigenetics*. Cold Spring Harbor, NY: Cold Spring Harbor Laboratory Press; 2007.
270. Sandman K, Reeve JN. Archaeal chromatin proteins: different structures but common function? *Curr Opin Microbiol* 2005;**8**:656–61.
271. Schuldiner M, Collins SR, Weissman JS, Krogan NJ. Quantitative genetic analysis in Saccharomyces cerevisiae using epistatic miniarray profiles (E-MAPs) and its application to chromatin functions. *Methods* 2006;**40**:344–52.
272. Heitman J. Evolution of eukaryotic microbial pathogens via covert sexual reproduction. *Cell Host Microbe* 2010;**8**:86–99.
273. Cuozzo C, Porcellini A, Angrisano T, Morano A, Lee B, Di Pardo A, et al. DNA damage, homology-directed repair, and DNA methylation. *PLoS Genet* 2007;**3**:e110.
274. Scott RJ, Spielman M. Genomic imprinting in plants and mammals: how life history constrains convergence. *Cytogenet Genome Res* 2006;**113**:53–67.

275. Renfree MB, Hore TA, Shaw G, Graves JA, Pask AJ. Evolution of genomic imprinting: insights from marsupials and monotremes. *Annu Rev Genomics Hum Genet* 2009;**10**:241–62.
276. Genevieve D, Sanlaville D, Faivre L, Kottler ML, Jambou M, Gosset P, et al. Paternal deletion of the GNAS imprinted locus (including Gnasxl) in two girls presenting with severe pre- and post-natal growth retardation and intractable feeding difficulties. *Eur J Hum Genet* 2005;**13**:1033–9.
277. Peters J, Wroe SF, Wells CA, Miller HJ, Bodle D, Beechey CV, et al. A cluster of oppositely imprinted transcripts at the Gnas locus in the distal imprinting region of mouse chromosome 2. *Proc Natl Acad Sci USA* 1999;**96**:3830–5.
278. Tamura K, Dudley J, Nei M, Kumar S. MEGA4: Molecular Evolutionary Genetics Analysis (MEGA) software version 4.0. *Mol Biol Evol* 2007;**24**:1596–9.
279. Price MN, Dehal PS, Arkin AP. FastTree: computing large minimum evolution trees with profiles instead of a distance matrix. *Mol Biol Evol* 2009;**26**:1641–50.

Natural History of the Eukaryotic Chromatin Protein Methylation System

L. Aravind,
Saraswathi Abhiman, and
Lakshminarayan M. Iyer

National Center for Biotechnology Information, National Library of Medicine, National Institutes of Health, Bethesda, Maryland, USA

Progress in Molecular Biology
and Translational Science, Vol. 101
DOI: 10.1016/B978-0-12-387685-0.00004-4

1877-1173/11 $35.00

In eukaryotes, methylation of nucleosomal histones and other nuclear proteins is a central aspect of chromatin structure and dynamics. The past 15 years have seen an enormous advance in our understanding of the biochemistry of these modifications, and of their role in establishing the epigenetic code. We provide a synthetic overview, from an evolutionary perspective, of the main players in the eukaryotic chromatin protein methylation system, with an emphasis on catalytic domains. Several components of the eukaryotic protein methylation system had their origins in bacteria. In particular, the Rossmann fold protein methylases (PRMTs and DOT1), and the LSD1 and jumonji-related demethylases and oxidases, appear to have emerged in the context of bacterial peptide methylation and hydroxylation systems. These systems were originally involved in synthesis of peptide secondary metabolites, such as antibiotics, toxins, and siderophores. The peptidylarginine deiminases appear to have been acquired by animals from bacterial enzymes that modify cell-surface proteins. SET domain methylases, which display the β-clip fold, apparently first emerged in prokaryotes from the SAF superfamily of carbohydrate-binding domains. However, even in bacteria, a subset of the SET domains might have evolved a chromatin-related role in conjunction with a BAF60a/b-like SWIB domain protein and topoisomerases. By the time of the last eukaryotic common ancestor, multiple SET and PRMT methylases were already in place and are likely to have mediated methylation at the H3K4, H3K9, H3K36, and H4K20 positions, and carried out both asymmetric and symmetric arginine dimethylation. Inference of H3K27 methylation in the ancestral eukaryote appears uncertain, though it was certainly in place a little later in eukaryotic evolution. Current data suggest that unlike SET methylases, which are universally present in eukaryotes, demethylases are not. They appear to be absent in the earliest-branching eukaryotic lineages, and emerged later along with several other chromatin proteins, such as the Dot1-methylase, prior to divergence of the kinetoplastid-heterolobosean lineage from the remaining eukaryotes. This period also corresponds to the point of origin of DNA cytosine methylation by DNMT1. Origin of major lineages of SET domains such as the Trithorax, Su(var)3-9, Ash1, SMYD, and TTLL12 and E(Z) might have played the initial role in the establishment of multiple distinct heterochromatic and euchromatic states that are likely to have been present, in some form, through much of eukaryotic evolution. Elaboration of these chromatin states might have gone hand-in-hand with acquisition of multiple jumonji-related and LSD1-like demethylases, and functional linkages with the DNA methylation and RNAi systems. Throughout eukaryotic evolution, there were several lineage-specific expansions of SET domain proteins, which might be related to a special transcription regulation process in trypanosomes, acquisition of new meiotic recombination hotspots in animals, and methylation and associated modifications of the diatom silaffin proteins involved in silica biomineralization. The use of specific domains to "read" the methylation marks appears to have been present in the ancestral eukaryote itself. Of these the chromo-like domains appear to have been acquired from bacterial secreted proteins that might have a role in binding cell-surface peptides or peptidoglycan. Domain architectures of the primary enzymes involved in the eukaryotic protein methylation system indicate key features relating to interactions with each other and other modifications in chromatin, such as acetylation. They also emphasize the profound functional distinction between the role of demethylation and deacetylation in regulation of chromatin dynamics.

I. Introduction

A. DNA Packaging and the Uniqueness of Eukaryotic Chromatin Organization

The three superkingdoms of life utilize very distinct strategies for packaging their genomic DNA. Most bacteria utilize members of the IHF/HU family as their primary DNA-packaging protein.[1] In addition, certain bacteria, such as chlamydiae, have specialized DNA-packaging proteins of the HC1/HC2 family that function in establishing the condensed chromatin that is typical of certain stages of their life cycle.[2,3] Archaea show a surprising diversity of DNA-packaging proteins that include members of the Alba, MC1, Sac7/Cren7/Sso7, and histone fold families.[4,5] The histone fold proteins, which they share with eukaryotes, are primarily observed only in two of the great divisions of archaea, namely the euryarchaea and the thaumarchaea. The archaeal histones represent a packaging strategy that appears to have been the precursor of the eukaryotic system. Currently characterized archaeal nucleosomes comprise a single or a pair of distinct histone subunits, assembling into a tetramer that wraps ~80 base-pairs (bp) of DNA around it (comparable to the eukaryotic histone H3–H4 tetrasomes).[4] The origin of the eukaryotes was accompanied by a dramatic development of this ancestral histone template. First, there was proliferation followed by divergence resulting in four distinct histones (H2A, H2B, H3, and H4) that are conserved throughout eukarya.[4] Second, these histones assembled into an octamer, as opposed to the archaeal tetrasome, and wrapped nearly twice as much DNA (~146 bp).[4] Third, the eukaryotic histones acquired extensions to the N-terminus and/or C-terminus of the globular DNA-binding histone fold, that are enriched in positively charged residues.[6] These extensions are known as the histone tails and provide additional surfaces that neutralize the negative charges of the DNA backbone.

Emergence of the histone octamer-based packaging in eukaryotes was also accompanied by several other major structural innovations pertaining to chromosomal organization.[5] Right in the common ancestor of all extant eukaryotes a transition was made from the predominantly circular chromosomes of prokaryotes to multiple linear ones whose ends are capped by telomeres. Further, the chromosomes were separated from the rest of the cell by a membrane bi-layer, resulting in the quintessential feature of the eukaryotes, the nucleus.[7–9] Emergence of the nucleus decoupled cytoplasmic translation from nuclear transcription and marked a major departure from the prokaryotic situation. This appears to have relaxed the constraints on the eukaryotic genes allowing them to be colonized by introns, as mRNA was no longer translated during transcription.[6] However, emergence of introns favored the origin of a new set of large protein complexes: the spliceosomal complexes that associated with transcribed genes, and acted on the intron-containing primary transcripts.[10] Emergence of the nucleus also appears to have favored the emergence of a distinct subnuclear organelle, the nucleolus, where the ribosomal proteins could be combined with the freshly synthesized rRNAs to generate functional ribosomal subunits.[11] Thus, the landscape of eukaryotic chromatin diverged considerably from that of the prokaryotes, with spliceosomal, rRNA processomal, and telomerase ribonucleoprotein complexes adding to the protein and nucleic acid mass of the chromosomes beyond just the genome and the histone octamers.

B. Covalent Modifications of Chromatin Proteins and Their Role as an Epigenetic Code

In terms of protein structure, the origin of the eukaryotes was characterized by an expansion of low-complexity sequences in proteins.[6,12] These form non-globular segments of proteins that typically exist as disordered or unstructured random coils, and tend to be enriched in a single or few amino acids. In addition to histone tails, such low-complexity regions are also abundant in eukaryotic nuclear proteins such as transcription factors (TFs) and spliceosomal proteins (e.g., RGG and SR repeats), and might play roles in protein–protein interactions and low-specificity nucleic acid interactions.[12] These low-complexity regions offered a niche for the diversification of a veritable ecosystem of enzymes in eukaryotes that catalyze addition of covalent modifications to the amino acid side chains or the N- and C-termini of polypeptides.[13–17] There also arose a corresponding array of enzymes that catalyzed the removal of such covalent modifications, to restore the given peptide to its unmodified state. In addition to histone tails, the other targets of this flux of modifications were peptides from proteins that are more transient or long-term residents of chromosomes. These modifications span a dramatic range in terms of molecular weight and biochemical diversity.[13–17] The simplest of these are low-molecular weight adducts (methyl, phosphate, and acetyl groups). Somewhat higher molecular weight modifications include mono-ADP-ribosylation, biotinylation, and spermidinylation. The largest modifications involve addition of whole biopolymers such as branched or linear polyADP-ribose (up to 200 ADP-ribose units), peptides such as polyglutamate or polyglycine (up to 20 amino acids) and polypeptides of the ubiquitin family such as ubiquitin (Ub) and Sumo. In addition to these adducts, there are covalent modifications that directly modify the amino acid side chain. These include citrullination that results from the deimidation of the guanidino group of arginine (releasing the ammonium ion) and ornithination that results from hydrolysis of the guanidino group (releasing urea).[18,19] Other direct modifications are hydroxylations of the side chains of proline, lysine, and asparagine which generate the corresponding hydroxy amino acids.[16,20,21]

Among chromatin proteins, the ε-amino group of lysine is the most prominent target for modification, and receives adducts such as methyl, acetyl, biotinyl, and ubiquitin-like polypeptides.[16] The target amino can accept up to 3 methyl groups, resulting in distinct mono-, di-, and tri- methyl forms of lysine. In contrast, the guanidino group of the other basic amino acid, arginine, is primarily the target for a single adduct, methylation. In this case, methylation can result in three distinct modifications namely monomethylarginine and either asymmetric dimethylarginine where both methyl groups are linked to a single nitrogen atom of the guanidine group or symmetric dimethylarginine with one methyl group on each of the two available nitrogens.[19] The alcoholic amino acids serine and threonine are the primary targets of phosphorylation, but tyrosine is also similarly modified, predominantly in the animal lineage.[15] Serine and threonine can also be glycanated by *N*-acetylglucosamine, the significance of which is only recently beginning to be understood.[22] The acidic side chain of glutamate is a target for several modifications such as mono- and poly-ADP-ribosylation, polyglutamylation, polyglycination, and potentially also methylation.[13,14] The amino termini of chromatin proteins are also often subject to processing followed by acetylation. These adducts, along with direct modifications of side chains (hydroxylation and citrullination), have a profound consequence on the biochemical properties of histones and other chromatin proteins. The most prevalent modifications of histones are acetylation and methylation.[15,17] The former has been observed on at least 13 lysine side chains distributed across the four standard octameric histones. Methylation target sites are also distributed across the four core histones, with six of those being arginine and the remaining seven being lysine.[15,17,23] These are followed by phosphorylation with at least six sites, ubiquitin-system modifications with at least five target sites and poly-ADP-ribosylation on a single site across the classic core histones or their variants like centromeric H3.[15,24,25] Other than the core histones, the linker histone H1 is also subject to various modifications, such as methylation (e.g., at H1.4K26).[26]

In a direct sense, all of these modifications can affect both the surface electrostatics and the net size of the modified polypeptide, and sterically affect its interactions with nucleic acids and proteins. For example, the acetylation of lysines can reduce the net positive charge, phosphorylation and polyglutamylation can increase the net negative charge, and ubiquitination and poly-ADP-ribosylation can drastically alter the size of the polypeptide.[13–15,17] Additionally, many of these modifications carry epigenetic information, commonly termed "the histone code." The introduction of these modifications by specific enzymes can be seen as a coding step, in which extragenetic information is "written" into the histones and transmitted through subsequent cell divisions.[15,17] Discrimination between modified and unmodified peptides by specific peptide-binding domains, which might then recruit other chromatin remodeling or modifying activities to chromatin, can be conceptualized as the "interpretation" of the epigenetic code.[17,27] Finally, the removal of these marks by other enzymes can be conceived as "resetting" of the epigenetic information and usually accompanies major differentiation events or transitions such as postzygotic development. These protein-based marks also functionally interact with both DNA modifications and the RNAi system to comprise the complete complex of epigenetic coding in eukaryotes.[17] Over the past two decades, biochemical and biological studies have unleashed an avalanche of information regarding the structural, mechanistic, and organismal dimensions of these systems of epigenetic information. In particular, a combination of computational analysis of proteins sequences and structures and experimental investigations have identified most of the major enzyme classes involved in the generation and erasure of epigenetic marks on proteins as well as the domains that discriminate among them.

C. Peptide-Modifying Enzymes in Chromatin and the Protein Methylation System

A key realization from the studies on chromatin protein modifications has been that, though most eukaryotes possess sizeable complements of proteins catalyzing the major modifications, they can all be unified into a relatively small set of protein superfamilies. Likewise, a relatively small set of structural scaffolds has been used repeatedly among the binding domains that discriminate modified from unmodified peptides in chromatin. Protein acetyltransferases can be unified as members of the GCN5-like acetyltransferase (GNAT) fold.[28,29] The deacetylases belong to two major folds, namely the HDAC-arginase-like fold that contains the prototypical histone deacetylase Rpd3, and the classical Rossmann fold which includes the deacetylases of the Sir2 superfamily.[30–32] Among kinases, most belong to the eukaryote-type protein kinase fold, though the recently characterized WSTF (that phosphorylates

H2A.X on tyrosine 142) appears to define a novel structural scaffold for protein kinases.[33] Ubiquitin- and SUMO-conjugating systems follow a three enzyme cascade (E1, E2, E3), of which all histone-modifying E3s contain a treble-clef domain of the RING finger superfamily as their catalytic element.[34–36] The deubiquitinating isopeptidases acting on histones contain a catalytic domain with either the papain-like fold, or the metal-dependent JAB domain of the deaminase-like fold. Likewise, the catalytic domains of methylating and demethylating enzymes, which are the focus of this chapter, belong to a small set of ancient structural scaffolds.

The chromatin protein methylation system can be defined as comprising lysine and arginine methylases, the corresponding demethylases, and the arginine deiminases that regulate arginine methylation by its conversion to citrulline. Domains that discriminate methylated peptides from their unmethylated counterparts (i.e., readers of the epigenetic code established by the above enzymes) may also be considered as an immediate extension of this system. In contrast to the several surveys that discuss chromatin protein methylation from a functional angle with a focus on human or yeast models, we adopt an evolutionary perspective and exploit the genomic information that has become available across the eukaryotic tree. We present a structural overview of the main types of protein methylases, demethylases and deiminases followed by an evolutionary consideration of each of the catalytic domains. We then briefly survey the structural diversity of the peptide-binding domains involved in discrimination of methyl marks and their potential role in recruiting other activities to chromatin. Thereafter, we consider the major trends in the domain architectures of enzymes belonging to the methylation system and discuss the emergent syntactical features in the context of the functions of these proteins. Finally, we try to place the evolutionary history of protein methylation in the context of the other major mediators of epigenetic information, namely the DNA methylation and the RNAi systems.

II. The Categories of Protein Methylases and Their Role in Chromatin Protein Methylation

Protein methylases have evolved among two structurally unrelated folds. The first group of protein methylases belongs to the classical methyltransferase superfamily along with numerous other methylases, and possesses the Rossmann fold.[37–39] The second group of currently known protein methylases, the SET domain superfamily, contains the β-clip fold.[40] Among the classical Rossmann fold-type methylases are several distinct protein methylase families, and two of these methylate histones and other chromatin proteins. In the SET domain superfamily, most characterized eukaryotic versions act on histones and probably other chromatin proteins, but a few also methylate cytoplasmic proteins such as cytochrome *C*, Rubisco and ribosomal proteins (L42, L12, L23), and the translation initiation factor eIF1A.[41–45]

A. The Rossmann Fold Chromatin Protein Methyltransferases

The Rossmann fold seen in methylases has a seven-stranded core sheet, in which the first six strands are in the parallel configuration as part of the regular succession of α/β units.[34,37,46] The topological crossover occurs between the 3rd and 4th strands, and juxtaposes the first strand (with the characteristic AdoMet-binding loop) and the 4th strand (with a key polar catalytic residue at its C-terminus) (Fig. 1).[38,39] However, there are variations on this basic structural template (see below). Protein methyltransferase activity appears to have arisen on multiple occasions within the Rossmann fold-containing superfamily of methylases. The most ancient of these is the HemK family of methylases that arose within the vast radiation of nucleic acid methylases with a [ND]PP[YFW] motif at the end of strand 4.[10,47,48] Members of this family catalyze the methylation of N(5) of specific glutamines in the ribosomal protein L3 (YfcB) and the class 1 peptide release factors RF1 and RF2 (HemK).[47,48] This family represents an early shift from nucleic acid- to protein-modifying activity, that appears to have occurred prior to the last universal common ancestor (LUCA) of extant life.[10] Another ancient family that was already present in the LUCA is the PrmA family, whose representatives are seen in bacteria, archaea, and eukaryotes. While the activity of the archaeal and eukaryotic members of this family remain poorly studied, the bacterial versions catalyze the trimethylation of the NH_2 terminus and the ε-amino groups of K3 and K39 of the ribosomal protein L11.[49] Thus, it appears likely that multiple protein methylation activities were present prior to the radiation of extant organisms from the LUCA, and that these acted primarily on proteins of the translation apparatus. Eukaryotic chromatin protein methylases belong to two distinct families, namely, the PRMT and the DOT1 families. Neither of these can be traced back to the LUCA, and appear to be late innovations relative to the above families.

B. The PRMT Family

1. Specificity and Functions of the PRMT Family

The PRMT family comprises dedicated arginine methyltransferases in eukaryotes that methylate specific arginines in histones and ribonucleoproteins[19,50,51] (Fig. 2). In vertebrates at least six distinct members of this family are active on nuclear proteins. Of these PRMT2, CARM1/PRMT4, PRMT5, and PRMT6 act on histone H3, with CARM1 targeting multiple arginines;

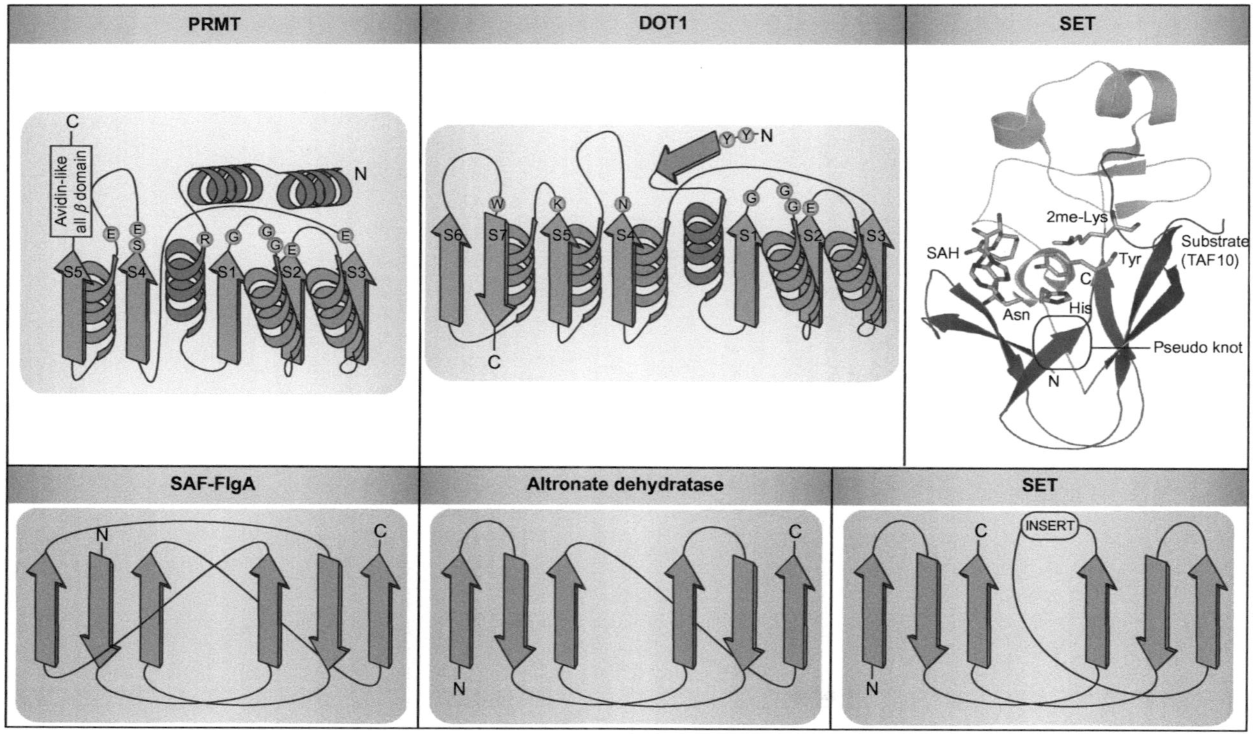

FIG. 1. Structure and sequence features of protein methyltransferases. The Rossmann fold protein methylases in the PRMT and DOT1 families are depicted as topology diagrams. Strands and helices of the Rossmannoid-fold core of the methylases are colored green and orange. Additional family-specific helices and strands are colored pink and blue, respectively. Strands of the core Rossmannoid fold are labeled S1-S5/S7. Key sequence features described in the text, including those involved in AdoMet-binding and catalysis, are shown in gray circles with the residue abbreviation at the corresponding structural

PRMT1, PRMT5, PRMT7, and probably PRMT2 act on an arginine in histone H4; and PRMT7 acts on an arginine in H2A.[52–58] PRMT6 also methylates arginines in the DNA-binding AT-hook motif of the chromatin protein HMG1A, whereas CARM1 methylates arginines in the transcription-factor-interacting adaptor module of the giant histone acetyltransferase protein p300.[52,59] Among the ribonucleoprotein complexes, PRMT1 methylates arginines in the glycine and arginine-rich low-complexity regions of hnRNPK and hnRNPR, PRMT5 and PRMT7 methylate arginines in the tails of the Sm proteins of multiple snRNPs, and PRMT3 methylates arginines in components of transcript cleavage stimulation factor (e.g., CstF-64) that processes pre-mRNAs for polyadenylation.[60–65] Additionally, PRMT3 also methylates the 40S ribosomal protein S2.[66] Genetic and biochemical evidence suggests that PRMT5 is also the primary methylase of arginines found in the glycine- and arginine-rich motifs of the components of the ribonucleoprotein nuage complex (e.g., Piwi), that associates with noncoding piwiRNAs to silence transposable elements in the germline of animals.[60,65] PRMT8 might modify the nuclear RNA-binding protein EWS, whose exact role remains unclear.[64] Beyond these, several other substrates, mainly nuclear RNA-associated proteins, have been recognized for these enzymes.[50] Given the widespread detection of methylated arginines in eukaryotic nuclear proteins, it is clear that the complete substrate map of the PRMTs still remains to be determined. PRMTs are categorized as either type I enzymes that catalyze sequential formation of monomethylarginine and then asymmetric dimethylarginine (CARM1, PRMT1, PMRT2, PMRT3, PRMT6, PRMT8), or type II enzymes that generate symmetric dimethylarginine (PRMT5 and PRMT7 and *Arabidopsis* PRMT10 which methylates H4 on arginine 3).[19,55,67] A distinctive lineage of PRMT enzymes from trypanosomes (TbPRMT7) has been described as a type III enzyme, as it only catalyzes arginine monomethylation.[68]

The action of the PRMTs is pleiotropic and appears to affect a wide-range of complexes that are permanent or transient components of chromatin as well as cytoplasmic complexes. First, the methylation of histones by PRMTs appears to have a major role in the positive regulation of transcription. Thus, CARM1 and several of the other PRMTs have been shown to function as coactivators for a diverse set of TFs.[51,52,56–58] The recruitment of enzymes such as PRMT2 to

element. The cartoon structure of the SET domain of SETD7 (PDB: 3M59), in complex with its substrate (TAF10) and AdoHcy, illustrates the 3-stranded repeat units within the β-clip fold, the pseudoknot, and key residues involved in AdoMet and substrate-lysine binding, and catalysis. Topology diagrams of the β-clip units of SAF-FlgA, altronate dehydratase, and the SET domain are shown in the bottom panels to illustrate the possible pathway of evolution of a SET-like β-clip unit from a SAF-FlgA-like unit through an altronate dehydratase like intermediate. The N-terminal unit of the β-clip is colored orange and the C-terminal one blue. (See Color Insert.)

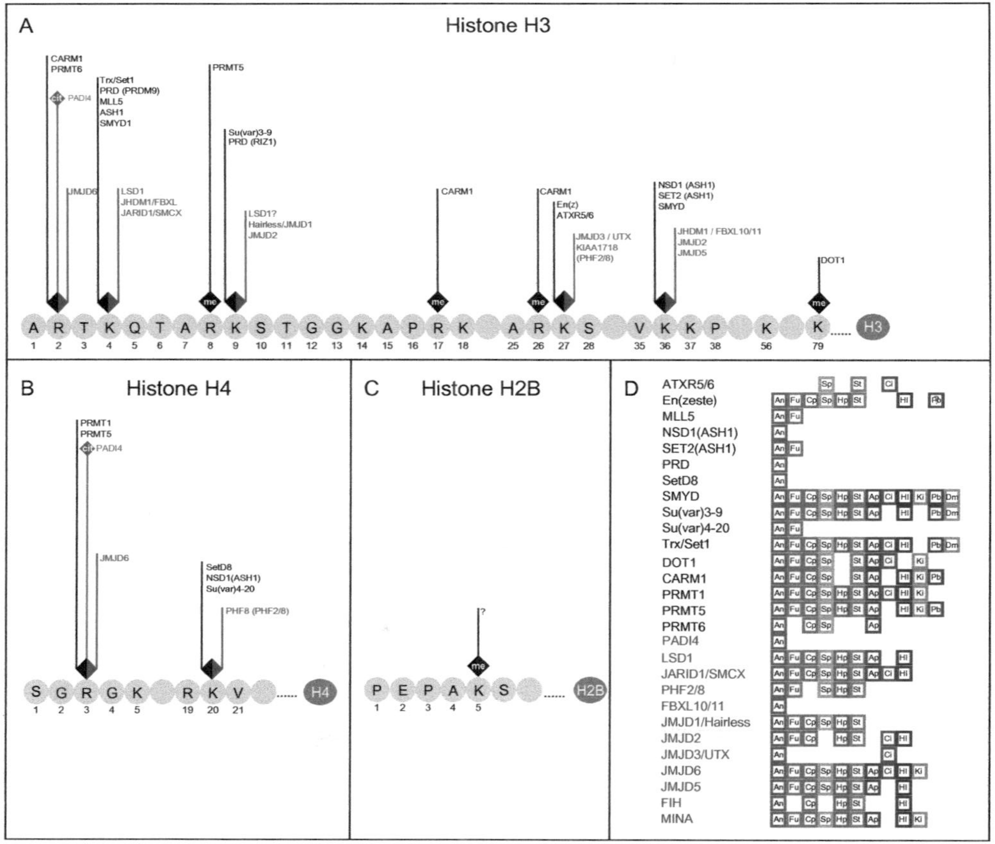

FIG. 2. Modification, location and phyletic patterns of enzymes involved in histone modifications. (A), (B), and (C) show the positions of histone residues modified by various methylases, demethylases and peptidylarginine deiminases. A rhombus is shown above the residue that is modified, and the extending line lists the enzymes involved in the modification. The human histone sequences were used as the reference sequence. Gaps of variable length

promoters by β-catenin is critical for the establishment of transcription programs associated with the Wnt signaling pathway in animals.[53] The observation that PRMT5 is stimulated by association with Swi2/Snf2 ATPases, such as BRG and BRM, suggests at least some of these enzymes might have a more general role in cooperating with chromatin-remodeling factors in establishing regions of open chromatin.[51] The association of PRMT5 with the transcription elongation factors Spt4 and Spt5 points to additional roles for methylation during elongation.[51] CARM1 also plays a role in coupling transcription to pre-mRNA splicing, whereas PRMT3 regulates its polyadenylation.[63] The presence of characteristic low-complexity repeats enriched in arginine in several eukaryotic RNA-associated proteins suggests that RNPs are a major target of the PRMTs, and that they have pervasive regulatory roles in processes such as splicing, which are yet to be fully understood. In particular the maturation of various RNP complexes appears to depend on the methylation of arginines in these low-complexity segments. DNA-repair proteins such as Mre11 and 53BP1, which transiently associate with chromatin, might also be targeted by PRMT1 and PRMT5, respectively, although the exact roles of these methylation events remain poorly understood.[50]

2. Structure and Evolutionary History of the PRMT Family

The PRMT family markedly departs from the standard methylase-type Rossmann fold in lacking the two C-terminal strands (strands 6 and 7) that form a β-hairpin[58,69] (Fig. 1). Instead, the 5th strand connects to a distinct β-barrel domain that adopts an avidin-like fold and appears to have displaced the two C-terminal strands, and this distinguishes the PRMT family from other methylase domains. Additionally, this family is characterized by three helices N-terminal to the initial strand of the Rossmann fold (Fig. 1). The first two of these helices are drawn over the AdoMet-binding site in the form a "cap" that largely shields the cofactor from solvent. The third N-terminal helix contains a highly conserved basic residue that binds the COOH group of the methionine moiety of AdoMet. Within the Rossmann fold domain, strand-2 bears a

between residues are shown as unmarked circles. Methylating enzymes are colored black and demethylating enzymes red. Citrullinating enzymes are colored blue. "me" represents methylation and "cit" citrullination. "?" depicts an uncertainty of the identity of the enzyme involved in that particular modification. The phyletic distribution of the histone modifying enzymes across 12 major eukaryotic lineages is shown in (D). The presence of an enzyme in a lineage is indicated by a colored box with an abbreviation of that eukaryotic lineage in the box. The eukaryotic lineages sampled for this analysis and the abbreviations used to depict them include: An, animals; Fu, fungi; Cp, chlorophyte algae; Sp, streptophyte plants; Hp, haptophytes; St, stramenopiles that includes oomycetes, phaeophytes, and diatoms; Ap, apicomplexa and *Perkinsus* (a sister group of dinoflagellates); Ci, ciliates; Hl, heterolobosea; Ki, kinetoplastids; Pb, parabasalids; Dm, diplomonads.

conserved acidic residue that forms hydrogen bonds with the hydroxyl groups of the sugar moiety of AdoMet, as is typical of methylases of this fold. A conserved acidic residue from strand-3 binds the amino group from the adenine moiety of AdoMet. The C-terminus of strand-4 bears an acidic residue that is central to the catalytic activity of these enzymes and plays a primary role in the transfer of the methyl group from AdoMet to the $-NH_2$ groups of arginine.[58,69] Thus, this residue is homologous and equivalent in function to the corresponding polar residue seen in the nucleic acid *N*-methylases. The substrate-binding pocket for arginine is formed by the combination of the N-terminal helices, the C-terminal avidin-fold domain, and an acidic residue from the loop just downstream of strand-4 of the Rossmann fold. The avidin-fold domain contributes a highly conserved histidine and tryptophan (HW motif) to form the "roof" of this substrate-binding pocket (Supplementary material).[58,69] The configuration of these elements appears to sterically restrict access to the active site, thereby allowing only certain peptides (especially those with glycine–arginine dyads) to serve as substrates.

Phylogenetic analysis suggests that the PRMT family had already diversified in the last eukaryotic common ancestor (LECA), as indicated by the presence of multiple members of this family in the early-branching eukaryote *Trichomonas vaginalis* (Supplementary material). Among the eukaryotic PRMTs, the primary split in the tree separates the PRMT5 clade from remaining PRMTs such as those of the CARM1 and PRMT1/PRMT3 clades. The versions in *T. vaginalis* belong to the PRMT5 clade, the CARM1-like clade and the PRMT1/PRMT3-like clades. This suggests that, prior to the LECA, the PRMT family had already radiated into at least one version that catalyzed symmetric dimethylation (PMRT5-like) and two distinct versions that catalyzed asymmetric dimethylation (CARM1-like and PRMT1/PRMT3-like). This would also imply that these two forms of methylation served as distinct signals right from the base of the eukaryotic tree (Supplementary material). By the time the trypanosomes branched off from the remaining eukaryotes, a distinct divergent group (the PRMT7 group) had emerged.[68] The exact affinities of this group within the PRMTs cannot be easily determined on account of its divergence; however, given its catalysis of symmetric methylation it is possible that it represents an offshoot of the ancestral PRMT5-like clade.[70] Subsequently in the course of eukaryotic evolution, the PRMT1/PRMT3-like clade split up into the distinct PRMT1- (including PRMT8), PRMT3-, and PRMT6-like clades. The enigmatic DmArt2/6/9, which is currently only known from animals and trypanosomes, also appears to be related to the above clade (Supplementary material). It might have emerged first in animals, followed by lateral transfer to the trypanosomes. The PRMT2-like clade appears to have emerged in the common ancestor of the unikonts (i.e., animals, fungi, and amoebozoans) as a branch of the CARM1-like clade, but was subsequently lost early in fungal evolution. Another offshoot of the CARM1-like clade is typified by *Arabidopsis* PRMT10,[67] found only in plants and stramenopiles. One or more of the PRMT methylases have been lost from several lineages of eukaryotes. Interestingly, the basal eukaryote *Giardia* and several microsporidians, which are highly reduced members of the fungal lineage, appear to have lost all members of the PRMT family. At face value, this would suggest that these organisms lack arginine methylation (at least of the conventional type observed in other eukaryotes), and indicates that eukaryotic life in principle could exist without it. Concomitant with their radiation, the PRMT proteins also acquired several distinct N-terminal domains: The PRMT5 clade acquired an N-terminal TIM-barrel domain related to the AP-endonucleases (Supplementary material). However, this domain lacks the catalytic residues of the AP-endonuclease and is likely to function in substrate recognition. The CARM1-clade acquired an N-terminal EVH1-like domain that adopts a PH-like β-barrel fold.[69] In animals, the PRMT2 clade acquired an N-terminal SH3 domain that recognizes proline-rich peptides. The PRMT3 clade acquired a C2H2 zinc finger domain in the common ancestor of the animals and the fungi.[71] The acquisition of these N-terminal domains appears to have played an important role in the emergence of distinct interactions and functional partners within the PRMT family.

Outside of eukaryotes, members of the PRMT family can be seen in several proteobacteria and members of the chlamydia lineages (Fig. 3, Supplementary material). These bacterial versions share both the N-terminal helices and the C-terminal avidin-fold domain with their eukaryotic counterparts, along with the above-described conserved motifs suggesting that they are also likely to methylate peptides. The bacterial versions are encoded by operons that also contain genes for multiple peptide dioxygenases of the double-stranded β-helix fold, such as the Jumonji-related enzymes, metal-binding cupins, and other 2-oxoglutarate Fe^{2+}-dependent dioxygenases (2OGFeDO). These operons may also encode genes for nonribosomal peptide synthetases with condensation domains and other potential peptide-oxidizing enzymes such as cytochrome P450 and LSD1-related amine oxidases (Supplementary material). All these enzymes are typical of systems that are involved in synthesis of secondary metabolites such as antibiotics, toxins, and siderophores from modified amino acids or peptides. Indeed, certain bacterial PRMTs are even incorporated as domains of the gigantic peptide antibiotic biosynthesis enzymes with multiple condensation domains that serially link amino acids in nonribosomal peptide synthesis (Fig. 3; e.g., anabaenopeptilide synthetase which synthesizes a toxic peptide metabolite in the cyanobacterium *Anabaena*; gi: 9715734).[72] Hence, it appears likely that the bacterial versions are involved in the synthesis of novel methylated peptide-derived secondary metabolites. At a higher level, the PRMT family can be unified with the ancient PrmA family of lysine

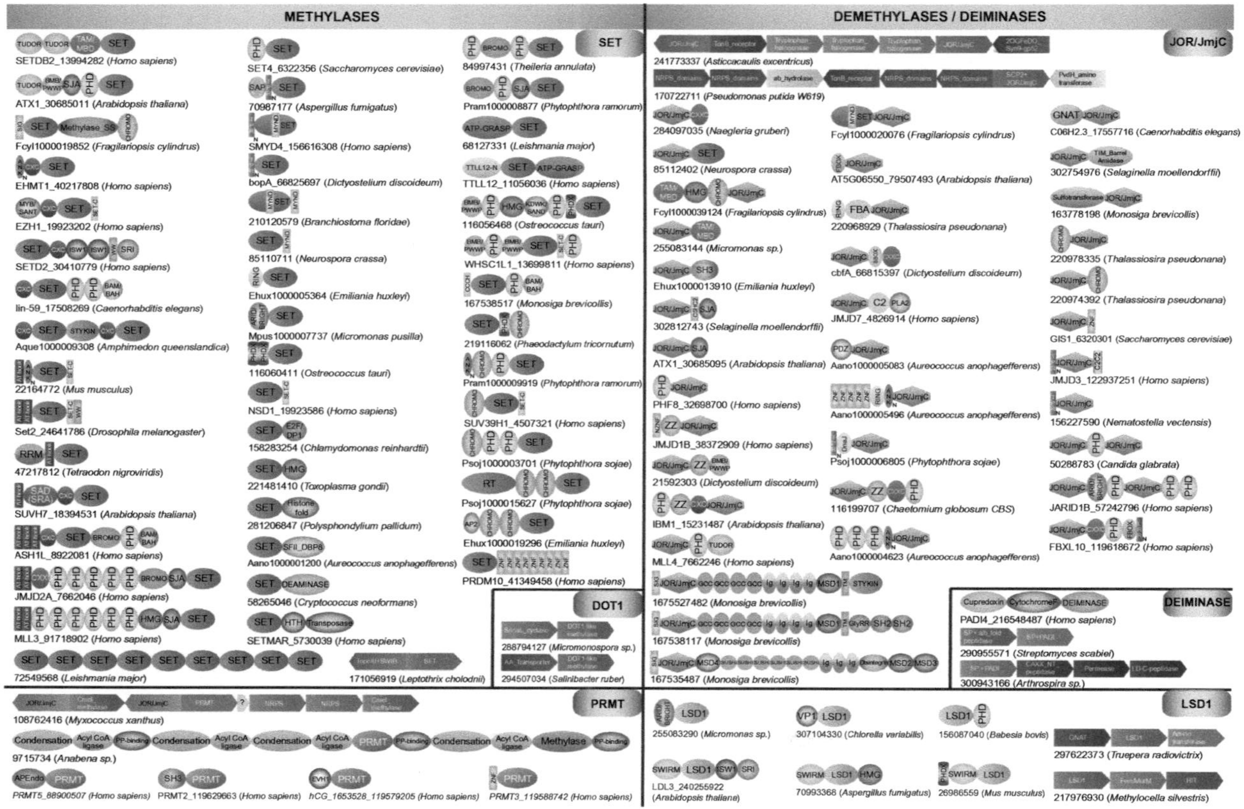

FIG. 3. Domain architectures and gene neighborhoods of various proteins related to protein methylation and demethylation. Super-secondary structure-forming repeats such as TPR, LRR, and Ankyrins are labeled with a "N" subscript to denote multiple (N) repeat units. Gene neighborhoods of prokaryotic homologs are shown as boxed arrows with the arrowhead pointing in the 3′ direction. Proteins are labeled with their Genbank index number

methylases.[49] This suggests that the PRMT family probably emerged first in bacteria from a PrmA-like precursor in the context of the diversification of secondary metabolite pathways. It was subsequently transferred to eukaryotes, where it radiated to give rise to diverse protein methylases.

B. The DOT1 Family

1. Specificity and Functions of the DOT1 Family

In eukaryotes members of the DOT1 family rather exclusively modify the H3K79, and processively methylate it to give rise to mono-, di-, and trimethylated forms.[73–75] DOT1-catalyzed methylation is rather distinctive in that it is a modification that targets a residue right within the globular histone fold, rather than lysines in low-complexity tails.[75] Unlike the histone methylations catalyzed by the PRMT family, methylation catalyzed by the DOT1 family appears to regulate gene expression primarily negatively across eukaryotes.[76–78] Studies in animals indicate that DOT1 is part of a large protein complex, including two pairs of paralogous proteins, all of which give rise to fusion proteins arising from chromosomal translocations in mixed lineage leukemia (MLL): (1) ENL and AF9/MLLT3, both similar to TAF14, with a N-terminal YEATS domain and a C-terminal BrC domain and (2) AF17/MLLT6 and AF10/MLLT10, both with two N-terminal PHD fingers and C-terminal AT-hook motifs.[74]

Studies in *Saccharomyces cerevisiae*, supported by studies in other eukaryotes, suggest distinct roles for the di- and trimethylated forms, which occur on largely mutually exclusive sets of genes.[77] H3K79me3 occurs predominantly within the gene body (i.e., protein-coding sequence), and is largely absent in promoters and intergenic regions. This form has been associated with genes that are transcriptionally less active, and is explicitly excluded from the nucleosomes associated with the most highly expressed genes: 50% of the genes generating just 1–4 mRNAs per hour are enriched in nucleosomes showing this modification, in contrast to just 2% of the genes giving rise to > 50 mRNAs per hour.[77] The increased processivity of DOT1 in catalyzing trimethylation appears to depend on prior monoubiquitination of histone H2BK123 by the Rad6/Bre1 ubiquitinating complex. Unlike H3K79me3 levels, which do not vary greatly over the cell cycle, H3K79me2 levels change significantly with the

and source species name. Gene names are also shown for proteins from model organisms. Standard abbreviations are used for most domains. The legend to Fig. 6 provides expansions for some nonstandard domain names used in this representation. Refer to the supplementary material for a comprehensive list of architectures and gene neighborhoods. Temporary gene names are used for proteins from the unpublished sequences of *Emiliania, Aureococcus anophagefferens*, and *Micromonas pusilla*. To access these protein sequences, refer to the Supplementary material in the FTP site.

cell-cycle, being lowest in G1 and elevated during the G2/M progression.[77,79] Further, H3K79me2 is not restricted to the gene bodies and is also seen in intergenic regions, including promoters. Moreover, the genes associated with this modification tend to be transcriptionally inactive during the G2/M phase, when its levels are elevated.[73,77] In trypanosomes, which possess three DOT1 paralogs, two have been functionally characterized.[80] The first, DOT1A, mainly catalyzes H3K79me2 formation in a cell-cycle dependent manner, whereas the other paralog DOT1B appears to be involved in subtelomeric gene-silencing associated with antigenic variation in trypanosomes.[80] In mammals, DOT1 appears to regulate heterochromatin formation at telomeric and centromeric regions, consistent with the observations in yeast and trypanosomes.[74,75,79] In addition to its role in silencing and heterochromatin organization, other observations suggest that DOT1 methylation regulates multiple aspects of DNA repair, such as base excision repair, Rad9-mediated checkpoint function, and negative regulation of the action of the translesion repair polymerases.[81,82]

2. Structure and Evolutionary History of the DOT1 Family

The core Rossmann fold domain of the DOT1 family does not deviate in any notable fashion from the scaffold typical of the methylase superfamily and is distinguished from other members by a unique N-terminal extension.[76,78] This comprises an element in an extended conformation, followed by a helix that leads to the first strand of the core Rossmann fold (Fig. 1). The extended element is drawn right over the AdoMet-binding site, comparable to the helical extension seen in the PRMT family, forming a “cap” over the active site. It contains two hydrophobic or aromatic residues that interact, respectively, with the methionine and adenine moieties of AdoMet (Fig. 1). In effect, this cap shields the cofactor from solvent, and forms part of the substrate recognition interface of these enzymes. Within the core Rossmann fold, as in other members of the methylase superfamily, the DOT1 family contains a conserved acidic residue that hydrogen bonds with the ribose moiety of AdoMet. The end of strand-4 is marked by a conserved asparagine that is critical for catalyzing the methyltransfer reaction.[78] The DOT1 family is characterized by a large loop between the strands 6 and 7 that forms the terminal hairpin (Fig. 1). This loop contains a large hydrophobic residue, which is usually a tryptophan in the classical eukaryotic DOT1 orthologs. The aromatic ring of this tryptophan is proximal to the strand-4 asparagine and is an important feature of the substrate recognition site.[78]

With the exception of the basal eukaryotes *Trichomonas* and *Giardia*, DOT1 orthologs are present in all other major eukaryotic lineages for which genome sequences are available (Fig. 2). However, within those lineages there are certain notable instances of gene loss—while the basal plant lineages such as the chlorophyte algae and lycopodiophytes have one or more DOT1 paralogs, they have been completely lost in the crown-group land plants such as angiosperms. Within animals and fungi, typically only a single DOT1 paralog is seen and they display a largely vertical pattern of evolution. However, in the caenorhabditiform nematodes there has been a notable lineage-specific expansion (LSE) of DOT1, with at least five paralogs in *Caenorhabditis elegans*. It seems important to study the potential functional compartmentalization of these newly emergent DOT1 versions in this organism. Phylogenetic analysis also suggests that the precursor of DOT1A and DOT1B in trypanosomes appears to have been acquired via lateral transfer from the animal lineage (Supplementary material). Following this transfer, it appears to have acquired an N-terminal Zn-chelating domain with four conserved cysteines, and was then duplicated to yield two functionally distinct paralogs. In microbial eukaryotes such as chlorophyte algae, stramenopiles, apicomplexans, ciliates, and trypanosomes, there appear to have been multiple lateral transfer events that have disseminated DOT1 paralogs between distantly related lineages. Consequently, some of these eukaryotes have multiple DOT1 paralogs, with particularly notable complements of three or more paralogs seen in certain stramenopiles and trypanosomes (the third trypanosome DOT1 paralog is distinct from the previously studied DOT1A and DOT1B). This presence of multiple DOT1 paralogs is rather different from the situation seen in most animals and fungi, raising the possibility that some of them might have evolved distinct substrate specificities or may regulate H3K79 methylation in alternative signaling or developmental contexts.

Outside of eukaryotes the closest homologs of DOT1 are seen in bacteria (Fig. 3, Supplementary material). Some of these bacterial forms, such as those seen in *Legionella*, myxobacteria, and *Protochlamydia*, are particular close to the eukaryotic forms and share conserved sequence motifs in both the N-terminal extended element and in the loop between strand-6 and strand-7 (e.g., the conserved aromatic residue). These versions are further nested within a large prokaryotic radiation of methylases that are involved in the synthesis of secondary metabolites such as mycolic acids in mycobacteria (including the mycolic acid cyclopropane synthases),[83] polyether antibiotics such as nigericin (e.g., NigE of *Streptomyces* sp. DSM4137) and as yet uncharacterized compounds in *Micromonospora* (gi: 288794127; in the same operon as a SnoaL-like polyketide cyclase).[84] Further, gene neighborhood analysis suggests that several members in this bacterial radiation (e.g., gi: 294507034 from *Salinibacter ruber*) are specified by a conserved operon along with an amino acid transporter. It is conceivable that these versions are involved in the utilization of particular amino acids, or metabolites derived from them (Fig. 3). Thus, it appears the DOT1-like group arose as part of the radiation of methylases involved in generating diversity among secondary metabolites by adding of specific methyl groups or utilization of amino acids. Interestingly, of the bacterial versions closest to the eukaryotic forms, some are encoded by intracellular pathogens (e.g., *Legionella*) or endosymbionts (*Protochlamydia*) of eukaryotes (Supplementary material). These versions have signal peptides that are likely to allow their secretion into the host cells. Hence, it would be of interest to investigate if they play a role in regulating host behavior via histone methylation comparable to the endogenous DOT1. This observation also suggests a possible route for the original acquisition of DOT1 by eukaryotes from their intracellular bacterial symbionts.

C. Methylases of the SET Domain Superfamily

1. Specificity and Functions of the SET Domain Superfamily

One of the great triumphs of genetics in the *Drosophila melanogaster* model was the discovery of the trithorax, polycomb, and position effect variegation (PEV) groups of genes.[17,85] Right from the early studies, it became clear that the protein products of these genes were key participants in the organization and dynamics of eukaryotic chromatin. A common denominator that spanned all these groups of proteins was the SET domain (also called the tromodomain), which was present in the trithorax group proteins, *Trithorax* and *Absent, small, or homeotic discs 1* (Ash1); the polycomb group protein of the PRC2 complex, *Enhancer of Zeste* [E(Z)]; and PEV gene, *Suppressor of Variegation 3–9* [Su(var)3–9].[17] These observations indicated that SET domain activity is required for the establishment of both repressive or heterochromatic states (the polycomb and PEV groups), as well as open chromatin associated with transcriptional activity (Trithorax group). Subsequently, sensitive sequence searches showed that the domain shared by these proteins was homologous to the experimentally characterized Rubisco and cytochrome *C* methylases.[45,86] This suggested that the SET domain was a methylase domain that was the long sought-after histone methylase. Subsequent experimental studies established that the SET domains are the primary methylases responsible for mono-, di-, and trimethylation of lysines in the tails of the histones H3 and H4.[87–89] Additionally, they methylate a number of other chromatin proteins such as the DNA methylase DNMT1 and the basal TF TAF10.[90,91] Of the currently known methylations catalyzed by SET domains, four are on lysines in the N-terminal tail of H3: K4, K9, K27, and K36 (Fig. 2).[15] H4 contains a single lysine target for methylation by SET domains—K20. The H3K4 methylation is catalyzed by the SET domains of the *D. melanogaster* trithorax group proteins, Trx and Ash1, their orthologs such as MLL1, MLL2, MLL3, MLL4, and MLL5 from mammals and Set1 from *S. cerevisiae*, and the SET-MYND (SMYD) proteins such as SMYD1.[87,88,92–94] H3K9 methylation is catalyzed primarily by the PEV group protein Su(var)3–9 and its orthologs such as Clr4 from

Schizosaccharomyces pombe, SUVH2 from plants and SUV39H1/SUV39H2 from mammals.[95–97] The same residue is also methylated by certain divergent SET domains of the PRD lineage (see below) such as RIZ1.[98] H3K27 methylation is primarily catalyzed by E(Z) and its orthologs such as EZH2 from mammals.[99–102] H3K36 is the target for the *S. cerevisiae* Set2-like, animal NSD1-like and the mammalian SMYD2-like proteins.[103,104] The H4K20 methylation is catalyzed by the *D. melanogaster* Su(var)4–20 and its orthologs, mammalian NSD1, and SETD8.[105,106] An as yet poorly understood modification, H2BK5me1, associated with transcriptionally active chromatin,[23] might also be catalyzed by a SET domain enzyme.

A recent study in *D. melanogaster* has identified five basic states of chromatin which were labeled BLACK, BLUE, and GREEN (heterochromatic) or YELLOW and RED (transcriptionally active euchromatin).[107] Given the conservation of the major fraction of chromatin proteins associated with these states and experimental evidence from other eukaryotes, there are indications that these are likely to represent key chromatin states widely present across the eukaryotic tree.[8,107] Many of these states are closely linked to the activities of particular types of SET domain proteins. H3K4me2 marks, synthesized primarily by the trithorax group SET domains like ASH1, are associated with both YELLOW and RED, defining a primary feature of active chromatin.[88,107,108] The ASH1 methylases function as part of a large protein complex that includes histone-binding proteins and a potential scaffold of noncoding RNAs, namely TRE1, TRE2, and TRE3.[109] In contrast, the H3K4me3 mark is associated with chromatin in the hotspots for initiation of meiotic recombination in several eukaryotic lineages. This modification is catalyzed by the PRDM9-like SET domains in animals.[110,111] The H3K36 mark is specifically associated with YELLOW chromatin, but not RED chromatin, and appears to define that subset of active genes that is typically associated with house-keeping and generic functions required across all cell-types.[107] The H3K36 marks are set up co-transcriptionally, at the time of the first transcription of the gene, due to association of these methylases with the transcription elongation complex.[112,113] These marks are also associated with facilitating nonhomologous end-joining DNA repair.[114] In RED chromatin, the trithorax group SET domains such as Trithorax appear to function in conjunction with other trithorax group proteins such as the SWI2/SNF2 ATPase Brahma, and are associated primarily with genes that are transcriptionally active in specific conditions or cell-types.[107] They are also enriched in the origin-recognition complex, and associated with the part of the genome that replicates earliest. The H3K9me2 mark, generated predominantly by the SET proteins belonging to the PEV group [e.g., Su(var)3–9] was localized to what was termed GREEN chromatin and corresponds to classical heterochromatin (which was recognized originally as the basis of PEV).[107] H3K27me3 was found to be associated with what was termed BLUE chromatin.[107] This predominantly repressive chromatin is enriched in the Polycomb group proteins, which include E(Z) that catalyzes this modification.[94,99,101,102] At least in animals, it is possible that H1K26 methylation of the linker histone by G9a/Glp SET domain proteins might play a role in BLACK chromatin.[26] The H4K20me2 appears to play a role in regulating DNA repair, whereas H4K20me3 might be associated with the classical heterochromatic state similar to H3K9me2.[105] The H4K20me1 modification is less understood and some studies suggest that it might be associated with actively replicating chromatin.[105] Consistent with this, SETD8 (which catalyzes this modification) associates with the DNA replication clamp, PCNA.[115] H4K20me2 also recruits the H4K16 acetylases like MSL3 that might activate transcription through acetylation.[116,117] Thus, histone methylation by SET domain proteins can both positively and negatively regulate gene expression, and are central to the establishment of the major chromatin states in a cell.

2. Structure and Active Site of the SET Domain Superfamily

The SET domain displays the peculiar β-clip fold, which is additionally found in the SAF domain, the dUTPases, the molybdenum cofactor biosynthesis protein MoeA C-terminal domain, urease β-subunits and some bacteriophage structural proteins[40] (Fig. 1). Of the β-clip domains, the SET domain is most closely related to the SAF domain, which binds low-molecular weight ligands such as sugars and their derivatives. The SAF domains display one of the simplest versions of the β-clip fold, and it is likely that the catalytic SET domain evolved from an ancestral sugar-binding precursor similar to the extant SAF domains.[40] The β-clip fold is a symmetric structure that has emerged through the duplication of a simple three-stranded unit.[118] The individual three-stranded units exist as an obligate dimer because one of the strands is swapped between the monomeric elements to yield two twisted 3-stranded sheets, that might further interact to result in a partial barrel configuration (Fig. 1). The successive strands associated with the topological swap are separated by a characteristic "hammer-head" or paper-clip handle-like structure that gives the fold its name (Fig. 1). Due to this property of obligate dimerization between the two 3-stranded units, different versions of the β-clip domain have repeatedly reemerged due to duplications of founder 3-stranded units. Additionally, the strands within a 3-stranded unit tend to undergo circular permutation resulting in configurational differences between different β-clip domains (Fig. 1). Within the SAF domain superfamily, such a circular permutation is seen in the N-terminal 3-stranded unit of hexuronate dehydratases, with respect to the ancestral state as typified by the bacterial flagellar FlgA proteins (Fig. 1).[40] Duplication of a founder unit based on such an

N-terminal 3-stranded unit could have given rise to the precursor of the SET domains. In the SET domains the end of C-terminal-most strand passes “under” the connector segment between the two 3-stranded units resulting a pseudoknot structure.[118]

Due to the internal duplication, the β-clip domains tend to have two symmetrically placed potential ligand binding pockets (Fig. 1). In the case of the classical SAF domains, such as that in the FlgA protein, the internal symmetry of the two pockets can be recognized at the sequence level and likely form the carbohydrate-binding sites.[40] In the case of the SET domain, the two pockets have greatly diverged at the sequence level. Further, an additional element is inserted between the two 3 stranded units that completely alters the environment of the two pockets.[118] This central element, unique to the SET domains, could have emerged through further duplication of one of the core 3-stranded elements. This central element is also the locus of large insertions such as an α-helical segment in the Rubisco methylases and the MYND finger, a binuclear Zn-chelating treble-clef domain, in the SET-MYND proteins (e.g., SMYD1, SMYD2).[92,119] The first of the two pockets of the SET β-clip appears to retain properties similar to the ancestral version seen in the SAF domain. It has a motif usually of the form GxG, whose backbone makes multiple contacts with AdoMet.[119–125] The second pocket in conjunction with the central elements forms the binding site for the modified peptide with the target lysine being deeply buried into it (Fig. 1). A highly conserved aromatic residue (typically tyrosine) from the end of the C-terminal strand, close to the pseudoknot, interacts with both the adenine moiety of AdoMet and the alkyl moiety of the target lysine side chain and guides it to the active site.[120,125] One of the key catalytic residues of the SET domain is a polar residue (typically asparagine), which is again associated with the pseudoknot structure in the connector segment between the two 3-stranded units of the β-clip (Fig. 1). This residue is often followed by a conserved histidine that interacts via π–π stacking with the aforementioned aromatic residue in the active site. It is likely that this residue functions as a “sensor” that detects and transmits the conformational change of the lysine interacting with the tyrosine to its N-terminal residue, which then mediates the transfer of the methyl group from AdoMet to the ε-NH_2 group on lysine. In several SET domains, there is a C-terminal extension that bears three conserved cysteines (usually a CXCXnC motif), which chelate a Zn^{2+} ion in conjunction with a conserved cysteine immediately after the histidine in the active site.[124] This motif has been erroneously defined as a distinct domain termed the “post-SET” domain. However, it merely represents an extension of the SET domain itself that uses metal-chelation to further stabilize the pseudoknot structure.

3. Evolutionary History of Conserved SET Domain Families

Analysis of the histone sequences indicates that the positions corresponding to H3K4, H3K9, H3K27, H3K36, and H4K20 are highly conserved throughout eukaryotes, and are confidently inferred as being lysine even in the LECA (Fig. 2, Supplementary material). Thus, the primary histone sites modified by the SET domain proteins were available in the ancestral eukaryote and are practically universally conserved, unlike the sites modified by the Rossmann fold methylases. All eukaryotic genomes sequenced to date include genes for at least one SET domain protein.[8] A single SET domain is specified by the highly reduced genome of the microsporidian *Encephalitozoon*, suggesting that a minimal histone methylation activity by these enzymes is essential for eukaryotic life. All other eukaryotes specify anywhere between 5 and > 80 SET domain proteins (e.g., the haptophyte alga *Emiliania huxleyii*), suggesting that histone methylation patterns comparable to the model systems are likely to be present across eukaryotes. Analysis of the proteomes of the basal eukaryotes *Giardia* and *Trichomonas* suggests that they contain 5 and 10 SET domain proteins, respectively. Of these at least five distinct versions can be interpreted as being present in the LECA, including members of the trithorax, Ash1/NSD1, Su(var)3-9, SMYD (Skm-Bop2), and TTLL12 clades. This picture is also corroborated by the evidence from another early-branching eukaryote, the heterolobosean *Naegleria*. Thus, enzymes capable of H3K4, H3K9, H3K36, and H4K20 methylation were present in the ancestral eukaryote. The TTLL12, which combines a N-terminal SET domain with a C-terminal ATP-grasp peptide-ligase domain that probably catalyzes polyglutamylation or aspartylation, is the most poorly studied of these ancient lineages[8,118] (Fig. 3). Given that polyglutamylation of histones has been previously observed, it would be most interesting to establish the dual modification catalyzed by this enzyme. These observations indicate a notable diversification of the SET domains both in structure and function even before the LECA.[8]

The nonchromatin protein methylases (i.e., those that methylate ribosomal proteins, translation factors, Rubisco, and cytochrome *C*) form a monophyletic assemblage.[118] Representatives of this assemblage are absent in diplomonads and parabasalids, but are present in most of the other major eukaryotic lineages such as the kinetoplastids, heteroloboseans, stramenopiles, plants, animals, and fungi. This clade is unified by a characteristic large insert within the SET domain and perhaps arose from the ancient SMYD group which has a similarly positioned large insert. Another related group includes a clade of poorly studied SET domains that are found in several animals, fungi, chlorophytes, stramenopiles, and kinetoplastids (the PolySET clade; Supplementary material).[118] The kinetoplastid versions of this clade (e.g., *Leishmania major* protein LmjF35.4550; gi: 72549568) are remarkable in having nine tandem SET domains, that could potentially modify multiple sites on one or more nucleosomes simultaneously. The E(Z) clade is currently detectable in animals, fungi, plants, stramenopiles and heteroloboseans, but is absent in more basal diplomonad and parabasalid lineages (Supplementary material). Two other SET domain clades, which have a similar phyletic spread, are those archtyped by the human protein C5orf35 (gi: 114600191) and *Ustilago maydis* UM04157.1 (gi: 71020147), although the latter clade is more prone to rampant gene loss. The C5orf35 clade also has a distinctive α-helical N-terminal domain and appears to be a particularly widespread clade whose catalytic specificity remains entirely unexplored (Supplementary material). Given the current state of the data, the above clades show a consistent pattern of being absent in the earliest-branching eukaryotes, diplomonads and parabasalids, but being present in the other early-branching clade such as the heteroloboseans or their sister group the kinetoplastids. This suggests that there was a second round of expansion of major SET domain clades after the divergence of the basal-most eukaryotes, but prior to the separation of the heterolobosean-kinetoplastid lineages from the other eukaryotes.[8]

The above observations imply that the ancestral eukaryote had a well-developed histone and chromatin protein methylation system but possibly lacked cytoplasmic SET domain-dependent systems for methylation of translation-related proteins and metabolic enzymes. Another possible corollary to the above observations is that the cognates of the two euchromatic states described as YELLOW and RED and the heterochromatic state GREEN established by the PEV proteins in the *Drosophila* model[107] are likely to have been present in this eukaryotic ancestor. However, the E(Z) clade of H3K27 methylating enzymes appear to have emerged slightly later, suggesting that the cognates of the polycomb PRC2-dependent or the BLUE chromatin state might have been an innovation after the divergence of the diplomonad and parabasalids. However, it cannot be ruled out if some other SET domains take up the role of the E(Z)-like proteins in these organisms (see below). It should also be noted that due to losses of particular SET domain clades in several eukaryotes these chromatin states might have been attenuated or lost in several individual lineages. For instance, in several fungi (e.g., *S. cerevisiae*), we see a loss of E(Z) and Clr4/Su(var)3–9 and also the demethylase LSD1 (see below) which are associated with different heterochromatinic states.[126] Consistent with this, endogenous H3K9 and H3K27 methylation are not observed in *S. cerevisiae*.[127] Another dramatic example of gene loss affecting the SET domains is seen in the genomes of the microsporidians, which are parasitic fungi that reside inside other eukaryotic host cells. Interestingly, unlike the yeasts, the only SET domain protein retained by these organisms is an ortholog of E(Z), which at face value would imply only the presence of H3K27 methylation and a polycomb-dependent repressive chromatin in these organisms. However, it would be worth investigating if this protein has adopted other methylation capabilities in the microsporidians.

Several additional clades of SET domains appear to have arisen relatively late and are restricted to particular eukaryotic lineages. The lineage archetyped by SETD1A and SETD1B, where the SET domain is combined with a RNA-recognition motif (RRM) domain, branched off from the Trithorax clade in the common ancestor of the animals, plants, and fungi (Fig. 3; Supplementary material). Around the same time, the SET2-like lineage appears to have branched off within the Ash1-like clade. Clades that emerged even later include the Su(var)4–20 and MLL5 clade, both of which arose in the common ancestor of animals and fungi. Certain clades emerged coeval with the origin of the animal (SETD8 and the PRD-SET clades) or the plant lineages (ATXR5/6). While the latter clade has been considered to be related to Trithorax,[128] neither the functions nor their phylogenetic affinities support such a specific relationship. Further, the ATXR5/6 clade is also found in the diatom *Thalassiosira* (gi: 220971332) and ciliates, into which it appears to have been transferred from the unicellular plant lineages. This pattern of transfer argues against a recent claim that these methylases were associated with the origin of multicellular land plants.[128] Of the newly emergent clades, the Su(var)4–20 and the SETD8 clade methylate H4K20, Mll5 methylates H3K4, and ATXR5/6 monomethylate H3K27.[15,115,128] The PRD-SET proteins include the distinctive SET domains of RIZ1 that methylates H3K4 and PRDM9 that trimethylates H3K9 positions[98,110,111] (Fig. 2). Concomitant with the emergence of both the animal and the plant lineages, the Su(var)3–9 family appears to have spawned specialized groups with distinctive domain architectures. In animals the G9a/Glp lineage emerged with fusion to distinctive methyllysine binding N-terminal ankyrin repeats.[129,130] In the plants, at least two distinct members of the Su(var)3–9 family emerged that respectively combine the SET domain with the methylated DNA-binding SAD/SRA domain (*Arabidopsis thaliana* SUVH5; gi: 15226918) and a C2H2 Zn-finger domain (*Physcomitrella*; gi: 162666292) (Fig. 3). Some innovations occurred even within the terminal branches of the tree of life. A striking instance of this is the fusion of a Su(var)3–9 clade SET domain with mariner-type transposase domain in the anthropoid primates resulting in the SETmar protein that function as a both an endonuclease and H3K36 methylase.[131,132] This represents a departure from the ancestral state in terms of the methylase target site and might have a role in DNA repair. Indeed, the transposase domain of SETmar appears to function as a potential DNA-repair enzyme that might help in reducing chromosomal translocations.[131,132] Several of these lineage-specific innovations represent alternative means of modifying the same residues that were already modified by the more ancient clades of SET domains. For example, the ATXR5/6 introduces a novel

H3K27me1 modification, which is distinct from the H3K27me3 introduced by the E(Z) clade, and is potentially critical for the establishment of highly condensed constitutive heterochromatin centers known as chromocenters, that are enriched in pericentromeric repeats, transposons, and ribosomal RNA genes.[128]

4. Lineage-Specific Expansions of SET Domains

The core representatives of most of the widespread clades of eukaryotic SET domains, which are associated with the generation of the primary histone marks typical of conserved chromatin states, show a largely vertical pattern of inheritance. However, some representatives of these clades, as well as certain more sporadically distributed groups, display LSEs with four more representatives limited to a particular branch of the eukaryotic tree (Supplementary material). We detail a few of these major LSEs and their possible biological significance. Among the earliest-branching eukaryotes, the parabasalid *Trichomonas* shows an LSE of nine distinctive SET domains and with N-terminal PHD finger (Fig. 3). While these SET domains have not been previously detected and do not group with any other lineage in trees, they show a few sequence features typical of the E(Z) family. Hence, it cannot be ruled out that they possess comparable catalytic activity. Kinetoplastids are remarkable in showing a large LSE of 25 SET domain proteins,[8] which is quite significant as it might explain certain important features of transcription regulation in kinteoplastids.[133] Kinetoplastids have a paucity of conventional TFs, and have large arrays of closely spaced genes that are cotranscribed into a polycistronic transcript by RNA polymerase II.[8,133] The transcription start sites of each of these gene arrays is marked by a specific type of nucleosome that is enriched in the H4K10ac mark, and the variant histones H2A.Z and H2B.V.[133,134] The production of the H4K10ac and the generation of specialized nucleosomes at the transcription start sites appear to depend on H3K4me3 marks. It is possible that the different expanded SET domains of kinetoplastids act as potential transcription regulators that introduce H3K4me3 and other methylated histone marks at specific start sites to activate transcription in a condition-specific manner. Some of these SET domains are also fused to ATP-grasp peptide-ligase domains that probably act as polyglutamylases or polyglycinases (Fig. 3).[8] These modifications might provide further regulatory marks beyond the methylation of histones.

Another extraordinary expansion of SET domains with 14–20 members per haploid genome is observed in the diatoms such as *Thalassiosira, Phaeodactylum*, and *Fragillariopsis* (Fig. 3, Supplementary material). These differ from all other eukaryotic SET domain families in bearing N-terminal signal peptides, implying that they are secreted proteins. Several members of this expansion show fusions to a spermidine synthase domain, which is a member of the Rossmann fold methylase superfamily (Fig. 3).[135] This fusion suggests that in addition to lysine methylation they might catalyze other modifications of the lysine side chain, such as generation of polyamine adducts analogous to spermidine from putrescine. Some of these proteins also contain methylated peptide-binding domains, such as the chromodomain (Fig. 3). These observations help solve a major mystery pertaining to biomineralization of silica by diatoms, which results in their intricately structured shells. Earlier studies have shown that silica biomineralization in diatoms is initiated by a class of lineage-specific secreted low complexity proteins known as silaffins.[136] These proteins contain repetitive sequences with lysines and serines; the side chains of the former are modified by dimethylation and extension through addition of two consecutive aminopropyl units to the ε-amino group, whereas the latter is modified by phosphorylation.[137,138] While the kinase catalyzing the latter reaction was recently identified,[139] the methylating and polyamine-forming enzymes that target most or all the lysines in a given silaffin have not yet been identified. Hence, we predict that this LSE of SET domains is involved in methylation of the silaffins, whereas the spermidine synthase domain is involved in addition of the aminopropyl units. Some of the lysines are also hydroxylated at the δ-position and this modification could be mediated by JOR/JmjC domain hydroxylases or other 2OGFeDOs (see below). These silaffin modifications establish a "code" comparable to the histone code and direct the species-specific morphology of the biomineralized silica.[137] Hence, it is apparent the diversity of the diatom SET domains and the inter-species differences in them might have a significant role in establishing distinct silaffin codes specific to each diatom. The expansion of the PRD-SET domains in the animal-lineage might be related to their recruitment for methylation of H3K9 at meiotic recombination initiation hotspots.[110,111] Several animal versions show over five paralogs of this family (e.g., 15 in *Homo sapiens* and 10 in the basal animal *Nematostella*). Previous studies have suggested that these hotspots are prone to loss due to mutant versions sweeping populations, through gene conversion known to occur in these regions.[140] The recruitment of alternative PRD-SET proteins from the LSEs, which widely differ in the DNA-recognizing C2H2 ZnF domains, could result in proteins which recognize new sequences thereby allowing emergence of new hotspots. Thus, the PRD-SET LSEs might provide potential backup against hotspot loss. This is in line with the observed positive selection of PRDM9 Zn-fingers, which allows rapid evolution of new recombination hotspots. Similarly, it is possible that other major expansions of SET domains such as one in the oomycetes (18–31 paralogs) or *Anopheles* (about 10 paralogs) provide diversity to recognize alternative recombination sites or mechanisms to silence diverging sequences associated with selfish elements (Supplementary material).

5. Prokaryotic SET Domains

The first prokaryotic SET domain was identified in the intracellular parasite *Chlamydia trachomatis*, which lead to the suggestion that it might have acquired it through lateral transfer from the host, just as with several other genes in this organism.[141] There is conflicting, but not mutually exclusive, evidence regarding the role of chlamydial SET proteins, with certain studies suggesting that it is secreted into host cells to modify histones in the nucleus[142] and others implicating it in the modification of the chlamydial chromatin proteins HC1/HC2.[143] Subsequently, the SET domain was recovered in organisms from all major prokaryotic lineages, which include parasitic bacteria, as well as free-living bacteria and methanogenic archaea[118] (Supplementary material). Studies on the archaeal SET domain suggest that it methylates the archaeal chromosome packaging protein MC1.[144] Analysis of phyletic profiles indicates that the SET domain often co-occurs in several parasitic as well as free-living bacteria with a protein containing the SWIB domain.[141] In certain bacteria, SWIB and SET domain genes are even in the same operon. Additionally, the SWIB domain is fused to topoisomerase-I in organisms such as *Chlamydia*, and topoisomerase-III in several β-proteobacteria (Fig. 3).[141] These observations suggest that at least a subset of the bacterial SET domains are likely to constitute a chromatin remodeling and modifying complex comprised of the SET, SWIB, and a topoisomerase protein. In eukaryotes, closely related SWIB domain proteins (e.g., human BAF60a/b) are part of a chromatin-remodeling coactivator complex, BAF, which they recruit to several unrelated TFs.[145–147] It is likely that the bacterial SWIB domains similarly play a role in recruiting this potential chromatin-modifying complex to specific chromosomal sites. However, some other bacterial SET proteins, such as CT0952 from *Chlorobium tepidum* (gi: 21673780) are fused to transmembrane segments, suggesting that they might modify membrane proteins.

While the initial hypothesis posited transfers of SET domains from eukaryotes to bacteria, their current widespread distribution in bacteria points in a different direction.[118] Like several other chromatin-related protein domains of eukaryotes, which have been acquired from bacteria, it is quite likely that the SET domain originally emerged in bacteria from the SAF carbohydrate-binding domain superfamily, which is greatly elaborated in bacteria (Fig. 1). Given that different lines of evidence support a chromatin-related role for at least a subset of the bacterial versions, it is probable that the SET domain had settled into such a role prior to its transfer to eukaryotes. Given the small size of the domain, with relatively few phylogenetically informative positions, it remains unclear if the bacterial versions were transferred on a single occasion or on multiple independent occasions to eukaryotes. Irrespective of this, it appears that at least one such transfer occurred prior to the LECA. It is possible that the five versions inferred as being present in the LECA emerged from the proliferation and diversification of a single bacterial progenitor. However, some of the later emergent clades were perhaps seeded by independent transfers from bacteria. Additionally, certain eukaryotic DNA viruses of nucleocytoplasmic large DNA virus clade (e.g., PBCV) have acquired SET domains that could have been transferred either from the host or bacterial endosymbionts that cohabit the same host cells.[148,149]

III. Enzymatic Mechanisms That Preempt or Reverse the Action of Protein Methylases in Chromatin

Protein methylation in chromatin proteins, both by Rossmann fold and SET domain methylases, is modulated by catalytic activities that either preempt or reverse the methyl marks.[15] The primary preemptive mechanism that has been characterized is citrullination of histone arginines. Demethylation affects both methylated arginines and lysines, and represents an important regulatory mechanism.

A. Citrullination and Arginine Deiminases

Arginine deiminases use water to catalyze the release of ammonia from the arginine guanidino group, yielding citrulline.[19] This reaction occurs in various organisms both on free arginines, such as in the urea cycle or during the production of nitric oxide by the nitric oxide synthase, as well as in polypeptides during protein modification.[18,19] The importance of this modification in proteins first became apparent with the identification of antibodies against citrullinated peptides in rheumatoid arthritis. Subsequently, it was shown to be a modification that occurs on multiple chromatin proteins such as histones and the histone chaperone nucleophosmin.[150] Follow-up studies established that citrullination was an important histone modification, that antagonizes arginine methylation by PRMT family methylases, by preemptive removal of the methyl-accepting sites.[151,152] All histone citrullination sites characterized to date correspond to arginines that are modified by the PRMT family methylases, namely H2AR3, H3R2, H3R8, H3R17, H3R26, and H4R3. The enzymes catalyzing arginine deimination in proteins are the peptidylarginine deiminases (PADIs) and, of the five paralogous PADIs characterized in humans, PADI4 has been identified as the primary histone-citrullinating enzyme.[19,153] Consistent with its role in countering the action of the PRMTs on histones, PADI4 acts as a transcription corepressor. PADI4 also citrullinates arginines in the adaptor module of the histone acetyltransferase p300.[154] This prevents the methylation of this module by CARM1, thereby blocking its interaction with

TFs via bridging proteins such as GRIP1, and providing a second means for PADI4 to act as a corepressor. There have been reports that PADI4 can also act on arginines already methylated by the PRMT family enzymes, catalyzing demethylimination.[151] However, biochemical characterization of the purified enzyme suggests that the demethylimination occurs at a very low rate and is hardly comparable to the rates of methylation by the PRMT enzymes.[155] In contrast, the *in vivo* studies supporting such an activity do not directly assay enzyme activity; instead they merely demonstrate that with increasing histone citrullination there is a fall in the level of asymmetric dimethylarginine on histones.[151,152] Hence, it appears likely that the demethylimination is not a physiologically relevant activity, but that histone methylation is precluded by deimination. The role of citrullination of other chromatin proteins still remains poorly understood, and it remains unclear whether there is catalytic restoration of arginine from citrulline as observed in the urea cycle.

The catalytic domain of the PADIs contains the unique pentein fold, which is comprised of five subdomains, each with a β–β–α–β secondary structure progression, arranged in a pentamerically symmetric fashion around a central aperture.[156] This fold is found in a superfamily of enzymes that operate on the guanidino group of arginine and its derivatives such as the *Porphyromonas*-type PADI, arginine deiminase (which acts on free arginine), succinylarginine dihydrolase, dimethylarginine dimethylaminohydrolase, L-arginine: glycine amidinotransferase and L-arginine: inosamine-phosphate amidinotransferase.[150,156,157] Additionally, a potentially inactive version is found in the archaeo-eukaryotic translation initiation factor eIF6. These observations suggest that the PADIs arose within this vast radiation of arginine metabolism enzymes, by acquiring specificity for peptidylarginine as against its free version. The PADIs are distinguished from all other members of this superfamily by fusion to N-terminal cupredoxin-like and cytochrome F-fold domains.[158] Additionally, their catalytic domain shows a circular permutation such that the first subdomain of the pentein scaffold includes elements from the extreme N- and C-termini of the domains. The active site of the PADIs is derived from three conserved motifs, respectively, associated with the first, second, and third subdomains of the pentein fold.[158,159] In the primary structure, the most N-proximal of these (a WxQD motif) is from the second subdomain, the [HRQ]xD motif is from the third subdomain, and the C-terminal most is an HC motif that belongs to the first subdomain because of the circular permutation in the PADI domain pentein fold. These residues project into the central aperture of the pentein fold and form a pocket to bind arginine.[158] In particular, the conserved aspartates in the first and second motifs are critical for arginine interaction through salt-bridge formation with the guanidino group. Their close approach to this group is another factor that argues against a physiological role for these enzymes in demethyimination, on the basis of steric hindrance. Catalysis is mediated primarily by the conserved cysteine in the third motif, which initiates cleavage of the C–N bond by forming an adduct with the guanidino carbon.[159] The histidines in either the second or the third motifs could be protonated and act as general acids in this reaction releasing ammonia. The subsequent hydrolytic step to generate citrulline could again be catalyzed by one of these histidines, or by the released ammonia acting to deprotonate water and release a hydroxyl anion.

Analysis of phyletic patterns of the PADI family suggests that the classical members are restricted to animals (Supplementary material). Hence, it is likely that chromatin protein citrullination by these enzyme is a late innovation that emerged only in the Metazoa. Within animals, the proliferation of PADIs was a late event that occurred in tetrapods, eventually resulting in the five paralogs observed in mammals.[153] Outside of animals, members of the PADI family are seen in bacteria and fungi (Supplementary material). The bacterial versions are secreted proteins specified by predicted operons, together with cell-surface peptidases (Fig. 3). Hence, it appears likely that the bacterial PADIs are part of a system that processes and modifies cell-surface proteins. The fungal and animal versions do not group together in phylogenetic trees (Supplementary material). Instead, each of them groups with distinct bacterial versions, suggesting that they were independently transferred from bacteria to the two eukaryotic lineages. Furthermore, the fungal versions are also secreted like the bacterial forms and are unlikely to modify chromatin proteins. It should be noted that the PADIs are part of a larger radiation of pentein fold enzymes (listed above) that occurred in bacteria. Some of these enzymes have independently acquired PADI activity (e.g., *Porphyromonas*-type PADI),[157] and might have been acquired by other eukaryotic lineages through independent transfers. Indeed, in the basal eukaryote, *Giardia*, there is evidence that the metabolic arginine deiminase plays a second role in citrullination of peptides in the cytoplasmic tails of variant surface proteins of this organism.[160] Hence, it cannot be ruled out that there are independent innovations of chromatin protein citrullination in some of the other eukaryotic lineages.

B. Chromatin Protein Demethylation

The demethylation of histones and arginines in chromatin proteins is currently known to be catalyzed by enzymes belonging to two distinct folds that use very different oxidative mechanisms to remove the methyl group. These are the LSD1-like demethylases with a flavin adenosine dinucleotide-binding Rossmann fold, and the jumonji-like demethylases with a 2-oxoglutarate-binding double-stranded β-helix fold (Fig. 4).

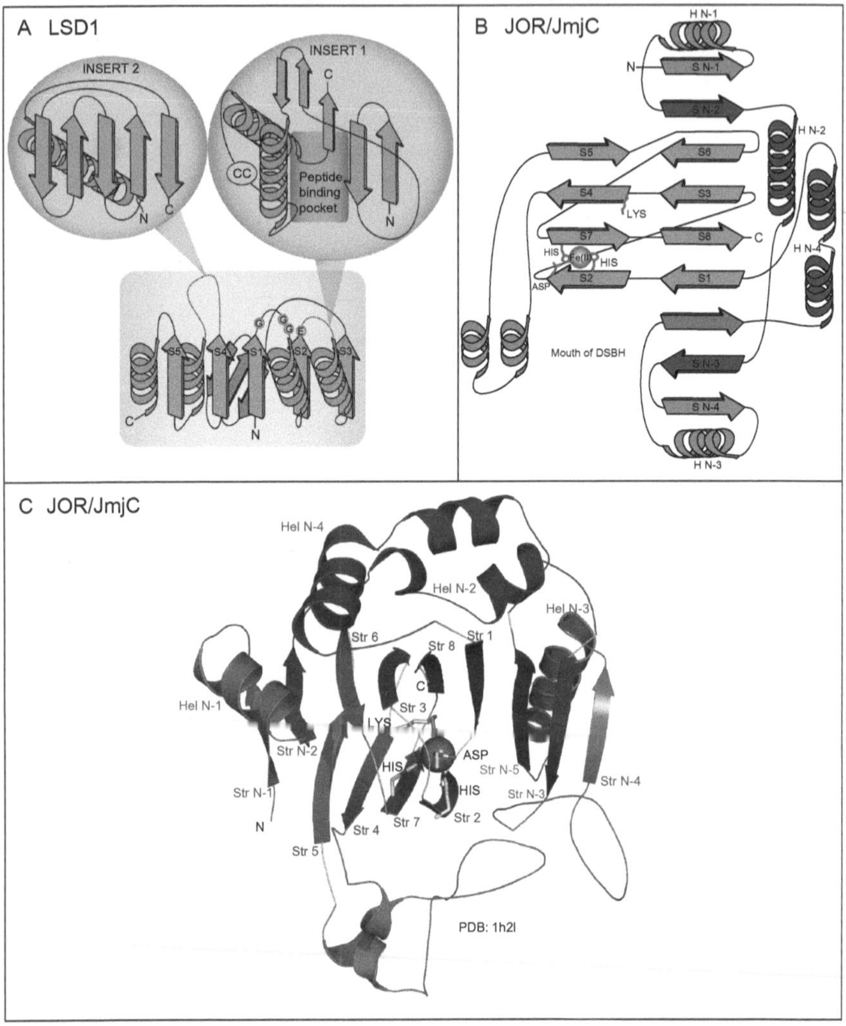

FIG. 4. Structure and sequence features of protein demethylases. (A) Topology diagrams of the LSD1 demethylase. The core secondary structure elements of the Rossmann fold are colored as in Fig. 1. The distinct substrate and flavin-binding domain (insert 1), and the FAD-binding cap domain (insert 2) are shown separately. CC refers to the coiled-coil motif found in insert 1. Shown in (B) and (C) are the topology and cartoon structure diagrams of the JOR/JmjC domain. The coloring scheme illustrates various shared features of the JOR/JmjC superfamily with related DSBH domains. Thus, the conserved strands of the DSBH core are colored green, and the additional strand shared with the metal-binding cupin is colored blue. Secondary structure elements additionally shared with all 2-oxoglutarate Fe^{2+}-dependent dioxygenases are colored pink. Shown in orange are structural elements and inserts that are only present in the JOR/JmjC superfamily. The core of the DSBH unit is labeled strands 1–8. Secondary structure elements N-terminal to the DSBH unit are labeled with a prefix "N" (e.g., Str N-5). The histidine and aspartate residues coordinate a Fe(II), whereas the lysine residue forms a salt bridge with the 2-oxoglutarate. (See Color Insert.)

C. LSD1-Like Demethylases

1. Specificity and Functions of the LSD1 Demethylases

The LSD1-like proteins were first identified as components of the corepressor complex, coREST, and subsequently predicted to function as chromatin-protein-modifying enzymes on the basis of their combining a Rossmann fold oxidoreductase domain to the SWIRM domain that is found in several distinct chromatin-associated proteins.[161,162] Subsequent studies demonstrated that LSD1-like enzymes are protein demethylases that specifically target mono- and dimethylated lysines.[163,164] They catalyze a distinctive reaction using the FAD cofactor, along with molecular oxygen and water, to oxidize the methyl groups and release formaldehyde and hydrogen peroxide.[165,166] This release of hydrogen peroxide might be one of the factors that accounts for the genetic evidence for superoxide dismutase having key chromatin-related functions. Consistent with the association of LSD1 with the CoREST complex, it demethylates H3K4me1 and H3K4me2, and thereby fosters negative regulation of gene expression.[164,167–170] This is also in line with the observation that the *Drosophila* PEV gene Su(var)3–3 is an LSD1 demethylase that acts on methylated H3K4, and that the *S. pombe* LSD1 demethylases strongly localize to heterochromatin.[167,171] However, in mammalian systems LSD1 also functions as a coactivator in conjunction with the sex hormone receptors and some other TFs, such as the POU-homeodomain protein PIT1.[169] This is also in line with the observation that LSD1-like proteins in *S. pombe* localize to start sites of certain active genes.[171] In mammals, the activating function has been attributed to a specific H3K9me2-demethylating activity of LSD1 when combined with these TFs.[163,166] However, *in vitro* studies do not currently provide clear evidence for H3K9me2 demethylation.[163,166] In addition to histones, LSD1 also demethylates methylated lysines on p53 and the DNA methylase DNMT1.[163,172] LSD1 is part of a large complex scaffolded by a noncoding RNA (HOTAIR), whose 5′ end binds the PRC2 polycomb complex with E(Z) while the 3′ end binds coREST with LSD1.[168] This assembly links demethylation of H3K4me2 by LSD1 with the methylation of H3K27 by E(Z) to establish repressive chromatin. Studies in *S. pombe* also suggest that LSD1-might be critical for establishing boundaries between euchromatin and heterochromatin, whereas studies in *C. elegans* have implicated it in erasure of epigenetic marks in the germline.[171,173,174]

2. Structural Features and Active Site of the LSD1-Demethylases

The LSD1 catalytic domain contains a Rossmann fold domain that is typical of the vast superfamily of NAD(P)/FAD-dependent oxidoreductases (Fig. 4).[161] Within this superfamily, LSD1 belongs to a specific clade of FAD-binding versions defined by peculiar inserts not found in those oxidoreductases having the standard form of the Rossmann fold. Most representatives of this clade catalyze the FAD-dependent oxidation of low-molecular weight substrates such as polyamines, L-amino acids, monoamines, porphyrinoids, and metabolites such as glycerol-3-phosphate.[161] The core β-sheet of the Rossmann fold of these proteins often has five parallel strands, the first of which is part of the β–α unit, bracketing the glycine-rich loop that forms the primary binding site for the phosphate backbone of FAD.[175–177] The C-terminus of the second strand contains a conserved acidic residue that forms hydrogen bonds with the hydroxyl groups of the sugar of FAD, just as the equivalent residue in the Rossmann methyltransferases interacts with the sugar in AdoMet (Fig. 4). This is followed by a glycine-rich patch that makes additional contacts with the nucleotide phosphate. The group of FAD-dependent oxidoreductases to which LSD1 belongs contains a large insert after the second strand, the core of which is 3-stranded sheet that binds the flavin moiety of FAD. This element also bears variable decorations that form the primary substrate-binding sites of these enzymes. In LSD1 it is greatly elaborated through insertion of a long coiled-coil element that sticks out of the core catalytic domain.[175–177] The fourth strand of the core sheet is placed adjacent to the first strand after a topological crossover typical of the Rossmann fold. However, the crossover element leading to it is rather different from the standard Rossmann fold in comprising of a β-meander of three strands (Fig. 4). Between the fourth strand and the helix leading to the last strand of the core β-sheet, the above clade of enzymes contains another large insert in the form of the "cap" domain that adopts an α+β fold with a 5-stranded sheet stacked against a single helix. This domain cooperates with the insert after strand-2 to form a cap over the active site to bind FAD.[175–177]

The generation of both hydrogen peroxide and formaldehyde in LSD1-catalyzed demethylation points to a two step reaction.[165,178] The first step involves FAD-dependent abstraction of hydrogen and electrons from the N–CH_3 of methylated lysine, resulting in an imine linkage ($N^+{=}CH_2$). Molecular oxygen is used as the terminal acceptor and is reduced to hydrogen peroxide. The imine linkage is then hydrolyzed by water to release formaldehyde and remove the methyl group. The peptide-binding site of LSD1 differs from those of all other peptide demethylases in that it makes extensive contacts with the tail of H3, which is buried deep within a pocket in the enzyme.[175–177] This pocket is primarily formed by the specific decorations in the insert after strand-2, supplemented by few positions from the cap domain and the C-terminal part of the Rossmann fold. Together, they make several contacts with specific side chains of the H3 tail, encompassing the whole region from residues 1–13, with the methylated H3K4 being precisely directed toward the flavin moiety of FAD.[175–177] In this configuration, H3K9 does not lie in the vicinity of the flavin moiety, raising questions about the H3K9me2 demethylation activity of LSD1. Given the extensive specific contacts, it is not clear how H3K9me2 could be accommodated in the same pocket even if it were indeed a genuine target.

3. Evolutionary History of LSD1 Demethylases

The LSD1 demethylases are found in all major eukaryotic lineages, with the exception of the basal eukaryotes *Giardia* and *Trichomonas*.[8] It is also absent in trypanosomes, but present in the sister lineage, the heteroloboseans (*Naegleria*), suggesting that it may have emerged relatively early in eukaryotic evolution, after the divergence of the parabasalid and diplomonad lineages from the remaining eukaryotes. However, there are several instances of loss of LSD1 within the remaining eukaryotes. A striking example is its loss in saccharomycete yeasts, such as *S. cerevisiae*, along with several other chromatin-associated proteins such as an HP1 homolog, E(Z)-like SET domains and a jumonji-related demethylase.[161] Hence, in these fungi there could be a general reduction of the heterochromatic states comparable to the GREEN and BLUE types observed in *Drososphila*.[107] Within eukaryotes, LSD1 proteins show some diversity in terms of domain architectures (Fig. 3). The classical LSD1 domains are fused to the N-terminal SWIRM domain which adopts the HTH fold. Additionally, in vertebrates a derived version of the PHD finger (PHD-X or ZF-CW), that binds methylated peptides, is found N-terminal to the SWIRM domain. LSD1 also shows fusions to a methylated peptide-binding C-terminal PHD finger in apicomplexan domains. LSD1 also shows multiple fusions to different DNA-binding domains: in fungi it is fused to a C-terminal DNA-binding HMG domain, while in the chlorophyte algae *Micromonas* and *Chlorella* it is fused, respectively, to an N-terminal BRIGHT (ARID) or VP1 (B3) domain in place of the SWIRM domain (Fig. 3, Supplementary material). These fusions might recruit LSD1 to specific DNA sequences or structures in these organisms. In land plants there is a small LSE of LSD1 paralogs (*Arabidopsis* ATPAO2, ATPAO3, and ATPAO4) that lack any N-terminal domains typical of chromatin proteins.

In eukaryotes, the closest relatives of LSD1 is a clade of proteins, typified by spermine oxidase (SMOX) and the peroxisomal polyamine oxide (PAOX), which is found in plants, animals, and fungi. These enzymes are major players in the catabolism of the polyamines spermidine and spermine through an oxidative process that appears to be active in the peroxisomes.[179,180] SMOX acts directly on spermine to oxidize it to 3-aminopropanal and spermidine. In an alternative pathway of spermine catabolism, one of the free NH_2 groups is first acetylated by an acetyltransferase of the GNAT superfamily, followed by oxidation by PAOX to release spermidine and 3-acetamidopropanal.[179,180] Both of these reactions are very similar to the lysine demethylation reaction and release hydrogen peroxide as a byproduct. Outside of eukaryotes, orthologs of LSD1, SMOX, and PAOX are found in several bacterial lineages such as proteobacteria, chloroflexi, and actinobacteria (Supplementary material). These are further nested within a large assemblage of prokaryotic L-amino acid and amine oxidases, suggesting that the precursors of the LSD1/SMOX/PAOX subgroup emerged as part of an adaptive radiation of these enzymes in bacteria. The bacterial orthologs of the LSD1/SMOX/PAOX subgroup are often specified by conserved operons that also contain adjacent genes for acetyltransferases of the GNAT superfamily (Fig. 3). This suggests that they may function in the oxidative metabolism of polyamines like spermine, similar to PAOX. Some others are combined in operons with Fem/MurM-like peptide ligases of the GNAT superfamily, and might represent alternative mechanisms for polyamine utilization through peptide ligation.[181] More divergent bacterial homologs of LSD1/SMOX/PAOX subgroup are in operons with genes for jumonji-related dioxygenases, and are predicted to participate in modification of nonribosomally synthesized peptides (Fig. 3).[182] Thus, it appears that the closely related activities of polyamine catabolism and peptide modification were already prevalent in these proteins even in the bacteria. Upon acquisition by eukaryotes, one of the lineages appears to have specialized in peroxisomal polyamine catabolism, whereas the second one acquired fusions to chromatin-associated proteins domains and was recruited as a histone demethylase. In light of the close relationship between these activities, it remains to be seen if LSD1 might have a role in oxidative modification of polyamines, which have been anecdotally reported to play a role in chromatin structure.[183,184]

D. Jumonji-Related Demethylases

1. Specificity and Functions of Jumonji-Related Demethylases

The *J*um*o*nji-*r*elated (JOR) or JmjC enzymes were first identified as potential chromatin-modifying enzymes using comparative genomics and sequence analysis, in a study on gene loss in *S. cerevisiae.*[126] Subsequent studies showed that these domains contained a double-stranded β-helix fold (DSBH), similar to the classical 2-oxoglutarate-dependent dioxygenases (2OGFeDO) and metal-dependent cupins, and suggested that they are likely to modify chromatin proteins via a dioxygenase reaction.[161,185,186] Eventually, biochemical studies confirmed the computational prediction that these domains were indeed dioxygenases and established that they were peptide demethylases.[187–189] JOR/JmjC catalyze demethylation of both lysines and arginines at almost all of the currently studied positions in histones, as well as carrying out hydroxylation of asparagine in proteins such as the hypoxia-induced factor-α and of the modified base wybutosine in tRNA.[20,182,189–192] Unlike the LSD1-demethylases, these enzymes can demethylate mono-, di-, and tri-methylated lysines and also the different forms of methylated arginines.[189,190,193] Hence, these enzymes are by far the most prevalent demethylases active in eukaryotic chromatin.

Members of this family, such as JARID1A/B/C/D, and *Little imaginal discs* (Lid), demethylate H3K4, the mono- and dimethylated forms of which are also targets of the LSD1-demethylase[194–199] (Fig. 2). However, enzymes such as Lid, might compete with LSD1, resulting in antagonistic rather than cooperative effects.[200] Demethylases targeting the H3K9 position include the JMDJ2A/B/C/D, JHDM2a/b, PHF8/JHDM1F proteins, and JHDM1d (KIAA1718) in humans.[193,201] JMJD1a is distinctive in that it only demethylates H3K9me1 and H3K9me2 but not H3K9me3.[193,202] These enzymes play a critical role in maintaining the appropriate levels of repressive chromatin established by H3K9 methylation. Mutations in one of these demethylases, PHF8, are associated with X-linked mental retardation and cleft lip/cleft palate.[203] Demethylases that counter H3K27 methylation are typified by human UTX, JMJD3, and JHDM1d (KIAA1718)[201,204,205]; these play a central role in modulating the action of the polycomb complexes and other histone modifications, such as H2AK120 ubiquitination by Bmi1 or *Posterior sex combs*. Removal of H3K36 methyl marks is catalyzed by some of the same enzymes as those involved in demethylation at the H3K9 position, such as human JMJD2A/B/C and *S. cerevisiae* Rph1.[127,206,207] Additionally, JMJD5 is also believed to demethylate H3K36me2 and regulates cell-cycle progression,[207] but lacks the features that are typical of histone demethylases[182] (see below). Similarly, the demethylases targeting the H4K20 position overlap with those involved in H3K9 demethylation (e.g., PHF8).[201,208] Coordinating the action of these demethylases with multiple alternative substrate sites might be an important regulatory theme that occurs across eukaryotes.

The JMJD6 proteins had been reported to demethylate methylarginines on histones, and thereby potentially counter the effects of the PRMT methylases.[190] However, other studies have suggested that they are not methylarginine demethylases, but lysine-5 hydroxylases that generate hydroxylysine in RNA-associated proteins such as U2AF65.[20] Domain architectures indicate that certain members of the JMJD6 family are fused to the methylated peptide-binding chromodomain (Figs. 3 and 5). These domains are usually not found in spliceosomal RNPs,[10] suggesting that at least a subset of the JMJD6 family might indeed modify histones. Alternatively, they possess a C-terminal helix-turn-helix domain which might bind single-stranded RNA, thereby enabling their association with spliceosomal complexes.[20] Hence, it remains unclear if JMJD6 is the primary demethylase of methylated arginines in histones. The demethylases, if any, which operate on H2BK5 and H3K79 remain as yet unknown. They might be recruited from some of the families of JOR/JmjC domains whose substrates remain to be determined. Unlike JMJD6 enzymes,

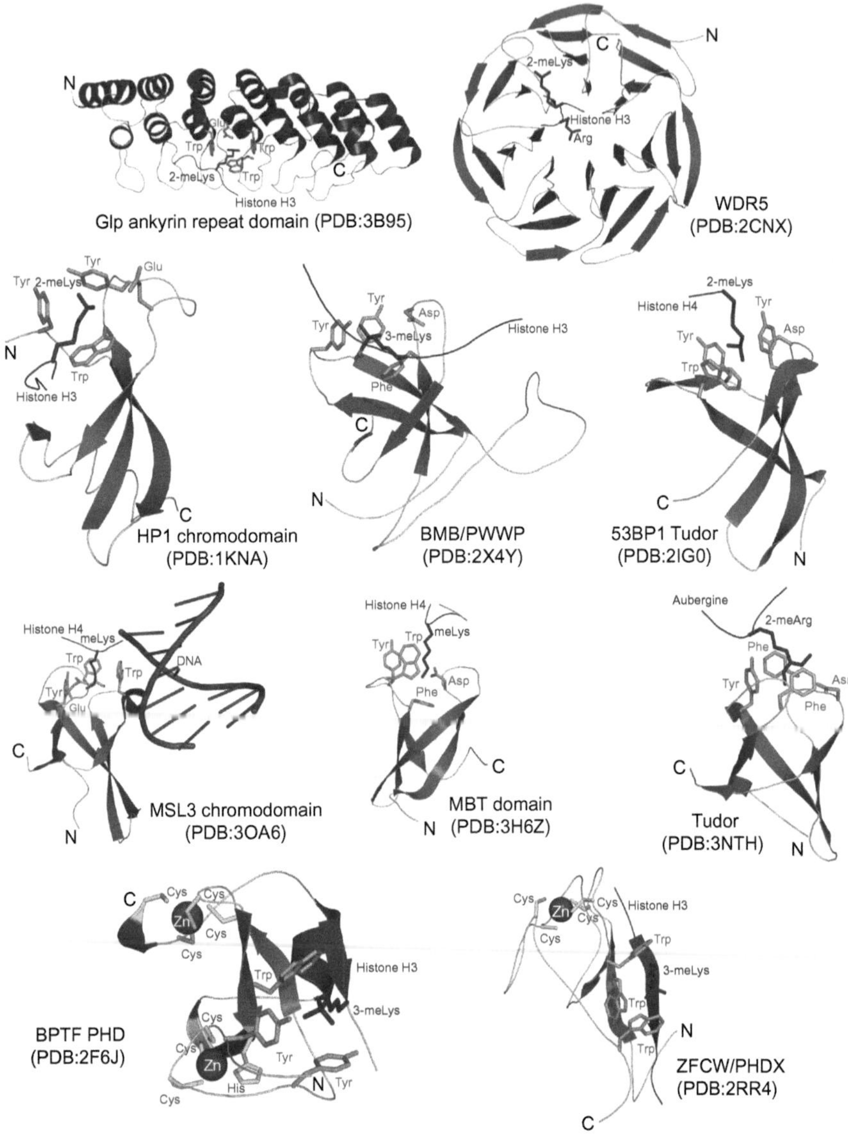

FIG. 5. Structures of various protein sidechain-methyl binding domains. The structures depict various protein side chain methyl binding domains, their binding modes and substrate contacting residues. Strands are colored green, and helices are shown in brick-red. Substrate polypeptides and amino acid side chains are colored blue and the binding residues are shown as stick models in orange. (See Color Insert.)

the JOR/JmjC domain of FIH1 catalyzes the hydroxylation of chromatin proteins such as HIF, whereas the related wybutosine (yW) hydroxylases modify tRNA.[182,191,192] The MINA/No66-type JOR/JmjC domains have been proposed to function as H3K9me3 demethylases.[209,210] However, it is debatable if this is their primary activity. First, these proteins lack the features typical of the classical histone demethylases, and are phylogenetically distant from them.[182] Second, they localize to the nucleolus, associate with ribonucleoprotein complexes related to ribosomal assembly, and cofractionate with ribosomal particles.[211] Hence, they might have a primary role in demethylation of ribosomal proteins or modifying RNA.

2. The Structure and Action of JOR/JmjC Domains

JOR/JmjC domains belong to a specific clade containing the DSBH fold, which also includes the cNMP-binding domains, domains traditionally labeled as the cupins, and classical 2OGFeDOs.[182,212] Within the above clade, the domains traditionally considered cupins, the classical 2OGFeDO superfamily, and the JOR/JmjC-like superfamily are further unified into a monophyletic clade by the presence of highly conserved histidines in strand-2 and strand-7 of the eight-stranded core of the DSBH.[182] Of these the metal-binding cupins, the classical 2OGFeDOs and the JOR/JmjC share a further modification in the ancestral active site comprising the two histidines—they possess a conserved residue downstream of the first ancestral conserved histidine, resulting in a signature of the form HXH or HXD (where X is any amino acid; Fig. 4). This modification converts the active site into a metal-chelating configuration, and the chelated ion (usually iron, zinc, nickel, or manganese) is critical for binding oxygen and catalysis of the dioxygenase reaction.[182] They also share a conserved strand that is antiparallel to strand-1 of the core DSBH. Members of the classical 2OGFeDO superfamily and JOR/JmjC superfamily further share several unique features that distinguish them from the metal-binding cupins.[182] One of these is the “straddling region” that crosses over from one end of the DSBH to the other, and is either a single long helix or a helical region interrupted by multiple loop-like regions (Fig. 4). This region is also prone to independent insertions of distinct domains in members of both these superfamilies, such as the insertion of a cysteine-rich metal-chelating cluster in the metazoan TET proteins (classical 2OGFeDO superfamily) and the Bright and PHD finger domains in the Jumonji/SMCX/JARID1 proteins (JOR/JmjC superfamily).[182] These superfamilies are also united by their dependence on both 2-oxoglutarate and iron for catalysis. The JOR/JmjC domains are further distinguished from the classical 2OGFeDO domains by several sequence and structure features. One main structural distinction is an N-terminal strand–helix extension, which in combination with the following strand has been erroneously defined as a distinct domain called the JmjN domain in several representatives such as Jumonji, JARID1, SMCX, and JMJD2.[182] However, it has no independent existence and simply represents a structural extension of the DSBH core. Another distinction is a helical hairpin with an extended connector loop that is inserted between strands 4 and 5 of the core DSBH fold and packs against the side of the core barrel (Fig. 4). This loop might bear additional inserts, such as a metal-cluster in the JMJD2-type proteins. In most JOR/JmjC families, the 2-oxoglutarate is kept in place in the cavity of the DSBH fold by a conserved basic residue in strand 4 which forms a salt bridge with the cofactor[182] (Fig. 4). In some JOR/JmjC families (namely CBF and Jmjd3), this basic residue has been substituted by lineage-specific conserved residues that are predicted to play a similar role. In contrast, the classical 2OGFeDO superfamily has a highly conserved arginine on strand 8, which performs an equivalent role.[21,182]

The 2-oxoglutarate cofactor enables the JOR/JmjC domain to distribute the two oxygens that catalyze the dioxygenase reaction. Unlike the metal-dependent cupin dioxygenases that insert both the oxygen atoms from molecular oxygen into the substrate, the JOR/JmjC enzymes insert one oxygen atom into 2-oxoglutarate, converting it to succinate+CO_2, and the second one into the demethylation substrate via hydroxylation. Hydroxylation of the N-linked methyl groups in methylated lysines and arginines results in the release of methyl groups as formaldehyde.[182] On the other hand, if the hydroxylation occurs on a carbon atom in the aliphatic side chain of asparagine, lysine, or the base wybutosine in tRNA, it results in a stable hydroxylated form of substrate.[20,182,191,192] Within the DSBH fold, the metal ion is situated close to strand-1 of the core near the mouth of the internal cavity of the domain (Fig. 4). Thus, the substrate accesses the active site via the mouth, which is bounded by the key inserts that differentiate the JOR/JmjC from other DSBH fold domains.[182] Hence, inserts appear to be central to the specificity of these enzymes. Further, the configuration of the approach to the active site primarily engages the modified amino acid side chain, and does not accommodate a large portion of the modified peptide (Fig. 4). This feature has allowed versions of JOR/JmjC to be adapted to demethylate most of the methylated lysines and arginines, unlike LSD1 which has a pocket rather specific for H3K4 due to engagement of multiple residues in the modified peptide.[175–177]

3. Evolutionary History of the JOR/JmjC Proteins

Unlike the SET domain methylases, JOR/JmjC domains are absent from the earliest-branching eukaryotes such as *Giardia* and *Trichomonas*, and also from parasites such as the microsporidian *Encephalitozoon* and *Entamoeba*.[8] Further, kinetoplastids, which are predicted to possess a robust histone methylation system, have representatives of only four major JOR/JmjC clades[182]

(Supplementary material). Of these, the wybutosine hydroxylases, JMJD6 and MINA/No66 clades, either do not (or have, at best, been dubiously proposed to) demethylate residues on histones. This leaves only the JMJD4 clade that could potentially function as a histone demethylase in kinetoplastids, although this activity has not been reported to date among proteins of this clade. JMJD4 also does not display any domain architectures (see below) typical of chromatin-modifying JOR/JmjC domains.[182] The same organisms also lack LSD1-like demethylases, suggesting that conventional chromatin protein demethylases are entirely absent.[8] Hence, in contrast to chromatin protein methylation, demethylation is probably not absolutely necessary for eukaryotic life. In particular, parasitic organisms may not tightly regulate the dynamic balance of different methylated and unmethylated states of histones. The complete absence of demethylases in diplomonads and parabasalids suggests that resetting of methyl marks on histones was not a major feature of the ancestral eukaryote. Of the early-branching eukaryotes a well-developed histone demethylation system is found only in the heterolobosean *Naegleria*,[8] implying that eukaryotes probably first acquired histone demethylation prior to the separation of this lineage from the remaining eukaryotes but after the diplomonads and parabasalids had diverged.

A comprehensive phylogenetic analysis of the JOR/JmjC domains revealed the presence of at least 17 distinct lineages in eukaryotes, divided into three major clades: histone demethylase-like, FIH1/yW-hydroxylase-like, and MINA/No66-like.[8,182] The first of these clades includes all well-established histone demethylases in addition to several poorly studied lineages: JMJD6, JMJD4, *Dictyostelium* CBF-like, Jumonji/JARID1, JMJD1/hairless, JMJD2, JMJD3, FBXL10/11, PHF2/8, JMJD8, and *Drosophila* CG2211-like. The FIH1/yW-hydroxylase-like clade includes the FIH1, yW-hydroxylase, JMJD5, JMJD7, HSPBAP1, and *S. pombe* Jmj4-like lineages. The MINA/No66-like is a unitary clade with no further differentiation of lineages within it. Heteroloboseans show at least nine of these lineages: JMDJ6, *Dictyostelium* CBF-like, Jumonji/JARID1, JMJD2, yW-hydroxylase, FIH1, JMJD5, JMJD7, and MINA/No66[182] (Supplementary material). Hence, a biochemically diverse group of JOR/JmjC proteins, that modify RNA, histones and other peptides, had emerged prior to the divergence of the heteroloboseans. Of the histone demethylation events, H3K4 and H3K9 demethylation can be inferred as having been the first to emerge in eukaryotes. Interestingly, JMJD3 acts on the H3K27 position and is found only in animals, choanoflagellates, and ciliates. This phyletic pattern does not match that of the corresponding histone methyltransferase E(Z), which is absent from ciliates but present in plants and fungi (which lack JMJD3). This suggests that there could be diversification of JMJD3 specificity in ciliates, and that other JOR/JmjC proteins might perform the role of methylated H3K27 demethylation in the lineages in which E(Z) is present.

A notable innovation of multiple lineages within the histone demethylase-like clade appears to have occurred in the common ancestor of animals, fungi and plants, resulting in the lineages JMJD1/hairless, PHF2/8, and JMJD8 (containing an HxH motif instead of the usual HxD in the active site) (Supplementary material). The PHF2/8 lineage further diversified at the time of the origin of animals, giving rise to the FBXL10/11 lineage, where the JOR/JmjC domain is fused to a C-terminal F-box domain (Fig. 3). This suggests that it is likely to recruit an ubiquitinating SCP-type E3 complex, whose targets remain poorly understood. A comparable fusion to F-box domains is also seen in certain members of the JMJD6 lineage, although here the fusion is at the N-terminus. Hence, the coupling of hydroxylation or demethylation with ubiquitination might be widely distributed in eukaryotes. In vertebrates, the PHF2/8 triplicated giving rise to the PHF2, PHF8, and JHDM1d, whereas the FBXL10/11 clade duplicated to give rise to FBXL10 and FBXL11 (Supplementary material).

Certain versions of the Jumonji/JARID1 lineage, such as Jumonji itself, appear to have lost their catalytic residues.[182] Hence, these forms might function as regulatory proteins that bind the methylated sites on histones, but do not catalyze demethylation. In animals, the *Drosophila* CG2211-like lineage of JOR/JmjC domains arose in the histone demethylase-like clade, and are typified by the presence of an N-terminal signal peptide and seven conserved cysteines flanking and within the catalytic domain (that are likely to form disulfide bridges).[182] Similar secreted forms are also observed in the JMJD5, FIH1, and MINA/No66 lineages, which might be fused to other domains involved in modification of surface proteins (e.g., sulfotransferases) or adhesion (e.g., immunoglobulin domains). These are hence likely to modify cell-surface proteins via hydroxylation (e.g., at lysine or asparagine hydroxylation) rather than demethylate histones. Secreted members of the MINA/No66 lineage in choanoflagellates (e.g., *Monosiga* gi: 167527482) also contain intracellular modules with protein kinase and SH2 domains[182] (Fig. 3), and might also function as sensors of redox potential and consequently transduce intracellular signals.[182] The secreted FIH1 clade members from diatoms (e.g., gi: 220973585 from *Thalassiosira pseudonana*), along with other secreted 2OGFe-DOs such as asparagine-β-hydroxylase family (e.g., gi: 220972489 from *Thalassiosira pseudonana*), and the Syn9-gp54 family hydroxylases (e.g. gi: 223994107 from *T. pseudonana*) are predicted to hydroxylate lysines in silaffins. Unlike representatives of the histone demethylase-like clade, those of the two other major clades of JOR/JmjC (see above) do not show fusions to chromatin-associated protein domains such as PHD fingers, chromo-like domains, or various DNA-binding domains (Fig. 3). This raises questions about their potential role as histone demethylases. Instead they could function as peptide hydroxylases that act on proteins in several distinct subcellular contexts such as

RNA–protein complexes. Members of the *S. pombe* Jmj4-like lineage are fused to a DNAJ domain in certain organisms (Fig. 3), suggesting that they might act in conjunction with Hsp70 to regulate protein stability.

Outside of eukaryotes, JOR/JmjC proteins are found in several bacteria. Phylogenetic analysis suggests that there are bacterial versions corresponding to each of the three major clades of JOR/JmjC domain found in eukaryotes.[182] Thus, the major groups of JOR/JmjC appear to have diversified in bacteria, followed by a transfer of at least one member from each of the three clades to the eukaryotes, prior to the divergence of the heterolobosean-kinetoplastid clade and the remaining eukaryotes.[182] This scenario implies that, after the acquisition from bacteria of ancestral members of the three clades of JOR/JmjC superfamily, they extensively radiated within eukaryotes. The bacterial representatives are coded by gene clusters involved in biosynthesis of several secondary metabolites, such as pyoverdine-like peptide-derived siderophores and possible peptide antibiotics. These gene clusters often specify multiple functionally linked dioxygenases of either the JOR/JmjC superfamily or the 2OGFeDO superfamily, tryptophan halogenase-like oxidoreductases and LSD1-like amino acid oxidases, and other enzymes involved in nonribosomal peptide biosynthesis and modification. Thus, the JOR/JmjC-like enzymes appear to have first emerged in the context of the secondary metabolism of peptide derivatives in bacteria.[182] In this respect, they mirror the PRMT and DOT1-like methylases and the LSD1-type oxidoreductases, but not the SET domains, which show no apparent connections to secondary metabolism in bacteria (see above).

IV. Domains Involved in Discrimination of Methylated Peptides

The discrimination of the methylation states of modified peptides in chromatin proteins is mediated by a number of structurally diverse domains. A comprehensive discussion of the proteins containing these domains is beyond the scope of this work, as it would involve most major groups of chromatin proteins. Hence, in this chapter, we briefly discuss the major structural scaffolds involved in binding modified peptides and discriminating their methylation status. Currently, methylated peptide binding is understood to have evolved independently in three distinct structural classes of domains[129] (Fig. 5). It is quite likely that future studies will uncover more such domains. The structural classes to which these domains belong include: (1) simple globular domains. These are classical folded domains, and chiefly include the chromo-like domains with the SH3 fold and possibly the catalytically inactive version of JOR/JmjC domain (e.g., in the animal Jumonji protein). (2) Metal-chelation supported domains. These are structures that depend primarily on chelation of Zn^{2+} ions for their stability, and include versions of the treble-clef fold typified by the PHD-finger domain and its structural derivatives. (3) Superstructure-forming repeats. These domains are composed of small repeated units that rarely exist by themselves as a globular structure. Instead, multiple repeats form globular units with a periodic structure that serve as peptide-binding sites. Among such structures, versions of the WD40 and ankyrin repeats bind methylated histones. An interesting unifying theme that has been frequently observed in several structurally distinct peptide-binding domains, is the use of a cage of aromatic residues to sense methylated lysine.

A. The Chromo-Like Domains with an SH3-Like Fold

These domains are archetyped by the chromodomain that was originally identified in archetypal chromatin proteins, such as polycomb and HP1. Subsequent sequence and structure analysis revealed that the tudor, BMB (PWWP), MBT, agenet, and BAM/BAH domains, that had been considered distinct, are actually related to the chromodomain and form a monophyletic assemblage.[8,213–215] The traditional classification into the above domains is not entirely reliable, due to the small size and extreme sequence divergence of these chromo-like domains. Through sequence and structure comparisons, it becomes apparent that the classical chromodomains BAM(BAH) and the BMB (PWWP) are well-defined families that retain their own distinctive features. The tudor, agenet, and MBT domains are likely to be unified into a higher order clade within which the boundaries are not clear at all. Hence, all these domains might be referred to simply as tudor-like domains.[216–219] This assemblage of tudor-like domains also includes the domains typified by the SAGA-associated factor 29, the "domains of unknown function" DUF3590 and DUF1325 defined by the PFAM database, the *S. cerevisiae* RAD9 RAD53BP domain and the so-called "tudor-knot" domain found in the histone acetyltransferase Esa1.[217,218,220] A previously undetected member of the tudor-like assemblage is found C-terminal to the Hsp90-like ATPase module in the eukaryotic Morc proteins.[221] Sequence profile analysis revealed that a conserved domain typified by the central conserved region in the plant auxin-response TF, the Aux-RF domain (PFAM: PF06507) is yet another previously unknown chromo-like domain which is distantly related to the tudor-like assemblage (Supplementary material). All these domains are unified by the presence of a core 5-stranded SH3-like β-barrel fold, which in some cases is partially open (Fig. 5). In these open forms, there is one region on the surface of the barrel where the adjacent strands do not come close enough to form hydrogen bonds. This feature allows a peptide to stack with one or both of the strands in the open region in an extended configuration.[222] Further, like many other β-barrels, such as the OB-fold domain, the barrel has an open mouth at one of the ends, thereby providing another location to accommodate a peptide ligand.[223] The strands forming the surface of the barrel are deployed to bind DNA or RNA in several representatives of the chromo-like fold.[116,220] By far, the chromo-like domains constitute the most versatile class of methylated histone binding proteins, in that they can recognize methylated peptides corresponding to most of the known histone methylation sites and also those in nonhistone proteins.

Currently available evidence suggests that the classical chromodomain can recognize methylation at H3K4, H3K9, H3K27, and H3K36.[116,222,224] Members of the tudor-like assemblage target methylated forms of H3K4, H3K9, and H4K20.[215–218,225] Different versions of these domains display alternative use of distinct binding interfaces. The N-terminally truncated versions of the classical chromodomain, such as those seen in the HP1 protein have a markedly open form of the barrel which allows them to accommodate the modified peptide (e.g., the H3K9me2-bearing peptide) in the extended configuration between the two strands flanking the opening in barrel, with the modified residue being accommodated in the mouth of the barrel (Fig. 5).[222] The complete versions of the classical chromodomain, such as that seen in the acetylase MSL3 (which binds H4K20me1), have a relatively closed barrel.[224] Hence, the H4K20me1-bearing peptide is bound only via the open mouth of the barrel, while the sides of the barrel bind DNA (Fig. 5). The BMB(PWWP) domain displays a partially open state (Fig. 5). Accordingly, it binds peptides such as the H3K36me3 via partial accommodation of the peptide in the open region of the barrel and accommodation of the trimethylated peptide in the mouth of the barrel.[215] The tudor-like domains, including the classical tudor domains that bind methylated arginine, the versions of the tudor domain that bind methylated lysine (e.g., H4K20me2), and the "MBT" domains that bind peptides such as the one containing H4K20me1, display completely closed barrels.[216–218,226,227] As a consequence they only accommodate the modified residue directly via the open mouth of the barrel (Fig. 5). The contextual specificity for different types of modified peptides in these cases and also in the case of the classical chromodomain and the BMB(PWWP) domain is achieved in large part by the residues in the loops that surround the mouth of the barrel. Thus, the different classical chromodomains are able to specifically recognize methylated peptides containing H3K4 (the two N-terminal chromodomains of the SWI2/SNF2 ATPase CHD1), H3K9 (HP1), and H3K27 (Polycomb).[116,222,224] The discrimination of the methylated residue in most of these cases occurs primarily via the aromatic cage buried within the mouth of the barrel. Additionally, all methyllysine-binding members of this fold have a conserved polar residue (typically acidic or asparagine) that also participates in interaction with the lysine (Fig. 5). However, differences in this cage are critical for the discrimination of different methylation states. For example, the tudor domain of the

DNA-repair regulator 53BP1 has a cage with five residues that sterically disfavors binding of H4K20me3 but is large enough to allow H4K20me2.[218] The binding mode of the BAM(BAH) domain is currently poorly understood. Studies in yeast suggest that it recognizes the H3K79 which is methylated by DOT1, but apparently it has preference for the unmethylated or hypomethylated version of H3K79.[214] It remains to be seen if versions of this domain might possess alternative specificities.

In eukaryotes, the explosive radiation of the chromo-like family appears to have occurred independently in several lineages. In particular, multiple such radiations are seen within the tudor-like: the so-called plant agenet domains represent one such example.[213] They were already present in the common ancestor of animals, plants, and fungi, but in the plant lineage they appear to have greatly expanded and diversified in their architectures (Fig. 3, Supplementary material).[8] In land plants (e.g., *Arabidopsis* gi: 75266597), they are combined with BAM(BAH) domains while in chlorophyte algae such as they are combined with the FHA domain (*Micromonas*; gi: 226463152), or the SWI2/SNF2 ATPases (*Chlamydomonas*; gi: 158276786) suggesting that these proteins might recognize multiple peptide positions simultaneously, or recruit chromatin-remodeling activities to nucleosomes with specific methyl marks. The Aux-RF domain has also undergone a massive expansion in the plant lineage (Supplementary material). However, its presence in animals and apicomplexans suggests that it might have emerged earlier. In plants, it is fused to an N-terminal VP1(B3) DNA binding domain, and is found in a large family of TFs that respond to various signals such as the plant hormone auxin. The AUX-RF domains display an aromatic cage typical of the methyllysine-binding versions of this domain, suggesting that they are likely to function similarly to other chromo-like domains and probably link methyllysine marks to transcription regulation. The animal and choanoflagellate version of the AUX-RF domain are fused to bromodomains and WD40 repeats (BRWD1/3). Thus, they might function as multivalent modified-histone-binding proteins. The AUX-RF domain in apicomplexans is fused to the RFD module, which is also found in the DNA methylase DNMT1 and other chromatin proteins, and might play a role in establishing repressive chromatin (see Chapter by Lakshminarayan M. Iyer *et al.*). Throughout eukaryotic evolution, the classical chromodomain appears to have been picked up by multiple transposons, possibly due to selection for avoiding integration into active gene bodies.[228] The recognition of specific methyl-marks in chromatin by transposons during integration might help them to integrate upstream rather than within genes, or in heterochromatic regions where they are unlikely to disrupt any gene function.

The classical SH3 and chromo-like domains had diverged from each other in bacteria, and were independently acquired by eukaryotes from bacterial precursors.[229] Bacterial homologs of the SH3 and chromo-like domains are found in secreted or periplasmic proteins associated with peptidoglycan, along with the bacterial precursor of another eukaryotic peptide-binding domain called the Slap homology domain 1 (SHD1).[8,229] This suggests that these domains probably acquired their peptide-recognition functions in the context of peptides found on the bacterial cell-surface. Their exact target peptides in bacteria, however, remain as yet unknown. The initial radiation of the chromo-like domains in eukaryotes might coincide with the expansion of protein methylases that happened prior to the LECA. This was paralleled by the radiation of the classical SH3 domains in eukaryotic cytoplasmic proteins in relation to peptide recognition in the context of the cytoskeleton.[8,229] Thus, different ancestral SH3-fold domains acquired from bacteria appear to have been recruited for distinct nuclear and cytoskeletal peptide interactions, probably concomitant with the origin of eukaryotic nucleo-cytoplasmic compartmentalization. Analysis of the early-branching eukaryotes suggests that the chromo, tudor-like, and BAM(BAH) domains had already separated from each other in the LECA itself, and that the BMB (PWWP) domain emerged just prior to divergence of the kinetoplastid-heterolobosean clade. At least three distinct versions of the chromo domain (including a HP1-like protein), one BAM(BAH) domain and one version of the chromatin-associated tudor-like domain, can be extrapolated as having been present in the LECA.[8,229]

B. The PHD-Like Domains

The PHD-finger contains a binuclear version of the treble-clef fold.[230] The basic mononuclear form of this fold has the characteristic Zn-chelating site comprised of one pair of cysteines from the initial flap and another pair from the beginning of the C-terminal helix of the domain (Fig. 5). The binuclear versions bind a second Zn^{2+} ion by means of one pair of chelating residues derived from the core “finger” region of the treble-clef fold and another pair derived from the C-terminal helix or a further C-terminal extension.[230] The PHD finger shares its binuclear organization with a functionally diverse set of treble-clef domains belong to the RING (the Ubiquitin E3-ligase), ZZ, MYND, B-box, C1, AN1, UBP, UPF1, and the FYVE superfamilies.[231] PHD fingers can bind methylated peptides (H3K4me2/H3K4me3), acetylated peptides (H3K14ac and H4S acetylated N-terminus), and nonacetylated peptides.[232–234] Versions involved in methylated peptide recognition appear to constitute a numerically dominant group of PHD fingers, suggesting that they are one of the primary domains involved in recognition of methylated H3K4.[233] H3K4me2/H3K4me3 or unmethylated H3K4 peptides stack in an antiparallel manner with the first strand of the core finger region of the PHD finger, in the extended configuration, via strand-like hydrogen bonding (Fig. 5). Interestingly, as in the case of the chromodomain, recognition of the H3K4 position appears to depend in large part on an aromatic cage, formed by aromatic residues from the

N-terminal flap and the end of the second strand of the core finger region (Fig. 5).[129,233] In versions that bind unmethylated histones or other peptides, these residues appear to be absent.[232] Several PHD fingers are indiscriminate in binding either H3K4me2 or H3K4me3.[232,233] Other more subtle differences in the peptide-binding site might help certain PHD fingers to differentiate between these states. For example, the divergent Pygopus-type PHD finger specifically binds H3K4me2, while the TAF3 PHD finger is apparently specific for H3K4me3.[235,236] The PHD-X/ZF-CW is a structurally degenerate version of the PHD domain that has lost the N-terminal flap of the treble-clef domain and the associated Zn-chelating site (Fig. 5). However, it retains the H3-binding interface and binds H3K4me3 in a manner similar to the conventional PHD domains.[237] Another possible derivative of the PHD finger is the mononuclear treble-clef seen C-terminal to the JOR/JmjC domain in the JMJD3 lineage—its binding specificity remains to be determined.[182]

Currently, the PHD finger is not known to occur outside of eukaryotes.[6,8] At least a single copy of the PHD finger is inferred as being present in the LECA making it a candidate for a potential early eukaryotic innovation. Its origin could be linked to the pre-LECA expansion of protein methylation that resulted in the emergence of H3K4 methylation. In course of eukaryotic evolution, the PHD finger was linked to several other domains involved in modified peptide recognition such as the bromodomain and chromo-like domains, several catalytic domains involved in histone modification, ubiquitination, and remodeling.[8] There is a LSE in *Trichomonas* of SET domain proteins with two N-terminal PHD fingers, at least one of which is capable of binding methylated H3K4, suggesting that such associations had emerged even in the earliest-branching eukaryotes. Like the chromodomain, the PHD finger has also been acquired by eukaryotic transposons,[228] suggesting that it might be similarly used to guide integration to sites away from gene bodies. The PHD-X/ZF-CW domain shows a remarkable expansion in the haptophyte alga *Emiliania huxleyi*. This organism specifies at least 19 distinct versions, many of which display specialized architectures including associations with other methylated histone recognition domains, as well as DNA-binding domains such as HMG, AP2, and homeodomains (Supplementary material). Thus, in haptophytes it appears to have become the dominant domain involved in discrimination of H3K4 methylation.

C. Superstructure-Forming Domains

Of the superstructure-forming repeats, the WD40 domain of the WDR5 protein was initially shown to bind H3K4me1, H3K4me2, and H3K4me3,[238–241] and later identified to be a peptidyl arginine recognition factor that recognizes arginines such H3R2 and MLL1R3765.[242] WDR5 is a subunit of the histone-methylating complex that includes the trithorax-like SET domains such as

MLL1. In mammalian models, loss of WDR5 results in global loss of H3K4 methylation. Structural analysis of the WDR5 suggests that it differs from other methylated histone-binding domains in being involved in presentation of the target H3K4 to SET domain methylases rather than reading of the methylated histone code.[241] This is achieved by the relatively superficial location of the H3K4 position when the corresponding peptide is bound by WDR5, allowing it be further modified by the SET domain.[241] It also differs from other methyl-lysine-binding domains in that it lacks a distinct aromatic cage to accommodate the methylated residue. The WD40 domain of WDR5 is a seven-bladed β-propeller with a circular permutation (i.e., the first blade is comprised of sequences from the extreme N- and C-termini of the protein) similar to that seen in the heterotrimeric GTPase β-subunits (Fig. 5). The substrate peptide is accommodated with the N-terminus and the second residue being buried deep within the mouth of the central aperture of the β-propeller.[238–241] Orthologs of WDR5 are seen in animals, fungi, slime molds, plants, stramenopiles, and heteroloboseans, suggesting that it was a fairly ancient component of the trithorax-like H3K4 methylating complexes, though there is no evidence for it having been present in the LECA. Other WD40 domains such as those in EED/Esc and RbBP4, which are a polycomb group proteins and required for E(Z) function could also potentially act similar to WDR5 in presenting peptides for methylation by the associated SET domain.[241,243] Other WD40 proteins such as groucho (associated with the histone deacetylase RPD3) and RbBP4 and RbBP7 (associated with the histone acetylase HAT1) interact with H3 and H4 tails, respectively. Similarly, the WD40 histone chaperone proteins such as HIRA associate with the H3 tail and also interact with the histone deacetylase HDAC1.[240] The WD40 repeats found in the animal BRWD1/3 proteins (associated with the methylated histone-binding Aux-RF protein), and in the WDSUB1 protein in conjunction with a SAM (sterile alpha motif) domain, might possibly bind histones.

The second superstructure-forming repeats confirmed to bind methylated histones are the ankyrin repeats of the animal SET domain methylases G9a/GLP, which recognize H3K9me1 and H3K9me2.[130,244] The ankyrin-repeat is a bihelical unit, followed by a peculiar hairpin that connects it to the first helix of the next repeat (Fig. 5). The repeats assume a slightly curved superstructure with a binding shelf formed between the bihelical units and the hairpins.[129] In the case of the GLP ankyrin domain, there are seven repeats with the H3 peptide being primarily bound by the fourth and fifth repeats.[130] Strikingly, despite displaying a completely distinct structural scaffold, the GLP ankyrin domain recognizes methyllysine by means of an aromatic cage formed by three tryptophan residues. Additionally, a glutamate residue forms a salt bridge with the methylated lysine (Fig. 5). Similar potential methylated histone binding sites have been proposed to exist in some other ankyrin repeat proteins such as

HIP14, but this interaction still remains to be confirmed.[245] G9a/GLP SET domains modify the same residue as the one bound by their Ankyrin repeats, and this phenomenon has been linked to spreading of methylated H3K9 marks.[244] The G9a/GLP lineage branched off from the Su(var)3–9 lineage in animals. This suggests that the methylated H3K9-recognizing-ankyrin repeats might be a relatively recent innovation. However, it is conceivable that independent innovation of histone recognition might have occurred in ankyrin-repeat chromatin proteins from other eukaryotes. One example is a predicted apicomplexan chromatin protein that combines a bromodomain with N-terminal ankyrin repeats (e.g., *Plasmodium* PF10_0328; gi: 23508131). Similar to the WD40 and ankyrin repeats, other superstructure forming repeats, such as the TPR and HEAT repeats could also be involved in histone recognition. For example the histone chaperone subunit HIR3 has multiple TPR repeats that could play a role in nucleosome interactions.[246] In this vein, it is worth determining whether the TPR repeats in the N-termini of the JMJD3 lineage and several SET domain proteins of the SMYD lineage (e.g., Human SMYD4, gi: 156616308) play a role in histone recognition.

V. Domain Architectures of Enzymes in the Chromatin Protein Methylation Network

A. Associations with DNA-Binding and Modified-Peptide-Recognition Domains

Analysis of domain architectures, and the network representation of the total set of architectures that are found among enzymes involved in the methylation system, reveal certain interesting patterns with considerable functional significance (Fig. 6). First, there is a striking difference in the architectures of the Rossmann fold methylases and deiminases on one side, and the SET domain methylases and JOR/JmjC and LSD1 demethylases on the other side (Fig. 6). The former show practically no fusions to conventional DNA-binding domains, modified-peptide- recognition domains or other chromatin-associated domains. The latter, on the other hand, show a rich diversity of domain architectures. This difference suggests that the Rossmann fold protein methylases and deiminases are not strongly regulated by means of their reading of preexisting epigenetic marks. This is consistent with the more generic roles of the PRMTs with a clear involvement in non-chromatin-related methylation.[50] With the exception of the SMYD-like and the RMT-like lineages of SET domains, most others show fusions to domains that are typical of chromatin proteins suggesting that a major subset of the SET domains are primarily involved in chromatin modifications. Both the SET and JOR/JmjC domains

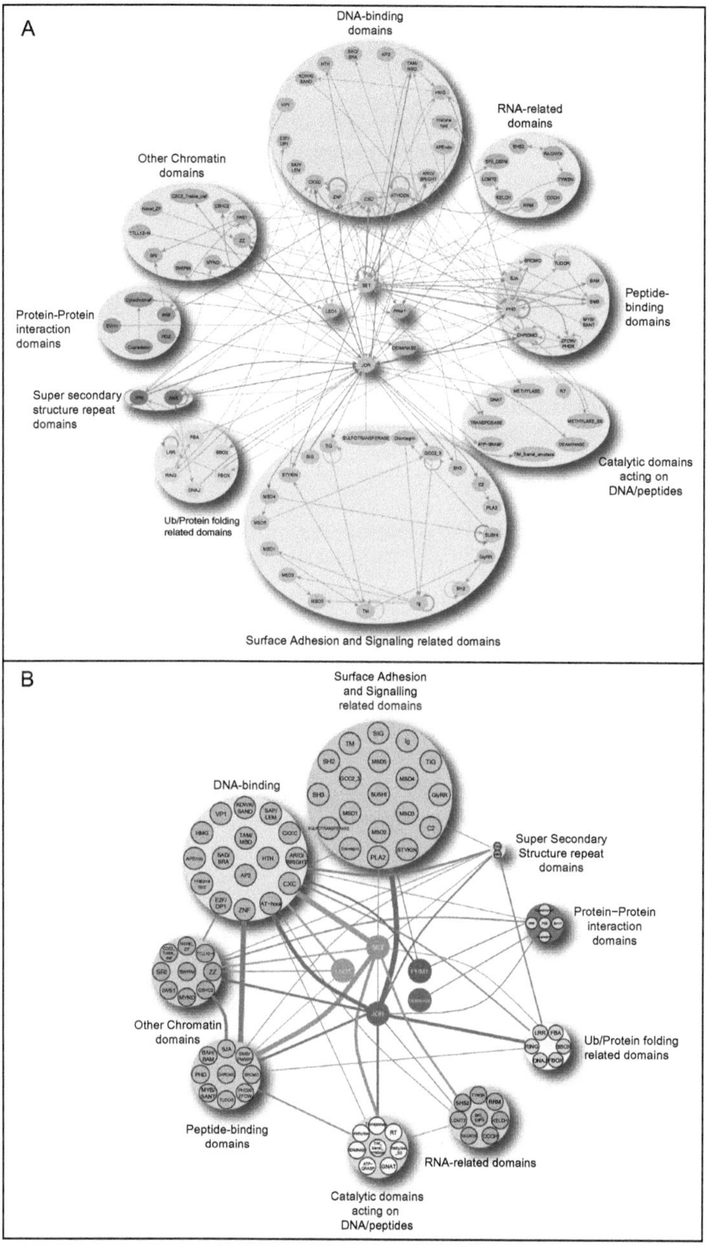

FIG. 6. Domain architectural network. This is shown as a network graph with nodes representing domains related to protein-methylation or -demethylation, and edges indicating their physical connectivity in a polypeptide. The meta-network (B) is used to highlight the overall trends

show a considerable tendency to associate with multiple DNA-binding domains as well as modified histone recognition domains (Fig. 6). In several cases, these combinations occur in the same polypeptide—for example, in MLL4, the SET domain is combined with both DNA-binding CXXC and AT-hook domains and the modified peptide-binding bromo and PHD domains (see Chapter by Lakshminarayan M. Iyer *et al.*). Thus, signals emanating from the recognition of both features of DNA and histone modifications appear to be integrated by these enzymes. Both the SET and JOR/JmjC domain are linked to the CXXC domain, which recognizes unmethylated CpG-containing DNA, while the SET is connected to SAD/SRA and AP2 domains and the JOR/JmjC to TAM/MBD (Fig. 6). This suggests that, in the course of eukaryotic evolution, both methylases and demethylases of chromatin have been recruited to recognize the methylation status of DNA.[108,206] These architectures appear to have been central to the integration of the DNA- and protein-based epigenetic marks and have repeatedly arisen in animals, amoebozoans, land plants, chlorophytes, and stramenopiles.[8]

SET, JOR/JmjC, and LSD1 proteins show linkages to PHD fingers, whereas only SET and JOR/JmjC are additionally linked to various chromo-like domains (Fig. 6). Thus, an important feature of both the SET methylases and demethylases appears to be recognition of preexisting methyl marks. This feature is important because recognition of preexisting marks might help in: (1) directing antagonistic modifications that either remove the preexisting mark or mediate a modification that functionally counters the original modification[193]; (2) directing a modification that further expands the functional significance of the original mark. For example, generation of bivalent histone methylations could proceed through such a process[94,175]; or (3) spreading the originally established marks. An example of this mechanism is the Su(var)3–9

of associations between different functional types of domains involved in protein methylation. The arrowheads depict directionality and point from the N-terminal to the C-terminal domain. Domains with similar functional roles are in the same color and further grouped into meta-nodes in the meta-network. Edges are colored based on the principal domain of an association; SET, magenta; PRMT, green; LSD1, orange; JOR/JmjC, blue; and deiminase, purple. Edges not involving these principal domains are colored gray. The edge thickness is proportional to the relative frequency with which linkages between two domains or metanodes recur in distinct polypeptides. Conventional abbreviations are used for domain nomenclature. Other domains with nonstandard abbreviations include: TTLL12-N, An alpha-helical domain found N-terminal to TTLL12; C2C2_Treble_clef, A distinct treble-clef fold domain with a C2C2 conservation; GlyRR, glycine rich repeat domain; MSD1-5, distinct *Monosiga*-specific extracellular domains; SIG, signal peptide; RT, reverse transcriptase; methylase_SS, spermidine synthase-like methylase; SFII_DBP8, superfamily II helicase of the DBP8 family; LCMT2, leucine carboxymethyl transferase family 2; APEndo, AP-endonucleases. (For interpretation of the references to color in this figure legend, the reader is referred to the Web version of this chapter.)

methylase which spreads the modification catalyzed by its SET domain by recognizing prior methyl marks.[95,167] These associations emerged even in the early-branching eukaryotes but considerably expanded over the course of eukaryotic evolution, especially in the plants, stramenopiles, haptophytes, and animals.[8]

While associations with methylated-peptide-recognition domains are common among the SET methylases and demethylases, associations with the bromodomain are relatively rare, with few SET proteins displaying fusions to the acetylated peptide-binding bromodomains (Fig. 6). Thus, recognition of preexisting acetyl marks appears to be of lower significance for direct methylation or demethylation events. However, analysis of the acetyltransferase architectures reveals that they have a much higher propensity to combine with both methyl and acetyl mark recognition domains. This would suggest that chromatin protein acetylation is, to a certain extent, hierarchically subservient to chromatin protein methylation. Another striking difference is seen when the architectures of the deacetylases are compared to the demethylases.[8] The latter are architecturally much richer than the former, suggesting that cross talk with preexisting modifications is of far greater significance in directing demethylation than in directing deacetylation. A domain associated with both JOR/JmjC and SET domains is the SET-JOR associated (SJA) domain also called the FYR domain.[8,247] This domain adopts a fold similar to the Polo-box domain which has been implicated in binding phosphoserine/phosphothreonine-containing peptides.[248] Hence, by comparison, it is possible that the SJA/FYR domain is involved in phosphopeptide recognition.

B. Associations with Other Domains Catalyzing and Directing Other Modifications

Combination of methylase and demethylase activities in the same polypeptide is extremely rare, with the only currently known example being a version from filamentous fungi (*Neurospora* NCU01973 gi: 85112402; Fig. 3). This enzyme might function as an editing enzyme that removes one methyl mark while concurrently adding another. Both demethylase and methylase domains also show no (or at best very rare) fusion to acetylase or deacetylase domains. The only currently known case is a fusion of a JOR/JmjC domain to an acetylase domain in nematodes (*C. elegans* C06H2.3; gi: 17557716). These observations suggest that in general, methylation and demethylation are in large part temporally decoupled from each, and also probably from acetylation and deacetylation. This does not imply that they are not functionally connected, but merely that there is no strong requirement for the modifications to happen concurrently. While versions of JOR/JmjC mediating hydroxylation of the RNA base yW show elaborate architectures typical of the RNA-modifying enzyme, no such architectures are seen among paralogs that modify chromatin proteins.[182] Both SET and JOR/JmjC have occasional fusions to the RING domain and other components of the E3 ubiquitin ligases (such as the F-box) or Ub-associated domains (such as the B-box) (Figs. 3 and 6). This suggests that concurrent methylation/demethylation and ubiquitination might have evolved in multiple lineages such as animals, plants, and stramenopiles. The association with another peptide-modifying enzymatic domain (polyglutamylase or polyglycinase) is observed in the case of the SET domains (Figs. 3 and 6; see above), and represents a potentially important modification that has been largely ignored. As noted above, the SET domain is also combined with the spermidine synthase domain in the diatoms. In fungi, the SET domain is fused to a nucleic acid deaminase domain related to the Tad3 tRNA-specific adenosine deaminase subunit (e.g., *Ustilago maydis* UM00758.1; gi: 71004478; Figs. 3 and 6).[8] It would be of interest to investigate if this deaminase domain might direct chromatin-associated RNA editing or perhaps even DNA modifications.

C. Associations with Other Domains and Their Possible Functional Implications

Certain SET domains contain an N-terminal domain that is rich in cysteines. This domain has been sometimes called the "pre-SET" domain. It is actually a version of the CxC domain and chelates 3 Zn^{2+} ions that are arranged as an equilateral triangle.[124,249] The classical CXC domains bind nucleic acids.[250–252] Consistent with this, at least certain members of the SET-associated version have been shown to bind single-stranded nucleic acids, suggesting that the domain might play a role in the previously observed association of RNA with the SET proteins.[253] Domain architectural analysis also points to the previously unexpected diversity of architectures of the SRI domains and their role in coupling histone methylation, demethylation, and other chromatin modifications to transcription. The SRI domain was first identified as a trihelical domain, at the extreme C-terminus of the animal and fungal proteins of the SET2 lineage, that mediates its interactions with the phosphorylated C-terminal domain of the RNA polymerase II (RNAPII-CTD).[254,255] This interaction is critical for SET2 to localize to gene bodies and locally methylate H3K36. Subsequent studies have identified a SRI domain at the C-terminus of the DNA helicase RecQL5, that mediates its binding to the RNAPII-CTD, and thereby helps maintain genomic stability during transcription.[256,257] Thus, the SRI domain appears to be a general module that links different activities to the gene bodies via association with the RNAP-CTD. Previously unknown SRI domains were detected at the C-termini of the plant LSD1 demethylase (e.g., *Arabidopsis* LDL3, gi: 240255922) and the PHD-RING finger protein PHRF1 (gi: 219518215), the animal splicing factors

SCAF1 (gi: 10440402) and SCAF11 (gi: 117676384), and the plant splicing regulator FES1 (gi: 30685865). These architectures suggest that not just methylases but also demethylases, E3 Ub-ligases and splicing factors might be localized to the RNAPII-CTD by the SRI domain (Figs. 3 and 6, Supplementary material). It is possible that co-transcriptional demethylation in plants might have a role in silencing or modulating active chromatin by associating with RNAPII. Similarly, the PHRF1 protein might recognize co-transcriptionally established methyl marks (perhaps by SET2 methylases) via its PHD finger, and ubiquitinate proteins associated with chromatin. This ubiquitination might have a role in organization of splicing complexes which also might be co-transcriptionally localized to gene bodies by the SRI domain. The exact role of the treble-clef fold MYND domain remains unclear, but the current evidence suggests that it might provide an extended interface for substrate recognition through peptide-binding.[92,103] Another domain that might have a role in recognition of unmodified peptides is the WW domain, which is found in methylases such as SET2 (Figs. 3 and 6).[258]

VI. Evolutionary Considerations

Complementary evidences from comparative genomics, sequence analysis, and structural biology have uncovered several key aspects of the provenance and the history of the eukaryotic protein methylation system and its integration with other regulatory mechanisms such as DNA methylation and RNAi. The evidence from comparative genomics indicates that many key players in each of these mechanisms have emerged in the bacterial world, as a part of different systems that were under selection for diversification (see also Chapter by Lakshminarayan M. Iyer *et al.*). However, there is little evidence for any cross talk between the precursors of these systems in the bacteria themselves. Thus, the major players of the regulatory system centered on DNA methylation and associated modifications arose predominantly in bacterial restriction–modification (R–M) systems and the DNA-modification operons of selfish elements such as viruses and transposons (Chapter by Lakshminarayan M. Iyer *et al.*). In sharp contrast, many of the precursors of the system centered on protein methylation arose in bacterial nonribosomal peptide synthesis systems associated with the production of diverse secondary metabolites such antibiotics, siderophores, toxins, and intermediates for amino acid utilization.[181,182] However, the SET domain alone appears to have had a distinct origin, with at least some of them already active as primitive chromatin-remodeling enzymes in prokaryotes. The chromo-like domains had yet another distinct role, probably in binding peptides on the cell-surface and in peptidoglycan. In the RNAi systems, the precursor of the RNA-dependent RNA polymerase (RDRP) appears to have been derived from bacterial selfish elements such as viruses and certain transposons, whereas the argonaute-like proteins were probably derived from bacterial RNA-based defense mechanisms that act against prokaryotic selfish elements.[10,259,260] However, beyond these components much of the RNAi system was put together from innovations that were specific to the eukaryotic lineage. Thus, the emerging picture is that the precursors of the individual components were already present in bacteria, but did not interact to result in regulatory systems such as those seen in eukaryotes. Their integration into coherent systems happened only in eukaryotes. Despite the functional distinctness of the precursors of these systems in bacteria, there appears to be a common denominator for several of them—most were under selection for diversification. Several precursors of the protein methylation system are part of the secondary metabolite biosynthesis pathways which are under strong selection for diversification due to resistance to antibiotics and siderophore "stealing."[181,182] Likewise, in the case of the precursors of the DNA methylation systems, the R–M systems are under strong selection for diversification due to the "arms-race" situation arising from resistance against restriction (see Chapter by Lakshminarayan M. Iyer *et al.*). Among precursors of the RNAi system, diversifying tendencies might be present due to similar resistance developed by the selfish elements against which they are deployed. These disparate diversifying processes in bacteria appear to have generated enormous amounts of raw material that could then be used by eukaryotes "for free."

Integration of the precursors into unified systems appears to have happened in multiple steps over the course of eukaryotic evolution. Importantly, the emergence of eukaryotes was accompanied by loss of the operonic regulatory systems typical of prokaryotes. This phenomenon seems to have broken up the old functional associations between genes and recombined the starting material acquired from the prokaryotic ancestors in several novel ways. The very first elements to be acquired and integrated appear to be the SET and PRMT protein methyltransferases, which had already considerably diversified in the LECA itself. This indicates that they were probably acquired as a part of the initial symbiogenic origin of eukaryotes, from either the bacterial or archaeal partner of the symbiosis.[6] Between the initial symbiotic event and the LECA they appear to have undergone multiple rounds of duplication and specialization. Hence, it is likely that they, along with the similarly acquired GCN5 and ELP3-like acetyltransferases and HDAC and Sir2 deacetylases, were critical to the very origin of the distinctive eukaryotic chromatin structure even as the positively charged histone-tails were acquired.[8] Peptide-binding domains secreted by the original bacterial endosymbiont appear to have provided the SH3 and chromo-like domains that were now recruited as peptide-binding domains in cytoskeletal and chromatin-related contexts. This innovation probably, for the first time, allowed recognition of modified peptides to be used as a form of epigenetic information. This early phase of eukaryotic evolution also saw the emergence of an incipient RNAi system that integrated the RNA-cleaving argonaute proteins and an amplificatory arm of the system based on the RDRPs. However, there are no clear signs of strong integration of the RNAi system with the protein-methylation system in these early-branching eukaryotes. The current data suggest that DNA cytosine methylation was also absent in these early-branching eukaryotes (see Chapter by Lakshminarayan M. Iyer *et al.*). Likewise, one might infer that after the basal lineages (the diplomonads and parabasalids) diverged from the remaining eukaryotes, there was a massive phase of new innovations again driven in large part by new transfers acquired from bacteria. First, this appears to be the time that DNA methylation was acquired as a chromatin-regulatory mechanism by eukaryotes. Second, the protein demethylases, both of the LSD1 and JOR/JmjC type appear to have emerged around this period. Also characteristic of this phase was the emergence of several SET domain lineages, DOT1-like methylases and the BMB(PWWP) domain. Thus, the systems and the chromatin structure typical of the eukaryotic "crown" group comprised of animals, fungi, amoebozoan, and plants might have had their origins during this phase.

Analysis of domain architectures reveals that there has been a trend toward increased complexity in chromatin proteins over the course of eukaryotic evolution. Diplomonads and parabasalids have the least complex domain architectures.[8] After their divergence, there appears to have been a phase of increased architectural diversity. However, the highest architectural complexity is observed in certain crown-group clades and, among those, animals are unparalleled in the complexity of their domain architecture networks.[7,8] This tendency is particularly prominent in the SET methylases and the JOR/JmjC demethylases. In many cases, this increase in architectural complexity occurs via "domain accretion" or fusion of new domains around an ancient orthologous core of the polypeptide (e.g., in the trithorax lineage of SET domains).[7,8,86] Despite having large absolute numbers of chromatin proteins, ciliates, and parabasalids tend to have much lower domain architectural complexity. Mere increase in proteome size without increase in architectural complexity of chromatin proteins, as seen in these organisms, might be sufficient to achieve relatively complex organization within a single cell. In contrast, the high complexity of animal, plant and stramenopile proteins points to a possible relationship between architectural complexity of chromatin proteins and emergence of a greater degree of cellular differentiation and multicellular or colonial states.[7,8,86] This suggests that increasing organizational complexity might have gone hand-in-hand with increased cross talk between epigenetic marks both at the protein and DNA level. Evidence from ciliates suggests that the RNAi system was functionally integrated with the protein methylation system by the time of the divergence of the alveolates from the remaining eukaryotes.[17,261,262] Studies on the patterns of gene loss in fungi suggest that this integration increased in the eukaryotic crown group with both the DNA- and protein-methylation systems acquiring extensive functional links to the RNAi system.[95,96,126,263] It is also possible that this phase was marked by the incorporation of noncoding RNAs such as the precursors of HOTAIR and the trithorax RNAs into the protein methylation complexes[109,168] (see Chapter by Anton Wutz). The strong functional collusion between DNA methylation and protein methylation systems probably first emerged in the common ancestor of the eukaryotic crown group.[90,108,206,264–266] However, these connections were considerably attenuated in fungi, probably as a result of gene loss associated with their saprophytic lifestyle and degeneration to the yeast-like morphotypes.[126] In contrast, in animals, plants, stramenopiles, and haptophytes these connections developed even further.

VII. General Conclusions

The past 15 years have witnessed an extraordinary expansion of studies pertaining to chromatin protein methylation and its functional significance.[17] In face of the enormous literature that has accumulated in this field, it is difficult to discern key new directions that might help in filling major lacunae. However, one important direction of study would be to create comprehensive regulatory networks that link all methylated proteins to their corresponding methylating or demethylating enzymes and methylated peptide-binding domains. The creation of such networks for multiple organisms is a rather daunting task, but at least the initial steps appear within grasp given the success of high-resolution mass spectrometry and related techniques in identifying such modifications.[23,77,267] Generation of such regulatory networks would enable "systems biology" type approaches to understanding the global aspects of protein methylation. This effort would also require a more comprehensive analysis of the targets of the enzymes of the methylation systems at a biochemical level. Integration of such data with studies like those recently conducted in the *Drosophila* system,[107] that have characterized the major cellular chromatin states, would be tremendously important in uncovering the basic regulatory processes behind major events in the nucleus. In terms of structural biology, the discovery of novel methylated peptide recognition interfaces beyond the PHD and chromo-like folds have opened up the possibility that other such interfaces remain to be discovered.[129,130,241] Finally, the current chapter attempts to emphasize the importance of the evolutionary framework in understanding protein methylation and its integration with DNA methylation. In light of this, studies in poorly explored organisms such as diatoms and haptophyte algae might reveal important aspects of these systems that have no direct cognates in current model systems. However, just as early studies in ciliates informed the work in model systems with respect to protein acetylation and RNAi,[17] it is likely that such studies of eukaryotic microbes will feed back new insights regarding traditional model systems. We hope that the synthesis offered here provides an evolutionary framework for such explorations.

Acknowledgments

Work by the authors is supported by the intramural funds of the National Library of Medicine, National Institutes of Health, USA. We would like to acknowledge the numerous contributions of various researchers in the protein methylation and chromatin field, which we were regrettably unable to cite due to sheer enormity of the literature under review.

Supplementary Information

A systematic collection of the different protein methylases, demethylases, and functionally related enzymes and other domains discussed in the text can be found at the following FTP site: ftp://ftp.ncbi.nih.gov/pub/aravind/chromatin/protein_methylase/supplementary.html

References

1. Dorman CJ. Nucleoid-associated proteins and bacterial physiology. *Adv Appl Microbiol* 2009;**67**:47–64.
2. Pedersen LB, Birkelund S, Christiansen G. Purification of recombinant Chlamydia trachomatis histone H1-like protein Hc2, and comparative functional analysis of Hc2 and Hc1. *Mol Microbiol* 1996;**20**:295–311.
3. Perara E, Ganem D, Engel JN. A developmentally regulated chlamydial gene with apparent homology to eukaryotic histone H1. *Proc Natl Acad Sci USA* 1992;**89**:2125–9.
4. Reeve JN, Bailey KA, Li WT, Marc F, Sandman K, Soares DJ. Archaeal histones: structures, stability and DNA binding. *Biochem Soc Trans* 2004;**32**:227–30.
5. Aravind L, Iyer LM, Anantharaman V. The two faces of Alba: the evolutionary connection between proteins participating in chromatin structure and RNA metabolism. *Genome Biol* 2003;**4**:R64.
6. Aravind L, Iyer LM, Koonin EV. Comparative genomics and structural biology of the molecular innovations of eukaryotes. *Curr Opin Struct Biol* 2006;**16**:409–19.
7. Anantharaman V, Iyer LM, Aravind L. Comparative genomics of protists: new insights into the evolution of eukaryotic signal transduction and gene regulation. *Annu Rev Microbiol* 2007;**61**:453–75.
8. Iyer LM, Anantharaman V, Wolf MY, Aravind L. Comparative genomics of transcription factors and chromatin proteins in parasitic protists and other eukaryotes. *Int J Parasitol* 2008;**38**:1–31.
9. Mans BJ, Anantharaman V, Aravind L, Koonin EV. Comparative genomics, evolution and origins of the nuclear envelope and nuclear pore complex. *Cell Cycle* 2004;**3**:1612–37.

10. Anantharaman V, Koonin EV, Aravind L. Comparative genomics and evolution of proteins involved in RNA metabolism. *Nucleic Acids Res* 2002;**30**:1427–64.
11. Staub E, Fiziev P, Rosenthal A, Hinzmann B. Insights into the evolution of the nucleolus by an analysis of its protein domain repertoire. *Bioessays* 2004;**26**:567–81.
12. Liu J, Tan H, Rost B. Loopy proteins appear conserved in evolution. *J Mol Biol* 2002;**322**:53–64.
13. Rouleau M, Aubin RA, Poirier GG. Poly(ADP-ribosyl)ated chromatin domains: access granted. *J Cell Sci* 2004;**117**:815–25.
14. van Dijk J, Miro J, Strub JM, Lacroix B, van Dorsselaer A, Edde B, et al. Polyglutamylation is a post-translational modification with a broad range of substrates. *J Biol Chem* 2008;**283**:3915–22.
15. Kouzarides T. Chromatin modifications and their function. *Cell* 2007;**128**:693–705.
16. Walsh CT, Garneau-Tsodikova S, Gatto GJ. Protein posttranslational modifications: the chemistry of proteome diversifications. *Angew Chem* 2005;**2005**:7342–72.
17. Allis CD, Jenuwein T, Reinberg D. *Epigenetics.* Cold Spring Harbor, N.Y: Cold Spring Harbor Laboratory Press; 2007.
18. Pereira CI, San Romao MV, Lolkema JS, Crespo MT. Weissella halotolerans W22 combines arginine deiminase and ornithine decarboxylation pathways and converts arginine to putrescine. *J Appl Microbiol* 2009;**107**:1894–902.
19. Thompson PR, Fast W. Histone citrullination by protein arginine deiminase: is arginine methylation a green light or a roadblock? *ACS Chem Biol* 2006;**1**:433–41.
20. Webby CJ, Wolf A, Gromak N, Dreger M, Kramer H, Kessler B, et al. Jmjd6 catalyses lysyl-hydroxylation of U2AF65, a protein associated with RNA splicing. *Science* 2009;**325**:90–3.
21. Aravind L, Koonin EV. The DNA-repair protein AlkB, EGL-9, and leprecan define new families of 2-oxoglutarate- and iron-dependent dioxygenases. *Genome Biol* 2001;**2**: RESEARCH0007.
22. Fujiki R, Chikanishi T, Hashiba W, Ito H, Takada I, Roeder RG, et al. GlcNAcylation of a histone methyltransferase in retinoic-acid-induced granulopoiesis. *Nature* 2009;**459**:455–9.
23. Barski A, Cuddapah S, Cui K, Roh TY, Schones DE, Wang Z, et al. High-resolution profiling of histone methylations in the human genome. *Cell* 2007;**129**:823–37.
24. Hewawasam G, Shivaraju M, Mattingly M, Venkatesh S, Martin-Brown S, Florens L, et al. Psh1 is an E3 ubiquitin ligase that targets the centromeric histone variant Cse4. *Mol Cell* 2010;**40**:444–54.
25. Ranjitkar P, Press MO, Yi X, Baker R, MacCoss MJ, Biggins S. An E3 ubiquitin ligase prevents ectopic localization of the centromeric histone H3 variant via the centromere targeting domain. *Mol Cell* 2010;**40**:455–64.
26. Trojer P, Zhang J, Yonezawa M, Schmidt A, Zheng H, Jenuwein T, et al. Dynamic Histone H1 Isotype 4 Methylation and Demethylation by Histone Lysine Methyltransferase G9a/KMT1C and the Jumonji Domain-containing JMJD2/KDM4 Proteins. *J Biol Chem* 2009;**284**:8395–405.
27. Jenuwein T, Allis CD. Translating the histone code. *Science* 2001;**293**:1074–80.
28. Dutnall RN, Tafrov ST, Sternglanz R, Ramakrishnan V. Structure of the yeast histone acetyltransferase Hat1: insights into substrate specificity and implications for the Gcn5-related N-acetyltransferase superfamily. *Cold Spring Harb Symp Quant Biol* 1998;**63**:501–7.
29. Neuwald AF, Landsman D. GCN5-related histone N-acetyltransferases belong to a diverse superfamily that includes the yeast SPT10 protein. *Trends Biochem Sci* 1997;**22**:154–5.
30. Burroughs AM, Allen KN, Dunaway-Mariano D, Aravind L. Evolutionary genomics of the HAD superfamily: understanding the structural adaptations and catalytic diversity in a superfamily of phosphoesterases and allied enzymes. *J Mol Biol* 2006;**361**:1003–34.
31. Blander G, Guarente L. The Sir2 family of protein deacetylases. *Annu Rev Biochem* 2004;**73**:417–35.

32. Leipe DD, Landsman D. Histone deacetylases, acetoin utilization proteins and acetylpolyamine amidohydrolases are members of an ancient protein superfamily. *Nucleic Acids Res* 1997;**25**:3693–7.
33. Xiao A, Li H, Shechter D, Ahn SH, Fabrizio LA, Erdjument-Bromage H, et al. WSTF regulates the H2A.X DNA damage response via a novel tyrosine kinase activity. *Nature* 2009;**457**:57–62.
34. Burroughs AM, Iyer LM, Aravind L. Natural history of the E1-like superfamily: implication for adenylation, sulfur transfer, and ubiquitin conjugation. *Proteins* 2009;**75**:895–910.
35. Burroughs AM, Jaffee M, Iyer LM, Aravind L. Anatomy of the E2 ligase fold: implications for enzymology and evolution of ubiquitin/Ub-like protein conjugation. *J Struct Biol* 2008;**162**:205–18.
36. Burroughs AM, Balaji S, Iyer LM, Aravind L. Small but versatile: the extraordinary functional and structural diversity of the beta-grasp fold. *Biol Direct* 2007;**2**:18.
37. Bujnicki JM. Comparison of protein structures reveals monophyletic origin of the AdoMet-dependent methyltransferase family and mechanistic convergence rather than recent differentiation of N4-cytosine and N6-adenine DNA methylation. *In Silico Biol* 1999;**1**:175–82.
38. Schubert HL, Blumenthal RM, Cheng X. Many paths to methyltransfer: a chronicle of convergence. *Trends Biochem Sci* 2003;**28**:329–35.
39. Fauman EB, Blumenthal RM, Cheng X. Structure and evolution of AdoMet-dependent methyltransferases. In: Cheng X, Blumenthal RM, editors. *S-adeno-sylmethionine-dependent methyltransferases: Structures and functions*. River Edge: World Scientific; 1999. pp. 1–54.
40. Iyer LM, Aravind L. The emergence of catalytic and structural diversity within the beta-clip fold. *Proteins* 2004;**55**:977–91.
41. Lipson RS, Webb KJ, Clarke SG. Two novel methyltransferases acting upon eukaryotic elongation factor 1A in *Saccharomyces cerevisiae*. *Arch Biochem Biophys* 2010;**500**:137–43.
42. Porras-Yakushi TR, Whitelegge JP, Clarke S. Yeast ribosomal/cytochrome c SET domain methyltransferase subfamily: identification of Rpl23ab methylation sites and recognition motifs. *J Biol Chem* 2007;**282**:12368–76.
43. Porras-Yakushi TR, Whitelegge JP, Miranda TB, Clarke S. A novel SET domain methyltransferase modifies ribosomal protein Rpl23ab in yeast. *J Biol Chem* 2005;**280**:34590–8.
44. Sadaie M, Shinmyozu K, Nakayama J. A conserved SET domain methyltransferase, Set11, modifies ribosomal protein Rpl12 in fission yeast. *J Biol Chem* 2008;**283**:7185–95.
45. Martzen MR, McCraith SM, Spinelli SL, Torres FM, Fields S, Grayhack EJ, et al. A biochemical genomics approach for identifying genes by the activity of their products. *Science* 1999;**286**:1153–5.
46. Kozbial PZ, Mushegian AR. Natural history of S-adenosylmethionine-binding proteins. *BMC Struct Biol* 2005;**5**:19.
47. Heurgue-Hamard V, Champ S, Engstrom A, Ehrenberg M, Buckingham RH. The hemK gene in Escherichia coli encodes the N(5)-glutamine methyltransferase that modifies peptide release factors. *EMBO J* 2002;**21**:769–78.
48. Nakahigashi K, Kubo N, Narita S, Shimaoka T, Goto S, Oshima T, et al. HemK, a class of protein methyl transferase with similarity to DNA methyl transferases, methylates polypeptide chain release factors, and hemK knockout induces defects in translational termination. *Proc Natl Acad Sci USA* 2002;**99**:1473–8.
49. Demirci H, Gregory ST, Dahlberg AE, Jogl G. Multiple-site trimethylation of ribosomal protein L11 by the PrmA methyltransferase. *Structure* 2008;**16**:1059–66.
50. Lee YH, Stallcup MR. Minireview: protein arginine methylation of nonhistone proteins in transcriptional regulation. *Mol Endocrinol* 2009;**23**:425–33.
51. Pal S, Sif S. Interplay between chromatin remodelers and protein arginine methyltransferases. *J Cell Physiol* 2007;**213**:306–15.

52. An W, Kim J, Roeder RG. Ordered cooperative functions of PRMT1, p300, and CARM1 in transcriptional activation by p53. *Cell* 2004;**117**:735–48.
53. Blythe SA, Cha SW, Tadjuidje E, Heasman J, Klein PS. beta-Catenin primes organizer gene expression by recruiting a histone H3 arginine 8 methyltransferase, Prmt2. *Dev Cell* 2010;**19**:220–31.
54. Chen SL, Loffler KA, Chen D, Stallcup MR, Muscat GE. The coactivator-associated arginine methyltransferase is necessary for muscle differentiation: CARM1 coactivates myocyte enhancer factor-2. *J Biol Chem* 2002;**277**:4324–33.
55. Lakowski TM, Frankel A. Kinetic analysis of human protein arginine N-methyltransferase 2: formation of monomethyl- and asymmetric dimethyl-arginine residues on histone H4. *Biochem J* 2009;**421**:253–61.
56. Strahl BD, Briggs SD, Brame CJ, Caldwell JA, Koh SS, Ma H, et al. Methylation of histone H4 at arginine 3 occurs in vivo and is mediated by the nuclear receptor coactivator PRMT1. *Curr Biol* 2001;**11**:996–1000.
57. Wang H, Huang ZQ, Xia L, Feng Q, Erdjument-Bromage H, Strahl BD, et al. Methylation of histone H4 at arginine 3 facilitating transcriptional activation by nuclear hormone receptor. *Science* 2001;**293**:853–7.
58. Yue WW, Hassler M, Roe SM, Thompson-Vale V, Pearl LH. Insights into histone code syntax from structural and biochemical studies of CARM1 methyltransferase. *EMBO J* 2007;**26**:4402–12.
59. Miranda TB, Webb KJ, Edberg DD, Reeves R, Clarke S. Protein arginine methyltransferase 6 specifically methylates the nonhistone chromatin protein HMGA1a. *Biochem Biophys Res Commun* 2005;**336**:831–5.
60. Anne J, Ollo R, Ephrussi A, Mechler BM. Arginine methyltransferase Capsuleen is essential for methylation of spliceosomal Sm proteins and germ cell formation in Drosophila. *Development* 2007;**134**:137–46.
61. Chan JY, Hsieh TY, Liu ST, Chou WY, Chung MH, Huang SM. Physical and functional interactions between hnRNP K and PRMT family proteins. *FEBS Lett* 2009;**583**:281–6.
62. Gonsalvez GB, Tian L, Ospina JK, Boisvert FM, Lamond AI, Matera AG. Two distinct arginine methyltransferases are required for biogenesis of Sm-class ribonucleoproteins. *J Cell Biol* 2007;**178**:733–40.
63. Kim C, Lim Y, Yoo BC, Won NH, Kim S, Kim G. Regulation of post-translational protein arginine methylation during HeLa cell cycle. *Biochim Biophys Acta* 2010;**1800**:977–85.
64. Pahlich S, Zakaryan RP, Gehring H. Identification of proteins interacting with protein arginine methyltransferase 8: the Ewing sarcoma (EWS) protein binds independent of its methylation state. *Proteins* 2008;**72**:1125–37.
65. Vagin VV, Wohlschlegel J, Qu J, Jonsson Z, Huang X, Chuma S, et al. Proteomic analysis of murine Piwi proteins reveals a role for arginine methylation in specifying interaction with Tudor family members. *Genes Dev* 2009;**23**:1749–62.
66. Bachand F, Silver PA. PRMT3 is a ribosomal protein methyltransferase that affects the cellular levels of ribosomal subunits. *EMBO J* 2004;**23**:2641–50.
67. Niu L, Lu F, Pei Y, Liu C, Cao X. Regulation of flowering time by the protein arginine methyltransferase AtPRMT10. *EMBO Rep* 2007;**8**:1190–5.
68. Fisk JC, Sayegh J, Zurita-Lopez C, Menon S, Presnyak V, Clarke SG, et al. A type III protein arginine methyltransferase from the protozoan parasite Trypanosoma brucei. *J Biol Chem* 2009;**284**:11590–600.
69. Troffer-Charlier N, Cura V, Hassenboehler P, Moras D, Cavarelli J. Functional insights from structures of coactivator-associated arginine methyltransferase 1 domains. *EMBO J* 2007;**26**:4391–401.

70. Lee JH, Cook JR, Yang ZH, Mirochnitchenko O, Gunderson SI, Felix AM, et al. PRMT7, a new protein arginine methyltransferase that synthesizes symmetric dimethylarginine. *J Biol Chem* 2005;**280**:3656–64.
71. Frankel A, Clarke S. PRMT3 is a distinct member of the protein arginine N-methyltransferase family. *Conferral of substrate specificity by a zinc-finger domain. J Biol Chem* 2000;**275**:32974–82.
72. Rouhiainen L, Paulin L, Suomalainen S, Hyytiainen H, Buikema W, Haselkorn R, et al. Genes encoding synthetases of cyclic depsipeptides, anabaenopeptilides, in Anabaena strain 90. *Mol Microbiol* 2000;**37**:156–67.
73. Mellor J. Linking the cell cycle to histone modifications: DOT1, G1/S, and cycling K79me2. *Mol Cell* 2009;**35**:729–30.
74. Okada Y, Feng Q, Lin Y, Jiang Q, Li Y, Coffield VM, et al. hDOT1L links histone methylation to leukemogenesis. *Cell* 2005;**121**:167–78.
75. Feng Q, Wang H, Ng HH, Erdjument-Bromage H, Tempst P, Struhl K, et al. Methylation of H3-lysine 79 is mediated by a new family of HMTases without a SET domain. *Curr Biol* 2002;**12**:1052–8.
76. Dlakic M. Chromatin silencing protein and pachytene checkpoint regulator DOT1p has a methyltransferase fold. *Trends Biochem Sci* 2001;**26**:405–7.
77. Schulze JM, Jackson J, Nakanishi S, Gardner JM, Hentrich T, Haug J, et al. Linking cell cycle to histone modifications: SBF and H2B monoubiquitination machinery and cell-cycle regulation of H3K79 dimethylation. *Mol Cell* 2009;**35**:626–41.
78. Zhang W, Hayashizaki Y, Kone BC. Structure and regulation of the mDOT1 gene, a mouse histone H3 methyltransferase. *Biochem J* 2004;**377**:641–51.
79. Oh S, Jeong K, Kim H, Kwon CS, Lee D. A lysine-rich region in DOT1p is crucial for direct interaction with H2B ubiquitylation and high level methylation of H3K79. *Biochem Biophys Res Commun* 2010;**399**:512–7.
80. Frederiks F, van Welsem T, Oudgenoeg G, Heck AJ, Janzen CJ, van Leeuwen F. Heterologous expression reveals distinct enzymatic activities of two DOT1 histone methyltransferases of Trypanosoma brucei. *J Cell Sci* 2010;**123**:4019–23.
81. Levesque N, Leung GP, Fok AK, Schmidt TI, Kobor MS. Loss of H3 K79 trimethylation leads to suppression of Rtt107-dependent DNA damage sensitivity through the translesion synthesis pathway. *J Biol Chem* 2010;**285**:35113–22.
82. Conde F, San-Segundo PA. Role of DOT1 in the response to alkylating DNA damage in *Saccharomyces cerevisiae*: regulation of DNA damage tolerance by the error-prone polymerases Polzeta/Rev1. *Genetics* 2008;**179**:1197–210.
83. Boissier F, Bardou F, Guillet V, Uttenweiler-Joseph S, Daffe M, Quemard A, et al. Further insight into S-adenosylmethionine-dependent methyltransferases: structural characterization of Hma, an enzyme essential for the biosynthesis of oxygenated mycolic acids in Mycobacterium tuberculosis. *J Biol Chem* 2006;**281**:4434–45.
84. Harvey BM, Mironenko T, Sun Y, Hong H, Deng Z, Leadlay PF, et al. Insights into polyether biosynthesis from analysis of the nigericin biosynthetic gene cluster in Streptomyces sp. DSM4137. *Chem Biol* 2007;**14**:703–14.
85. Richards EJ, Elgin SC. Epigenetic codes for heterochromatin formation and silencing: rounding up the usual suspects. *Cell* 2002;**108**:489–500.
86. Koonin EV, Aravind L, Kondrashov AS. The impact of comparative genomics on our understanding of evolution. *Cell* 2000;**101**:573–6.
87. Nakamura T, Mori T, Tada S, Krajewski W, Rozovskaia T, Wassell R, et al. ALL-1 is a histone methyltransferase that assembles a supercomplex of proteins involved in transcriptional regulation. *Mol Cell* 2002;**10**:1119–28.

88. Beisel C, Imhof A, Greene J, Kremmer E, Sauer F. Histone methylation by the Drosophila epigenetic transcriptional regulator Ash1. *Nature* 2002;**419**:857–62.
89. Wang H, Cao R, Xia L, Erdjument-Bromage H, Borchers C, Tempst P, et al. Purification and functional characterization of a histone H3-lysine 4-specific methyltransferase. *Mol Cell* 2001;**8**:1207–17.
90. Esteve PO, Chang Y, Samaranayake M, Upadhyay AK, Horton JR, Feehery GR, et al. A methylation and phosphorylation switch between an adjacent lysine and serine determines human DNMT1 stability. *Nat Struct Mol Biol* 2011;**18**:42–8.
91. Couture JF, Collazo E, Hauk G, Trievel RC. Structural basis for the methylation site specificity of SET7/9. *Nat Struct Mol Biol* 2006;**13**:140–6.
92. Sirinupong N, Brunzelle J, Ye J, Pirzada A, Nico L, Yang Z. Crystal structure of cardiac-specific histone methyltransferase SmyD1 reveals unusual active site architecture. *J Biol Chem* 2010;**285**:40635–44.
93. Briggs SD, Bryk M, Strahl BD, Cheung WL, Davie JK, Dent SY, et al. Histone H3 lysine 4 methylation is mediated by Set1 and required for cell growth and rDNA silencing in *Saccharomyces cerevisiae*. *Genes Dev* 2001;**15**:3286–95.
94. Saleh A, Al-Abdallat A, Ndamukong I, Alvarez-Venegas R, Avramova Z. The Arabidopsis homologs of trithorax (ATX1) and enhancer of zeste (CLF) establish 'bivalent chromatin marks' at the silent AGAMOUS locus. *Nucleic Acids Res* 2007;**35**:6290–6.
95. Zhang K, Mosch K, Fischle W, Grewal SI. Roles of the Clr4 methyltransferase complex in nucleation, spreading and maintenance of heterochromatin. *Nat Struct Mol Biol* 2008;**15**:381–8.
96. Nakayama J, Rice JC, Strahl BD, Allis CD, Grewal SI. Role of histone H3 lysine 9 methylation in epigenetic control of heterochromatin assembly. *Science* 2001;**292**:110–3.
97. Naumann K, Fischer A, Hofmann I, Krauss V, Phalke S, Irmler K, et al. Pivotal role of AtSUVH2 in heterochromatic histone methylation and gene silencing in Arabidopsis. *EMBO J* 2005;**24**:1418–29.
98. Derunes C, Briknarova K, Geng L, Li S, Gessner CR, Hewitt K, et al. Characterization of the PR domain of RIZ1 histone methyltransferase. *Biochem Biophys Res Commun* 2005;**333**:925–34.
99. Margueron R, Li G, Sarma K, Blais A, Zavadil J, Woodcock CL, et al. Ezh1 and Ezh2 maintain repressive chromatin through different mechanisms. *Mol Cell* 2008;**32**:503–18.
100. Nekrasov M, Wild B, Muller J. Nucleosome binding and histone methyltransferase activity of Drosophila PRC2. *EMBO Rep* 2005;**6**:348–53.
101. Muller J, Hart CM, Francis NJ, Vargas ML, Sengupta A, Wild B, et al. Histone methyltransferase activity of a Drosophila Polycomb group repressor complex. *Cell* 2002;**111**:197–208.
102. Jacob Y, Stroud H, Leblanc C, Feng S, Zhuo L, Caro E, et al. Regulation of heterochromatic DNA replication by histone H3 lysine 27 methyltransferases. *Nature* 2010;**466**:987–91.
103. Brown MA, Sims 3rd RJ, Gottlieb PD, Tucker PW. Identification and characterization of Smyd2: a split SET/MYND domain-containing histone H3 lysine 36-specific methyltransferase that interacts with the Sin3 histone deacetylase complex. *Mol Cancer* 2006;**5**:26.
104. Rao B, Shibata Y, Strahl BD, Lieb JD. Dimethylation of histone H3 at lysine 36 demarcates regulatory and nonregulatory chromatin genome-wide. *Mol Cell Biol* 2005;**25**:9447–59.
105. Balakrishnan L, Milavetz B. Decoding the histone H4 lysine 20 methylation mark. *Crit Rev Biochem Mol Biol* 2010;**45**:440–52.
106. Rayasam GV, Wendling O, Angrand PO, Mark M, Niederreither K, Song L, et al. NSD1 is essential for early post-implantation development and has a catalytically active SET domain. *EMBO J* 2003;**22**:3153–63.

107. Filion GJ, van Bemmel JG, Braunschweig U, Talhout W, Kind J, Ward LD, et al. Systematic protein location mapping reveals five principal chromatin types in Drosophila cells. *Cell* 2010;**143**:212–24.
108. Milne TA, Briggs SD, Brock HW, Martin ME, Gibbs D, Allis CD, et al. MLL targets SET domain methyltransferase activity to Hox gene promoters. *Mol Cell* 2002;**10**:1107–17.
109. Sanchez-Elsner T, Gou D, Kremmer E, Sauer F. Noncoding RNAs of trithorax response elements recruit Drosophila Ash1 to Ultrabithorax. *Science* 2006;**311**:1118–23.
110. Baudat F, Buard J, Grey C, Fledel-Alon A, Ober C, Przeworski M, et al. PRDM9 is a major determinant of meiotic recombination hotspots in humans and mice. *Science* 2010;**327**:836–40.
111. Myers S, Bowden R, Tumian A, Bontrop RE, Freeman C, MacFie TS, et al. Drive against hotspot motifs in primates implicates the PRDM9 gene in meiotic recombination. *Science* 2010;**327**:876–9.
112. Xu L, Zhao Z, Dong A, Soubigou-Taconnat L, Renou JP, Steinmetz A, et al. Di- and tri- but not monomethylation on histone H3 lysine 36 marks active transcription of genes involved in flowering time regulation and other processes in Arabidopsis thaliana. *Mol Cell Biol* 2008;**28**:1348–60.
113. Yoh SM, Lucas JS, Jones KA. The Iws1:Spt6:CTD complex controls cotranscriptional mRNA biosynthesis and HYPB/Setd2-mediated histone H3K36 methylation. *Genes Dev* 2008;**22**:3422–34.
114. Fnu S, Williamson EA, De Haro LP, Brenneman M, Wray J, Shaheen M, et al. Methylation of histone H3 lysine 36 enhances DNA repair by nonhomologous end-joining. *Proc Natl Acad Sci USA* 2011;**108**:540–5.
115. Huen MS, Sy SM, van Deursen JM, Chen J. Direct interaction between SET8 and proliferating cell nuclear antigen couples H4-K20 methylation with DNA replication. *J Biol Chem* 2008;**283**:11073–7.
116. Kim D, Blus BJ, Chandra V, Huang P, Rastinejad F, Khorasanizadeh S. Corecognition of DNA and a methylated histone tail by the MSL3 chromodomain. *Nat Struct Mol Biol* 2010;**17**:1027–9.
117. Larschan E, Alekseyenko AA, Gortchakov AA, Peng S, Li B, Yang P, et al. MSL complex is attracted to genes marked by H3K36 trimethylation using a sequence-independent mechanism. *Mol Cell* 2007;**28**:121–33.
118. Aravind L, Iyer LM. Provenance of SET-domain histone methyltransferases through duplication of a simple structural unit. *Cell Cycle* 2003;**2**:369–76.
119. Trievel RC, Beach BM, Dirk LM, Houtz RL, Hurley JH. Structure and catalytic mechanism of a SET domain protein methyltransferase. *Cell* 2002;**111**:91–103.
120. Jacobs SA, Harp JM, Devarakonda S, Kim Y, Rastinejad F, Khorasanizadeh S. The active site of the SET domain is constructed on a knot. *Nat Struct Biol* 2002;**9**:833–8.
121. Kwon T, Chang JH, Kwak E, Lee CW, Joachimiak A, Kim YC, et al. Mechanism of histone lysine methyl transfer revealed by the structure of SET7/9-AdoMet. *EMBO J* 2003;**22**:292–303.
122. Manzur KL, Farooq A, Zeng L, Plotnikova O, Koch AW, Sachchidanand , et al. A dimeric viral SET domain methyltransferase specific to Lys27 of histone H3. *Nat Struct Biol* 2003;**10**:187–96.
123. Wilson JR, Jing C, Walker PA, Martin SR, Howell SA, Blackburn GM, et al. Crystal structure and functional analysis of the histone methyltransferase SET7/9. *Cell* 2002;**111**:105–15.
124. Zhang X, Tamaru H, Khan SI, Horton JR, Keefe LJ, Selker EU, et al. Structure of the Neurospora SET domain protein DIM-5, a histone H3 lysine methyltransferase. *Cell* 2002;**111**:117–27.
125. Xiao B, Jing C, Wilson JR, Walker PA, Vasisht N, Kelly G, et al. Structure and catalytic mechanism of the human histone methyltransferase SET7/9. *Nature* 2003;**421**:652–6.
126. Aravind L, Watanabe H, Lipman DJ, Koonin EV. Lineage-specific loss and divergence of functionally linked genes in eukaryotes. *Proc Natl Acad Sci USA* 2000;**97**:11319–24.

127. Klose RJ, Gardner KE, Liang G, Erdjument-Bromage H, Tempst P, Zhang Y. Demethylation of histone H3K36 and H3K9 by Rph1: a vestige of an H3K9 methylation system in *Saccharomyces cerevisiae*? *Mol Cell Biol* 2007;**27**:3951–61.
128. Jacob Y, Feng S, LeBlanc CA, Bernatavichute YV, Stroud H, Cokus S, et al. ATXR5 and ATXR6 are H3K27 monomethyltransferases required for chromatin structure and gene silencing. *Nat Struct Mol Biol* 2009;**16**:763–8.
129. Brent MM, Marmorstein R. Ankyrin for methylated lysines. *Nat Struct Mol Biol* 2008;**15**:221–2.
130. Collins RE, Northrop JP, Horton JR, Lee DY, Zhang X, Stallcup MR, et al. The ankyrin repeats of G9a and GLP histone methyltransferases are mono- and dimethyllysine binding modules. *Nat Struct Mol Biol* 2008;**15**:245–50.
131. Miskey C, Papp B, Mates L, Sinzelle L, Keller H, Izsvak Z, et al. The ancient mariner sails again: transposition of the human Hsmar1 element by a reconstructed transposase and activities of the SETMAR protein on transposon ends. *Mol Cell Biol* 2007;**27**:4589–600.
132. Shaheen M, Williamson E, Nickoloff J, Lee SH, Hromas R. Metnase/SETMAR: a domesticated primate transposase that enhances DNA repair, replication, and decatenation. *Genetica* 2010;**138**:559–66.
133. Siegel TN, Hekstra DR, Kemp LE, Figueiredo LM, Lowell JE, Fenyo D, et al. Four histone variants mark the boundaries of polycistronic transcription units in Trypanosoma brucei. *Genes Dev* 2009;**23**:1063–76.
134. Mandava V, Janzen CJ, Cross GA. Trypanosome H2Bv replaces H2B in nucleosomes enriched for H3 K4 and K76 trimethylation. *Biochem Biophys Res Commun* 2008;**368**:846–51.
135. Lu PK, Tsai JY, Chien HY, Huang H, Chu CH, Sun YJ. Crystal structure of Helicobacter pylori spermidine synthase: a Rossmann-like fold with a distinct active site. *Proteins* 2007;**67**:743–54.
136. Kroger N, Deutzmann R, Sumper M. Polycationic peptides from diatom biosilica that direct silica nanosphere formation. *Science* 1999;**286**:1129–32.
137. Sumper M, Hett R, Lehmann G, Wenzl S. A code for lysine modifications of a silica biomineralizing silaffin protein. *Angew Chem Int Ed Engl* 2007;**46**:8405–8.
138. Kroger N, Lorenz S, Brunner E, Sumper M. Self-assembly of highly phosphorylated silaffins and their function in biosilica morphogenesis. *Science* 2002;**298**:584–6.
139. Sheppard V, Poulsen N, Kroger N. Characterization of an endoplasmic reticulum-associated silaffin kinase from the diatom Thalassiosira pseudonana. *J Biol Chem* 2010;**285**:1166–76.
140. Boulton A, Myers RS, Redfield RJ. The hotspot conversion paradox and the evolution of meiotic recombination. *Proc Natl Acad Sci USA* 1997;**94**:8058–63.
141. Stephens RS, Kalman S, Lammel C, Fan J, Marathe R, Aravind L, et al. Genome sequence of an obligate intracellular pathogen of humans: *Chlamydia trachomatis*. *Science* 1998;**282**:754–9.
142. Pennini ME, Perrinet S, Dautry-Varsat A, Subtil A. Histone methylation by NUE, a novel nuclear effector of the intracellular pathogen Chlamydia trachomatis. *PLoS Pathog* 2010;**6**: e1000995.
143. Murata M, Azuma Y, Miura K, Rahman MA, Matsutani M, Aoyama M, et al. Chlamydial SET domain protein functions as a histone methyltransferase. *Microbiology* 2007;**153**:585–92.
144. Manzur KL, Zhou MM. An archaeal SET domain protein exhibits distinct lysine methyltransferase activity towards DNA-associated protein MC1-alpha. *FEBS Lett* 2005;**579**:3859–65.
145. Hsiao PW, Fryer CJ, Trotter KW, Wang W, Archer TK. BAF60a mediates critical interactions between nuclear receptors and the BRG1 chromatin-remodeling complex for transactivation. *Mol Cell Biol* 2003;**23**:6210–20.
146. Ito T, Yamauchi M, Nishina M, Yamamichi N, Mizutani T, Ui M, et al. Identification of SWI. SNF complex subunit BAF60a as a determinant of the transactivation potential of Fos/Jun dimers. *J Biol Chem* 2001;**276**:2852–7.

147. Oh J, Sohn DH, Ko M, Chung H, Jeon SH, Seong RH. BAF60a interacts with p53 to recruit the SWI/SNF complex. *J Biol Chem* 2008;**283**:11924–34.
148. de Souza RF, Iyer LM, Aravind L. Diversity and evolution of chromatin proteins encoded by DNA viruses. *Biochim Biophys Acta* 2010;**1799**:302–18.
149. Mujtaba S, Manzur KL, Gurnon JR, Kang M, Van Etten JL, Zhou MM. *Epigenetic transcriptional repression of cellular genes by a viral SET protein.* Nat Cell Biol; 2008.
150. Hagiwara T, Nakashima K, Hirano H, Senshu T, Yamada M. Deimination of arginine residues in nucleophosmin/B23 and histones in HL-60 granulocytes. *Biochem Biophys Res Commun* 2002;**290**:979–83.
151. Wang Y, Wysocka J, Sayegh J, Lee YH, Perlin JR, Leonelli L, et al. Human PAD4 regulates histone arginine methylation levels via demethylimination. *Science* 2004;**306**:279–83.
152. Cuthbert GL, Daujat S, Snowden AW, Erdjument-Bromage H, Hagiwara T, Yamada M, et al. Histone deimination antagonizes arginine methylation. *Cell* 2004;**118**:545–53.
153. Vossenaar ER, Zendman AJ, van Venrooij WJ, Pruijn GJ. PAD, a growing family of citrullinating enzymes: genes, features and involvement in disease. *Bioessays* 2003;**25**:1106–18.
154. Lee YH, Coonrod SA, Kraus WL, Jelinek MA, Stallcup MR. Regulation of coactivator complex assembly and function by protein arginine methylation and demethylimination. *Proc Natl Acad Sci USA* 2005;**102**:3611–6.
155. Kearney PL, Bhatia M, Jones NG, Yuan L, Glascock MC, Catchings KL, et al. Kinetic characterization of protein arginine deiminase 4: a transcriptional corepressor implicated in the onset and progression of rheumatoid arthritis. *Biochemistry* 2005;**44**:10570–82.
156. Shirai H, Blundell TL, Mizuguchi K. A novel superfamily of enzymes that catalyze the modification of guanidino groups. *Trends Biochem Sci* 2001;**26**:465–8.
157. McGraw WT, Potempa J, Farley D, Travis J. Purification, characterization, and sequence analysis of a potential virulence factor from Porphyromonas gingivalis, peptidylarginine deiminase. *Infect Immun* 1999;**67**:3248–56.
158. Arita K, Hashimoto H, Shimizu T, Nakashima K, Yamada M, Sato M. Structural basis for Ca(2+)-induced activation of human PAD4. *Nat Struct Mol Biol* 2004;**11**:777–83.
159. Lu X, Galkin A, Herzberg O, Dunaway-Mariano D. Arginine deiminase uses an active-site cysteine in nucleophilic catalysis of L-arginine hydrolysis. *J Am Chem Soc* 2004;**126**:5374–5.
160. Touz MC, Ropolo AS, Rivero MR, Vranych CV, Conrad JT, Svard SG, et al. Arginine deiminase has multiple regulatory roles in the biology of Giardia lamblia. *J Cell Sci* 2008;**121**:2930–8.
161. Aravind L, Iyer LM. The SWIRM domain: a conserved module found in chromosomal proteins points to novel chromatin-modifying activities. *Genome Biol* 2002;**3**: RESEARCH0039.
162. Humphrey GW, Wang Y, Russanova VR, Hirai T, Qin J, Nakatani Y, et al. Stable histone deacetylase complexes distinguished by the presence of SANT domain proteins CoREST/kiaa0071 and Mta-L1. *J Biol Chem* 2001;**276**:6817–24.
163. Nicholson TB, Chen T. LSD1 demethylates histone and non-histone proteins. *Epigenetics* 2009;**4**:129–32.
164. Shi Y, Lan F, Matson C, Mulligan P, Whetstine JR, Cole PA, et al. Histone demethylation mediated by the nuclear amine oxidase homolog LSD1. *Cell* 2004;**119**:941–53.
165. Culhane JC, Cole PA. LSD1 and the chemistry of histone demethylation. *Curr Opin Chem Biol* 2007;**11**:561–8.
166. Forneris F, Binda C, Battaglioli E, Mattevi A. LSD1: oxidative chemistry for multifaceted functions in chromatin regulation. *Trends Biochem Sci* 2008;**33**:181–9.
167. Rudolph T, Yonezawa M, Lein S, Heidrich K, Kubicek S, Schafer C, et al. Heterochromatin formation in Drosophila is initiated through active removal of H3K4 methylation by the LSD1 homolog SU(VAR)3-3. *Mol Cell* 2007;**26**:103–15.

168. Tsai MC, Manor O, Wan Y, Mosammaparast N, Wang JK, Lan F, et al. Long noncoding RNA as modular scaffold of histone modification complexes. *Science* 2010;**329**:689–93.
169. Wang J, Scully K, Zhu X, Cai L, Zhang J, Prefontaine GG, et al. Opposing LSD1 complexes function in developmental gene activation and repression programmes. *Nature* 2007;**446**:882–7.
170. Wang Y, Zhang H, Chen Y, Sun Y, Yang F, Yu W, et al. LSD1 is a subunit of the NuRD complex and targets the metastasis programs in breast cancer. *Cell* 2009;**138**:660–72.
171. Gordon M, Holt DG, Panigrahi A, Wilhelm BT, Erdjument-Bromage H, Tempst P, et al. Genome-wide dynamics of SAPHIRE, an essential complex for gene activation and chromatin boundaries. *Mol Cell Biol* 2007;**27**:4058–69.
172. Huang J, Sengupta R, Espejo AB, Lee MG, Dorsey JA, Richter M, et al. p53 is regulated by the lysine demethylase LSD1. *Nature* 2007;**449**:105–8.
173. Katz DJ, Edwards TM, Reinke V, Kelly WG. A *C. elegans* LSD1 demethylase contributes to germline immortality by reprogramming epigenetic memory. *Cell* 2009;**137**:308–20.
174. Swigut T, Wysocka J. Fallen immortals. *Cell* 2009;**137**:203–5.
175. Yang M, Gocke CB, Luo X, Borek D, Tomchick DR, Machius M, et al. Structural basis for CoREST-dependent demethylation of nucleosomes by the human LSD1 histone demethylase. *Mol Cell* 2006;**23**:377–87.
176. Forneris F, Binda C, Adamo A, Battaglioli E, Mattevi A. Structural basis of LSD1-CoREST selectivity in histone H3 recognition. *J Biol Chem* 2007;**282**:20070–4.
177. Chen Y, Yang Y, Wang F, Wan K, Yamane K, Zhang Y, et al. Crystal structure of human histone lysine-specific demethylase 1 (LSD1). *Proc Natl Acad Sci USA* 2006;**103**:13956–61.
178. Cook T, Roos D, Morada M, Zhu G, Keithly JS, Feagin JE, et al. Divergent polyamine metabolism in the Apicomplexa. *Microbiology* 2007;**153**:1123–30.
179. Vujcic S, Liang P, Diegelman P, Kramer DL, Porter CW. Genomic identification and biochemical characterization of the mammalian polyamine oxidase involved in polyamine back conversion. *Biochem J* 2003;**370**:19–28.
180. Seiler N. Catabolism of polyamines. *Amino Acids* 2004;**26**:217–33.
181. Iyer LM, Abhiman S, Maxwell Burroughs A, Aravind L. Amidoligases with ATP-grasp, glutamine synthetase-like and acetyltransferase-like domains: synthesis of novel metabolites and peptide modifications of proteins. *Mol Biosyst* 2009;**5**:1636–60.
182. Iyer LM, Abhiman S, de Souza RF, Aravind L. Origin and evolution of peptide-modifying dioxygenases and identification of the wybutosine hydroxylase/hydroperoxidase. *Nucleic Acids Res* 2010;**38**:5261–79.
183. Murray-Stewart T, Wang Y, Goodwin A, Hacker A, Meeker A, Casero Jr. RA. Nuclear localization of human spermine oxidase isoforms - possible implications in drug response and disease etiology. *FEBS J* 2008;**275**:2795–806.
184. Larsson J, Zhang J, Rasmuson-Lestander A. Mutations in the Drosophila melanogaster gene encoding S-adenosylmethionine synthetase [corrected] suppress position-effect variegation. *Genetics* 1996;**143**:887–96.
185. Anantharaman V, Koonin EV, Aravind L. Regulatory potential, phyletic distribution and evolution of ancient, intracellular small-molecule-binding domains. *J Mol Biol* 2001;**307**:1271–92.
186. Clissold PM, Ponting CP. JmjC: cupin metalloenzyme-like domains in jumonji, hairless and phospholipase A2beta. *Trends Biochem Sci* 2001;**26**:7–9.
187. Whetstine JR, Nottke A, Lan F, Huarte M, Smolikov S, Chen Z, et al. Reversal of histone lysine trimethylation by the JMJD2 family of histone demethylases. *Cell* 2006;**125**:467–81.
188. Tsukada Y, Fang J, Erdjument-Bromage H, Warren ME, Borchers CH, Tempst P, et al. Histone demethylation by a family of JmjC domain-containing proteins. *Nature* 2006;**439**:811–6.

189. Klose RJ, Kallin EM, Zhang Y. JmjC-domain-containing proteins and histone demethylation. *Nat Rev* 2006;**7**:715–27.
190. Chang B, Chen Y, Zhao Y, Bruick RK. JMJD6 is a histone arginine demethylase. *Science* 2007;**318**:444–7.
191. Elkins JM, Hewitson KS, McNeill LA, Seibel JF, Schlemminger I, Pugh CW, et al. Structure of factor-inhibiting hypoxia-inducible factor (HIF) reveals mechanism of oxidative modification of HIF-1 alpha. *J Biol Chem* 2003;**278**:1802–6.
192. Dann 3rd CE, Bruick RK, Deisenhofer J. Structure of factor-inhibiting hypoxia-inducible factor 1: an asparaginyl hydroxylase involved in the hypoxic response pathway. *Proc Natl Acad Sci USA* 2002;**99**:15351–6.
193. Krishnan S, Horowitz S, Trievel RC. Structure and function of histone h3 lysine 9 methyltransferases and demethylases. *Chembiochem* 2011;**12**:254–63.
194. Seward DJ, Cubberley G, Kim S, Schonewald M, Zhang L, Tripet B, et al. Demethylation of trimethylated histone H3 Lys4 in vivo by JARID1 JmjC proteins. *Nat Struct Mol Biol* 2007;**14**:240–2.
195. Secombe J, Li L, Carlos L, Eisenman RN. The Trithorax group protein Lid is a trimethyl histone H3K4 demethylase required for dMyc-induced cell growth. *Genes Dev* 2007;**21**:537–51.
196. Iwase S, Lan F, Bayliss P, de la Torre-Ubieta L, Huarte M, Qi HH, et al. The X-linked mental retardation gene SMCX/JARID1C defines a family of histone H3 lysine 4 demethylases. *Cell* 2007;**128**:1077–88.
197. Eissenberg JC, Lee MG, Schneider J, Ilvarsonn A, Shiekhattar R, Shilatifard A. The trithorax-group gene in Drosophila little imaginal discs encodes a trimethylated histone H3 Lys4 demethylase. *Nat Struct Mol Biol* 2007;**14**:344–6.
198. Tahiliani M, Mei P, Fang R, Leonor T, Rutenberg M, Shimizu F, et al. The histone H3K4 demethylase SMCX links REST target genes to X-linked mental retardation. *Nature* 2007;**447**:601–5.
199. Secombe J, Eisenman RN. The function and regulation of the JARID1 family of histone H3 lysine 4 demethylases: the Myc connection. *Cell Cycle* 2007;**6**:1324–8.
200. Di Stefano L, Walker JA, Burgio G, Corona DF, Mulligan P, Naar AM, et al. Functional antagonism between histone H3K4 demethylases in vivo. *Genes Dev* 2011;**25**:17–28.
201. Yokoyama A, Okuno Y, Chikanishi T, Hashiba W, Sekine H, Fujiki R, et al. KIAA1718 is a histone demethylase that erases repressive histone methyl marks. *Genes Cells* 2010;**15**:867–73.
202. Liu Z, Zhou S, Liao L, Chen X, Meistrich M, Xu J. Jmjd1a demethylase-regulated histone modification is essential for cAMP-response element modulator-regulated gene expression and spermatogenesis. *J Biol Chem* 2010;**285**:2758–70.
203. Laumonnier F, Holbert S, Ronce N, Faravelli F, Lenzner S, Schwartz CE, et al. Mutations in PHF8 are associated with X linked mental retardation and cleft lip/cleft palate. *J Med Genet* 2005;**42**:780–6.
204. Lee MG, Villa R, Trojer P, Norman J, Yan KP, Reinberg D, et al. Demethylation of H3K27 regulates polycomb recruitment and H2A ubiquitination. *Science* 2007;**318**:447–50.
205. Agger K, Cloos PA, Christensen J, Pasini D, Rose S, Rappsilber J, et al. UTX and JMJD3 are histone H3K27 demethylases involved in HOX gene regulation and development. *Nature* 2007;**449**:731–4.
206. Blackledge NP, Zhou JC, Tolstorukov MY, Farcas AM, Park PJ, Klose RJ. CpG islands recruit a histone H3 lysine 36 demethylase. *Mol Cell* 2010;**38**:179–90.
207. Hsia DA, Tepper CG, Pochampalli MR, Hsia EY, Izumiya C, Huerta SB, et al. KDM8, a H3K36me2 histone demethylase that acts in the cyclin A1 coding region to regulate cancer cell proliferation. *Proc Natl Acad Sci USA* 2010;**107**:9671–6.

208. Suganuma T, Workman JL. Features of the PHF8/KIAA1718 histone demethylase. *Cell Res* 2010;**20**:861–2.
209. Sinha KM, Yasuda H, Coombes MM, Dent SY, de Crombrugghe B. Regulation of the osteoblast-specific transcription factor Osterix by NO66, a Jumonji family histone demethylase. *Embo J* 2010;**29**:68–79.
210. Okamoto M, Van Stry M, Chung L, Koyanagi M, Sun X, Suzuki Y, et al. Mina, an Il4 repressor, controls T helper type 2 bias. *Nat Immunol* 2009;**10**:872–9.
211. Eilbracht J, Reichenzeller M, Hergt M, Schnolzer M, Heid H, Stohr M, et al. NO66, a highly conserved dual location protein in the nucleolus and in a special type of synchronously replicating chromatin. *Mol Biol Cell* 2004;**15**:1816–32.
212. Chen Z, Zang J, Whetstine J, Hong X, Davrazou F, Kutateladze TG, et al. Structural insights into histone demethylation by JMJD2 family members. *Cell* 2006;**125**:691–702.
213. Maurer-Stroh S, Dickens NJ, Hughes-Davies L, Kouzarides T, Eisenhaber F, Ponting CP. The Tudor domain 'Royal Family': Tudor, plant Agenet, Chromo, PWWP and MBT domains. *Trends Biochem Sci* 2003;**28**:69–74.
214. Norris A, Boeke JD. Silent information regulator 3: the Goldilocks of the silencing complex. *Genes Dev* 2010;**24**:115–22.
215. Wang Y, Reddy B, Thompson J, Wang H, Noma K, Yates 3rd JR, et al. Regulation of Set9-mediated H4K20 methylation by a PWWP domain protein. *Mol Cell* 2009;**33**:428–37.
216. Sathyamurthy A, Allen MD, Murzin AG, Bycroft M. Crystal structure of the malignant brain tumor (MBT) repeats in Sex Comb on Midleg-like 2 (SCML2). *J Biol Chem* 2003;**278**:46968–73.
217. Lancelot N, Charier G, Couprie J, Duband-Goulet I, Alpha-Bazin B, Quemeneur E, et al. The checkpoint *Saccharomyces cerevisiae* Rad9 protein contains a tandem tudor domain that recognizes DNA. *Nucleic Acids Res* 2007;**35**:5898–912.
218. Huyen Y, Zgheib O, Ditullio Jr. RA, Gorgoulis VG, Zacharatos P, Petty TJ, et al. Methylated lysine 79 of histone H3 targets 53BP1 to DNA double-strand breaks. *Nature* 2004;**432**:406–11.
219. Callebaut I, Mornon JP. The human EBNA-2 coactivator p100: multidomain organization and relationship to the staphylococcal nuclease fold and to the tudor protein involved in Drosophila melanogaster development. *Biochem J* 1997;**321**(Pt 1):125–32.
220. Shimojo H, Sano N, Moriwaki Y, Okuda M, Horikoshi M, Nishimura Y. Novel structural and functional mode of a knot essential for RNA binding activity of the Esa1 presumed chromodomain. *J Mol Biol* 2008;**378**:987–1001.
221. Iyer LM, Abhiman S, Aravind L. MutL homologs in restriction-modification systems and the origin of eukaryotic MORC ATPases. *Biol Direct* 2008;**3**:8.
222. Jacobs SA, Khorasanizadeh S. Structure of HP1 chromodomain bound to a lysine 9-methylated histone H3 tail. *Science* 2002;**295**:2080–3.
223. Murzin AG. OB(oligonucleotide/oligosaccharide binding)-fold: common structural and functional solution for non-homologous sequences. *EMBO J* 1993;**12**:861–7.
224. Flanagan JF, Mi LZ, Chruszcz M, Cymborowski M, Clines KL, Kim Y, et al. Double chromodomains cooperate to recognize the methylated histone H3 tail. *Nature* 2005;**438**:1181–5.
225. Grimm C, Matos R, Ly-Hartig N, Steuerwald U, Lindner D, Rybin V, et al. Molecular recognition of histone lysine methylation by the Polycomb group repressor dSfmbt. *EMBO J* 2009;**28**:1965–77.
226. Huang Y, Fang J, Bedford MT, Zhang Y, Xu RM. Recognition of histone H3 lysine-4 methylation by the double tudor domain of JMJD2A. *Science* 2006;**312**:748–51.
227. Liu H, Wang JY, Huang Y, Li Z, Gong W, Lehmann R, et al. Structural basis for methylarginine-dependent recognition of Aubergine by Tudor. *Genes Dev* 2010;**24**:1876–81.

228. Perez-Alegre M, Dubus A, Fernandez E. REM1, a new type of long terminal repeat retrotransposon in Chlamydomonas reinhardtii. *Mol Cell Biol* 2005;**25**:10628–38.
229. Ponting CP, Aravind L, Schultz J, Bork P, Koonin EV. Eukaryotic signalling domain homologues in archaea and bacteria. Ancient ancestry and horizontal gene transfer. *J Mol Biol* 1999;**289**:729–45.
230. Krishna SS, Majumdar I, Grishin NV. Structural classification of zinc fingers: survey and summary. *Nucleic Acids Res* 2003;**31**:532–50.
231. Aravind L, Iyer LM, Koonin EV. Scores of RINGS but no PHDs in ubiquitin signaling. *Cell Cycle* 2003;**2**:123–6.
232. Chakravarty S, Zeng L, Zhou MM. Structure and site-specific recognition of histone H3 by the PHD finger of human autoimmune regulator. *Structure* 2009;**17**:670–9.
233. Li H, Ilin S, Wang W, Duncan EM, Wysocka J, Allis CD, et al. Molecular basis for site-specific read-out of histone H3K4me3 by the BPTF PHD finger of NURF. *Nature* 2006;**442**:91–5.
234. Zeng L, Zhang Q, Li S, Plotnikov AN, Walsh MJ, Zhou MM. Mechanism and regulation of acetylated histone binding by the tandem PHD finger of DPF3b. *Nature* 2010;**466**:258–62.
235. Miller TC, Rutherford TJ, Johnson CM, Fiedler M, Bienz M. Allosteric remodelling of the histone H3 binding pocket in the Pygo2 PHD finger triggered by its binding to the B9L/BCL9 co-factor. *J Mol Biol* 2010;**401**:969–84.
236. van Ingen H, van Schaik FM, Wienk H, Ballering J, Rehmann H, Dechesne AC, et al. Structural insight into the recognition of the H3K4me3 mark by the TFIID subunit TAF3. *Structure* 2008;**16**:1245–56.
237. He F, Umehara T, Saito K, Harada T, Watanabe S, Yabuki T, et al. Structural insight into the zinc finger CW domain as a histone modification reader. *Structure* 2010;**18**:1127–39.
238. Wysocka J, Swigut T, Milne TA, Dou Y, Zhang X, Burlingame AL, et al. WDR5 associates with histone H3 methylated at K4 and is essential for H3 K4 methylation and vertebrate development. *Cell* 2005;**121**:859–72.
239. Schuetz A, Allali-Hassani A, Martin F, Loppnau P, Vedadi M, Bochkarev A, et al. Structural basis for molecular recognition and presentation of histone H3 by WDR5. *EMBO J* 2006;**25**:4245–52.
240. Couture JF, Collazo E, Trievel RC. Molecular recognition of histone H3 by the WD40 protein WDR5. *Nat Struct Mol Biol* 2006;**13**:698–703.
241. Ruthenburg AJ, Wang W, Graybosch DM, Li H, Allis CD, Patel DJ, et al. Histone H3 recognition and presentation by the WDR5 module of the MLL1 complex. *Nat Struct Mol Biol* 2006;**13**:704–12.
242. Trievel RC, Shilatifard A. WDR5, a complexed protein. *Nat Struct Mol Biol* 2009;**16**:678–80.
243. Tie F, Stratton CA, Kurzhals RL, Harte PJ. The N terminus of Drosophila ESC binds directly to histone H3 and is required for E(Z)-dependent trimethylation of H3 lysine 27. *Mol Cell Biol* 2007;**27**:2014–26.
244. Collins R, Cheng X. A case study in cross-talk: the histone lysine methyltransferases G9a and GLP. *Nucleic Acids Res* 2010;**38**:3503–11.
245. Gao T, Collins RE, Horton JR, Zhang X, Zhang R, Dhayalan A, et al. The ankyrin repeat domain of Huntingtin interacting protein 14 contains a surface aromatic cage, a potential site for methyl-lysine binding. *Proteins* 2009;**76**:772–7.
246. Balaji S, Iyer LM, Aravind L. HPC2 and ubinuclein define a novel family of histone chaperones conserved throughout eukaryotes. *Mol Biosyst* 2009;**5**:269–75.
247. Garcia-Alai MM, Allen MD, Joerger AC, Bycroft M. The structure of the FYR domain of transforming growth factor beta regulator 1. *Protein Sci* 2010;**19**:1432–8.
248. Elia AE, Rellos P, Haire LF, Chao JW, Ivins FJ, Hoepker K, et al. The molecular basis for phosphodependent substrate targeting and regulation of Plks by the Polo-box domain. *Cell* 2003;**115**:83–95.

249. Min J, Zhang X, Cheng X, Grewal SI, Xu RM. Structure of the SET domain histone lysine methyltransferase Clr4. *Nat Struct Biol* 2002;**9**:828–32.
250. Brzeska K, Brzeski J, Smith J, Chandler VL. Transgenic expression of CBBP, a CXC domain protein, establishes paramutation in maize. *Proc Natl Acad Sci USA* 2010;**107**:5516–21.
251. Fauth T, Muller-Planitz F, Konig C, Straub T, Becker PB. The DNA binding CXC domain of MSL2 is required for faithful targeting the Dosage Compensation Complex to the X chromosome. *Nucleic Acids Res* 2010;**38**:3209–21.
252. Andersen SU, Algreen-Petersen RG, Hoedl M, Jurkiewicz A, Cvitanich C, Braunschweig U, et al. The conserved cysteine-rich domain of a tesmin/TSO1-like protein binds zinc in vitro and TSO1 is required for both male and female fertility in Arabidopsis thaliana. *J Exp Bot* 2007;**58**:3657–70.
253. Krajewski WA, Nakamura T, Mazo A, Canaani E. A motif within SET-domain proteins binds single-stranded nucleic acids and transcribed and supercoiled DNAs and can interfere with assembly of nucleosomes. *Mol Cell Biol* 2005;**25**:1891–9.
254. Vojnic E, Simon B, Strahl BD, Sattler M, Cramer P. Structure and carboxyl-terminal domain (CTD) binding of the Set2 SRI domain that couples histone H3 Lys36 methylation to transcription. *J Biol Chem* 2006;**281**:13–5.
255. Kizer KO, Phatnani HP, Shibata Y, Hall H, Greenleaf AL, Strahl BD. A novel domain in Set2 mediates RNA polymerase II interaction and couples histone H3 K36 methylation with transcript elongation. *Mol Cell Biol* 2005;**25**:3305–16.
256. Kanagaraj R, Huehn D, MacKellar A, Menigatti M, Zheng L, Urban V, et al. RECQ5 helicase associates with the C-terminal repeat domain of RNA polymerase II during productive elongation phase of transcription. *Nucleic Acids Res* 2010;**38**:8131–40.
257. Islam MN, Fox 3rd D, Guo R, Enomoto T, Wang W. RecQL5 promotes genome stabilization through two parallel mechanisms–interacting with RNA polymerase II and acting as a helicase. *Mol Cell Biol* 2010;**30**:2460–72.
258. Ingham RJ, Colwill K, Howard C, Dettwiler S, Lim CS, Yu J, et al. WW domains provide a platform for the assembly of multiprotein networks. *Mol Cell Biol* 2005;**25**:7092–106.
259. Iyer LM, Koonin EV, Aravind L. Evolutionary connection between the catalytic subunits of DNA-dependent RNA polymerases and eukaryotic RNA-dependent RNA polymerases and the origin of RNA polymerases. *BMC Struct Biol* 2003;**3**:1.
260. Muljo SA, Kanellopoulou C, Aravind L. MicroRNA targeting in mammalian genomes: genes and mechanisms. *Wiley Interdiscip Rev Syst Biol Med* 2010;**2**:148–61.
261. Juranek SA, Rupprecht S, Postberg J, Lipps HJ. snRNA and heterochromatin formation are involved in DNA excision during macronuclear development in stichotrichous ciliates. *Eukaryot Cell* 2005;**4**:1934–41.
262. Liu Y, Taverna SD, Muratore TL, Shabanowitz J, Hunt DF, Allis CD. RNAi-dependent H3K27 methylation is required for heterochromatin formation and DNA elimination in Tetrahymena. *Genes Dev* 2007;**21**:1530–45.
263. Gerace EL, Halic M, Moazed D. The methyltransferase activity of Clr4Suv39h triggers RNAi independently of histone H3K9 methylation. *Mol Cell* 2010;**39**:360–72.
264. Feng S, Cokus SJ, Zhang X, Chen PY, Bostick M, Goll MG, et al. Conservation and divergence of methylation patterning in plants and animals. *Proc Natl Acad Sci USA* 2010;**107**:8689–94.
265. Zemach A, McDaniel IE, Silva P, Zilberman D. Genome-wide evolutionary analysis of eukaryotic DNA methylation. *Science* 2010;**328**:916–9.
266. Wang J, Hevi S, Kurash JK, Lei H, Gay F, Bajko J, et al. The lysine demethylase LSD1 (KDM1) is required for maintenance of global DNA methylation. *Nat Genet* 2009;**41**:125–9.
267. Choudhary C, Kumar C, Gnad F, Nielsen ML, Rehman M, Walther TC, et al. Lysine acetylation targets protein complexes and co-regulates major cellular functions. *Science* 2009;**325**:834–40.

DNA Methylation in *Drosophila*—A Critical Evaluation

Veiko Krauss
and Gunter Reuter

Developmental Genetics, Institute of Biology, Martin Luther University Halle-Wittenberg, Weinbergweg, Halle/S, Germany

Drosophila belongs to the so-called "Dnmt2 only" organisms, and does not contain any of the canonical DNA methyltransferases (Dnmt1 and Dnmt3). Furthermore, no functional homologs of known 5-methylcytosine reader proteins are found. Nevertheless, there is strong evidence for DNA methylation in this organism. It has been suggested that DNA methylation in *Drosophila* is simply a byproduct of Dnmt2, which is a DNA methyltransferase (Dnmt) according to structure and type of catalysis but functions *in vivo* as a tRNA methyltransferase. However, concerning the very specific timing of cytosine methylation in *Drosophila*, their suggested functions in control of retrotransposon silencing and genome stability, and the obvious DNA methylation activity of Dnmt2 enzymes in the protozoans *Dictyostelium discoideum* and *Entamoeba histolytica*, we tend to disagree with this notation. Dnmt2 probably serves, and not only in *Drosophila*, as a methyltransferase of both specific DNA and tRNA targets.

Progress in Molecular Biology
and Translational Science, Vol. 101
DOI: 10.1016/B978-0-12-387685-0.00003-2

I. Evidence for 5-Methyl Cytosine in *Drosophila*

In *Drosophila*, the search for DNA methylation has a long and controversial history. Although a small amount of 5-methylcytosine was already reported in an early study using specific antibodies,[1] other studies using methyl-sensitive restriction enzymes (MSREs), nearest neighbor analysis (NNA), or high-pressure liquid chromatography (HPLC) did not provide strong evidence for DNA methylation.[2–6] [For this chapter, "mC" refers to 5mC and not N4mC.]

During the past 10 years, additional studies detected DNA methylation within the genome of *Drosophila melanogaster*, applying altogether 10 different experimental techniques (Table I). Using mC-specific antibodies for immunocytology, genome-wide DNA methylation in early *Drosophila* embryos was reported.[9,11] However, these studies did not include Dnmt2 null mutant controls, which are essential to avoid false-positive signals.[19] Recently, a genome-wide bisulfite analysis revealed that if ubiquitous DNA methylation is present in *Drosophila*, it occurs at a level, which is below the detection limit of this technique.[18]

More often, evidence for DNA methylation in *Drosophila* was found at restricted genomic locations, as at retrotransposons[10,12,14,16,17], at promoters for specific genes,[13,17] or at individual enhancer elements.[15] At endogenous genes, aberrant DNA hypermethylation was detected at promoters silenced during tumorigenesis.[13,15] Thus, DNA methylation appears to be more involved in transposon control than that in developmental gene regulation. Further, bisulfite studies at genome scale[18] would surely have been detected methylation of single copy sequences, whereas methylation at only some copies of middle repetitive elements as retrotransposons could easily be overlooked using this technique (see Chapter by Eleanor Wong and Chia-Lin Wei).

All studies, with the exception of those connected with tumor formation,[13,15] agreed that DNA methylation in *Drosophila* occurs preferentially during early embryogenesis. This is also consistent with two studies analyzing nucleotide substitution patterns that suggest DNA methylation occurs in the germ line.[12,14] However, in early embryos of inbred laboratory strains, the relative amount of mC is below 0.5% (Table I), but could be even much lower (0.08% or 0.11%; Refs. 18 and 8). Again, genome-wide bisulfite sequence analysis[18] might underestimate the total amount of mC, as reads could only be mapped to the assembled 84% (168 of about 200 Gb) of the total *D. melanogaster* genome.[20]

For *Drosophila*, preferential occurrence of mC at CpG or at any other specific sequence motif has never been demonstrated firmly (Table I), and likely there is no preference for any specific nucleotide at the 3′ position. This

TABLE I

EVIDENCE OF CYTOSINE METHYLATION IN *DROSOPHILA MELANOGASTER*

Genomic location of mC	Developmental timing of mC	mC/C (%)	Sequence context of mC	Methods used	References
n.d.	Embryos and larvae	n.d.	n.d.	DB	1
n.d.	Early embryo (0–4 h)	0.35 (2D-TLC), 0.25 (BS)	CpN with preference for CpT	BS, 2D-TLC, HPLC	7
n.d.	All stages, less abundant in adults	0.08	n.d.	2D-TLC, HPLC	8
IC: ubiquitous	Embryos	CE: 0.37 ± 0.07	n.d.	CE, IC, SB	9
Gag ORF of rover	Adults	n.d.	insignificant	AC, BS	10
IC: ubiquituous	Embryos	n.d.	n.d.	IC	11
DNAREP1_DM	Germ line	n.d.	CpN	NSP	12
Rbf promoter and exon 1, pathogenic	Eye disc	n.d.	CpN, with preference for CpG	BS, MSRE	13
Retrotransposons	Germ line	n.d.	CpN	NSP	14
Ftz enhancer, pathogenic	Germ line, embryos, adults	BS: 8.4	CpN	BS, MSRE, SB	15
Invader 4 LTR	Embryos	BS: up to 57.9	CpN	BS, MSRE	16
Gene promoters (e.g., Rb, Antp, CG2316) and retrotransposons (HeT-A and Rt1b{})	Soma	BS: 7.0/8.0 (Rb promoter/exon1 in S2 cell expressing transgenic SETDB)	CpN	BS, ChIP, MSRE	17
Genome-wide	0–3 h embryos	0.11	CpN	BS	18

n.d., not determined; AC, affinity chromatography; BS, bisulfite sequencing; CE, capillary electrophoresis; ChIP, chromatin immunoprecipitation using mC antibodies; 2D-TLC, 2D thin layer chromatography; DB, dot blot; HPLC, high pressure liquid chromatography; IC, immunocytology; MSRE, methylation-sensitive restriction endonucleases; NSP, nucleotide substitution pattern; SB, slot blot.

could be explained most simply by the evolutionary loss of the maintenance enzyme of CpG methylation, the DNA methyltransferase 1 (Dnmt1) (but see Chapter by Željko M. Svedružić).

II. Evidence for DNA Methylation Activity of Dnmt2 Enzymes

Three types of cytosine-5 DNA methyltransferases (Dnmts) have been found in metazoans: Dnmt1, Dnmt2, and Dnmt3 (for review, see Ref. 21, and Chapters by Lakshminarayan M. Iyer *et al.*; Željko M. Svedružić; and Frédéric Chédin). Dnmt3 enzymes methylate cytosine within unmethylated dsDNA, and can thus serve as *de novo* methyltransferases. Dnmt1 enzymes are maintenance methyltransferases because they show clear preference for hemimethylated DNA under some conditions. The smallest eukaryotic methyltransferase, Dnmt2, is most widely distributed in animals, fungi, protists, and plants,[22] but its catalytic activity on DNA is very weak according to analyses of the human and the *Entamoeba* enzymes[23,24] (Table II). This enzyme is the only Dnmt found in dipterans including *Drosophila* (see Chapter by Lakshminarayan M. Iyer *et al.*).

Several attempts were made to demonstrate a DNA methylation activity of Dnmt2 in *D. melanogaster.*[16,25–27] All these studies revealed (indirectly) low, but significant activity using different methods. Overexpression of *Dnmt2* in *Drosophila* appears to increase genome-wide DNA methylation from 0.2% to 0.4–0.7% as determined by capillary electrophoresis.[26,27] Limited bisulfite sequencing data confirmed these results, at least qualitatively.[25,26] One study reported a clear-cut difference of methylation within *Invader 4* LTRs between wild type and *Dnmt2*$^{-/-}$flies.[16] However, the bisulfite and MSRE data were generated using a wild-type strain with high early embryonic *Dnmt2* expression, which is not found regularly in inbred laboratory strains. There is thus highly varied *Dnmt2* expression between different wild-type strains (V. Krauss and G. Reuter, unpublished data). In summary, Dnmt2 appears to methylate actually the *Drosophila* genome, but this activity and its functional consequences are probably variable between different strains.

As in *Drosophila*, in the social amoeba *Dictyostelium discoideum* the Dnmt2 ortholog (DnmA) is the only putative Dnmt, and has been evaluated thoroughly for Dnmt activity. In DnmA knockout cells, DNA methylation was reduced according to dot blot (DB) and MSRE analysis by about half, whereas immunocytology and capillary electrophoresis could not detect significant differences from the wild-type controls.[29,30] However, bisulfite analysis demonstrated a nearly complete loss of DNA methylation at *Skipper* and *DIRS-1* retrotransposons in the mutants. This suggests that Dnmt2 may not be the only functional Dnmt in *Dictyostelium*, and may be involved specifically in suppression of retrotransposon activity.[29]

TABLE II

EVIDENCE OF DNA METHYLATION BY DNMT2 (ORDERED ACCORDING TO THE PHYLOGENETIC RELATIONSHIP TO *DROSOPHILA MELANOGASTER*)

Genomic location of mC	Species and developmental timing	mC/C (%)	Sequence context of mC	Methods used	References
Two genes (*CG8547*, *CG8553*) and a cotransfected plasmid	*Drosophila melanogaster*, 0–6 h embryos, S2 cells	n.d.	CpN	BS, EE	25
(Insufficient data)	*Drosophila melanogaster*, Embryos (IC), adults (BS, CE)	WT: 0.2% Overexpression: 0.6–0.7% (CE)	CpW (BS)	BS, CE, EE, IC	26
n.d.	UAS-driven mouse transgenic in *Drosophila melanogaster*	BS: 0.0% (WT), 1.1% (Dnmt2-transgenic flies) CE: 0.2–0.4% more mC to background	CpH	CE, EE	27
Invader 4 LTR	*Drosophila melanogaster*, embryos	BS: up to 57.9	CpN	BS, MSRE	16
558 bp PCR fragment with 34 CpG sites (*in vitro*)	*Homo sapiens*	0.4% (BS), 0.7% (2D-TLC)	ttnCGgar	BS, 2D-TLC, IVM	23
n.d.	*Entamoeba histolytica*	n.d.	n.d.	EAT	24
Scaffold/matrix attachment region (MRS2)	*Entamoeba hystolytica*	BS: 51.5	CpN	BS, OHS, SW	28
BS: *DIRS-1* LTR, *Skipper* RT ORF	*Dictyostelium discoideum*	CE: 0.2, BS: DIRS-1 LTR: 23.1, BS: Skipper RT: 33.3	CpN	BS, CE, IC	29
DIRS-1, *guaB* gene	*Dictyostelium discoideum*	n.d.	n.d.	DB, MSRE	30

n.d., not determined; BS, bisulfite sequencing; CE, capillary electrophoresis; 2D-TLC, 2D thin layer chromatography; DB, dot blot; EAT, enzyme assay using AdoMet with tritium; EE, ectopic expression; IC, immunocytology; IVM, *in vitro* mutagenesis; MSRE, methylation-sensitive restriction endonucleases; OHS, one-hybrid system; SW, south-western.

DNA methylation was also shown in other "Dnmt2-only" organisms, especially in other dipterans, although direct evidence for involvement of Dnmt2 is still missing. About 1 in 600 (0.17%) cytosines are methylated in the DNA of the mosquito *Aedes albopictus*[31] (Table III). Data for five other dipteran species were obtained by immunocytology, slot blot analysis, and capillary electrophoresis.[11] Recently, female-specific DNA methylation was reported for *Drosophila willistoni* and its relatives[35,36]; however, this evidence is based exclusively on MSRE analysis using restriction enzymes that do not exhibit the sensitivity or resistance to mC suggested by the authors.[37] DNA methylation was also reported for the fungus gnat *Sciara coprophila*, based on immunocytological analysis with anti-mC antibodies.[32] Here again, a negative control is not available and the validity of this report remains questionable because nonspecific signals could not be excluded.[19]

There is substantial evidence for DNA methylation in two "Dnmt2-only" protists: *Entamoeba* and *Dictyostelium* (Table III). In *Entaemoba*, evidence for methylation of repetitive sequences (rDNA, SAR/MAR, RT-LINE retrotransposons) is strong and validated by several methods, including affinity chromatography, DB analysis, and bisulfite sequencing.[24,28,33,34] Similarly, DNA methylation was detected in *Dictyostelium* by several methods, in three different retrotransposons and in one gene (Table III).[29,30] As in *Drosophila*, a preferred sequence context was never established in these "Dnmt2-only" protists.[24,29,30,33,34]

III. Methylation of tRNA by Dnmt2

Prompted by the rather weak Dnmt activity and the predominantly cytoplasmic localization of the enzyme, Goll *et al.*[38] demonstrated that human and *Drosophila* Dnmt2 enzymes methylate $tRNA^{Asp}$, specifically from Dnmt2-less cells of mouse, *D. melanogaster*, and *Arabidopsis thaliana*. These experiments identified cytosine 38 in the anticodon loop of $tRNA^{Asp}$ as the Dnmt2 methylation target.[38] Conservation of Dnmt2 and the C38-containing anticodon loop of $tRNA^{Asp}$ in several organisms, including the bacterium *Geobacter sulferreducens* but not *Saccharomyces cerevisiae* or *Caenorhabditis elegans*, suggest a coevolution of Dnmt2 and the target tRNA[38] (see Chapter by Lakshminarayan M. Iyer *et al.*).

In *Dictyostelium*, however, $tRNA^{Asp}$ appears to be methylated weakly if at all by the Dnmt2 ortholog DnmA,[39] though this might depend on growth conditions or the cell cycle phase. Notably, for RNA methylation, Dnmt2 uses a Dnmt-like mechanism, in that similar residues from motifs IV, VI, and VIII are involved in catalysis that are functionally relevant in Dnmts.[39] Recently, it was reported that not only $tRNA^{Asp}$ but also $tRNA^{Val}$ and $tRNA^{Gly}$ are methylated by Dnmt2 in *Drosophila*.[40] Methylation of these tRNAs in their

TABLE III

Evidence of Cytosine Methylation in Other Dnmt2-Only Organisms (Ordered According to the Phylogenetic Relationship to *Drosophila Melanogaster*)

Genomic location of mC	Species and developmental timing	mC/C (%)	Sequence context of mC	Methods used	References
IC: ubiquitous	*Drosophila simulans*, embryos	n.d.	n.d.	IC, SB	11
IC: ubiquitous	*Drosophila pseudoobscura*, embryos	CE: 0.21 ± 0.05	n.d.	CE, IC, SB	11
IC: ubiquitous	*Drosophila virilis*, embryos	n.d.	n.d.	IC, SB	11
IC: ubiquitous	*Drosophila hydei*, embryos	n.d.	n.d.	IC, SB	11
Heterochromatin (L chromosomes)	*Sciara coprophila*	n.d.	n.d.	IC	32
n.d.	*Anopheles gambiae*, embryos	CE: 0.49 ± 0.09	n.d.	CE, SB	11
n.d.	*Aedes albopictus*, cell culture	0.17	n.d.	PC	31
rDNA	*Entamoeba histolytica*	n.d.	CpN	AC, BS, DB	24
Scaffold/matrix attachment region (MRS2)	*Entamoeba histolytica*	BS: 51.5	CpN	BS, OHS, SW	28
Transgenic *RT-LINE* copies from *E. invadens*	*Entamoeba histolytica*	BS: 100 (transgenic *RT-LINE* copy)	CpN	AC, BS	33
Consensus sequence with stretch of 10 adenines	*Entamoeba histolytica*	BS: 16.5 (fragments from 6.0 to 42.0)	CpN	AC, BS	34
Promoter and ORF from *RT LINE*	*Entamoeba invadens*	BS: 100 (*RT-LINE*)	CpN	AC, BS, COBRA, IC	33
BS: *DIRS-1* LTR, *Skipper* RT ORF	*Dictyostelium discoideum*	CE: 0.2, BS: DIRS-1 LTR: 23.1, Skipper RT: 33.3	CpN	BS, CE, IC	29
DIRS-1, *guaB* gene	*Dictyostelium discoideum*, developed cells	DP: 0.14	n.d.	CpGD, DB, MSRE	30

n.d., not determined; AC, affinity chromatography; BS, bisulfite sequencing; CE, capillary electrophoresis; COBRA, combined bisulfite restriction analysis; CpGD, CpG depletion; DB, dot blot; IC, immunocytology; MSRE, methylation-sensitive restiction endonucleases; OHS, one-hybrid system; PC, paper chromatography; SB, slot blot; SW, south-western.

anticodon loops seems to suppress their cleavage during cellular stress, which points to a potential biological function of Dnmt2-dependent tRNA methylation in stress response.

IV. Biological Function of Dnmt2 in *Drosophila*

At a cellular level, null mutant analysis reveals three processes in which Dnmt2 is involved, including retrotransposon silencing, stabilization of repetitive DNA, and cell homeostasis.

A. Retrotransposon Silencing

Dnmt2 plays a significant role in three of five at least partially independent silencing pathways of *Drosophila*.[16] Whereas *Dnmt2* null mutations showed no modifier effects on gene silencing induced by pericentromeric heterochromatin, a dominant suppressor effect has been observed on the *white*$^{+}$ reporter gene of P element constructs inserted into *Invader4* or *HetA* retrotransposons, and on repeat-dependent *white* gene silencing in T(2,3)V21ePlacW. Dnmt2-dependent retrotransposon silencing is only observed in somatic cells of *Drosophila*, independently whether the mutation was paternally or maternally inherited.[16] Thus, Dnmt2 is zygotically expressed early, as confirmed by RT-PCR analysis. According to chromatin immunoprecipitation (ChIP) analyses, Dnmt2 is associated not only with *Invader4* but also with *Rover* and *Copia* retrotransposons in early embryos[16] (see also Ref. 10). Transcripts of all tested retrotransposons are significantly increased in *Dnmt2* mutants. Loss of Dnmt2 eliminates first DNA methylation and, later on, H4K20 trimethylation at retrotransposons. This impairs maintenance of retrotransposon silencing in somatic but not in germline cells, where retrotransposon silencing depends on an RNAi mechanism.[41,42]

B. Stabilization of Repeats

Fluorescent *in situ* hybridization (FISH) analysis, of polytene chromosomes from Dnmt2 null strains, revealed reproducible loss of the subtelomeric *Invader4* clusters at 2R and 3R telomeres.[16] After 12 successive generations in a wild-type background, FISH analysis revealed that the *Invader4* clusters at 3R telomeres remain lost once they have been deleted. Analysis of other Dnmt2 mutant alleles and backgrounds revealed that the loss of subtelomeric *Invader4* repeats is stochastic but essentially inevitable.

These stable losses of subtelomeric *Invader4* clusters indicate that Dnmt2 is essential to genomic stability. This function is probably limited to early germ line cells. In contrast, impairment of satellite and rDNA repeat stability observed in *Su(var)3–9* and *ago2* null mutants seems to be restricted to somatic cells.[43] It remains to be determined whether Dnmt2 also stabilizes other repeat sequences.

C. Longevity and Stress Response

While *Dnmt2* mutants are viable and fertile,[38,39] such flies, on average, have shorter life spans than wild-type flies under standard laboratory conditions.[40,44] Conversely, *Dnmt2*-overexpressing flies live significantly longer. Specifically, two- to fourfold overexpression leads to a 58% longer life span on average, compared to nontransgenic flies with identical wild-type background. In addition, *Dnmt2*-overexpressing flies show greater resistance to starvation and to oxidative damage by paraquat and H_2O_2, whereas *Dnmt2* mutants show opposite phenotypes and lower thermotolerance at 29 °C.[40,44] Remarkably, expression of the small heat shock proteins Hsp22, Hsp23, and Hsp26 is positively correlated with *Dnmt2* expression.[44] Specifically, the overexpression of all three Hsps increase life span, while decreased Hsp26 shortened the life span of *Drosophila*. Thus, the improvement of stress resistance and the prolongation of life span by Dnmt2 may be mediated by these heat shock proteins. Dnmt2 may facilitate expression of these heat shock proteins, which in turn might prevent the aggregation of damaged proteins. It remains to be shown that the effect of *Dnmt2* on life span is indeed mediated by this pathway, for example, by analyzing genetic interactions between mutants of *Dnmt2* and of the small heat shock proteins.

With respect to tRNA stability, *Dnmt2* inhibits stress-induced cleavage of $tRNA^{Gly}$ and $tRNA^{Asp}$ by methylation of cytosine 38 within these tRNAs.[40] However, the possible causal connection between this effect of Dnmt2 on stress-induced tRNA cleavage and the stress-protecting impact of *Dnmt2* needs to be resolved.

V. Dinucleotide Abundances in "Dnmt2-only" Organisms: A Hint of DNA Methylation?

DNA methylation causes a specific mutation pressure: CpG dinucleotides are depleted because cytosines in a CpG context are preferentially methylated, and mC can be oxidatively deaminated to thymine.[45] As CpG is depleted, the mutation product TpG and its complement CpA are observed at elevated levels.[46] DNA methylation correlates with CpG depletion even in rather weakly methylated genomes, such as those of insects[47–49]. The resulting signal, therefore, can be used to investigate distribution and level of DNA methylation. A CpG level below 0.78 (observed/expected) and a concurrent enrichment of

TpG and CpA (each of above 1.23 [observed/expected]) are highly significant if found in whole genomes.[50] These are conservative estimates, so even values of 0.9 and 1.1, respectively, are noteworthy.[45]

Accordingly, we have computed these values for three "Dnmt2-only" organisms with independent records of DNA methylation (*D. discoideum, Entamoeba histolytica*, and *D. melanogaster*), and for two organisms that contain no known Dnmts and do not show DNA methylation (*C. elegans* and *S. cerevisiae*) (Fig. 1). Given that the CpG value is more important than the TpG and CpA values,[45] both *D. discoideum* and *E. histolytica* show imprints of genome-wide CpG methylation. In *D. melanogaster*, TpG and CpA are similarly enriched (1.11 for both TpG and CpA); however, the CpG depletion is weak

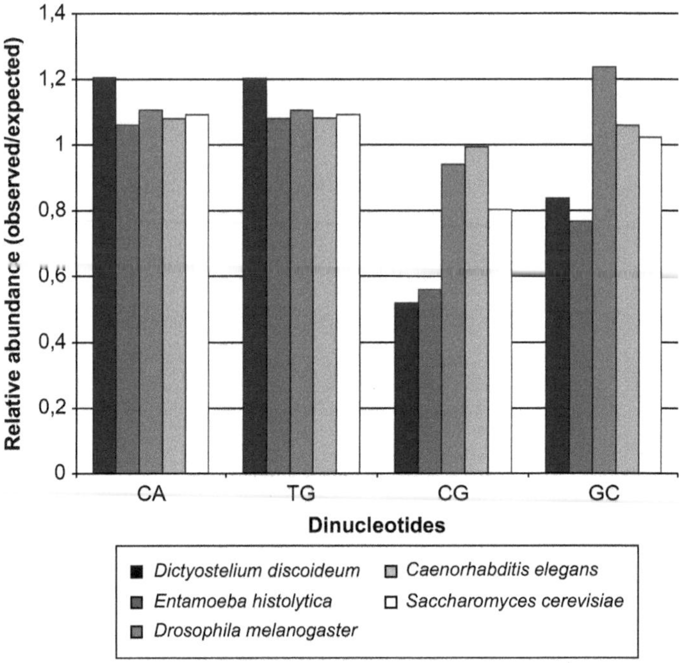

FIG. 1. *Dinucleotide frequency analysis*. A record of DNA methylation (*Dictyostelium, Entamoeba*, and *Drosophila*) is typically coupled with a depletion of CpG sites below 0.78 (*Dictyostelium* and *Entamoeba*) and an enrichment of TpG and CpA sites above 1.1 (*Dictyostelium* and *Drosophila*). Note that *D. discoideum* fulfills all three conditions, whereas *D. melanogaster* shows similar relative CpG, TpG, and CpA abundances as the Dnmt-less species *C. elegans* and *S. cerevisiae*. Frequencies of dinucleotides were determined from genome project data (Lechner, Marz, Ihling, Sinz, Stadler, and Krauss: Genome Size and Methylation Rate are positively correlated in Metazoans, unpublished).

(0.94) and not significantly different from organisms without CpG methylation. This is consistent with a CpN methylation as described for *Drosophila*, but also with a weak CpG methylation.

For *Dictyostelium* and *Entamoeba*, the support of CpG-oriented distribution of mC by CpG depletion is somewhat surprising, as nearly all analyses concerning Dnmt2-only organisms support cytosine methylation in the CpN context. A possible explanation may be that Dnmt2 is not the only Dnmt in Dnmt2-only organisms. Katoh *et al.*[30] have knocked out Dnmt2 in *D. discoideum*, and found only a 50% reduction of DNA methylation in their mutants. In the same species, Kuhlmann *et al.*[29] were unable to detect a genome-wide decrease of methylation in their knockout strains for Dnmt2. Both studies obtained no information about the genome-wide sequence context of mC in this species with the strongest CpG depletion. Thus, dinucleotide bias in Dnmt2-only organisms may result from a until now undescribed, CpG-oriented Dnmt activity. In *Drosophila*, sequence analyses on retrotransposons revealed about twofold more C:G → T:A than T:A → C:G substitutions.[12,14] Such an imbalance is consistent with a methylation bias by deamination of mCpN positions. In *Drosophila*, therefore, only Dnmt2 and no additional Dnmt activity is suggested by nucleotide bias.

VI. Methyl-Binding Proteins in *Drosophila*—Evidence, Concepts, and Inconsistencies

Epigenetic signaling requires specific enzymes, which are able to emplace signals ("writers") and other molecules that can detect these signals ("readers").[51] In this respect, DNA methylation represents an unusual form of epigenetic change: it could mediate functional consequences without the help of a specialized reader. Cytosine methylation results in compaction of nucleosomes,[52] could inhibit sequence-specific transcription factor binding, and increases the mutation rate through deamination of mC to thymine. Thus, it is conceivable that organisms maintain a DNA methylation apparatus to ensure transcriptional silencing and genome stability without specific "reader" proteins (see Chapter by Lakshminarayan M. Iyer *et al.*).

However, there is a candidate reader molecule of DNA methylation in *Drosophila*: the ortholog of the mammalian methyl-binding protein MBD2.[5] Two analyses of this protein, named MBD2/3, reported methyl-specific binding capacities on CpG[53] or CpA/T dinucleotides,[9] respectively. However, methodical issues rendered the significance of these findings unclear.[54] Other studies described MBD2/3 as a component of a transcriptional suppressor complex without methylation-specific DNA interaction.[5,54] An alignment of the methyl-binding domain (MBD) of *Drosophila* MBD2/3 with those of other MBDs

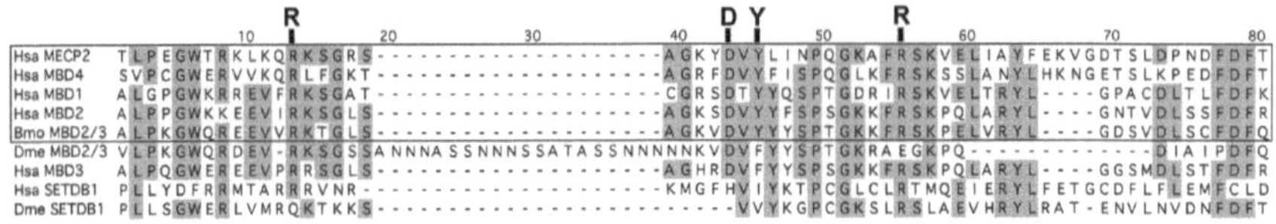

FIG. 2. *Alignment of selected MBD domains, ordered according to their mC binding.* Residues that contact the bases of the mCG motif[55] are shown above the alignment. Note that tyrosine 45 is necessary for mC binding.[55] For all cases within the frame (upper five sequences), mC-specific binding activity has been demonstrated.

reveals why the *Drosophila* ortholog probably could not recognize mC (Fig. 2). All methyl-specific binding proteins of this type contain arginine, aspartic acid, tyrosine, and arginine at the alignment positions 13, 43, 45, and 56, respectively. These four positions directly contact the bases of the mCpG dinucleotide, and are necessary to bind the mCpG double strand motif.[55] In addition, the CpN methylation pattern of *Drosophila* is more consistent with a mCpG-independent DNA-binding of *Drosophila* MBD2/3.

It was recently reported that the *Drosophila* MBD-containing histone H3K9 methyltransferase SETDB1 binds mC specifically, within a CpA context.[17] As MBDs interact with double-stranded DNA and mCpA could not be methylated symmetrically, this binding activity might not be mediated by the MBD domain. It remains to be shown that *Drosophila* indeed uses SETDB1 as methylation reader *in vivo*.

In summary, *Drosophila* exhibits neither a Dnmt of high efficiency nor a reader protein of the common MBD type. Nevertheless, the evidence for DNA methylation is overwhelming. It has been suggested that DNA methylation in *Drosophila* is simply a byproduct of a former Dnmt that becomes transformed to a tRNA methylase (Dnmt2).[56] Concerning the very specific temporal control of CpN methylation in *Drosophila*, and the obvious Dnmt activity of orthologous enzymes in *Dictyostelium* and *Entamoeba*, we tend to disagree with this notion. Dnmt2 probably serves, and not only in *Drosophila*, as a methylase of both DNA and RNA, as suggested by Jeltsch *et al.*[57]

Acknowledgment

Research on Dnmt2 was supported by a grant from Deutsche Forschungsgemeinschaft (DFG) to G. R. (Re911/9-1; FOR 1082).

References

1. Achwal CW, Iyer CA, Chandra HS. Immunochemical evidence for the presence of 5mC, 6mA and 7mG in human, Drosophila and mealybug DNA. *FEBS Lett* 1983;**158**:353–8.

2. Doerfler W. DNA methylation and gene activity. *Annu Rev Biochem* 1983;**52**:93–124.
3. Patel CV, Gopinathan KP. Determination of trace amounts of 5-methylcytosine in DNA by reverse-phase high-performance liquid chromatography. *Anal Biochem* 1987;**164**:164–9.
4. Tweedie S, Charlton J, Clark V, Bird AP. Methylation of genomes and genes at the invertebrate-vertebrate boundary. *Mol Cell Biol* 1997;**17**:1469–75.
5. Tweedie S, Ng HH, Barlow AL, Turner BM, Hendrich B, Bird AP. Vestiges of a DNA methylation system in Drosophila melanogaster? *Nat Genet* 1999;**23**:389–90.
6. Lyko F, Ramsahoye BH, Kashevsky H, Tudor M, Mastrangelo MA, Orr-Weaver TL, et al. Mammalian (cytosine-5) methyltransferases cause genomic DNA methylation and lethality in Drosophila. *Nat Genet* 1999;**23**:363–6.
7. Lyko F, Bernard Ramsahoye R, Jaenisch R. DNA methylation in Drosophila melanogaster. *Nature* 2000;**408**:538–40.
8. Gowher H, Leismann O, Jeltsch A. DNA of Drosophila melanogaster contains 5-methylcytosine. *EMBO J* 2000;**19**:6918–23.
9. Marhold J, Kramer K, Kremmer E, Lyko F. The Drosophila MBD2/3 protein mediates interactions between the MI-2 chromatin complex and CpT/A-methylated DNA. *Development* 2004;**131**:6033–9.
10. Salzberg A, Fisher O, Siman-Tov R, Ankri S. Identification of methylated sequences in genomic DNA of adult Drosophila melanogaster. *Biochem Biophys Res Commun* 2004;**322**:465–9.
11. Marhold J, Rothe N, Pauli A, Mund C, Kuehle K, Brueckner B, et al. Conservation of DNA methylation in dipteran insects. *Insect Mol Biol* 2004;**13**:117–23.
12. Singh ND, Arndt PF, Petrov DA. Genomic heterogeneity of background substitutional patterns in Drosophila melanogaster. *Genetics* 2005;**169**:709–22.
13. Ferres-Marco D, Gutierrez-Garcia I, Vallejo DM, Bolivar J, Gutierrez-Aviño FJ, Dominguez M. Epigenetic silencers and Notch collaborate to promote malignant tumours by Rb silencing. *Nature* 2006;**439**:430–6.
14. Keller I, Bensasson D, Nichols RA. Transition-transversion bias is not universal: a counter example from grasshopper pseudogenes. *PLoS Genet* 2007;**3**:E22.
15. Xing Y, Shi S, Le L, Lee CA, Silver-Morse L, Li WX. Evidence for transgenerational transmission of epigenetic tumor susceptibility in Drosophila. *PLoS Genet* 2007;**3**:1598–606.
16. Phalke S, Nickel O, Walluscheck D, Hortig F, Onorati MC, Reuter G. Retrotransposon silencing and telomere integrity in somatic cells of Drosophila depends on the cytosine-5 methyltransferase DNMT2. *Nat Genet* 2009;**41**:696–702.
17. Gou D, Rubalcava M, Sauer S, Mora-Bermudez F, Erdjument-Bromage H, Tempst P, et al. SETDB1 is involved in postembryonic DNA methylation and gene silencing in Drosophila. *PLoS One* 2010;**5**:E10581.
18. Zemach A, McDaniel IE, Silva P, Zilberman D. Genome-wide evolutionary analysis of eukaryotic DNA methylation. *Science* 2010;**328**:916–9.
19. Weyrich A, Tang X, Xu G, Schrattenholz A, Hunzinger C, Hennig W. Mammalian DNMTs in the male germ line DNA of Drosophila. *Biochem Cell Biol* 2008;**86**:380–5.
20. Bosco G, Campbell P, Leiva-Neto JT, Markow TA. Analysis of Drosophila species genome size and satellite DNA content reveals significant differences among strains as well as between species. *Genetics* 2007;**177**:1277–90.
21. Hermann A, Gowher H, Jeltsch A. Biochemistry and biology of mammalian DNA methyltransferases. *Cell Mol Life Sci* 2004;**61**:2571–87.
22. Ponger L, Li W-H. Evolutionary diversification of DNA methyltransferases in eukaryotic genomes. *Mol Biol Evol* 2005;**22**:1119–28.
23. Hermann A, Schmitt S, Jeltsch A. The human Dnmt2 has residual DNA-(cytosine-C5) methyltransferase activity. *J Biol Chem* 2003;**278**:31717–21.

24. Fisher O, Siman-Tov R, Ankri S. Characterization of cytosine methylated regions and 5-cytosine DNA methyltransferase (Ehmeth) in the protozoan parasite Entamoeba histolytica. *Nucleic Acids Res* 2004;**32**:287–97.
25. Tang L-Y, Reddy MN, Rasheva V, Lee T-L, Lin M-J, Hung M-S, et al. The eukaryotic DNMT2 genes encode a new class of cytosine-5 DNA methyltransferases. *J Biol Chem* 2003;**278**:33613–6.
26. Kunert N, Marhold J, Stanke J, Stach D, Lyko F. A Dnmt2-like protein mediates DNA methylation in Drosophila. *Development* 2003;**130**:5083–90.
27. Mund C, Musch T, Strödicke M, Assmann B, Li E, Lyko F. Comparative analysis of DNA methylation patterns in transgenic Drosophila overexpressing mouse DNA methyltransferases. *Biochem J* 2004;**378**:763–8.
28. Banerjee S, Fisher O, Lohia A, Ankri S. Entamoeba histolytica DNA methyltransferase (Ehmeth) is a nuclear matrix protein that binds EhMRS2, a DNA that includes a scaffold/matrix attachment region (S/MAR). *Mol Biochem Parasitol* 2005;**139**:91–7.
29. Kuhlmann M, Borisova BE, Kaller M, Larsson P, Stach D, Na J, et al. Silencing of retrotransposons in Dictyostelium by DNA methylation and RNAi. *Nucleic Acids Res* 2005;**33**:6405–17.
30. Katoh M, Curk T, Xu Q, Zupan B, Kuspa A, Shaulsky G. Developmentally regulated DNA methylation in Dictyostelium discoideum. *Eukaryot Cell* 2006;**5**:18–25.
31. Adams RL, McKay EL, Craig LM, Burdon RH. Methylation of mosquito DNA. *Biochim Biophys Acta* 1979;**563**:72–81.
32. Greciano PG, Ruiz MF, Kremer L, Goday C. Two new chromodomain-containing proteins that associate with heterochromatin in Sciara coprophila chromosomes. *Chromosoma* 2009;**118**:361–76.
33. Harony H, Bernes S, Siman-Tov R, Ankri S. DNA methylation and targeting of LINE retrotransposons in Entamoeba histolytica and Entamoeba invadens. *Mol Biochem Parasitol* 2006;**147**:55–63.
34. Lavi T, Siman-Tov R, Serge Ankri S. Insights into the mechanism of DNA recognition by the methylated LINE binding protein EhMLBP of Entamoeba histolytica. *Mol Biochem Parasitol* 2009;**166**:117–25.
35. Nunes Garcia R, D'Avila MF, Robe LJ, da Silva Loreto EL, Panzera Y, de Heredia FO, et al. First evidence of methylation in the genome of Drosophila willistoni. *Genetica* 2007;**131**:91–105.
36. D'Avila MF, Nunes Garcia R, Panzera Y, da Silva Valente VL. Sex-specific methylation in Drosophila: an investigation of the Sophophora subgenus. *Genetica* 2010;**138**:907–13.
37. REBASE. The Restriction Enzyme Database. http://rebase.neb.com/rebase/rebase.html.
38. Goll MG, Kirpekar F, Maggert KA, Yoder JA, Hsieh C-L, Zhang X, et al. Methylation of tRNAAsp by the DNA methyltransferase homolog Dnmt2. *Science* 2006;**311**:395–8.
39. Jurkowski TP, Meusburger M, Phalke S, Helm M, Nellen W, Reuter G, et al. Human DNMT2 methylates tRNA(Asp) molecules using a DNA methyltransferase-like catalytic mechanism. *RNA* 2008;**14**:1663–70.
40. Schaefer M, Pollex T, Hanna K, Tuorto F, Meusburger M, Helm M, et al. RNA methylation by Dnmt2 protects transfer RNAs against stress-induced cleavage. *Genes Dev* 2010;**24**:1590–5.
41. Vagin VV, Sigova A, Li C, Seitz H, Gvozdev V, Zamore PD. A distinct small RNA pathway silences selfish genetic elements in the germline. *Science* 2006;**313**:320–4.
42. Yin H, Lin H. An epigenetic activation role of Piwi and a Piwi-associated piRNA in Drosophila melanogaster. *Nature* 2007;**450**:304–8.
43. Peng JC, Karpen GH. H3K9 methylation and RNA interference regulate nucleolar organization and repeated DNA stability. *Nat Cell Biol* 2007;**9**:25–35.

44. Lin M-J, Tang L-Y, Reddy MN, Shen C-KJ. DNA methyltransferase gene dDnmt2 and longevity of Drosophila. *J Biol Chem* 2005;**280**:861–4.
45. Simmen MW. Genome-scale relationships between cytosine methylation and dinucleotide abundances in animals. *Genomics* 2008;**92**:33–40.
46. Bird AP. DNA methylation and the frequency of CpG in animal DNA. *Nucleic Acids Res* 1980;**8**:1499–504.
47. Elango N, Hunt B, Goodisman M, Yi S. DNA methylation is widespread and associated with differential gene expression in castes of the honeybee, Apis mellifera. *Proc Natl Acad Sci USA* 2009;**106**:11206–11.
48. Krauss V, Eisenhardt C, Unger T. The genome of the stick insect Medauroidea extradentata is strongly methylated within genes and repetitive DNA. *PLoS One* 2009;**4**:E7223.
49. Walsh TK, Brisson JA, Robertson HM, Gordon K, Jaubert-Possamai S, Tagu D, et al. A functional DNA methylation system in the pea aphid, Acyrthosiphon pisum. *Insect Mol Biol* 2010;**19**(Suppl. 2):215–28.
50. Karlin S, Cardon LR. Computational DNA sequence analysis. *Annu Rev Microbiol* 1994;**48**:619–54.
51. Prohaska SJ, Stadler PF, Krakauer DC. Innovation in gene regulation: the case of chromatin computation. *J Theor Biol* 2010;**265**:27–44.
52. Choy JS, Wei S, Lee JY, Tan S, Chu S, Lee T-H. DNA methylation increases nucleosome compaction and rigidity. *J Am Chem Soc* 2010;**132**:1782–3.
53. Roder K, Hung M-S, Lee TL, Lin TY, Xiao H, Isobe KI, et al. Transcriptional repression by Drosophila methyl-CpG-binding proteins. *Mol Cell Biol* 2000;**20**:7401–9.
54. Ballestar E, Pile LA, Wassarman DA, Wolffe AP, Wade PA. A Drosophila MBD family member is a transcriptional corepressor associated with specific genes. *Eur J Biochem* 2001;**268**:5397–406.
55. Ho KL, McNae IW, Schmiedeberg L, Klose RJ, Bird AP, Walkinshaw MD. MeCP2 binding to DNA depends upon hydration at methyl-CpG. *Mol Cell* 2008;**29**:525–31.
56. Motorin Y, Lyko F, Helm M. 5-Methylcytosine in RNA: detection, enzymatic formation and biological functions. *Nucleic Acids Res* 2010;**38**:1415–30.
57. Jeltsch A, Nellen W, Lyko F. Two substrates are better than one: dual specificities for Dnmt2 methyltransferases. *Trends Biochem Sci* 2006;**31**:306–8.

DNA Methylation in Zebrafish

Mary G. Goll* and Marnie E. Halpern†

*Developmental Biology Program, Sloan-Kettering Institute, New York, USA

†Carnegie Institution for Science, Department of Embryology, Baltimore, Maryland, USA

DNA methylation is crucial for normal development and cellular differentiation in many large-genome eukaryotes. The small tropical freshwater fish *Danio rerio* (zebrafish) has recently emerged as a powerful system for the study of DNA methylation, especially in the context of development. This review summarizes our current knowledge of DNA methylation in zebrafish and provides evidence for the general conservation of this system with mammals. In addition, emerging strategies are highlighted that use the fish model to address some of the key unanswered questions in DNA methylation research.

Progress in Molecular Biology
and Translational Science, Vol. 101
DOI: 10.1016/B978-0-12-387685-0.00005-6

1877-1173/11 $35.00

I. Introduction

DNA methylation is essential for normal development in plants and some animals, but it is absent or found at very low levels in classic genetic models such as *Saccharomyces cerevisiae*, *Schizosaccharomyces pombe*, *Caenorhabditis elegans*, and *Drosophila melanogaster* (reviewed in Ref. 1). Currently, much of our understanding of DNA methylation has been derived from studies in mouse, the plant *Arabidopsis thaliana*, and the fungus *Neurospora crassa*. While each of these organisms offers advantages, the zebrafish is becoming a compelling alternative for the study of DNA methylation.

On the basis of gene homology and genome-wide profiling, the zebrafish methylation system appears more similar to mammals than to plants or fungi. Therefore, it is an appealing option for probing vertebrate-specific aspects of the DNA methylation machinery and the evolution of methylation in the vertebrate lineage. Moreover, because zebrafish do not require imprinting of genes or sex chromosomes for viability,[2] as is the case in mammals, they provide a simplified system for exploring the roles of DNA methylation in vertebrate development.

In recent years, zebrafish have become widely accepted as a useful model for studying development. Many of the same attributes that have made them appealing for these studies are also relevant to the study of epigenetic regulation. Zebrafish are easy to maintain in large numbers and each female can lay hundreds of eggs per week. Such large numbers of embryos facilitate quantitation of non-Mendelian inheritance patterns, which frequently arise in epigenetic studies. Zebrafish embryos develop rapidly and externally, which allow access to stages of development that can be difficult to study in mammals. Moreover, haploid and uniparental diploid embryos can be produced readily,[2] allowing analysis of methylation patterns inherited specifically from the maternal or paternal genome. Importantly, the developing embryos are transparent, permitting the transcriptional status of fluorescent reporter genes to be monitored *in vivo* and across generations. Finally, zebrafish are highly amenable to forward genetic screens that have the potential to identify new factors not previously implicated in epigenetic regulation. Collectively, these advantages provide the framework for new and unique approaches to the study of DNA methylation.

Despite its advantages, the potential for epigenetic research in zebrafish has not been fully realized. In part, this is due to the belated application of technologies for targeted deletion of specific genes and the reliance on antisense morpholino oligonucleotides injected at the one-cell stage. As morpholinos reduce protein by blocking translation or splicing of mRNAs, they are not always effective and rarely result in 100% depletion of protein. In addition, up to 18% of morpholinos appear to have "off-target" effects that include p53-mediated cell death in the central nervous system and somites.[3,4] Although rescue of phenotypes by coinjection of mRNA can control these secondary effects,[5] the analysis of morpholino-injected embryos remains challenging. Fortunately, the number of available zebrafish mutants is rapidly growing, and methods for targeted gene mutation such as zinc finger nucleases have now been validated. Thus, while information gained from morpholino-based experiments provides valuable initial insights into DNA methylation in zebrafish, it will be important to revisit these experiments using genetic mutations.

II. Methylation Profile of the Zebrafish Genome

A. Early Evidence for DNA Methylation

The first clues that zebrafish might have an efficient DNA methylation system came from early studies of integrated transgenes. Initial approaches to transgenesis in zebrafish relied on injection of naked, linearized DNA into the one-cell stage embryo.[6] In a small percentage of embryos, the linearized DNA would integrate randomly into the genome, usually as a large concatemeric array containing many copies of the transgene. Despite having numerous copies, these early zebrafish lines were often reported to have low or variegating transgene expression.[7–10] Silencing of transgene expression was reminiscent of the DNA methylation and transcriptional silencing that had been described in mouse lines carrying integrated transgene arrays[11–16] and suggested a similar mechanism might be operating in zebrafish.

The presence of DNA methylation at zebrafish transgene arrays was confirmed by Gibbs et al., using integrated luciferase reporters that were strongly silenced.[9] High levels of DNA methylation were found at the transgenes by digestion with methylation-sensitive restriction enzymes, and modest increases in luciferase expression were achieved after exposing embryos to the DNA methyltransferase inhibitor 5-azacytidine.[9] The role of DNA methylation in transgene silencing was further substantiated by Martin and McGowan, who likewise demonstrated the accumulation of DNA methylation at transgene arrays with variegated expression of the chloramphenicol acetyltransferase resistance gene *cat*.[10,17]

Curiously, methylation levels at the *cat* resistance transgene array appeared to be somewhat plastic, with modest differences in the extent of methylation reported among individuals that depended on genetic background, rearing temperature, sex, and parent of origin.[17] While these early studies provided the first evidence that DNA methylation could play a role in transcriptional silencing in zebrafish, they revealed little about the overall DNA methylation landscape or how DNA methylation systems in zebrafish might compare to other organisms.

B. Global Methylation Profiles in Fish

Genome-wide sequencing technologies have now provided a wealth of data, enabling comparison of DNA methylation systems in diverse organisms (see Chapter by Eleanor Wong and Chia-Lin Wei). Completely sequenced genomes allow comparison of DNA methyltransferase homologs present in different lineages, while recent high-throughput methods of shotgun bisulfite sequencing allow the methylation profile of many organisms to be characterized at the level of the whole genome.[18,19]

Bisulfite sequencing analysis of the zebrafish genome has been reported for pooled larvae at 5 days postfertilization (dpf).[18] Similar to mouse embryos, zebrafish larvae have high levels of DNA methylation (80%) at CpG dinucleotides in virtually all sequence contexts, except CpG islands where it is low.[18] Although levels of DNA methylation are otherwise uniformly high, a mild enrichment can nonetheless be detected at repetitive elements and gene bodies.[18] The level and distribution of CpG methylation observed in zebrafish is on par with mouse (74% methylation) and significantly higher than the levels observed in plants (e.g., *Arabidopsis*, 22%) or insects (e.g., honey bee, 0.93%; Table I).[18]

While non-CpG methylation in the forms of CpHpG and CpHpH can be readily detected in plant and fungal sequences undergoing *de novo* methylation, its role in mammals is less clear[1] (H = A, C, or T). Similar to mouse, in zebrafish both CpHpG and CpHpH methylation are present at relatively low levels and evenly distributed across the zebrafish genome.[18] The amount of non-CpG methylation in the zebrafish embryo is somewhat higher than that reported for mouse embryos (2.13% compared to 0.59%; Table I), although further analysis will be required to determine whether this difference is significant.[18]

So far, dense non-CpG methylation has only been described once in the zebrafish literature, with injected plasmid DNA as the substrate.[22] In contrast, a number of other experiments examining specific genomic loci by bisulfite

TABLE I
METHYLATION OF THE ZEBRAFISH GENOME COMPARED TO OTHER MODEL ORGANISMS

	Zebrafish	Mouse	Honey bee	*Arabidopsis*	*Neurospora*
CpG	80.3%[18]	74.2%[18]	0.93%[18]	22.3%[18]	0.37%[20]
CpNpG	1.22%[18]	0.30%[18]	0.26%[18]	5.92%[18]	NA
CpHpH	0.91%[18]	0.29%[18]	0.17%[18]	1.51%[18]	1.2%[20]
Transposons	Mildly enriched[18]	Mildly enriched[18]	Not enriched[18]	Strongly enriched[18]	Strongly enriched[21]

NA, not assessed. H, not G (i.e., A, C, or T).

sequencing have reported only background levels of non-CpG methylation.[23–25] As in mammals, the biological significance of non-CpG methylation in zebrafish has not been determined.

Studies of other species suggest DNA methylation is widespread in fish. DNA methylation has been reported in *Oryzias latipes* (medaka), *Gasterosteus aculeatus* (stickleback), *Tetraodon nigroviridis* (green spotted puffer), and a number of other fish species.[19,26–28] Detailed analysis of DNA methylation at the level of shotgun bisulfite sequencing has also been performed on DNA from *Tetraodon*, which has a small genome containing very few transposable elements.[29] Similar to zebrafish, the *Tetraodon* genome was heavily methylated at CpG dinucleotides (66%) with enrichment of methylation in genes and the few present transposons.[19] Curiously, non-CpG methylation was not detected in the *Tetraodon* genome.[19] However, because zebrafish and *Tetraodon* methylation levels were determined in two different studies, direct comparison is difficult. It is possible that these differences are due to the developmental stage at which methylation was examined. Methylation in zebrafish was examined in larvae, while the status of a more mature juvenile *Tetrodon* was reported. Given that in mouse, highest levels (3.1%) of non-CpG methylation are present in embryonic stem (ES) cells,[30] it may be informative to reexamine non-CpG methylation in zebrafish and *Tetraodon* at the same early and late developmental stages.

III. The Zebrafish Methylation Machinery

Based on sequence similarity in 10 diagnostic amino acid motifs, the zebrafish genome codes for eight DNA methyltransferase orthologs, which include at least one member of the Dnmt1, Dnmt2, and Dnmt3 families found in mammals[22] (Fig. 1). Not surprisingly, homologs of the plant-specific chromomethylase and fungal-specific Dim-2 families were not detected in the zebrafish genome. The presence of mammalian-like Dnmt1 and Dnmt3 proteins suggests that pathways involved in maintenance and *de novo* methylation are generally conserved in vertebrate species (see also Chapter by Lakshminarayan M. Iyer *et al.*), and that zebrafish may be well suited for their study.

A. Maintenance of Methylation by Dnmt1 and Uhrf1 Proteins

In order to be maintained over the lifetime of an organism, methylated cytosine residues must be propagated at every cell division. Thus far, two proteins have been identified that act cooperatively to maintain methylation through DNA replication. In mouse, hemimethylated DNA is recognized and

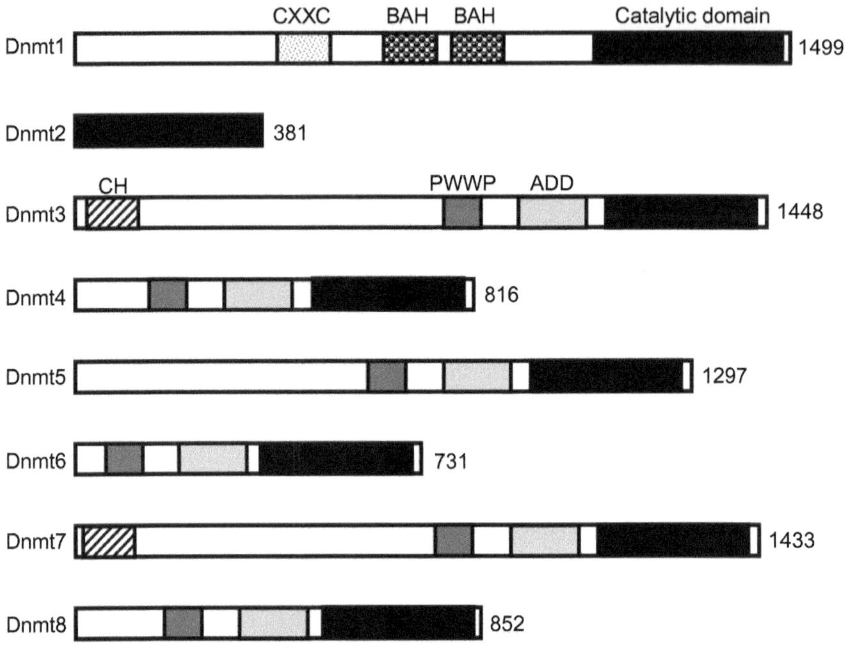

FIG. 1. Schematic of the eight DNA methyltransferase orthologs present in the zebrafish genome (adapted from Ref. 22). All orthologs share a C-terminal catalytic domain containing amino acid motifs that are conserved in all 5mC-generating DNA methyltransferases. In addition, these proteins exhibit distinct motifs in the N-terminus. BAH: bromo-adjacent homology domain, CH: calponin homology, CXXC: cysteine-rich domain, PWWP: conserved proline-tryptophan-tryptophan-proline domain, and ADD: ATRX-DNMT3-DNMT3L domain.

bound by a SET- and RING-associated (SRA) domain containing protein called UHRF1 that recruits the DNA methyltransferase DNMT1 to the replication fork[31,32] (see Chapter by Jafar Sharif and Haruhiko Koseki). A methyl group is then added to the newly synthesized DNA strand, allowing for passive transfer of previously established methylation patterns. A version of this simple, yet elegant mechanism for maintaining DNA methylation appears to be widely used in plants, vertebrates, and possibly some insect lineages (reviewed in Ref. 33).

1. Dnmt1 and Uhrf1 Conservation

Dnmt1 was the first DNA methyltransferase identified in zebrafish, providing an early indication that fish and mammals might share common mechanisms of DNA methylation.[34] Zebrafish and human orthologs are 73% identical overall, with 89% identity in the C-terminal catalytic domain. Zebrafish Dnmt1 retains the N-terminal bromo-adjacent homology (BAH) and CXXC zinc finger domains found in mammalian Dnmt1 homologs.

In mouse, the *Dnmt1* locus includes 5′ sex-specific exons that are expressed specifically in either the testes or ovaries. In the ovary, an alternative 5′exon leads to production of high levels of active DNMT1 protein, which becomes sequestered in the cytoplasm during later stages of oocyte growth.[35] This transcript is important for the proper maintenance of imprinted genes during oogenesis.[36] In contrast, the spermatocyte-specific 5′ exon interferes with translation and decreases DNMT1 levels during male meiosis.[35] A directed search for sex-specific *dnmt1* transcripts has yet to be performed in zebrafish, and available expressed sequence tags (ESTs) do not provide evidence for oocyte-specific exons. However, one EST (BC057461) cloned from an adult zebrafish male does code for a nonfunctional version of Dnmt1 protein, suggesting that alternative splicing could also play a regulatory role in zebrafish gametes. A definitive answer to this question will require careful 5′ RACE studies on dissected gonadal tissues.

As expected, the zebrafish genome also includes an ortholog of *uhrf1* that is 66% identical to the human protein, and contains the SRA, ubiquitin-like, and plant homeodomain (PHD) domains present in the mammalian orthologs.[37] Similar to mouse and human, the zebrafish genome also contains a second related gene, *uhrf2*.[38] However, data linking Uhrf2 proteins and DNA methylation have yet to be reported in any organism.

2. Dnmt1 and Uhrf1 Are Required for DNA Methylation

As would be predicted for proteins with a generalized role in maintenance of DNA methylation, expression of both *dnmt1* and *uhrf1* appears to be fairly ubiquitous in the developing embryo. Highest levels of expression are observed throughout early development with later widespread expression in zones of proliferation during organogenesis.[39–42] *Dnmt1* expression has not been examined in adults, but *uhrf1* mRNA was detected in proliferating adult tissues by RT-PCR.[37] There is also indirect evidence to suggest that maternal Uhrf1 and Dnmt1 proteins may be maternally deposited in the oocytes.[23,42,43]

Mutations in the zebrafish *dnmt1* and *uhrf1* genes support a conserved role for this pair of proteins in DNA methylation. When present in homozygous form, null mutations in the zebrafish *dnmt1* gene lead to decreased levels of DNA methylation.[42] This finding is consistent with an earlier report describing a 40% reduction in DNA methylation following depletion of Dnmt1 protein by injection of translation-blocking antisense morpholinos into the one-cell embryo.[40] Similarly, zebrafish homozygous for a hypomorphic mutation in the *uhrf1* gene have a 75% reduction in *uhrf1* mRNA, and an approximately ~50% reduction in global DNA cytosine methylation.[18,37,44] Consistent with a generalized role in maintenance of DNA methylation, methylation in *uhrf1* mutants is decreased throughout the genome including at intergenic regions, exons, and introns.[18]

3. Maintenance Methylation Is Essential for Normal Zebrafish Development

In initial experiments to address the role of DNA methylation in zebrafish development, embryos were exposed to the DNA methyltransferase inhibitor 5-azacytidine.[39] Following exposure, a fraction of the developing embryos exhibited defects including a shortened trunk and abnormal development of the somites, notochord, and muscle. However, the timing of administration was critical and only 20–30% of embryos treated during the blastula period showed these phenotypes.[39]

Somewhat different phenotypes were reported upon depletion of Dnmt1 using antisense morpholinos. At 24 hpf, 40% of injected embryos were dead, while the remaining ones exhibited developmental abnormalities including curled tails, pericardial edema, and jaw defects.[40] Most strikingly, knockdown of Dnmt1 caused loss of expression of terminal differentiation markers in the intestine, exocrine pancreas, and retina, while markers of the liver and endocrine pancreas appeared normal.[40]

Two alleles of *dnmt1*, identified in a screen for mutations that affect zebrafish liver and pancreas development, recapitulated many of the phenotypes observed in morpholino experiments.[42] Mutants initially developed normally, but by 4 dpf had a markedly smaller exocrine pancreas as well as liver, eye, and jaw defects. *Dnmt1* mutant zebrafish also showed increased p53-mediated apoptosis similar to that observed for *Dnmt1* mutant mouse fibroblasts.[42,45] At the morphological level, *uhrf1* mutants and *dnmt1* mutants have very similar phenotypes,[37] supporting the hypothesis that the primary role of Uhrf1 in development is to target Dnmt1 to the replication foci for maintenance of DNA methylation.

While all of these experiments point to a fundamental role for DNA methylation in development, the phenotypes observed following 5-azacytidine treatment are distinct from those in *dnmt1* mutants and morphants. There are several possibilities for the discrepancy. First, treatment with 5-azacytidine is expected to block activity of all DNA methyltransferases, which might cause a more severe phenotype than depletion of Dnmt1 alone. Second, maternal Dnmt1 protein deposited in the yolk would not be depleted in mutant or morphant embryos, but would be inactivated by azacytidine treatment. Therefore, 5-azacytidine phenotypes may be due to earlier loss of Dnmt1 activity. However, 5-azacytidine is also highly unstable in water, raising the potential for nonspecific effects. It is difficult to rule out this possibility because of the crude assay used to measure methylcytosine depletion in this early experiment. Analyses of maternal zygotic *dnmt1* mutants may aid in clarifying the consequence of early inhibition of Dnmt1 activity.

The tissue-specific differences in sensitivity following loss of Dnmt1 are also intriguing. Given that the tissues undergoing the highest levels of proliferation at 3–4 dpf were also the most sensitive to loss of Dnmt1 protein, it is possible that proliferation itself caused susceptibility. Maternal protein stores should be depleted by cell division and, thus, cells undergoing more rounds of division might have increased sensitivity to loss of zygotic Dnmt1. Alternatively, Rai et al. have suggested that differences in tissue responses could indicate a role for DNA methylation in terminal differentiation of some organs but not others. While further studies will be required to distinguish between these two possibilities, characterization of *dnmt1* morpholino-injected embryos and mutants supports the general conclusion that Dnmt1 function is important for growth or maintenance of endodermal organs.

4. Posttranslational Regulation of Dnmt1

Given the widespread conservation of the maintenance methylation machinery, it is not surprising that Dnmt1 and Uhrf1 functions are conserved in zebrafish. However, it is not yet known whether other proteins involved in the regulation of maintenance methylation in mammals have the same function in fish. Recent reports have suggested that regulation of DNMT1 protein is achieved, at least in mammals, via the antagonistic action of a lysine-specific protein methyltransferase and demethylase. The methyltransferase SET7 is thought to lead to protein degradation through direct methylation of specific lysine residues in DNMT1[46] (see Chapter by Shannon R. Morey Kinney and Sriharsa Pradhan). Conversely, the demethylase LSD1 has been hypothesized to prevent this degradation by removing the methyl mark.[47] Although methylation of Dnmt1 has not been demonstrated in zebrafish, the N-terminal lysine and flanking amino acid signature targeted by SET7 are conserved, and orthologs for *set7* and *lsd1* are present in the zebrafish genome.

B. Conserved Role for Zebrafish Dnmt2 in RNA Methylation

Dnmt2 proteins represent the most widely conserved family of DNA methyltransferase paralogs in eukaryotes, but their function has been the most controversial. Although several reports have described very low levels of Dnmt2-dependent DNA methylation in *Dictyostelium discoideum* and *Drosophila*,[48–50] additional reports have failed to identify any DNA methylation activity for this family of proteins.[48–50] In contrast, Dnmt2 shows robust methyltransferase activity on specific tRNA substrates and this activity is important for protecting RNAs against stress-induced cleavage.[51–53]

The zebrafish genome includes a single copy of *dnmt2* (Fig. 1), which is ubiquitously expressed and has activity on $tRNA^{Asp}$.[54] Surprisingly, while mutations in *dnmt2* in a variety of organisms including mouse, *Drosophila*, *Arabidopsis*, and *Schizosaccaromyces* do not result in any obvious phenotypes, depletion of Dnmt2 in zebrafish embryos using an antisense morpholino affected the development of the retina, liver, and brain.[54] No detectable changes in DNA methylation were observed and Dnmt2 protein targeted to the cytoplasm, but not the nucleus, could rescue the phenotypes.[54] Although it is unclear why loss of Dnmt2 should cause developmental abnormalities specifically in zebrafish, these results suggest that Dnmt2-mediated RNA methylation rather than DNA methylation is required for normal development in this organism.[54]

C. Dnmt3 *De Novo* Methyltransferases

The establishment of DNA methylation has not been extensively explored in zebrafish. To date, *de novo* methylation has been reported at the zebrafish promoter of the *no tail* gene, as well as at newly integrated transgenes.[9,17,23,55] More comprehensive efforts to identify targets of *de novo* methylation have not been described. In mammals, the closely related proteins DNMT3A and DNMT3B are primarily responsible for the establishment of DNA methylation[56,57] (see Chapter by Frédéric Chédin). Based on sequence comparisons, the zebrafish genome specifies six likely orthologs of DNMT3A and DNMT3B that could potentially serve as *de novo* methyltransferases.[22] Similar to mouse, the zebrafish dnmt3 family members (*dnmt3, dnmt4, dnmt5, dnmt6, dnmt7*, and *dnmt8*) all contain conserved PWWP (pro-trp-trp-pro) and cysteine-rich ADD (ATRX-DNMT3-DNMT3L) motifs near the N-terminus (Figs. 1 and 2).

1. DNMT3A-LIKE PROTEINS

Zebrafish possesses two highly related proteins, Dnmt6 and Dnmt8 (Fig. 1), that are the most related to mouse DNMT3A (Fig. 2). The presence of two DNMT3A orthologs appears to be conserved in many fish species, as two genes specifying Dnmt3a-like proteins are also present in the genomes of *Tetraodon*, fugu, and stickleback (Fig. 2). Dnmt6 and Dnmt8 show 77% identity overall, with 89% identity in the catalytic domain. In mouse, *Dnmt3a* is expressed as two different isoforms, with the short form being highly expressed in germ cells and ES cells. While EST data do not show evidence for alternative splicing of *dnmt6* or *dnmt8* in zebrafish, it is of note that the most N-terminal portions of the proteins are the most divergent, and vary in length by 150 amino acids. It is possible that, instead of employing different isoforms of *Dnmt3a*, fish achieve similar results by using two independent genes. Although high sequence similarity makes gene expression analyses challenging, knowing whether the zebrafish Dnmt3a homologs show tissue-specific differences in their expression, particularly in germ cells, would be informative.

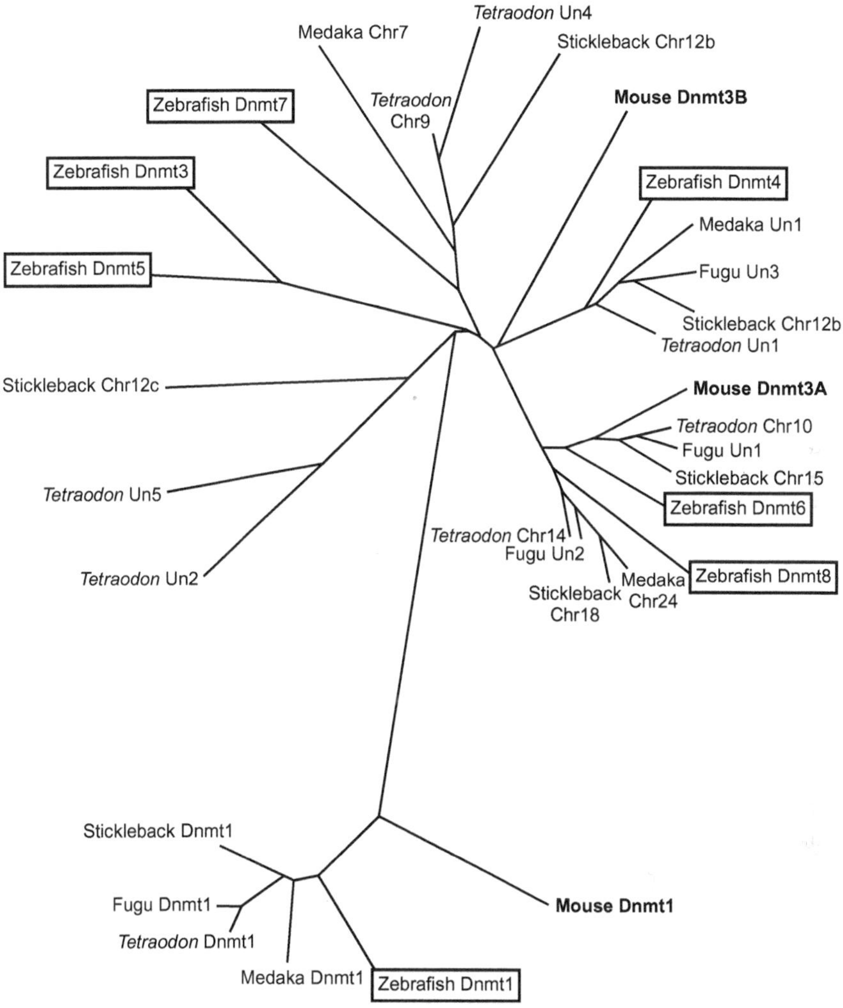

FIG. 2. Phylogenetic tree of Dnmt1 and Dnmt3 families from mouse, zebrafish, fugu, *Tetraodon*, medaka, and stickleback. Alignments were performed on full-length amino acid sequences via ClustalW and DRAWTREE (http://mobyle.pasteur.fr/cgi-bin/portal.py). The multiple alignment parameters were as follows: Gonnet 250 matrix with gap open penalty 10, gap extension penalty 0.2, and gap separation penalty 4 sequences. Mouse and zebrafish protein accessions are as follows: mouse Dnmt1 NP_034196.3, Dnmt3A NP_031898.1, Dnmt3B NP_001003960.1 and zebrafish Dnmt1 NP_571264.1, Dnmt3 NP_571461.1, Dnmt4 NP_001020621.1, Dnmt5 NP_001018315.1, Dnmt6 NP_001018150, Dnmt7 NP_001018312.1, Dnmt8 NP_001018144.1. Fugu, *Tetraodon*, medaka, and stickleback homologs represent predicted proteins identified by homology to zebrafish proteins using BLAT (http://genome.ucsc.edu). Ensemble predictions numbers are: Medaka Chr24 ENSORLP00000022087, Chr Un1 ENSORLP00000025108, Chr7 ENSORLP00000018521; *Tetraodon* Chr10 ENSTNIP00000022492, Chr14

2. Dnmt3B-Like Proteins

In mammals, DNMT3B is responsible for *de novo* methylation of diverse sequences. Mutations in the mouse gene result in embryonic lethality, while the immune centromere and facial abnormalities (ICF) syndrome is caused by mutations in the human gene. The zebrafish protein Dnmt4 shows highest similarity to mouse DNMT3B, with 68% identity in amino acid sequences (Fig. 2). However, injection of *dnmt4* antisense morpolinos did not cause any developmental phenotypes, and targets of Dnmt4 methylation have not yet been described. Nonetheless, dynamic expression of *dnmt4* in early embryos and tissue-specific expression in larvae imply that this protein may have important roles during zebrafish development.[41]

3. Dnmt3 Orthologs Unique to Fish

Fish-specific Dnmt3 family members can also be detected in the genomes of zebrafish, *Tetraodon*, fugu, and stickleback. While these proteins have substantial sequence similarity to mouse DNMT3A and DNMT3B, they do not appear to be any more related to one protein than the other (Fig. 2). Moreover, these additional orthologs appear to be diverging from each other much faster than DNMT3A or DNMT3B family members (Fig. 2). It is most odd that zebrafish Dnmt3 and Dnmt7 both contain an N-terminal calponin homology domain[22] (Fig. 1). Calponin homology domains are generally associated with actin binding, and are typically found in cytoskeletal and signal transduction proteins.[58] Thus far, calponin homology domains have not been reported in any methyltranferase protein outside of fish and their functional significance remains to be determined. Dnmt5 is most closely related to Dnmt3, but lacks the N-terminal calponin domain. *dnmt3* and *dnmt5* reside in a gene cluster with *dnmt4*, suggesting they may have arisen from a recent gene duplication event followed by functional diversification.[59]

4. DNMT3L Orthologs Are Absent in Fish

Mammalian genomes encode for an additional Dnmt3 homolog, DNMT3L, which has lost most catalytic motifs associated with methylation. Despite lacking these motifs, DNMT3L is essential for methylation at monoalleleically expressed imprinted genes in the maternal genome.[60] Unlike

ENSTNIP00000022331, Un1 ENSTNIP00000009615, Un2 ENSTNIP00000009617, Chr9 ENSTNIP00000003297, Chr4 ENSTNIP00000003297, Un5 ENSTNIP00000009617; Fugu Un1 ENSTRUP00000024461, Un2 ENSTRUP00000009050, Un3 ENSTRUP00000031152; Stickleback Chr15 ENSGACP00000013579, Chr18 ENSGACP00000009407, Chr12a ENSGACP00000013579, Chr12b ENSGACP00000015822, Chr12c ENSGACP00000013567. Un indicates undesignated chromosome.

mammals, zebrafish do not require monoalleleic expression of imprinted genes for survival, as both androgenetic and parthogenetic zebrafish are viable and fertile in adulthood.[2,61] DNMT3L orthologs have not been detected in the zebrafish genome, and it has been hypothesized that DNMT3L may have been acquired during the evolution of imprinting.[62] However, DNMT3L is also required for the methylation of transposable elements in the paternal mouse genome,[63] and methylation has been detected at transposable elements in zebrafish.[18] It will be important to determine whether *de novo* methylation of transposable elements occurs in the zebrafish germ line and, if so, whether any of the fish-specific Dnmt3 orthologs substitute for DNMT3L in this process.

5. Evidence for Methylation by Dnmt3 Family Members

Although all five *dnmt3* genes were shown to be transcribed in 1-day-old embryos by 5′- and 3′-RACE, so far there is only limited functional evidence for their activity.[22] Significant effects on global methylation were not observed following injection of antisense morpholinos against each mRNA.[22] At the level of individual genes, injection of antisense morpholino against *dnmt7*, but not other *dnmt3* family members, was reported to decrease CpG methylation at the *ntl* gene. However, embryos developed normally and no differences in *ntl* gene expression could be detected.[22] In contrast, another study reported modest decreases in methylation and correlating increases in gene expression at the *no tail* and *lef1* genes using a *dnmt3* antisense morpholino.[64] It is likely that the zebrafish Dnmt3 family members share at least some overlapping sequence targets and that combinatorial knockdown of Dnmt3 proteins may reveal more pronounced effects on methylation.

6. Phenotypes of Dnmt3 Mutants

Thus far, Dnmt3 is the only member of the zebrafish *de novo* methyltransferase family for which a phenotype has been described. *Dnmt3* morpholino-injected embryos die at 96 hpf with multiple abnormalities including very small brains, defective pharyngeal arch formation, and abnormal retinal neural epithelial differentiation.[64] Although overall patterning of the developing brain appeared normal, several proneural markers were not expressed. The authors attribute the loss of these makers to decreased DNA methylation and ectopic expression of the *lef1* gene, a transcription factor that is critical for neurogenesis in the hypothalamus.[65]

It is noteworthy that the phenotypes reported for *dnmt3* morphants are very distinct from those observed in *dnmt1* morphants, and that overexpression of either mRNA cannot rescue phenotypes associated with the other.[64] Given that DNMT1 is typically thought to maintain methylation established by DNMT3s, it is curious that the neuronal defects seen in *dnmt3* morpholino-injected embryos were not observed in *dnmt1* morphants or mutants. However, expression of *dnmt3* mRNA was detected in the brain, eyes, and somites between 1 and 3 dpf[64] and it is possible that continuous *de novo* methylation by Dnmt3 might perpetuate DNA methylation in the absence of Dnmt1.

D. Conservation of Additional Proteins Associated with Methylation

The zebrafish genome also codes for orthologs of additional proteins that have been associated with DNA methylation in mammals. Although these proteins have not been extensively studied in fish, their apparent presence is consistent with broad conservation of the methylation machinery between fish and mammals.

1. CFP1

In mouse, the CXXC finger protein 1 (CFP1) specifically binds to unmethylated CpG dinucleotides, and has been demonstrated to influence chromatin structure at CpG islands by recruitment of the H3K4 methyltransferase Setd1.[66] In addition, mouse ES cells lacking CFP1 fail to differentiate, and exhibit a 60–80% reduction in global genomic cytosine methylation.[67] Although the mechanism by which CFP1 influences global methylation levels has not been fully elucidated, it appears that this role is conserved among vertebrates: as depletion of the zebrafish Cfp1 protein by morpholino injection results in a 60% reduction in genome-wide methylation.[68]

2. Methyl-Binding Proteins

Mammalian proteins that show selective binding to methylated CpG sites have been proposed to inhibit the transcription of methylated promoters (see Chapter by Pierre-Antoine Defossez and Irina Stancheva). Orthologs of the most prominent of the proteins, MeCP2 and Mbd2, are also found in the zebrafish genome.[69]

3. Piwi Proteins

The mouse Piwi-related proteins MIWI, MIWI2, and MILI are important for the *de novo* methylation of transposons in germ cells and likely mediate this activity through interactions with small RNA molecules called piRNAs (see Chapter by Anton Wutz). The zebrafish genome encodes for two *piwi* homologs: *ziwi* and *zili*.[70,71] Both proteins and piRNAs with homology to transposons can be found specifically in germ cells, and increased transposon expression was reported in the gonads of *zili* homozygous mutants.[71] However, because Piwi proteins are essential for germ cell development, it has been technically challenging to assess the methylation status of transposable elements in mutants. In contrast to mouse, *Drosophila* Piwi proteins mediate silencing of transposons independent of DNA methylation.[72] It is therefore of interest to determine whether zebrafish Piwi proteins facilitate transposon constraint through DNA methylation or other mechanisms.

IV. Link Between DNA and Histone Modification in Zebrafish

As the primary protein component of chromatin, histones have been attributed roles in both packaging of DNA into higher order structures and in regulation of transcription, in both cases through modification of their N-terminal tails (see Chapter by Taiping Chen). Despite extensive study, it remains unclear how DNA and histone modifications are targeted to particular regions of the genome. However, emerging data suggest that there is coordination between DNA methylation and histone modifications in many organisms including zebrafish.

While dependence of DNA methylation on histone H3 lysine 9 (H3K9) methylation has been clearly demonstrated at non-CpG dinucleotides in plants and fungi,[73,74] the relationship between these modifications has been less clear in mammals. Decreased DNA methylation was observed at pericentric satellite DNA from ES cells lacking the H3K9 methyltransferases SUV39h1/SUV39h2, suggesting that in some situations, mammalian DNA methylation is also dependent on H3K9 methylation. In contrast, H3K9 methylation of pericentric satellite DNA was normal in *Dnmt1* mutant ES cells.[75]

Surprisingly, evidence for the opposite relationship has been reported in zebrafish, with H3K9 methylation being dependent on DNA methylation. *Dnmt1* morpholino-injected embryos were reported to have a 70% reduction in H3K9 methylation, whereas Suv39h1-depleted embryos had normal levels of cytosine methylation.[40] Remarkably, developmental defects observed in *dnmt1* morpholino-injected embryos could be rescued by overexpression of *Suv39h1* mRNA.[40] These results suggest that, in contrast to other systems, zebrafish CpG methylation may direct H3K9 methylation and even more surprisingly, H3K9 methylation may be sufficient for normal development in the absence of Dnmt1-mediated DNA methylation.

Additional work in zebrafish has also suggested a link between the euchromatic H3K9 methyltransferase G9a and methylation by Dnmt3. Global changes in methylation of DNA or H3K9 were not detected following injection of morpholinos for either gene. However, both decreased H3K9 methylation and DNA methylation were observed at the *lef1* promotor in Dnmt3 morphant embryos ("morphants" are organisms treated with morpholinos). In addition, injection of G9a mRNA into Dnmt3 morphant embryos was reported to rescue phenotypes associated with *lef1* dysregulation.[64] Although it will be important to extend these observations to additional genes, this work raises the interesting possibility that distinct pairs of DNA and histone methyltransferases might act upon unique subsets of genes.[64]

V. DNA Methylation Dynamics in Early Development

In mammals, erasure and reestablishment of global methylation patterns have been described during gametogenesis and early embryogenesis.[76] While the functional significance of these dynamic events has not been confirmed, they are thought to reset gametic methylation patterns, thereby reprogramming the embryonic genome for somatic development.[77] In mouse, the process of erasure and reestablishment in the early embryo has been well studied. Demethylation of the paternal pronucleus occurs almost immediately after fertilization. Additional stepwise decreases in embryonic methylation then continue until the morula stage when *de novo* methylation is initiated.[78]

Because zebrafish embryos should also need to erase gametic methylation patterns prior to somatic development, one might predict conservation of these early reprogramming events. However, there are conflicting reports on whether a similar phenomenon occurs in zebrafish. Initial studies by Macleod et al. did not detect changes in DNA methylation over the course of development, at DANA repetitive elements or three randomly selected zebrafish genes.[24] Similarly, an early study of methylation dynamics in medaka detected constant levels of global methylation between 16-cell and larval stages of development.[26]

In contrast, other studies have provided evidence for dynamic methylation in the early embryo. Using methyl-sensitive restriction digests of genomic DNA, Mhanni et al. reported very low levels of global methylation in one to two cell stage embryos, compared to blastula stage embryos.[79] These observations were corroborated by an additional report using a 5-methyl cytosine antibody to monitor methylation in the early embryo. In this study, robust methylation was detected in blastulae by Southwestern blots and immunocytochemistry, but little to no methylation was detected before this stage.[80]

It is difficult to fully reconcile the results of these reports. The study by Macleod et al. omitted very early embryonic stages from analysis, and only examined a selected group of sequences. Thus, demethylation occurring immediately after fertilization or at other unexamined sequences would have been missed. However, while the report from Mhanni et al. described the most severe demethylation in one to two cell embryos, substantial demethylation of both genomic DNA and DANA elements was also described at 2.2 hpf, a time at which Macleod et al. reported DANA elements to be fully methylated.

Perhaps one reason for the discrepancies is the substantial technical hurdles associated with measuring DNA methylation in the early embryo, where small amounts of starting material, a high yolk to DNA ratio, and abundant stores of maternal RNA complicate analyses. Given that a large amount of high-quality genomic DNA is essential for accurate assessment of DNA methylation by restriction enzyme digestion, alternative methods such as bisulfite sequencing may be better suited to the early embryo. Many different genomic loci will need to be examined to characterize genome-wide methylation dynamics but, given its importance from an evolutionary standpoint, a more comprehensive assessment to clarify methylation dynamics in zebrafish should be well worth the effort involved.

VI. Evidence for Active DNA Demethylation in Zebrafish

If DNA methylation patterns during development are dynamic, a means for demethylation is required. In animals, DNA demethylation has a long and controversial history and, despite the efforts of multiple laboratories, mechanisms for its control remain unclear (see Chapter by Taiping Chen). One mechanism termed "passive demethylation" involves failure to methylate CpG dinucleotides in the newly synthesized DNA strand during replication. Accordingly, after two cycles of replication, 50% of DNA duplexes would be fully demethylated. However, some experiments have detected DNA demethylation in the absence of replication, a scenario that would require "active" enzymatic removal of methylated cytosine. Over the years, many different mechanisms have been proposed and discounted in animals,[81] and to date active demethylation remains a contentious issue.

A. Early Evidence for Active Demethylation in Zebrafish

There is some evidence for active demethylation in zebrafish. Initial studies reported demethylation of *in vitro* methylated plasmid DNA injected into the one-cell stage embryo, even in the presence of the replication inhibiting drug camptothecin.[82] Supporting the DNA replication-independent nature of the demethylase activity, extracts from embryos collected at 4 hpf were also capable of demethylating plasmid DNA *in vitro*. Surprisingly, this activity was abolished upon treatment with RNase A, suggesting that the activity was RNA dependent. While these early studies favor the presence of an active demethylase in zebrafish embryos, *in vivo* targets of the demethylase activity have not been described nor the mechanism of demethylation identified.

B. Demethylation by Aid and Mbd4

In animals, activation-induced cytosine deaminase (AID) is currently at the epicenter of demethylation research. AID has been implicated in the erasure of DNA methylation in primordial germ cells and in fibroblasts undergoing reprogramming. Recent work suggests that Aid is likewise capable of mediating DNA demethylation in zebrafish.[25,83,84]

Provocative work in the zebrafish system suggests a coupled mechanism of demethylation, in which Aid-mediated deamination of methylated cytosine residues is followed by base excision through the glycosylase activity of the methyl-binding domain protein Mbd4. Supporting this model, overexpression of *aid* and *mbd4* mRNAs resulted in a 30% reduction in genomic DNA methylation, as assayed by mass spectrometry and *Hpa*II digestion.[25] In addition, the growth arrest and DNA damage protein Gad45, which had been previously implicated in DNA demethylation, stimulated Mbd4- and Aid-mediated demethylation.[25] It remains to be determined whether the same demethylase system is utilized for demethylation in the early zebrafish embryo.

C. Potential Roles for Aid Demethylase Activity in Development

The Aid demethylase machinery has been implicated in the maintenance of intestinal cells and possibly the progression of colon cancer. At 72 hpf *aid*, *mbd4* and *gadd45* mRNAs were all shown to be upregulated in zebrafish homozygous for mutations in the *adenomatous polyposis coli* (*apc*) tumor suppressor gene.[85] Genome-wide changes in DNA methylation were not detected in *apc* mutants; however, methylated DNA immunoprecipitation (MeDIP) revealed decreases in DNA methylation at a number of genes many of which are involved in intestinal differentiation and colorectal cancer progression.[85] In particular, the authors observed high expression and hypomethylation of a marker of colon crypt progenitor cells. Increased expression of the progenitor cell marker correlated with decreased expression of an intestinal differentiation marker. Proper expression of both genes could be partially rescued by injection of an anti-Aid morpholino, implicating Aid-mediated demethylation in their regulation. These observations lead to the hypothesis that APC downregulates the demethylase machinery to facilitate progenitor commitment to differentiation during normal development and, conversely, increased expression of *aid*, *mbd4*, and *gadd45* in *apc* mutants may drive the hypomethylation frequently observed in colon cancer.[85]

The developmental defects described in this study are intriguing although somewhat surprising given that overall development appears normal in *Aid* and *Mbd4* mutant mice.[83,86] To clarify the roles of Aid in normal zebrafish development, it will be important to examine its direct role independent of the *apc* mutant background. Intriguingly, zebrafish embryos injected with antisense morpholinos against either *aid* or *mbd4* were reported to show reduced expression of the neural marker *neurogenin-1*,[25] suggesting AID could have more widespread functions in normal development.

VII. Perspectives and Future Directions

The initial studies presented in this review provide a fundamental understanding of DNA methylation in zebrafish and demonstrate conservation of the basic methylation machinery with mammals. These studies provide a framework for new research that exploits the unique features of the zebrafish system for the study of DNA methylation, particularly in the context of development.

A. DNA Methylation in Development

An important question in the field of methylation research centers around why DNA methylation is required for development in vertebrates, when many other organisms survive without methylated genomes.[1] Mammals require methylation at imprinted genes, whose proper regulation is necessary for viability.[1] Thus, it has been hypothesized that dysregulation of imprinting could explain the lethality observed in mammalian *Dnmt1* mutant embryos. Zebrafish do not require imprinting for normal development, yet similar to mice, *Dnmt1* mutant zebrafish are inviable and show significant levels of p53-mediated apoptosis. These observations suggest that there are other underlying reasons for hypomethylation-induced lethality.[40,42] The accessibility of the developing zebrafish embryo, the wide array of characterized developmental markers, and the lack of imprinting should make zebrafish a simple and well-defined system for probing the additional requirements for DNA methylation in vertebrates.

B. *In Vivo* Monitoring of Methylation

To date, much of our understanding of methylation has come from biochemical assays on DNA extracted from whole animals or tissues. However, these snapshots of methylation may overlook dynamic changes during development or differences in methylation among distinct cell populations in heterogeneous tissues. Moreover, biochemical approaches prove challenging for multigenerational studies as animal sacrifice is frequently required for sample collection. The ease of creating transgenic zebrafish, coupled with the clarity of the developing larvae, makes zebrafish an ideal system for *in vivo* monitoring of methylation via fluorescent reporters.

This strategy has recently been validated in transgenic zebrafish carrying the green fluorescent protein (GFP) under the control of 14 copies of the yeast upstream activation sequence (14X UAS). In the absence of DNA methylation, the yeast transcriptional activator Gal4 binds to the 14X UAS to drive transcription of GFP. However, the UAS is rich in CpG dinucleotides and the 14 copies constitute a repetitive sequence, which likely attracts methylation.[23] Over several generations, methylation accumulates at the 14X UAS leading to increasing variegation in GFP labeling and finally complete silencing of the transgene.[23] Consistent with this, reactivation of GFP expression is observed in *dnmt1* and *uhrf1* zebrafish mutants that have reduced methylation.[18,23] Thus, transgenic lines can be used to continuously monitor methylation in the developing zebrafish at the resolution of single cells.

Somewhat surprisingly, when reactivation of the 14X UAS was examined in *dnmt1* mutant larvae using a brain-specific driver, not all regions of the brain showed equivalent reactivation of GFP expression. While robust labeling was observed in the mid- and forebrain regions, little to no reactivation was observed in the hindbrain.[23] The mechanisms contributing to this difference have not been identified, although it is possible that other DNA methyltransferases compensate for Dnmt1 in the hindbrain. The observation of regional differences in reactivation underscores the importance of monitoring DNA methylation *in situ* and in the context of development.

C. Reprogramming and Cellular Differentiation

Additional transgenic lines may reveal important information about the roles of DNA methylation in cellular differentiation and regeneration.[87] Transgenic lines carrying GFP reporters that are silenced in adult tissue, but reactivated in dividing cells of the regenerating fin, have recently been described.[88] The same transgenes were also found to be reexpressed during retinal regeneration and could potentially serve as reporters in a variety of regeneration models.[85] Intriguingly, DNA methylation may control the expression status of these transgenes.[88] Although additional research will be required to validate these tools, they offer enormous potential for monitoring dynamic changes in methylation during the regeneration process.[85] Given that roles for Dnmt1 and Uhrf1 have recently been described in zebrafish pancreas and liver regeneration,[37,42] and that Aid-mediated demethylation has been shown to be important for the maintenance of intestinal stem cells, the zebrafish is likely to become a widely used system for understanding the roles of DNA methylation in cellular differentiation and regeneration.

D. Environmental Influences on DNA Methylation

A new and intriguing research direction has emerged, which is focused on the effects of dietary and environmental factors on DNA methylation patterns, and their relevance to the etiology of human disease.[89] For example, bisphenol A (BPA), an endocrine-disrupting compound used in the production of plastics and resins, causes hypomethylation of several genetic loci in mouse.[89]

Additional studies in rodents and humans have also suggested that exposure to a number of other chemicals including metals, peroxisome proliferators, air pollutants, and endocrine-disrupting toxins may alter methylation patterns.[90] Rearing adult or larval zebrafish in water containing chemicals would provide a simple way to probe how exposure impacts DNA methylation. Changes in methylation can be assayed by direct examination of harvested genomic DNA, or *in vivo* using transgenic reporters of methylation such as those described above. Preliminary evidence for changes in methylation at several zebrafish genes following exposure to either 17α-ethinylestradiol or arsenic validates the potential of these approaches.[91,92]

E. Genetic Screens to Identify Factors Important for Methylation

Despite extensive research, the complete machinery required for the establishment, maintenance, and interpretation of methylation marks has not been fully elucidated in vertebrates. Forward genetic screens in *Arabidopsis* have provided a number of mutations that affect methylation; however, many genes identified from plant screens are not conserved in animals.[93–95] A screen for dominant modifiers of an epigenetically regulated transgene in mouse has provided mutants demonstrating the importance of animal screens.[96] However, to date, an unbiased screen for recessive mutations that affect DNA methylation in vertebrates has not been described.

One of the most powerful features of the zebrafish system is its potential for unbiased genetic and chemical screens. Screens for components required for DNA methylation in zebrafish could be performed by direct analysis of genomic DNA, as has been described in plants.[97] Alternatively, transgenic reporters of methylation such as those described above may offer an efficient means for rapid screening at multiple stages of development. Reactivation of transgenic reporters in zebrafish mutants with hypomethylated genomes provides proof of principle for using genetic screens to identify new mutations that influence gene silencing.

VIII. Conclusions

Although there are nearly 300 million years separating the last common fish and human ancestor, DNA methylation pathways remain well conserved. Given the similarities between DNA methylation in mammals and zebrafish, the zebrafish model holds great promise for providing new insights into the mechanisms that regulate DNA methylation in vertebrates and for better understanding its roles in normal development and disease.

References

1. Goll MG, Bestor TH. Eukaryotic cytosine methyltransferases. *Annu Rev Biochem* 2005;**74**:481–514.
2. Streisinger G, Walker C, Dower N, Knauber D, Singer F. Production of clones of homozygous diploid zebra fish (Brachydanio rerio). *Nature* 1981;**291**:293–6.
3. Ekker SC, Larson JD. Morphant technology in model developmental systems. *Genesis* 2001;**30**:89–93.
4. Robu ME, Larson JD, Nasevicius A, Beiraghi S, Brenner C, Farber SA, et al. p53 activation by knockdown technologies. *PLoS Genet* 2007;**3**:e78.
5. Bill BR, Petzold AM, Clark KJ, Schimmenti LA, Ekker SC. A primer for morpholino use in zebrafish. *Zebrafish* 2009;**6**:69–77.
6. Stuart GW, McMurray JV, Westerfield M. Replication, integration and stable germ-line transmission of foreign sequences injected into early zebrafish embryos. *Development* 1988;**103**:403–12.
7. Culp P, Nusslein-Volhard C, Hopkins N. High-frequency germ-line transmission of plasmid DNA sequences injected into fertilized zebrafish eggs. *Proc Natl Acad Sci USA* 1991;**88**:7953–7.
8. Amsterdam A, Lin S, Hopkins N. The Aequorea victoria green fluorescent protein can be used as a reporter in live zebrafish embryos. *Dev Biol* 1995;**171**:123–9.
9. Gibbs PD, Peek A, Thorgaard G. An in vivo screen for the luciferase transgene in zebrafish. *Mol Mar Biol Biotechnol* 1994;**3**:307–16.
10. Stuart GW, Vielkind JR, McMurray JV, Westerfield M. Stable lines of transgenic zebrafish exhibit reproducible patterns of transgene expression. *Development* 1990;**109**:577–84.
11. Linn F, Heidmann I, Saedler H, Meyer P. Epigenetic changes in the expression of the maize A1 gene in Petunia hybrida: role of numbers of integrated gene copies and state of methylation. *Mol Gen Genet* 1990;**222**:329–36.
12. Mittelsten Scheid O, Paszkowski J, Potrykus I. Reversible inactivation of a transgene in Arabidopsis thaliana. *Mol Gen Genet* 1991;**228**:104–12.
13. Davis BP, MacDonald RJ. Limited transcription of rat elastase I transgene repeats in transgenic mice. *Genes Dev* 1988;**2**:13–22.
14. Sharpe JA, Wells DJ, Whitelaw E, Vyas P, Higgs DR, Wood WG. Analysis of the human alpha-globin gene cluster in transgenic mice. *Proc Natl Acad Sci USA* 1993;**90**:11262–6.
15. Robertson G, Garrick D, Wu W, Kearns M, Martin D, Whitelaw E. Position-dependent variegation of globin transgene expression in mice. *Proc Natl Acad Sci USA* 1995;**92**:5371–5.
16. Garrick D, Fiering S, Martin DI, Whitelaw E. Repeat-induced gene silencing in mammals. *Nat Genet* 1998;**18**:56–9.
17. Martin CC, McGowan R. Genotype-specific modifiers of transgene methylation and expression in the zebrafish, Danio rerio. *Genet Res* 1995;**65**:21–8.
18. Feng S, Cokus SJ, Zhang X, Chen PY, Bostick M, Goll MG, et al. Conservation and divergence of methylation patterning in plants and animals. *Proc Natl Acad Sci USA* 2010;**107**:8689–94.
19. Zemach A, McDaniel IE, Silva P, Zilberman D. Genome-wide evolutionary analysis of eukaryotic DNA methylation. *Science* 2010;**328**:916–9.
20. Borkovich KA, Alex LA, Yarden O, Freitag M, Turner GE, Read ND, et al. Lessons from the genome sequence of Neurospora crassa: tracing the path from genomic blueprint to multicellular organism. *Microbiol Mol Biol Rev* 2004;**68**:1–108.
21. Selker EU, Tountas NA, Cross SH, Margolin BS, Murphy JG, Bird AP, et al. The methylated component of the Neurospora crassa genome. *Nature* 2003;**422**:893–7.

22. Shimoda N, Yamakoshi K, Miyake A, Takeda H. Identification of a gene required for de novo DNA methylation of the zebrafish no tail gene. *Dev Dyn* 2005;**233**:1509–16.
23. Goll MG, Anderson R, Stainier DY, Spradling AC, Halpern ME. Transcriptional silencing and reactivation in transgenic zebrafish. *Genetics* 2009;**182**:747–55.
24. Macleod D, Clark VH, Bird A. Absence of genome-wide changes in DNA methylation during development of the zebrafish. *Nat Genet* 1999;**23**:139–40.
25. Rai K, Huggins IJ, James SR, Karpf AR, Jones DA, Cairns BR. DNA demethylation in zebrafish involves the coupling of a deaminase, a glycosylase, and gadd45. *Cell* 2008;**135**:1201–12.
26. Walter RB, Li HY, Intano GW, Kazianis S, Walter CA. Absence of global genomic cytosine methylation pattern erasure during medaka (Oryzias latipes) early embryo development. *Comp Biochem Physiol B Biochem Mol Biol* 2002;**133**:597–607.
27. Aniagu SO, Williams TD, Allen Y, Katsiadaki I, Chipman JK. Global genomic methylation levels in the liver and gonads of the three-spine stickleback (Gasterosteus aculeatus) after exposure to hexabromocyclododecane and 17-beta oestradiol. *Environ Int* 2008;**34**:310–7.
28. Varriale A, Bernardi G. DNA methylation and body temperature in fishes. *Gene* 2006;**385**:111–21.
29. Jaillon O, Aury JM, Brunet F, Petit JL, Stange-Thomann N, Mauceli E, et al. Genome duplication in the teleost fish Tetraodon nigroviridis reveals the early vertebrate proto-karyotype. *Nature* 2004;**431**:946–57.
30. Ramsahoye BH, Biniszkiewicz D, Lyko F, Clark V, Bird AP, Jaenisch R. Non-CpG methylation is prevalent in embryonic stem cells and may be mediated by DNA methyltransferase 3a. *Proc Natl Acad Sci USA* 2000;**97**:5237–42.
31. Bostick M, Kim JK, Esteve PO, Clark A, Pradhan S, Jacobsen SE. UHRF1 plays a role in maintaining DNA methylation in mammalian cells. *Science* 2007;**317**:1760–4.
32. Sharif J, Muto M, Takebayashi S, Suetake I, Iwamatsu A, Endo TA, et al. The SRA protein Np95 mediates epigenetic inheritance by recruiting Dnmt1 to methylated DNA. *Nature* 2007;**450**:908–12.
33. Ooi SK, Bestor TH. Cytosine methylation: remaining faithful. *Curr Biol* 2008;**18**:R174–6.
34. Mhanni AA, Yoder JA, Dubesky C, McGowan RA. Cloning and sequence analysis of a zebrafish cDNA encoding DNA (cytosine-5)-methyltransferase-1. *Genesis* 2001;**30**:213–9.
35. Mertineit C, Yoder JA, Taketo T, Laird DW, Trasler JM, Bestor TH. Sex-specific exons control DNA methyltransferase in mammalian germ cells. *Development* 1998;**125**:889–97.
36. Howell CY, Bestor TH, Ding F, Latham KE, Mertineit C, Trasler JM, et al. Genomic imprinting disrupted by a maternal effect mutation in the Dnmt1 gene. *Cell* 2001;**104**:829–38.
37. Sadler KC, Krahn KN, Gaur NA, Ukomadu C. Liver growth in the embryo and during liver regeneration in zebrafish requires the cell cycle regulator, uhrf1. *Proc Natl Acad Sci USA* 2007;**104**:1570–5.
38. Bronner C, Achour M, Arima Y, Chataigneau T, Saya H, Schini-Kerth VB. The UHRF family: oncogenes that are drugable targets for cancer therapy in the near future? *Pharmacol Ther* 2007;**115**:419–34.
39. Martin CC, Laforest L, Akimenko MA, Ekker M. A role for DNA methylation in gastrulation and somite patterning. *Dev Biol* 1999;**206**:189–205.
40. Rai K, Nadauld LD, Chidester S, Manos EJ, James SR, Karpf AR, et al. Zebra fish Dnmt1 and Suv39h1 regulate organ-specific terminal differentiation during development. *Mol Cell Biol* 2006;**26**:7077–85.
41. Thisse B, Heyer V, Lux A, Alunni V, Degrave A, Seiliez I, et al. Spatial and temporal expression of the zebrafish genome by large-scale in situ hybridization screening. *Methods Cell Biol* 2004;**77**:505–19.

42. Anderson RM, Bosch JA, Goll MG, Hesselson D, Dong PD, Shin D, et al. Loss of Dnmt1 catalytic activity reveals multiple roles for DNA methylation during pancreas development and regeneration. *Dev Biol* 2009;**334**:213–23.
43. Mhanni AA, McGowan RA. Variations in DNA (cytosine-5)-methyltransferase-1 expression during oogenesis and early development of the zebrafish. *Dev Genes Evol* 2002;**212**:530–3.
44. Amsterdam A, Nissen RM, Sun Z, Swindell EC, Farrington S, Hopkins N. Identification of 315 genes essential for early zebrafish development. *Proc Natl Acad Sci USA* 2004;**101**:12792–7.
45. Jackson-Grusby L, Beard C, Possemato R, Tudor M, Fambrough D, Csankovszki G, et al. Loss of genomic methylation causes p53-dependent apoptosis and epigenetic deregulation. *Nat Genet* 2001;**27**:31–9.
46. Esteve PO, Chin HG, Benner J, Feehery GR, Samaranayake M, Horwitz GA, et al. Regulation of DNMT1 stability through SET7-mediated lysine methylation in mammalian cells. *Proc Natl Acad Sci USA* 2009;**106**:5076–81.
47. Wang J, Hevi S, Kurash JK, Lei H, Gay F, Bajko J, et al. The lysine demethylase LSD1 (KDM1) is required for maintenance of global DNA methylation. *Nat Genet* 2009;**41**:125–9.
48. Kuhlmann M, Borisova BE, Kaller M, Larsson P, Stach D, Na J, et al. Silencing of retrotransposons in Dictyostelium by DNA methylation and RNAi. *Nucleic Acids Res* 2005;**33**:6405–17.
49. Kunert N, Marhold J, Stanke J, Stach D, Lyko F. A Dnmt2-like protein mediates DNA methylation in Drosophila. *Development* 2003;**130**:5083–90.
50. Phalke S, Nickel O, Walluscheck D, Hortig F, Onorati MC, Reuter G. Retrotransposon silencing and telomere integrity in somatic cells of Drosophila depends on the cytosine-5 methyltransferase DNMT2. *Nat Genet* 2009;**41**:696–702.
51. Goll MG, Kirpekar F, Maggert KA, Yoder JA, Hsieh CL, Zhang X, et al. Methylation of tRNAAsp by the DNA methyltransferase homolog Dnmt2. *Science* 2006;**311**:395–8.
52. Jurkowski TP, Meusburger M, Phalke S, Helm M, Nellen W, Reuter G, et al. Human DNMT2 methylates tRNA(Asp) molecules using a DNA methyltransferase-like catalytic mechanism. *RNA* 2008;**14**:1663–70.
53. Schaefer M, Lyko F. Solving the Dnmt2 enigma. *Chromosoma* 2010;**119**:35–40.
54. Rai K, Chidester S, Zavala CV, Manos EJ, James SR, Karpf AR, et al. Dnmt2 functions in the cytoplasm to promote liver, brain, and retina development in zebrafish. *Genes Dev* 2007;**21**:261–6.
55. Yamakoshi K, Shimoda N. De novo DNA methylation at the CpG island of the zebrafish no tail gene. *Genesis* 2003;**37**:195–202.
56. Okano M, Xie S, Li E. Cloning and characterization of a family of novel mammalian DNA (cytosine-5) methyltransferases. *Nat Genet* 1998;**19**:219–20.
57. Okano M, Bell DW, Haber DA, Li E. DNA methyltransferases Dnmt3a and Dnmt3b are essential for de novo methylation and mammalian development. *Cell* 1999;**99**:247–57.
58. Bu W, Su LK. Characterization of functional domains of human EB1 family proteins. *J Biol Chem* 2003;**278**:49721–31.
59. Smith TH, Dueck CC, Mhanni AA, McGowan RA. Novel splice variants associated with one of the zebrafish dnmt3 genes. *BMC Dev Biol* 2005;**5**:23.
60. Bourc'his D, Xu GL, Lin CS, Bollman B, Bestor TH. Dnmt3L and the establishment of maternal genomic imprints. *Science* 2001;**294**:2536–9.
61. Corley-Smith GE, Lim CJ, Brandhorst BP. Production of androgenetic zebrafish (Danio rerio). *Genetics* 1996;**142**:1265–76.
62. Yokomine T, Hata K, Tsudzuki M, Sasaki H. Evolution of the vertebrate DNMT3 gene family: a possible link between existence of DNMT3L and genomic imprinting. *Cytogenet Genome Res* 2006;**113**:75–80.

63. Bourc'his D, Bestor TH. Meiotic catastrophe and retrotransposon reactivation in male germ cells lacking Dnmt3L. *Nature* 2004;**431**:96–9.
64. Rai K, Jafri IF, Chidester S, James SR, Karpf AR, Cairns BR, et al. Dnmt3 and G9a cooperate for tissue-specific development in zebrafish. *J Biol Chem* 2010;**285**:4110–21.
65. Lee JE, Wu SF, Goering LM, Dorsky RI. Canonical Wnt signaling through Lef1 is required for hypothalamic neurogenesis. *Development* 2006;**133**:4451–61.
66. Thomson JP, Skene PJ, Selfridge J, Clouaire T, Guy J, Webb S, et al. CpG islands influence chromatin structure via the CpG-binding protein Cfp1. *Nature* 2010;**464**:1082–6.
67. Carlone DL, Lee JH, Young SR, Dobrota E, Butler JS, Ruiz J, et al. Reduced genomic cytosine methylation and defective cellular differentiation in embryonic stem cells lacking CpG binding protein. *Mol Cell Biol* 2005;**25**:4881–91.
68. Young SR, Mumaw C, Marrs JA, Skalnik DG. Antisense targeting of CXXC finger protein 1 inhibits genomic cytosine methylation and primitive hematopoiesis in zebrafish. *J Biol Chem* 2006;**281**:37034–44.
69. Coverdale LE, Martyniuk CJ, Trudeau VL, Martin CC. Differential expression of the methyl-cytosine binding protein 2 gene in embryonic and adult brain of zebrafish. *Brain Res Dev Brain Res* 2004;**153**:281–7.
70. Houwing S, Kamminga LM, Berezikov E, Cronembold D, Girard A, van den Elst H, et al. A role for Piwi and piRNAs in germ cell maintenance and transposon silencing in Zebrafish. *Cell* 2007;**129**:69–82.
71. Houwing S, Berezikov E, Ketting RF. Zili is required for germ cell differentiation and meiosis in zebrafish. *EMBO J* 2008;**27**:2702–11.
72. Sarot E, Payen-Groschene G, Bucheton A, Pelisson A. Evidence for a piwi-dependent RNA silencing of the gypsy endogenous retrovirus by the Drosophila melanogaster flamenco gene. *Genetics* 2004;**166**:1313–21.
73. Tamaru H, Zhang X, McMillen D, Singh PB, Nakayama J, Grewal SI, et al. Trimethylated lysine 9 of histone H3 is a mark for DNA methylation in Neurospora crassa. *Nat Genet* 2003;**34**:75–9.
74. Jackson JP, Lindroth AM, Cao X, Jacobsen SE. Control of CpNpG DNA methylation by the KRYPTONITE histone H3 methyltransferase. *Nature* 2002;**416**:556–60.
75. Lehnertz B, Ueda Y, Derijck AA, Braunschweig U, Perez-Burgos L, Kubicek S, et al. Suv39h-mediated histone H3 lysine 9 methylation directs DNA methylation to major satellite repeats at pericentric heterochromatin. *Curr Biol* 2003;**13**:1192–200.
76. Morgan HD, Santos F, Green K, Dean W, Reik W. Epigenetic reprogramming in mammals. *Hum Mol Genet* 2005;**14**(1):R47–58.
77. Monk M, Boubelik M, Lehnert S. Temporal and regional changes in DNA methylation in the embryonic, extraembryonic and germ cell lineages during mouse embryo development. *Development* 1987;**99**:371–82.
78. Santos F, Hendrich B, Reik W, Dean W. Dynamic reprogramming of DNA methylation in the early mouse embryo. *Dev Biol* 2002;**241**:172–82.
79. Mhanni AA, McGowan RA. Global changes in genomic methylation levels during early development of the zebrafish embryo. *Dev Genes Evol* 2004;**214**:412–7.
80. MacKay AB, Mhanni AA, McGowan RA, Krone PH. Immunological detection of changes in genomic DNA methylation during early zebrafish development. *Genome* 2007;**50**:778–85.
81. Ooi SK, Bestor TH. The colorful history of active DNA demethylation. *Cell* 2008;**133**:1145–8.
82. Collas P. Modulation of plasmid DNA methylation and expression in zebrafish embryos. *Nucleic Acids Res* 1998;**26**:4454–61.
83. Popp C, Dean W, Feng S, Cokus SJ, Andrews S, Pellegrini M, et al. Genome-wide erasure of DNA methylation in mouse primordial germ cells is affected by AID deficiency. *Nature* 2010;**463**:1101–5.

84. Bhutani N, Brady JJ, Damian M, Sacco A, Corbel SY, Blau HM. Reprogramming towards pluripotency requires AID-dependent DNA demethylation. *Nature* 2010;**463**:1042–7.
85. Rai K, Sarkar S, Broadbent TJ, Voas M, Grossmann KF, Nadauld LD, et al. DNA demethylase activity maintains intestinal cells in an undifferentiated state following loss of APC. *Cell* 2010;**142**:930–42.
86. Millar CB, Guy J, Sansom OJ, Selfridge J, MacDougall E, Hendrich B, et al. Enhanced CpG mutability and tumorigenesis in MBD4-deficient mice. *Science* 2002;**297**:403–5.
87. Khavari DA, Sen GL, Rinn JL. DNA methylation and epigenetic control of cellular differentiation. *Cell Cycle* 2010;**9**:3880–3.
88. Thummel R, Burket CT, Hyde DR. Two different transgenes to study gene silencing and re-expression during zebrafish caudal fin and retinal regeneration. *ScientificWorldJournal* 2006;**6** (Suppl. 1):65–81.
89. Dolinoy DC, Huang D, Jirtle RL. Maternal nutrient supplementation counteracts bisphenol A-induced DNA hypomethylation in early development. *Proc Natl Acad Sci USA* 2007;**104**:13056–61.
90. Baccarelli A, Bollati V. Epigenetics and environmental chemicals. *Curr Opin Pediatr* 2009;**21**:243–51.
91. Stromqvist M, Tooke N, Brunstrom B. DNA methylation levels in the 5′ flanking region of the vitellogenin I gene in liver and brain of adult zebrafish (Danio rerio)—sex and tissue differences and effects of 17alpha-ethinylestradiol exposure. *Aquat Toxicol* 2010;**98**:275–81.
92. Li D, Lu C, Wang J, Hu W, Cao Z, Sun D, et al. Developmental mechanisms of arsenite toxicity in zebrafish (Danio rerio) embryos. *Aquat Toxicol* 2009;**91**:229–37.
93. Chan SW, Henderson IR, Jacobsen SE. Gardening the genome: DNA methylation in Arabidopsis thaliana. *Nat Rev Genet* 2005;**6**:351–60.
94. Kanno T, Mette MF, Kreil DP, Aufsatz W, Matzke M, Matzke AJ. Involvement of putative SNF2 chromatin remodeling protein DRD1 in RNA-directed DNA methylation. *Curr Biol* 2004;**14**:801–5.
95. Murfett J, Wang XJ, Hagen G, Guilfoyle TJ. Identification of Arabidopsis histone deacetylase HDA6 mutants that affect transgene expression. *Plant Cell* 2001;**13**:1047–61.
96. Blewitt ME, Vickaryous NK, Hemley SJ, Ashe A, Bruxner TJ, Preis JI, et al. An *N*-ethyl-*N*-nitrosourea screen for genes involved in variegation in the mouse. *Proc Natl Acad Sci USA* 2005;**102**:7629–34.
97. Vongs A, Kakutani T, Martienssen RA, Richards EJ. Arabidopsis thaliana DNA methylation mutants. *Science* 1993;**260**:1926–8.

Section II

Mammalian DNA Methyltransferases — Structure and Function

Dnmt1: Structure and Function

Željko M. Svedružić

Medical Biochemistry, PB Rab, Faculty of Medicine, University of Rijeka, Rab, Croatia

Dnmt1, the principal DNA methyltransferase in mammalian cells, is a large and a highly dynamic enzyme with multiple regulatory features that can control DNA methylation in cells. This chapter highlights how insights into Dnmt1 structure and function can advance our understanding of DNA methylation in cells. The allosteric site(s) on Dnmt1 can regulate processes of *de novo* and maintenance DNA methylation in cells. Remaining open questions include which molecules, by what mechanism, bind at the allosteric site(s) in cells? Different phosphorylation sites on Dnmt1 can change its activity or ability to bind DNA target sites. Thirty-one different molecules are currently known to have physical and/or functional interaction with Dnmt1 in cells. The Dnmt1 structure and enzymatic mechanism offer unique insights into those interactions. The interacting molecules are involved in chromatin organization, DNA repair, cell cycle regulation, and apoptosis and also include RNA polymerase II, some RNA-binding proteins, and some specific Dnmt1-inhibitory RNA molecules. Combined insights from studies of different enzymatic features of Dnmt1 offer novel ideas for development of drug candidates, and can be used in selection of promising drug candidates from more than 15 different compounds that have been identified as possible inhibitors of DNA methylation in cells.

Progress in Molecular Biology
and Translational Science, Vol. 101
DOI: 10.1016/B978-0-12-387685-0.00006-8

I. Introduction

DNA methylation is a fundamental mechanism in functional organization of the human genome. Studies of DNA methylation can help us to tackle some of the key questions in the current biomedical sciences, such as carcinogenesis,[1] host infection by different viruses,[2–7] cell differentiation,[8] autoimmune diseases,[9] different types of mental illness and neurological disorders,[10–12] and environmental toxicology.[13] Unfortunately, effective therapies and preventive treatments targeting DNA methylation are still very underdeveloped. This is in large part due to lack of knowledge about the molecular mechanism of DNA methylation, starting with the principal DNA methyltransferase in human cells, Dnmt1 (DNA methyltransferase 1).

The first evidence for DNA methylation was reported in 1948,[14] and 27 years later, the first attempts to purify mammalian DNA methyltransferase were described.[15] Over the next 15 years, Dnmt1 was found to be a very large and a complex enzyme. Dnmt1 has catalytic preference for hemimethylated sites,[16,17] and its activity can be regulated by a specific group of DNA and RNA molecules.[18] Dnmt1 has high kinetic preference for poly(dI-dC) substrate, and a distinct allosteric site.[16] The molecular biology era of Dnmt1 began with cloning of its cDNA[19] and the full-length gene in 1992.[20] The first complete purification from mammalian cells[21] laid ground for more than a dozen mechanistic studies (reviewed in Ref. 22). Mechanistic studies of Dnmt1 were greatly aided by the first crystal structure of a flipped-out target base by bacterial enzyme M.HhaI[23] and the numerous subsequent mechanistic studies that were inspired by that landmark achievement.

Dnmt1 expression using a baculovirus vector[24–26] allowed preparation and characterization of different fragments of Dnmt1,[27–29] and made purification of Dnmt1 almost routine for many research groups (today, purified Dnmt1 can be also purchased from several commercial sources). The majority of Dnmt1 studies have been focused on the function of its allosteric site,[22,30–32] on selectivity between *de novo* and maintenance methylation,[30,33] and on characterization of different Dnmt1 fragments.[28,29,34,35] Relatively little work has been done on characterization of the catalytic mechanism in the active site,[36,37] on the mechanisms of action for proposed inhibitors of Dnmt1,[38] and on the multiple phosphorylations of Dnmt1.[39,40] A large research effort has been devoted to search for molecules that interact with Dnmt1 in cells. At this writing (December, 2010), 31 different molecules have been found to interact with Dnmt1 (Table I). Many of those interactions have been mapped to different parts of Dnmt1 sequence,[35,72] only a small fraction of those interactions have been functionally characterized. The majority of Dnmt1 studies

TABLE I

Summary of Molecules Currently Known to Interact Physically and/or Functionally with Dnmt1

Core chromatin replication complex	DNA repair, cell cycle control, and regulation of apoptosis	RNA-directed DNA methylation
– Dnmt3a and Dnmt3b[41] – SNF2h-containing chromatin-remodeling complex NoRC[42] – LSH protein (lymphoid-specific helicases) protein related to the SNF2 family of chromatin-remodeling ATPases[43] – PCNA, DNA clamp processivity-promoting factor[44] – UHRF1[45,46] – HP1β, heterochromatin protein 1β isoform (chromobox protein)[47,48] – SUV39H1, histone-lysine *N*-methyltransferase[48] – G9a histone methyltransferase[49] – HDAC1 and HDAC2, histone deacetylase 1/2[50–52] – PML-RAR promyelocytic leukemia-retinoic acid receptor, oncogene transcription factor[53] – RIP140, metabolic repressor, also known as NRIP1 (nuclear receptor interacting protein 1)[51] – CFP1 CXXC finger protein 1 (PHD domain)[54] – MBD2/MBD3, methyl-CpG-binding domain protein[55] – PcG-EZH2 Polycomb-group proteins enhancer of Zeste homolog 2[56]	– PARP-1 (poly(ADP-ribose) polymerase 1) and poly(ADP-ribose)[57,58] – pRb/E2F1, Retinoblastoma tumor suppressor protein, control of G1/S transition and S-phase[29,59,60] – p53, tumor suppressor, regulation of cell cycle and apoptosis[61] – DMAP1, DNA methyltransferase 1-associated protein 1[50,62,63] – RGS6 member of mammalian RGS (regulator of G-protein signaling) proteins[62] – CK1δ/ϵ kinase that phosphorylates Dnmt1[40] – Annexin V, scaffolding proteins that anchors other proteins to the cell membrane and participates in apoptosis[64] – Hsp90, chaperon[65] – p23, cochaperone[66] – SET 7, protein lysine methyltransferase (Chapter by Shannon R. Morey Kinney and Sriharsa Pradhan) – ATK1, serine/threonine protein kinase (Chapter by Shannon R. Morey Kinney and Sriharsa Pradhan)	– MeCP2, methyl-CpG binding domain protein 2 (Rett syndrome)[67,68] – RNA Pol II[69–71] – Specific tRNA and mRNA[18,24]

The interacting molecules are arranged in different classes in an attempt to highlight the various physiological processes that can regulate DNA methylation in cells.

used human[27–29,31,32] and murine enzymes[30,33,34,36,39,73,74]; the two enzymes share 78% sequence identity and many of the fine details in catalytic mechanism.[36]

Today, 35 years after the first reported study of mammalian DNA methyltransferase,[15] the results from different studies of Dnmt1 structure and function have been summarized in a number of excellent review articles. These cover distinct areas, including a general description of different DNA methyltransferases and DNA methylation,[75–77] different Dnmt1 fragments and their expression,[35,77] enzymatic properties and enzymatic assays of Dnmt1,[22] and functional interactions with molecules involved in DNA methylation.[72,78,79] This chapter is written as an extension of the earlier review articles in the desire to motivate use of the knowledge about Dnmt1 structure and function to advance our understanding of DNA methylation in cells. One of the major limiting factors in current DNA methylation research is a poor connection between cell-based, enzyme-based, medicinal chemistry, and pharmacological studies of DNA methylation. In Section II, I will highlight the major points about Dnmt1 structure and enzymatic function with a specific emphasis on development of Dnmt1 inhibitors. In Section II, I will discuss how Dnmt1 structure and its enzymatic mechanism can help us to understand its functional interactions with other molecules involved in DNA methylation in cells. Finally, in Section IV, I will summarize the results of a new Dnmt1 crystal structure, published as this chapter was being completed.

II. The Functional Domains of Dnmt1 and Dnmt1 Inhibitors

The first landmark study of the catalytic mechanism of bacterial DNA methyltransferase M.HhaI[80] was followed by more than a hundred crystallographic, enzyme kinetics, biophysical, and computational studies to reach the current knowledge about this relatively small enzyme. M.HhaI is about five times smaller than Dnmt1; nevertheless, the success of M.HhaI studies greatly advanced our understanding of Dnmt1 and showed how much work is needed to understand large complex enzymes like Dnmt1. Studies of Dnmt1 structure and function can answer many of the key questions in physiology of DNA methylation, such as: (i) design of Dnmt1 inhibitors and activators, (ii) interaction with other molecules, (iii) Dnmt1 phosphorylation and methylation, or (iv) the mechanism that controls the difference between *de novo* and maintenance methylation activity of Dnmt1.

We still do not know to what extent Dnmt1 can function as maintenance or *de novo* methyltransferase in cells.[78,79] This is one of the key questions in DNA methylation research since the changes in methylation patterns can trigger pathological events.[75,81,82] Dnmt1 is often described in the literature as the maintenance methyltransferase, based on the initial studies that reported higher activity on hemimethylated sites relative to unmethylated sites more than 20 years ago.[16,17] The crucial insights about Dnmt1 mechanism that have accumulated in the past 20 years have not yet been incorporated into the great majority of the current publications that have attempted to describe DNA methylation in cells.[1,72,75,81,83,84] The often quoted large preference for hemimethylated sites can be observed only in the early presteady state that represent only one segment of different features of allosteric regulation of Dnmt1.[22,30] In the case of allosteric regulation with fully methylated DNA, Dnmt1 is almost equally effective as a maintenance and a *de novo* methyltransferase[30]; under those conditions, the catalytic activity of Dnmt1 can be about 10-fold higher than that of the alleged *de novo* methyltransferases Dnmt3a and Dnmt3b.[32,33,36,85,86] Finally, in certain conditions, the allosteric regulation can completely inhibit Dnmt1 activity on any DNA substrate.[74,87] In sum, our ability to understand DNA methylation and Dnmt1 activity in cells directly depends on our ability to understand its allosteric regulation.[22,30] To complete our understanding of allosteric regulation of Dnmt1, we still have to answer several key questions: (i) How many different allosteric sites exist on Dnmt1?[29,34,88] (ii) Is there cooperativity between the different allosteric sites? (iii) How allosteric regulation affects the catalytic activity of Dnmt1 and its ability to bind the substrate DNA?[74] (iv) What is the binding specificity of the allosteric site(s) (i.e., DNA or RNA, unmethylated or premethylated, single stranded or double stranded)?[22]

Studies of Dnmt1 structure and function can also help us to understand its interaction with other molecules, or different phosphorylations that target Dnmt1. As noted above, 31 different molecules were reported to interact with Dnmt1 (Table I), and this number is likely to grow in the future. It is a delusion to think that we can understand 31 different interactions *in situ* in cells if we do not understand the enzymatic properties of purified Dnmt1. Purified Dnmt1 is first needed for description of different functional properties of this unique enzyme, and then the acquired insights can be used to understand the interaction with other molecules or the functional consequences of different phosphorylation events targeting Dnmt1. In essence, the interacting molecules or phosphorylations can (i) stimulate or inhibit Dnmt1 activity, (ii) guide Dnmt1 to methylation sites, (iii) facilitate dissociation from target-DNA sites, and (iv) modulate the ability of Dnmt1 to interact with other molecules. All four of these events can be measured with purified Dnmt1 using currently available methods.[22,28–30,34] The basic principles for building correlations between Dnmt1 activity in cell-based and enzyme-based assays have been described.[22] The studies of interaction between Dnmt1 and retinoblastoma protein (Rb) are one example of an admirable success in studies of Dnmt1 structure and function, its interaction with other molecules in cells, and the physiological significance of those events.[29,59,60]

The third key segment where studies of Dnmt1 structure and function can have a major impact is the development of Dnmt1 inhibitors or activators. Both, Dnmt1 inhibitors and activators, are acutely needed as a research tool for studies of different physiological processes that depend on DNA methylation, and ultimately in clinics for treatment of cancer, viral infections, mental illness, or autoimmune diseases. Both activators and the inhibitors are attractive, as the pathological processes related to DNA methylation depend on both an increase and a decrease in DNA methylation at specific genomic loci.[81,82] Dnmt1 has a number of regulatory mechanisms and a number of potential target sites that can be exploited in developing inhibitors and activators. Unfortunately, we still do not have a satisfactory inhibitor of Dnmt1. A number of compounds that inhibit or alter DNA methylation in cells have been described in the literature. Based on their structure and/or putative mechanism of action, the inhibitors can be divided to different groups: (i) cytosine analogues,[89] (ii) AdoMet analogues,[22,90] (iii) hydrazines,[91] (iv) phthalimides,[92] (v) polyphenols,[93] (vi) aminobenzoic derivatives,[94,95] (vii) modified oligonucleotides,[87] (viii) antisense oligos,[96] and (ix) disulfide analogues.[97]

A comprehensive analysis of different Dnmt1 inhibitors indicates that, despite the impressive number of inhibitors reported in the literature, all of those compounds are plagued with fundamental problems.[38,93] Many lack reproducibility,[98] some do not target-DNA methyltransferases and affect DNA methylation only indirectly, and many have unacceptable toxicity. Some compounds that target Dnmt1 have weak inhibition potency, so additional mechanistic studies are needed to increase the potency. Finally, some of the inhibitors share structural features with environmental toxins.[13] Very little of the current knowledge regarding the enzymatic mechanism of mammalian DNA methyltransferase has been used in inhibitor development. Some exceptions are studies of procainamide,[94] single-stranded oligonucleotide inhibitors,[87] or cytosine analogues.[89,99] Despite of described limitations, some of the compounds have entered clinical trials, some have even reached late stage clinical trials, while 5-aza-2′-deoxycytidine (Dacogen®) and 5-azacytidine (Vidaza®) have been approved for treatment of myelodysplastic syndrome.[38]

Dnmt1 has about 1620 amino acids; the actual length depends on the species and the expression of tissue-specific exons.[79] The studies of large enzymes like Dnmt1 can greatly benefit if the enzyme can be divided to smaller domains. Historically, Dnmt1 is divided into a smaller catalytic domain and the large N-terminal regulatory domain.[100] The separation into catalytic and the regulatory domains will be used in the next two sections, to highlight key aspects of Dnmt1 structure and function, with emphasis on development of Dnmt1 inhibitors.

A. The Catalytic C-Terminal Domain

Sequence analysis[101] and the crystal structures of bacterial DNA methyltransferases[23] showed that cytosine carbon 5 DNA methyltransferase, such as Dnmt1, needs at least three structural elements for catalytic activity: an

AdoMet-binding domain, a target base-binding cavity, and a target sequence recognition domain. The first two can be identified, based on conserved sequence motifs.[101] In Dnmt1, the AdoMet-binding site and the target base-binding cavity can be found within the C-terminal region after multiple Gly-Lys repeats.[20] The target recognition domain (TRD) is not precisely defined by the conserved sequence motifs, and cannot be assigned with certainty based on the DNA sequence. Historically, catalytic domain is defined as the C-terminal sequence after the Gly-Lys repeats.[100] This "catalytic domain" does not have catalytic activity.[29,34,102] For catalytic activity, Dnmt1 needs to have about 1000 amino acids from the C-terminal region; precisely the human Dnmt1 constructs missing the first 121, 501, 580, or 621 amino acids are still active, but those missing the first 672 amino acids or more are not active.[28,29,35]

Almost everything that is known today about the catalytic mechanism in the active site of cytosine carbon 5 DNA methyltransferases, such as Dnmt1, comes from studies of bacterial enzymes, most notably M.HhaI. Computational studies, enzyme kinetics, and protein crystallography have been used to describe fine details of its DNA binding, AdoMet binding, base flipping, and catalytic mechanism, encompassing more than 100 relevant studies. About 20 different crystal structures of M.HhaI can be used to trace different catalytic steps in atomic detail.[23,103–113] The crystal structures of M.HhaI also gave valuable insights into the mechanisms of inhibition by cytosine analogues.[89,99] In contrast to M.HhaI, only a handful of studies have analyzed the catalytic mechanism in the active site of Dnmt1. There is a general assumption that Dnmt1 and M.HhaI share the same mechanism in the active site based on the conserved sequence motifs,[20,101] ^{3}H exchange reaction,[36,114] and inhibition by 5-fluoro-cytosine[37] and other cytosine analogues.[38,89] Further comparisons between Dnmt1 and M.HhaI can be a productive strategy to advance our understanding of the catalytic mechanism in the active site of Dnmt1, and the mechanisms of inhibition for the inhibitors that target the catalytic domain.

Methylation and ^{3}H exchange reactions with different AdoMet analogues showed that Dnmt1 and M.HhaI share some very fine features in the catalytic steps that take place once the target base is positioned in the active site cavity.[36,114] In particular, M.HhaI and Dnmt1 have very similar catalytic processes at the carbon 5 of the activated target base including the methyltransfer step.[36,114] The observed similarity is encouraging for studies of Dnmt1 inhibitors, as many of the current inhibitors exploit catalytic processes at the carbon 5 of activated target base. The methyltransfer step is the rate-limiting step for both enzymes.[36,114] Interestingly, even though Dnmt1 and M.HhaI share the same rate-limiting step in methylation and ^{3}H exchange reaction, M.HhaI can have more than a 100-fold faster catalytic rate.[36,114] The difference in catalytic rates can be attributed to the difference in the rapid equilibrium between the initial steps that lead to the target base attack, that is, the recognition of the target base, the base-flipping steps, and formation of the unstable covalent adduct intermediate[36] (base-flipping experiments have never been reported for Dnmt1; however, it is very likely that Dnmt1 like other DNA methyltransferase flips the target base out of DNA helix[23]). Due to such rapid equilibrium, both M.HhaI and Dnmt1 have a unique feature that their catalytic rates can be controlled independently at two levels: (i) the rate-limiting methyltransfer step and (ii) the rapid equilibrium prior to the irreversible rate-limiting step.[22,36,114] Changes in the rapid equilibrium between the initial catalytic steps is the most likely explanation for the surprisingly high kinetic preference for poly(dI-dC) substrates that are unique to Dnmt1 and cannot be observed with M.HhaI.[16,36] Also it appears that the allosteric regulation of Dnmt1 leads to changes in the methylation rates as a result of changes in the rapid equilibrium between the early catalytic steps.[22,36,114]

The conserved sequence motifs defining the active site cavity translate into very fine similarities in the catalytic steps in the active sites of Dnmt1 and M.HhaI. Dnmt1 and M.HhaI also share conserved sequence motifs that define the AdoMet-binding site; however, the two enzymes show a number of significant differences in interaction with AdoMet. Dnmt1 has a relatively high K_m for AdoMet in comparison to the bacterial enzymes, up to 100-fold higher than the value measured with M.HhaI.[32,33,115,116] With Dnmt1, DNA binding does not lead to a large change in AdoMet-binding affinity as in M.HhaI.[28,36,115,116] However with Dnmt1, AdoMet binding at the start of catalysis can control enzyme activity through slow structural changes that appear to be coupled to allosteric inhibition.[22,29,36] Unlike Dnmt1, M.HhaI shows very rapid rates of target base attack in the absence of AdoMet. This is very surprising, as target base attack in the absence of AdoMet will greatly increase the chances of mutagenic deamination.[117–120] The ability to support mutagenic deamination might be an inherent deficiency of the bacterial enzymes, or a physiological adaptation that can be advantageous to bacterial enzymes but not to Dnmt1.[36,114] Briefly, under conditions of AdoMet insufficiency, mutations induced by the bacterial methyltransferase might slow digestion of host DNA by cognate endonucleases by changing occurrences of the recognition sequence. Deamination (affecting one strand) can subsequently be repaired effectively or, if not repaired, there is a good chance that it can result in a tolerable mutation.[121] The double-strand breaks caused by the nucleases are much more difficult to repair, cannot be tolerated, and therefore are much more lethal.[121] In contrast, AdoMet interaction with Dnmt1 has regulatory features that might prevent Dnmt1 from attacking the target base in the absence of the cofactor.[22,36] It remains to be seen if these regulatory features complicate or facilitate design of Dnmt1 inhibitors that target the active site and the AdoMet-binding site.

The comparisons between Dnmt1 and M.HhaI are often oversimplified, with arguments that Dnmt1 has a large preference for hemimethylated DNA (relative to unmethylated DNA) that cannot be observed with M.HhaI.[84,122]

The ability of M.HhaI to discriminate between hemimethylated and unmethylated sites is often forgotten,[84,122] even though it is well documented by kinetic and crystallographic studies.[108,115,116] Dnmt1 has much more complex interactions with DNA molecules than M.HhaI.[22,30] With Dnmt1, the reported difference in catalytic activity between hemimethylated and unmethylated substrates vary between 3- and 40-fold; some reports indicated that the difference may vary as much as 2- to 200-fold.[30,75,88,123] Such wide range can be confusing; to avoid misleading conclusions, it is necessary to understand the mechanism that can lead to such large variability. Briefly, Dnmt1 has at least two DNA-binding sites, the active site and the regulatory allosteric site.[22,28–30,88] Depending on the DNA bound to each site, Dnmt1 can show different catalytic activity.[22,30]

For example, a comparison of Dnmt1 methylation reactions on hemimethylated and unmethylated DNA revealed that in the early presteady state, the measured methylation rates can differ by 30-fold or more.[22,30] The large difference is due to allosteric inhibition that is caused by an excess of unmethylated substrate.[22,36] However, slow conformational changes at the start of catalysis lead to relief from allosteric inhibition,[36] so that the observed difference between the presteady state methylation rates for the two substrate is only about fivefold[22] (a slow relief from allosteric inhibition at the start of catalysis can be seen only if the catalytic rates on given substrates are faster that the rate of relief from allosteric inhibition). With further progress of the reaction to the steady state, the unmethylated and hemimethylated substrates show almost equal rates.[22,30] A decrease in Dnmt1 saturation with AdoMet results in slower relief from allosteric inhibition and therefore elevated selectivity for premethylated substrate.[22] The described features depend on DNA bound at the active site and the allosteric site[22,30,36]; therefore, different experimental setups will show variations as described in detail elsewhere.[22,30,88] In sum, Dnmt1 is a large and somewhat unique enzyme; design and the interpretation of its activity studies is a unique challenge.[22,36] The established enzyme assay and enzyme kinetics textbooks were written primarily for much faster enzymes that target simple small-molecule substrates. Therefore, established approaches for the studies of enzyme activity[124–126] have to be adapted for Dnmt1 studies[22,36] to avoid inconsistencies and confusion in the future activity studies.

A number of Dnmt1 inhibitors can probably be improved by exploiting details of the catalytic mechanism in the active site and the AdoMet-binding site. Initial design and optimization of such inhibitors can take advantage of the conserved features between Dnmt1 and M.HhaI.[99] Cytosine analogues are the best known active site inhibitors.[38,89,99] The cytosine analogues cannot inhibit Dnmt1 directly, but have to be incorporated into DNA, where they trap Dnmt1 in the process of DNA methylation by forming a covalent bond with the enzyme.[89] This results in a "suicide inhibition," but also excessive DNA damage and high toxicity. The cytosine analogues are most effective at low concentration that represents a balance between desired interference with DNA methyltransferase and acceptable toxicity.[38,99] The cytosine analogues are also valuable tool for cell-based studies of DNA methylation[127,128] and biochemical studies of DNA methyltransferase[37,89] (the cell-based studies that use cytosine analogues to study DNA methylation often do not differentiate between the inhibitor's effects on DNA methylation from the cellular response to DNA damage that is caused by the inhibitor). Different cytosine analogues show different toxicity, pharmacodynamics, and pharmacokinetic properties in clinical and cell-based studies.[38] Interestingly, the rate of inactivation[129–131] has never been reported for any of the analogues, so we cannot correlate the potency and the toxicity for any of the analogues with its ability to form the covalent adduct with the enzyme (the rate of inactivation by 5-fluoro-cytosine is likely to be comparable to the methylation rates, as 5-fluoro-cytosine is methylated only about two times slower than cytosine[132]).

Despite the continuing effort to improve the pharmacology of cytosine analogues, the reality is that none of the analogues in their current form can reach the true potential that inhibitors of DNA methylation can have. The cytosine analogues have a number of desirable features as lead compounds; however, they need to be modified to achieve the suicide inhibition without the need for toxic incorporation into genomic DNA. It is very likely that Dnmt1, like other DNA methyltransferases, flips the target base out of the DNA helix.[23] Thus, we need to design small-molecule lead compounds that can position cytosine analogues in the active site cavity, mimicking the flipped-out base. The base-flipping mechanism has been described in numerous studies. Design of a compound that can mimic the flipped-out base appears to be very challenging, as base flipping is a dynamic process that depends on an equilibrium between networks of competing interactions.[22,133,134] Stivers and colleagues reported small-molecule derivatives of uracyl that can inhibit uracyl-glycosidase,[135] another base-flipping enzyme that depends on dynamic equilibrium between base-flipping and base-restacking process.[133] However, the most favorable IC_{50} value is about 9 μM,[135] indicating that we are still far from design of a potent small-molecule inhibitor that can mimic the flipped-out base. An alternative, to incorporation of cytosine analogues into small molecules that can mimic the flipped-out base, is to attach cytosine analogue to a small molecule that can exploit binding interactions between Dnmt1 and AdoMet. [1,2-dihydropyrimidin-2-one]-5-methylene-(methylsulfonium)-adenosyl is currently in development as a potential lead compound for such class of Dnmt1 inhibitors.[22] Moreover, due to known action mechanism, [1,2-dihydropyrimidin-2-one]-5-methylene-(methylsulfonium)-adenosyl could be continually modified to increase its binding affinity and specificity.[22]

Procaine and procainamide are two other small molecules that have been reported to inhibit Dnmt1.[94] The two inhibitors are attractive due to known pharmacological properties and good tolerance. Procainamide inhibits DNA methylation in cells and purified Dnmt1, and the IC_{50} for purified Dnmt1 is about 10 μM.[94] This IC_{50} value can be a good starting point for a promising lead compound[94]; however, additional optimization and a higher potency are required for possible application. An elaborate mechanistic study suggested that procainamide binds at the AdoMet-binding site.[94] A fortunate situation, for all studies of Dnmt1 inhibitors that target the AdoMet site, is that Dnmt1–AdoMet interaction is the only Dnmt1 measurement that can use steady state approximations and related equations to determine competition and binding affinity.[36] Unlike substrate DNA, AdoMet is always in a large excess relative to Dnmt1, and catalytic rates with poly(dI-dC) substrates are fast enough to assure multiple turnovers and linear reaction profiles, the key requirements for every steady state analysis.[36,124]

RG108 is another small molecule that can inhibit DNA methylation in cells and purified Dnmt1.[93] The reported IC_{50} value for Dnmt1 inhibition in enzyme-based assay is a very impressive 115 nM.[93] However, RG108 has the basic structure of a phthalate, and phthalates are known to be genotoxic.[136] It remains to be seen if the genotoxic effects are a result of interference with DNA methylation only, or if other targets are involved. The initial results with RG108 are encouraging: the cell-based studies did not reveal multiple targets or excessive toxicity.[93] The binding site for RG108 has not been confirmed experimentally, though initial molecular modeling studies indicated that RG108 might target the active site.[93] This is consistent with the observation that both Dnmt1 and the bacterial enzyme M.SssI can be inhibited by RG108, as the two enzymes share conserved sequence motifs in the active site and the AdoMet-binding pocket.[92] Another still untested possibility is that both enzymes are inhibited by RG108 indirectly as a result of inhibitor's intercalation into substrate DNA. RG108 structure has planar conjugated hydrophobic rings that in principle could intercalate into DNA and interfere with all molecules that bind to DNA. In sum, additional mechanistic studies are needed to show how and if RG108 can directly bind and inhibit Dnmt1.

B. The Regulatory N-Terminal Domain

The bacterial enzyme M.HhaI shows that three basic domains and 327 amino acids are enough for cytosine methylation within a GCGC sequence, with preference for hemimethylated sites. Mammalian Dnmt1 has about 1620 amino acids, suggesting that Dnmt1 has evolved far beyond its basic function of DNA methylation. The N-terminal domain of Dnmt1 appears to have three functions: (i) allosteric regulation of the catalytic activity, (ii) multiple phosphorylation and methylation sites (see Chapter by Shannon R. Morey Kinney and Sriharsa Pradhan) that regulate catalytic activity of Dnmt1, and (iii) orchestration of interactions between Dnmt1 and other molecules that affect DNA methylation (of 31 molecules known to interact with Dnmt1 (Table I), only one of them, cochaperone p23, was found to bind the C-terminal domain[66]). In sum, the N-terminal domain has multiple regulatory mechanisms that control the activity and the specificity of DNA methylation. Interestingly, we still do not have a Dnmt1 inhibitor that targets any of the regulatory sites. One possible exception is a single-stranded DNA oligonucleotide with one 5-methyl-cytosine that has been developed as an allosteric inhibitor of mouse Dnmt1[87] (it remains to be seen if the same oligonucleotide inhibits human Dnmt1). Design of small-molecule inhibitors or activators that target the N-terminal domain is especially attractive. The inhibitors that target the catalytic domain can only inhibit catalytic activity of Dnmt1, the wide spectra of regulatory features in the N-terminal domain indicate a potential for design of Dnmt1 inhibitors and activators, or the compounds that destabilize the Dnmt1–target–DNA complex. Both activators and the inhibitors are attractive, as the pathological processes related to DNA methylation depend on both an increase and a decrease in methylation at specific genomic loci.[81,82]

Today, we think that the N-terminal domain has a loose structure and acts as a collection of interacting and yet somewhat separate subdomains that can even fold independently.[34,137] This thinking is based on the observations that isolated segments of the N-terminal domain can bind DNA,[28,29,34] or interact with proteins that bind to Dnmt1[35] (some of those interactions are listed in Section III of this chapter). The Dnmt1 fragments that can interact with other molecules range in length from 20 to more than 300 amino acids (the protein fragments bigger than 100 amino acids can be expected to form the secondary structure that is close to their native structures[124]). The isolated segments of the N-terminal domain can effectively mimic Dnmt1 in its interaction with other molecules only if they can form a structure that is very close to their native structures. (To remind the reader, a structural change of one hydrogen bond can lead to a decrease in binding energy that can correspond to a decrease in binding affinity of about an order of magnitude[138–140]; some nice experimental examples of that principle are shown in Ref. 141.) The multiple-independent domains connected by flexible linkers will make determination and interpretation of crystal structures of Dnmt1 very challenging. It appears that at this moment, one productive research strategy would be to continue characterizations of the functional fragments of Dnmt1. For example, crystal structure of human Dnmt1 replication foci-targeting sequence is available at 2.3 Å resolution (PDB code: 3epz,[142] 268 amino acids, residues 350–618), while the approximate structure for some parts of Dnmt1 can be deduced from the homology modeling of the conserved structural motifs.[77,143]

Different segments of murine Dnmt1 have been expressed and purified from *Escherichia coli* cells and to some extent functionally characterized.[34] Baculovirus expression was also used to prepare different segments of human

Dnmt1,[28,29,137] human Dnmt1 that had increasingly shorter N-terminal domain,[27–29,137] or full-length human Dnmt1 with specific mutations,[29] or human Dnmt1 missing sections of its sequence.[28] Yeast two-hybrid systems have been used to identify interactions between different fragments of Dnmt1.[102] The studies of enzymatic function of different Dnmt1 fragments have been focused on two major questions: (i) which parts of the N-terminal domain are required for the catalytic activity and (ii) which parts of the N-terminal domain represent the regulatory allosteric site(s). Steady, admirable progress has been achieved in addressing both questions, and the future prospect for additional insights look very promising.

The search for catalytically active Dnmt1 fragments started with the first report of a successful separation of functional fragments of Dnmt1 using partial proteolysis.[100] The proteolysis resulted in admirably clean cleavage of Dnmt1 into a C-terminal fragment that shows catalytic activity, and an N-terminal regulatory domain that binds radioactive zinc ions. The cleaved "catalytic domain" appeared to be free from allosteric inhibition at high substrate concentration, suggesting that the N-terminal domain contains the allosteric site that leads to substrate inhibition.[16,36] The initial success by Bestor inspired numerous studies; however, none of those studies could prepare the active catalytic domain that was described in the original Bestor's study.[27–29,34,137] The human Dnmt1 constructs missing the first 121, 501, or 580 amino acids are still active, and show the ability to discriminate between hemimethylated and unmethylated DNA substrates similar to the full-length enzyme.[29,35] Human Dnmt1 constructs missing the first 621 amino acids are still active, but constructs missing the first 672 amino acids are not.[28,35]

Interestingly, even though the first 580 amino acids are not necessary for activity, this segment has multiple phosphorylation sites that can control catalytic activity of Dnmt1.[39,40,144,145] Possibly, the phosphorylation sites are not directly involved in catalysis, but conformational changes in the flexible protein structure allow the phosphorylated sites to interact with different Dnmt1 parts and cause inhibition. This proposal is supported by the observation that a very specific peptide mimic of the phosphorylation site at Ser515 can inhibit activity of mouse Dnmt1, apparently by altering the interactions between different domains.[39] Dnmt1 phosphorylation at Ser515 is needed for enzyme activity.[39] Additional phosphorylation sites on mouse Dnmt1 are found within amino acid region 1–290 that binds casein kinase 1δ/ϵ (CK1δ/ϵ).[40] The first phosphorylation takes place at Ser146, and then additional phosphorylations spread to the surrounding amino acids.[40] Described phosphorylations appear to inhibit Dnmt1 as a result of destabilization of the Dnmt1–DNA complex.[40] Phosphorylated peptides that can control Dnmt1 activity are potentially interesting as Dnmt1 inhibitors,[39] though additional mechanistic studies are needed to show if there are practically feasible strategies. Interestingly, Dnmt1 missing the first 501 amino acids has higher activity than the full-length enzyme,[27,32] and the missing phosphorylation sites could be one of the reasons for this. Apart from the phosphorylation sites, the first 580 amino acids contain exons specific to mammalian enzymes, one of the regulatory DNA-binding sites,[28,29,34] exons specific for different tissues and cell types,[77,79,137] and phosphorylation and methylation sites that control Dnmt1 turnover in cells (Chapter by Shannon R. Morey Kinney and Sriharsa Pradhan).

Two studies, using different experimental approaches, showed that Dnmt1 has two DNA-binding sites in the N-terminal domain and one DNA-binding site in the C-terminal domain.[29,34] The first DNA-binding site is located in the human enzyme within the region 261–356, and more specifically amino acids Lys 284, Lys 285, His 286, and Arg 287.[29] The second DNA-binding site is located within the Zn-finger domain, in the amino acid segment 580–697. Interestingly, functional studies suggest that the two DNA-binding sites might function, at least to some degree, independently. Activation by fully methylated DNA can be observed with full-length Dnmt1 and Dnmt1 missing the first 121 amino acids; however, the activation cannot be observed with the constructs missing the first 501 or 580 amino acids.[29] Nevertheless, the constructs missing the first 501 or 580 amino acids can differentiate between hemimethylated and unmethylated substrates,[28,29] which is one of the features that depends on the allosteric inhibition at high substrate concentrations.[36] Thus, the allosteric activation by the fully methylated DNA and the allosteric inhibition by an excess of unmethylated DNA substrate appear to be at least partially independent. Strictly speaking, we do not know if fully methylated DNA activates Dnmt1, or if instead it leads to higher activity by reducing inhibition. The activation by fully methylated DNA is among the strongest evidence of allosteric regulation of Dnmt1,[30] a number of other more subtle evidences have been reported.[22]

The DNA-binding site between amino acids 261 and 356 does not show sequence homology with any of the known DNA-binding proteins; however, the Zn-finger domain has strong similarity to MBD1 protein, a chromatin modification protein that binds to unmethylated CpG sites (MBD proteins are covered in Chapter by Pierre-Antoine Defossez and Irina Stancheva and in Ref. 79). Human Dnmt1 missing the first 580 amino acids shows slightly higher binding affinity for unmethylated substrate relative to hemimethylated and fully methylated DNA, though interestingly the opposite preference is observed in the catalytic activity.[28] This is consistent with the well-known observation that unmethylated DNA can induce Dnmt1 inhibition by binding at the allosteric site.[16,36] The isolated Zn-finger (amino acids 651–697) shows obvious binding preference for unmethylated DNA, though its DNA-binding affinity is about 100-fold lower than the DNA affinity of full-length Dnmt1. This suggests that additional amino acids around the Zn finger also participate in DNA binding. Consistent with this proposal, full-length Dnmt1 that is missing the Zn-finger domain shows clear evidences of inhibition by increasing concentration of unmethylated substrate and about 10-fold reduced activity.[28] In sum, the accumulated evidence suggests that the Zn-finger domain could be the allosteric site that leads to inhibition by an excess of unmethylated DNA. The inhibition by excess unmethylated substrate was the first textbook evidence for allosteric regulation of Dnmt1.[16,124–126] Although somewhat subtle, the inhibition by excess unmethylated substrate has been reproduced in numerous studies over the last 30 years.[22]

A number of key questions remain open. First, there is a question about positive or negative cooperativity between the two allosteric sites on Dnmt1. Precisely, even though externally added fully methylated DNA can activate Dnmt1 in methylation reactions on unmethylated substrates,[29,30] Dnmt1 is not self-activated by the methylation sites produced in its methylation reaction on unmethylated substrate.[36] The evidence of self-activation in the enzymatic reaction is easy to detect. In the case of product activation, the reaction time profile after the first turnover shows a continual increase in the catalytic rates as the product accumulates with time.[124–126,140] Following the first turnover, the Dnmt1 methylation reactions show linear reaction time profiles and a decrease in catalytic rates.[30,32,33,36,74]

The allosteric decreases in catalytic activity can be rather drastic. In the early presteady state, the methylation reaction on unmethylated DNA can be more than 30-fold slower than that on hemimethylated DNA or on unmethylated DNA in the presence of allosteric regulation by the fully methylated DNA.[30] However, after the first turnover (i.e., the progress from presteady state to the steady state), the rate difference between different reactions drops drastically and the time profiles for all reactions show very similar rates.[22,30,33,36] This rate drop cannot be attributed to any of the common causes such as product inhibition, enzyme inactivation with time, or substrate depletion.[36,74] The whole process represents an outstanding display of allosteric regulation of Dnmt1.

The second group of interesting questions is which molecules, by what mechanism, bind at the allosteric site(s) of Dnmt1 in cells[22]? The answers to these questions are likely the key for understanding of Dnmt1 function in cells. Apart from the allosteric activation by fully methylated DNA, or the allosteric inhibition by unmethylated DNA, a number of other molecules were found to bind and/or inhibit Dnmt1, such as specific RNA and tRNA molecules,[18,24] single-stranded oligonucleotides with one 5-methyl-C site,[87] poly(ADP-ribose),[57] poly(dA)·poly(dT),[18] or poly(dA-dT).[74] It remains to be seen if any of these interactions compete for the same binding sites or cooperate in control of Dnmt1 activity in cells. Apart from their physiological significance, all insights into the mechanism of allosteric regulation could be exploited in development of Dnmt1 inhibitors and activators.

III. Structure and Function of Dnmt1 and Its Interaction with Other Molecules

In cells, DNA is organized in chromatin structures that have multiple functions, including DNA packing and protection from DNA damage,[121,146,147] and organization of dynamic physiological processes such as transcription, replication, repair, and recombination. DNA methyltransferases affect each of these processes by acting within different multimolecular structures. The interacting molecules can control the access of Dnmt1 to the target sites[148] and/or Dnmt1 activity at the potential target sites.[22,149,150] As noted above, about 31 different molecules have been reported to interact with Dnm1 (Table I); this number is likely to grow in the future. The interacting molecules include proteins, RNA molecules, and poly(ADP-ribose). Nice maps of Dnmt1 fragments that interact with other proteins are provided in earlier publications.[35,72] Interestingly, all of the interacting molecules are found to bind the regulatory N-terminal domain, except co-chaperone p23, that was found to bind the catalytic C-terminal domain.[66]

A full list of different molecules that are known to interact with Dnmt1 is summarized in Table I, the reader is also referred to Jafar Sharif and Haruhiko Koseki in this volume, and to some relevant publications.[18,35,47,72,78,79] Here, I want to point out that it is highly unlikely that all of the interacting molecules bind to Dnmt1 at the same time, thus it will be very useful to group the different interactions according to their physiological functions (Table I). It appears that Dnmt1 participates in four different yet functionally integrated interactions: (i) Dnmt1 dimerization; (ii) core chromatin replication complex; (iii) interaction with molecules involved in DNA repair, cell cycle control, and apoptosis; and (iv) interaction with RNA Pol II, RNA-binding proteins, and specific RNA molecules. The function of different Dnmt1 interactions overlaps to some extent, and this overlapping probably explains why inactivation of many of the interacting proteins results in only partial loss of DNA methylation.[84] In the coming paragraphs, I will discuss some examples of how insights in enzymatic properties of Dnmt1 can help us to understand the interaction between Dnmt1 and other molecules in cells.

A. Dnmt1 Dimerization

Dnmt1 dimers can be observed with purified Dnmt1 and in cells overexpressing Dnmt1.[151] Several bacterial enzymes including M.HhaI can also form dimers.[113] Dnmt1 dimers are formed by interaction between the regulatory N-terminal domains (amino acids 310–629). It has been proposed that dimerization increases enzyme specificity for hemimethylated sites, but unfortunately this has not been confirmed in studies of enzyme activity. Activity studies at different concentrations of Dnmt1 can reveal differences in the enzymatic properties of Dnmt1 monomers and dimers. I have observed that an increase in Dnmt1 concentration leads to partial inhibition of Dnmt1 activity on unmethylated DNA, independent of DNA concentration.[36] If similar inhibition cannot be observed with premethylated substrate, we could argue that Dnmt1 activity studies support the idea that its dimerization leads to increased selectivity for premethylated substrate. In sum, it is still unknown how, and if, dimerization can affect the enzymatic properties of Dnmt1; the experimental tools required to answer those questions in the future studies are available.[22,27,29,30,32,36]

B. Core Chromatin Replication Complex

During the process of DNA methylation in cells, Dnmt1 is a part of a large multiprotein complex that acts as *epigenetic code replication complex.*[47,149,150] The primary function of such complex is preservation of the existing epigenetic organization and chromatin structures. I will highlight some of the functions of the core chromatin replication complex that can directly depend on enzymatic properties of Dnmt1 (see also Chapters by Xiaodong Cheng and Robert M. Blumenthal; Jafar Sharif and Haruhiko Koseki).

It has been suggested that Dnmt1 can methylate DNA wrapped in nucleosomes, and that nucleosomes influence DNA methylation patterns.[152–155] Studies with purified Dnmt1 and mononucleosomes indicate that Dnmt1 activity on nucleosomes is as high as 20–40% of the activity observed with free DNA, while for some DNA sequences nucleosomes are methylated to a similar extent as the free DNA.[154] Using a very similar approach, another group showed that Dnmt1 and Dnmt3 can methylate nucleosomes at rates two times higher than the free DNA.[153] That study also indicated that Dnmt1 binds tightly to nucleosomes and, surprisingly, that the methylation sites can be found in different translational and rotational setting in the nucleosome.[153] In both studies, the bacterial enzyme M.SssI was inhibited by nucleosomes to a much higher degree than Dnmt1.[153,154]

Mapping of DNA methylation sites in *Arabidopsis thaliana* and human genomic DNA showed that nucleosomal DNA is methylated at a higher frequency and with 10 bp periodicity. The methylation sites are presumably facing away from the histone–DNA interface.[152] This prompted the authors to speculate that DNA might be methylated while it is still bound to the nucleosomes and that the rotational setting on nucleosomes surface could guide DNA methyltransferases. However, there are alternative explanations that have not been considered. First, 5-methyl-cytosine makes DNA more rigid especially for compressions in the major groove,[156,157] so that DNA methylation patterns might have evolved in cohort with the nucleosome positioning sequences to adapt to DNA bending and breading in nucleosomes.[158–161] Second, the 10-bp periodicity might not be a result of guided action of DNA methyltransferase, but an adaptation of methylation sites that can facilitate the methylation readout by the proteins that bind to methylated and hemimethylated sites while DNA is in compact chromatin structures.

To fully appreciate the unique features that Dnmt1 (and Dnmt3 enzymes) needs in order to methylate nucleosomes, we have to briefly describe the general principles of how nucleosomes affect DNA interaction with proteins. In nucleosomes, 147 bp of DNA is wrapped 1.67 times around a histone octamer so that the DNA sites are not freely accessible to DNA-binding proteins.[162–165] DNA sites are only transiently released from the surface of histone octamers[161] in correlation with DNA flexibility and the translational and the rotational setting of each site.[162–164,166] In a truly landmark study, Polach and Widom used about a dozen different restriction enzymes to show that nucleosomes can block enzyme activity by approximately factors of 10, 100, and 1000 relative to free DNA.[165] The progressive decrease in enzyme activity corresponds to different translational settings starting from the loose DNA ends toward the nucleosome dyad. An enzyme's ability to act on nucleosomes is directly proportional to the enzyme concentration, while the rate-limiting step appears to be the enzyme's k_{cat}/K_m ratio rather than nucleosome unwinding.[165] Dnmt1 has exceptionally slow turnover rates in comparison to other enzymes,[36] which would make Dnmt1 relatively inept to act on target sites positioned in nucleosomes.

Some DNA-binding proteins can access DNA sites on nucleosomes since they do not completely encircle the DNA helix upon binding. For example, uracyl-glycosidase binds DNA asymmetrically[167] so it can effectively access DNA repair sites even at the nucleosome dyad axis.[162] However, even the enzymes that bind DNA asymmetrically can be inhibited by different rotational settings of the DNA target site on the nucleosome surface.[162,164] With a molecular mass of 186 kDa, Dnmt1 is about 1.5 times bigger than nucleosomes. For all of us working on nucleosomes, it would be fascinating to understand how, and if, a large and slow enzyme like Dnmt1 can methylate and bind nucleosomes keeping the original rotation setting of the target sites as the initial studies suggested.[152–154] However, if future studies continue to compare total activity of Dnmt1 to that of M.SssI, the authors need to take into account that M.SssI is 100- to 200-fold faster than Dnmt1.[36,114] In the time it takes for Dnmt1 to methylate one of the most accessible sites in nucleosomes, M.SssI can methylate all of the easy and many not-so-easy accessible sites. This can give the false impression that in comparison to free DNA, the nucleosome-bound DNA affects the activity of M.SssI much more than that of Dnmt1. In sum, the key requirement in studies of enzyme activity on nucleosomes is the ability to quantitatively differentiate between the activity on the easily accessible site at the free DNA ends from the activity on the progressively less-accessible sites toward the nucleosome dyad.[161–166] This key requirement was never achieved in any of the Dnmt1 studies.

Independent of the possibility that Dnmt1 can act on nucleosomes, the reported interactions between Dnmt1 and proteins involved in chromatin remodeling indicate that some forms of chromatin relaxation take place before Dnmt1 can access its target sites. Dnmt1 is functionally associated with SNF2h-containing chromatin-remodeling complex NoRC,[42] and with LSH protein (lymphoid-specific helicases), another protein related to the SNF2 family of chromatin-remodeling ATPases.[43] LSH function showed that even inactive Dnmt1 is crucial for its function. Both studies indicated that functional interactions between Dnmt1 and chromatin-remodeling proteins are closely associated with the function of two histone deacetylases HDAC1 and HDAC2. The two deacetylases were also reported to interact directly with Dnmt1.[60] A number of other histone modifying proteins interact with Dnmt1, like histone-lysine *N*-methyltransferase SUV39H1,[48] G9a histone methyltransferase,[49] and heterochromatin-binding protein HP1β.[47,48] A close functional link between DNA methylation and histone modification is driven by the shared function; the main unanswered questions are as follows: (i) What is the precise order of these events? (ii) What type of cooperativity exists between DNA methylation and histone modification events?[45]

Many of the proteins involved in histone modification that interact with Dnmt1 also interact with DNA methyltransferases Dnmt3a and Dnmt3b[47,79] (Dnmt3 enzymes are described in Chapter by Frédéric Chédin). Dnmt1 also interacts with Dnmt3a and Dnmt3b.[41] Studies of the mechanism of interactions between Dnmt1 and Dnmt3 enzymes can offer answers to some of the key questions about the alleged physiological function of these three enzymes, that is, Dnmt1 as the maintenance methyltransferase and Dnmt3a and Dnmt3b as *de novo* methyltransferases. Many of the cell-based studies suggested that the classification of Dnmt1 as a maintenance methyltransferase and Dnmt3 enzymes as *de novo* methyltransferase is not justified, as both Dnmt1 and Dnmt3 enzymes participate in *de novo* methylation via unknown molecular mechanisms.[79]

The enzyme-based studies support this conclusion, as purified Dnmt1 and Dnmt3 enzymes have (within experimental error) equal catalytic rates as *de novo* methyltransferases.[32,33,36,85,86] However, it is very important to notice that in the presence of fully methylated DNA, Dnmt1 can have more than 10 times higher *de novo* catalytic activity than Dnmt3 enzymes. In those conditions, *de novo* catalytic activity of Dnmt1 is almost identical to its activity as a maintenance methyltransferase.[153] The allosteric effects appear to be the main regulator of Dnmt1 activity as *de novo* and maintenance methyltransferase, the remaining important question is "How do different DNA molecules bind at the allosteric site of Dnmt1 in cells?" It has been suggested that catalytic product of Dnmt3 can stimulate catalytic activity of Dnmt1, but the actual mechanism is not known.[168] It is possible that Dnmt3 can increase catalytic activity of Dnmt1 by feeding its methylated DNA directly to the allosteric site on Dnmt1. This proposal is supported by the observation that Dnmt3 binds the regulatory domain of Dnmt1.[41] Direct feeding of premethylated DNA to the allosteric site on Dnmt1 can be crucial, since in free solution different DNA molecules randomly compete for different DNA-binding sites on Dnmt1, a mere presence of methylated DNA does not lead to activation (as described in Section II.B).

Apart from Dnmt1, the methylated DNA sites are also recognized by other proteins, such as UHRF1 (ubiquitin-like with PHD and ring finger domain 1) and MBD (methyl-CpG-binding domain) proteins (see also Chapters by Jafar Sharif and Haruhiko Koseki; Pierre-Antoine Defossez and Irina Stancheva). The hemimethylated sites are small, so it is impossible for large proteins like Dnmt1, UHRF1, or MBD to bind such sites simultaneously. This raises the question of whether those molecules cooperate to increase the sensitivity of Dnmt1 for hemimethylated sites and maintenance methylation.[83,123,150] Here, I will discuss interaction between Dnmt1 and UHRF1 (the interaction with MBD proteins is described later in this text in the section on RNA-directed DNA methylation). UHRF1 protein is indispensable for Dnmt1 activity in cells (Ref. 45 and Chapter by Jafar Sharif and Haruhiko Koseki). UHRF1 is a large multidomain protein that appears to be the central molecule in the epigenetic code replication complex.[45] UHRF1 can recognize hemimethylated sites and histone methylation, and bind both Dnmt1 and histone methyltransferase G9a.[45,46,169,170] Thus, UHRF1 appears to link histone methylation and DNA methylation.[45]

The known enzymatic features of Dnmt1 can help us to address the debate how, and if, UHRF1 and Dnmt1 cooperate in preserving hemimethylated sites during replication.[83,123,150] Dnmt1 preference for the hemimethylated sites is primarily a result of kinetic preferences that depend on the allosteric regulation.[22,30,74] UHRF1 or any other molecule can affect the specificity of Dnmt1 for hemimethylated sites only if the process affects the allosteric regulation of Dnmt1 (Section II.B and Refs. 22,30). Dnmt1 preference for hemimethylated DNA sites is not a result of preferred binding equilibrium[28,171]; therefore, UHRF1 cannot increase Dnmt1's preference for hemimethylated sites by increasing its binding affinity for such sites.[83] UHRF1 and Dnmt1 are too big to bind hemimethylated sites simultaneously, and the binding affinity of Dnmt1 for hemimethylated site depends only on its interaction energy with that site. Thus, UHRF1 cannot preactivate Dnmt1 by binding at the hemimethylated site prior to Dnmt1[83] since such "preactivation" would be against the basic chemistry principle that the binding equilibrium does not depend on the path by which the equilibrium is achieved.[124,138–140]

C. DNA Repair, Cell Cycle Control, and Regulation of Apoptosis

The actions of multiprotein complex involved in epigenetic code replication have to be closely correlated with cell cycle regulation. The chromatin structures and the proteins involved in chromatin organization are known to block the access of DNA repair proteins to the damage sites.[121] Therefore, it can be expected that the molecules involved in response to DNA damage can stop DNA methylation or even facilitate dissociation of DNA methyltransferases from the target sites. The actual experiments have shown that the molecules involved in response to DNA damage can stop DNA methylation at the level of Dnmt1 transcription[172,173] or by direct interaction with Dnmt1,[40,57–59] or regulate apoptosis in complex with Dnmt1.[50,60–63] Probably, one of the best understood examples of such regulation is Dnmt1 interaction with the Rb that results in Dnmt1 inhibition and dissociation of Dnmt1–DNA complex.[29,59] The amino acid region 261–356 in the N-terminal domain of Dnmt1 binds to specific pockets on Rb.[29] The interaction between Dnmt1 and Rb protein could compete with the allosteric regulation of Dnmt1 by fully methylated DNA.[29]

Dnmt1 is also inhibited by poly(ADP-ribose) and PARP1.[57,58] The mechanism of Dnmt1 inhibition by poly(ADP-ribose) has not been investigated. It is not known if the interaction leads only to inhibition, or also to destabilization of Dnmt1 binding to its target site.[74] It is unknown if poly(ADP-ribose) binds at the regulatory N-terminal domain, and whether such binding interferes with other molecules that regulate Dnmt1 activity by binding to the regulatory domain.[22,28–30] Dnmt1 can be inhibited by poly(dA)·poly(dT) and poly(dA-dT), but not by poly(A) or poly(dA).[18,74] It remains to be seen if the inhibiting molecules bind at the same site as poly(ADP-ribose). The specificity of the poly (ADP-ribose)-binding site can be explored as a potential target for design of small-molecule Dnmt1 inhibitors as drug candidates. The experimental tools required to answer the questions related to Dnmt1 inhibition by poly(ADP-ribose) have been developed.[22,27,29,30,32,36,74]

Dnmt1 interacts with p53, and this interaction affects expression of the *survivin* gene, an inhibitor of apoptosis.[61] Dnmt1 and p53 appear to be involved in downregulation of protein phosphatases that regulate the cell cycle.[174] DNA repair and cell cycle regulation are two closely related processes; therefore, it is no surprise that Dnmt1 also interacts with some of the kinases involved in cell cycle regulation.[40,144] As noted above, Dnmt1 is phosphorylated at a number of sites in its regulatory N-terminal domain.[39,40,144,145] Decreased DNA-binding affinity is observed after phosphorylation of serine 146 and surrounding sites by the CK1δ/ϵ.[40] In sum, it appears that Dnmt1 interaction with the molecules involved in response to DNA damage leads to inhibition of DNA methylation (by the release of the DNA target sites), cell cycle arrest, and regulation of apoptosis.

D. RNA-Directed DNA Methylation

The interaction with specific RNA molecules was probably the first report of Dnmt1 interaction with other molecules.[18] The original study is almost 30 years old; nevertheless, many of the fine features that were reported by Weissbach and colleagues almost 30 years ago are fully reproducible.[22] Similar to Weissbach and colleagues, we routinely found during Dnmt1 purification that its activity is strongly inhibited in cell extracts.[24] The inhibition was found to be due to tightly bound RNA molecule(s).[18,24] RNA-directed DNA methylation has been described in plant cells, and a comparative analysis of plant and mammalian cells shows that mammalian cells have counterparts for the molecules involved in RNA-directed DNA methylation.[175] Growing evidence indicates that RNA molecules are involved in control of DNA methylation in mammalian cells, through the mechanism is poorly understood[176–180] (see Chapter by Anton Wutz).

Several studies reported evidence of physical or functional interaction between Dnmt1 and RNA Pol II,[69–71] but the actual mechanism and the consequences of that interaction are not understood. RNA Pol II is involved in synthesis of the noncoding RNA molecules.[181] Dnmt1 interacts with several members from the family of methyl-CpG-binding domain (MBD) proteins[55,79,182] that bind several types of RNA molecules[183] and interact with histone modifying components.[67] Most notable is MeCP2 (Rett syndrome protein[68]); MeCP2 participates in displacement of histone H1[184,185] and its functional interaction with Dnmt1 can be crucial for neuronal development.[68] Methylated DNA sites are too small to bind Dnmt1 and MBD proteins at the same time, so it is unclear how these proteins act together in the recognition of methylated sites. Interestingly, MBD proteins cannot bind methylated DNA sites when they bind RNA molecules,[183] suggesting that RNA molecules might coordinate the access to methylated DNA sites between MBD proteins and Dnmt1. Dnmt3a, another protein that can bind Dnmt1, also binds RNA molecules in cells.[186] Dnmt3a binds the antisense strand of siRNA molecules in cells, acting in cohort with RNA Pol II and histone methylation.[186]

Interestingly, Dnmt1 cannot be inhibited by every RNA molecule, only some mRNA and tRNA molecules can cause inhibition and it appears that some of them have higher potency than others, with poly(G) being the most preferred sequence known to date.[18] It remains to be seen if tRNA molecules are really involved in DNA methylation, or some tRNA molecules just mimic some of the noncoding RNA molecules that participate in DNA methylation. A puzzling link between DNA methylation and tRNA molecules is also an open question in Dnmt2 studies[187,188]. If RNA molecules control Dnmt1 activity by binding to an allosteric site, the sensitivity of allosteric regulation to 5-methylcytosine suggests that an RNA methyltransferase might methylate the regulatory RNA molecules and thus control Dnmt1 and DNA methylation.[22] In sum, the allosteric regulation of Dnmt1, and Dnmt1 inhibition by the RNA molecules, unfairly remains to this day the two least-explored enzymatic features of Dnmt1.[22] The two features are likely to be related, and likely to be the key for our understanding of differences between maintenance and *de novo* DNA methylation, the two processes that can drive the changes in methylation patterns that lead to diseases.[81,82] The future studies of Dnmt1 interaction with RNA molecules are likely to have a significant impact on our understanding of Dnmt1 action in cells, and on rational design of novel modulators of Dnmt1 activity in cells.[22]

IV. New Crystal Structures of Large C-Terminal Fragment of Mouse and Human Dnmt1

Just as I finished writing the final version of this chapter, new crystal structures of the large Dnmt1 fragment have been published.[189] The first structure shows mouse Dnmt1 (residues 650–1602) in complex with *S*-adenosyl homocysteine and a 19-bp DNA duplex that contains two unmethylated CpG sites separated by 8 bp (3.0 Å resolution; Fig. 1). The second crystal structure shows human Dnmt1 (residues 646–1600) in complex with the same 19-bp DNA and *S*-adenosyl homocysteine (3.6 Å resolution). The new structures complement the existing structure of human Dnmt1 replication foci-targeting

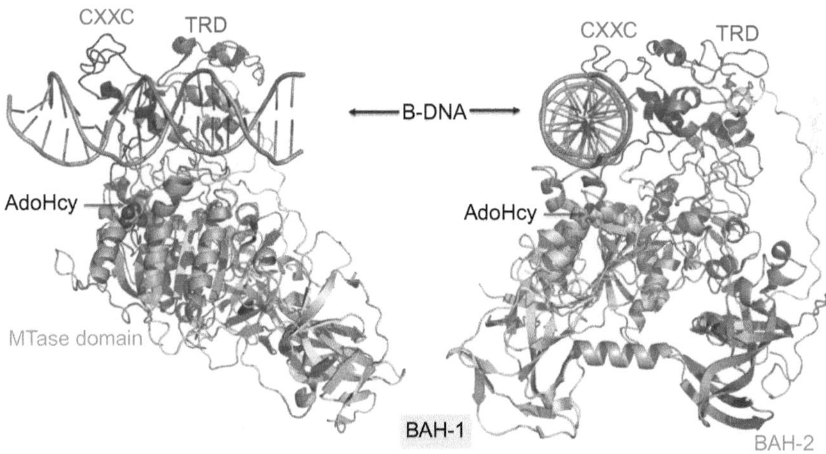

FIG. 1. Structure of mouse Dnmt1. Two orthogonal views of mouse Dnmt1 (residues 650–1602) in complex with unmethylated DNA substrate and cofactor *S*-adenosyl homocysteine (AdoHcy) (PDB 3PT6).[189] The domain structure from N-to-C terminal are Zn-finger CXXC domain (magenta), BAH-1 (yellow), BAH-2 (orange), methyltransferase (green), and target recognition domain (blue). (See Color Insert.)

domain (residues 350–618, PDB code: 3epz[142]). The new structures represent a landmark achievement that will accelerate the future studies of Dnmt1 and design of novel inhibitors or activators.

I will briefly discuss here how the new structural insights add to three decades of Dnmt1 studies that were summarized in the earlier sections. The crystal structures are consistent with Dnmt1 fragments that were prepared following activity studies,[28,29] and confirm predictions that Dnmt1 is composed of multiple subdomains that are connected with flexible regulatory loops. Flexible protein loops are known to develop in protein evolution to facilitate protein–protein interactions, protein phosphorylation, or dynamic structural changes that control enzyme activity.[190,191] Loops from BAH1 and BAH2 domain penetrate into the catalytic domain, suggesting a possible explanation for the earlier observations that separated catalytic domain is inactive,[28,29] but it can become active by binding the other domains.[34] Similar binding sites for penetrating loops might explain Dnmt1 inhibition by phosphorylated peptide with specific amino acid sequence.[39]

The crystal structure confirmed that Dnmt1 and M.HhaI share very similar structure in both the active site and the AdoMet-binding domain,[23] with one major difference. Specifically, the active site cysteine is in a retracted position in Dnmt1. This retracted position might explain why Dnmt1 (unlike M.HhaI) is very slow in attacking its target base in the absence of cofactor, and/or why Dnmt1 shows slow activation (i.e., enzyme hysteresis) at the start of catalysis on unmethylated DNA substrate.[22,36,114] Big differences between Dnmt1 and M.HhaI are observed in the target (DNA methylation site) recognition domain. For both enzymes, the TRDs are composed from flexible loops. The Dnmt1 TRD is about twice the size of the one in M.HhaI, even though its target sequence (CG) is only half the size of the M.HhaI target sequence (GCGC). The loop from the BAH2 domain penetrates the TRD, suggesting that the relatively large domain is an adaptation to regulatory interactions with other parts of Dnmt1. Finally, it appears that (unlike M.HhaI) the TRD of Dnmt1 does not completely encircle the target DNA, and molecular modeling studies might be used to test whether Dnmt1 can dock sideways onto DNA bound to nucleosomes[192] and methylate the target sites facing away from the nucleotide surface.[152–154]

The new structures offer insights into the earlier observations that Dnmt1 is inhibited by unmethylated substrate.[16,22,36,74,87] It appears that the Zn-finger CXXC domain binds DNA first, and this interaction depends on existing DNA methylation. In the case of hemimethylated sites, the Zn-finger CXXC domain will not bind DNA, and the active site should be open for DNA binding and maintenance methylation. In the case of unmethylated substrate, Zn-finger CXXC domain binds the unmethylated CpG site so that the loop between BAH1 and CXXC domain is in position where it can block the DNA binding at the active site, while the TRD is in retracted position bound to the loop from

BAH2 domain. The inhibitory structure must be unstable and dynamic, since Dnmt1 is active on unmethylated substrate. The slow relief from allosteric inhibition that can be observed at the start of catalysis with unmethylated substrate could be a direct consequence of the underlining structural changes.[22,36]

In sum, multiple domains connected by flexible loops, multiple phosphorylation sites,[39,40] and multiple interacting molecules (Table I) indicate that Dnmt1 has a very dynamic structure with many subtle regulatory features. Thus, we always have to bear in mind that the static crystal structures do not reflect many of the catalytic and regulatory features of the active enzyme. Future insights into Dnmt1 function, and design of its inhibitors and activators, will strongly depend on building a correlation between the available structures and the enzyme assays that are designed to capture apparently subtle and yet important regulatory features.[22] I am excited about the possibility that in the next decade, we could expect a burst in enzymatic studies of Dnmt1 that could surpass the burst in M.HhaI studies that came after the first structure with the flipped-out base.[23]

References

1. Sharma S, Kelly TK, Jones PA. Epigenetics in cancer. *Carcinogenesis* 2010;**31**:27–36.
2. Fernandez AF, Esteller M. Viral epigenomes in human tumorigenesis. *Oncogene* 2010;**29**:1405–20.
3. Bloom DC, Giordani NV, Kwiatkowski DL. Epigenetic regulation of latent HSV-1 gene expression. *Biochim Biophys Acta* 2010;**1799**:246–56.
4. Wei H, Zhou MM. Viral-encoded enzymes that target host chromatin functions. *Biochim Biophys Acta* 2010;**1799**:296–301.
5. Takacs M, Banati F, Koroknai A, Segesdi J, Salamon D, Wolf H, et al. Epigenetic regulation of latent Epstein-Barr virus promoters. *Biochim Biophys Acta* 2010;**1799**:228–35.
6. White MK, Safak M, Khalili K. Regulation of gene expression in primate polyomaviruses. *J Virol* 2009;**83**:10846–56.
7. Hoelzer K, Shackelton LA, Parrish CR. Presence and role of cytosine methylation in DNA viruses of animals. *Nucleic Acids Res* 2008;**36**:2825–37.
8. Sen GL, Reuter JA, Webster DE, Zhu L, Khavari PA. DNMT1 maintains progenitor function in self-renewing somatic tissue. *Nature* 2010;**463**:563–7.
9. Richardson B. DNA methylation and autoimmune disease. *Clin Immunol* 2003;**109**:72–9.
10. Miller G. Epigenetics. The seductive allure of behavioral epigenetics. *Science* 2010;**329**:24–7.
11. Miller G. Epigenetics. A role for epigenetics in cognition. *Science* 2010;**329**:27.
12. van Bokhoven H, Kramer JM. Disruption of the epigenetic code: an emerging mechanism in mental retardation. *Neurobiol Dis* 2010;**39**:3–12.
13. Skinner MK, Manikkam M, Guerrero-Bosagna C. Epigenetic transgenerational actions of environmental factors in disease etiology. *Trends Endocrinol Metab* 2010;**21**:214–22.
14. Weissbach A. A chronicle of DNA methylation (1948-1975). *EXS* 1993;**64**:1–10.
15. Roy PH, Weissbach A. DNA methylase from HeLa cell nuclei. *Nucleic Acids Res* 1975;**2**:1669–84.

16. Pedrali-Noy G, Weissbach A. Mammalian DNA methyltransferases prefer poly(dI-dC) as substrate. *J Biol Chem* 1986;**261**:7600–2.
17. Stein R, Gruenbaum Y, Pollack Y, Razin A, Cedar H. Clonal inheritance of the pattern of DNA methylation in mouse cells. *Proc Natl Acad Sci USA* 1982;**79**:61–5.
18. Bolden A, Ward C, Siedlecki JA, Weissbach A. DNA methylation. Inhibition of de novo and maintenance methylation in vitro by RNA and synthetic polynucleotides. *J Biol Chem* 1984;**259**:12437–43.
19. Bestor TH. Cloning of a mammalian DNA methyltransferase. *Gene* 1988;**74**:9–12.
20. Yen RW, Vertino PM, Nelkin BD, Yu JJ, el-Deiry W, Cumaraswamy A, et al. Isolation and characterization of the cDNA encoding human DNA methyltransferase. *Nucleic Acids Res* 1992;**20**:2287–91.
21. Xu G, Flynn J, Glickman JF, Reich NO. Purification and stabilization of mouse DNA methyltransferase. *Biochem Biophys Res Commun* 1995;**207**:544–51.
22. Svedruzic ZM. Mammalian cytosine DNA methyltransferase Dnmt1: enzymatic mechanism, novel mechanism-based inhibitors, and RNA-directed DNA methylation. *Curr Med Chem* 2008;**15**:92–106.
23. Klimasauskas S, Kumar S, Roberts RJ, Cheng X. HhaI methyltransferase flips its target base out of the DNA helix. *Cell* 1994;**76**:357–69.
24. Glickman JF, Flynn J, Reich NO. Purification and characterization of recombinant baculovirus-expressed mouse DNA methyltransferase. *Biochem Biophys Res Commun* 1997;**230**:280–4.
25. Glickman JF, Reich NO. Baculovirus-mediated high level expression of a mammalian DNA methyltransferase. *Biochem Biophys Res Commun* 1994;**204**:1003–8.
26. Pradhan S, Talbot D, Sha M, Benner J, Hornstra L, Li E, et al. Baculovirus-mediated expression and characterization of the full-length murine DNA methyltransferase. *Nucleic Acids Res* 1997;**25**:4666–73.
27. Bacolla A, Pradhan S, Larson JE, Roberts RJ, Wells RD. Recombinant human DNA (cytosine-5) methyltransferase. III. Allosteric control, reaction order, and influence of plasmid topology and triplet repeat length on methylation of the fragile X CGG.CCG sequence. *J Biol Chem* 2001;**276**:18605–13.
28. Pradhan M, Esteve PO, Chin HG, Samaranayke M, Kim GD, Pradhan S. CXXC domain of human DNMT1 is essential for enzymatic activity. *Biochemistry* 2008;**47**:10000–9.
29. Pradhan S, Esteve PO. Allosteric activator domain of maintenance human DNA (cytosine-5) methyltransferase and its role in methylation spreading. *Biochemistry* 2003;**42**:5321–32.
30. Goyal R, Reinhardt R, Jeltsch A. Accuracy of DNA methylation pattern preservation by the Dnmt1 methyltransferase. *Nucleic Acids Res* 2006;**34**:1182–8.
31. Bacolla A, Pradhan S, Roberts RJ, Wells RD. Recombinant human DNA (cytosine-5) methyltransferase. II. Steady-state kinetics reveal allosteric activation by methylated DNA. *J Biol Chem* 1999;**274**:33011–9.
32. Pradhan S, Bacolla A, Wells RD, Roberts RJ. Recombinant human DNA (cytosine-5) methyltransferase. I. Expression, purification, and comparison of de novo and maintenance methylation. *J Biol Chem* 1999;**274**:33002–10.
33. Flynn J, Glickman JF, Reich NO. Murine DNA cytosine-C5 methyltransferase: pre-steady- and steady-state kinetic analysis with regulatory DNA sequences. *Biochemistry* 1996;**35**:7308–15.
34. Fatemi M, Hermann A, Pradhan S, Jeltsch A. The activity of the murine DNA methyltransferase Dnmt1 is controlled by interaction of the catalytic domain with the N-terminal part of the enzyme leading to an allosteric activation of the enzyme after binding to methylated DNA. *J Mol Biol* 2001;**309**:1189–99.

35. Pradhan S, Esteve PO. Mammalian DNA (cytosine-5) methyltransferases and their expression. *Clin Immunol* 2003;**109**:6–16.
36. Svedruzic ZM, Reich NO. DNA cytosine C5 methyltransferase Dnmt1: catalysis-dependent release of allosteric inhibition. *Biochemistry* 2005;**44**:9472–85.
37. Yoder JA, Soman NS, Verdine GL, Bestor TH. DNA (cytosine-5)-methyltransferases in mouse cells and tissues. Studies with a mechanism-based probe. *J Mol Biol* 1997;**270**:385–95.
38. Yu N, Wang M. Anticancer drug discovery targeting DNA hypermethylation. *Curr Med Chem* 2008;**15**:1350–75.
39. Goyal R, Rathert P, Laser H, Gowher H, Jeltsch A. Phosphorylation of serine-515 activates the mammalian maintenance methyltransferase Dnmt1. *Epigenetics* 2007;**2**:155–60.
40. Sugiyama Y, Hatano N, Sueyoshi N, Suetake I, Tajima S, Kinoshita E, et al. The DNA-binding activity of mouse DNA methyltransferase 1 is regulated by phosphorylation with casein kinase 1delta/epsilon. *Biochem J* 2010;**427**:489–97.
41. Kim GD, Ni J, Kelesoglu N, Roberts RJ, Pradhan S. Co-operation and communication between the human maintenance and de novo DNA (cytosine-5) methyltransferases. *EMBO J* 2002;**21**:4183–95.
42. Zhou Y, Grummt I. The PHD finger/bromodomain of NoRC interacts with acetylated histone H4K16 and is sufficient for rDNA silencing. *Curr Biol* 2005;**15**:1434–8.
43. Myant K, Stancheva I. LSH cooperates with DNA methyltransferases to repress transcription. *Mol Cell Biol* 2008;**28**:215–26.
44. Chuang LS, Ian HI, Koh TW, Ng HH, Xu G, Li BF. Human DNA-(cytosine-5) methyltransferase-PCNA complex as a target for p21WAF1. *Science* 1997;**277**:1996–2000.
45. Bostick M, Kim JK, Esteve PO, Clark A, Pradhan S, Jacobsen SE. UHRF1 plays a role in maintaining DNA methylation in mammalian cells. *Science* 2007;**317**:1760–4.
46. Sharif J, Muto M, Takebayashi S, Suetake I, Iwamatsu A, Endo TA, et al. The SRA protein Np95 mediates epigenetic inheritance by recruiting Dnmt1 to methylated DNA. *Nature* 2007;**450**:908–12.
47. Cheng X, Blumenthal RM. Coordinated chromatin control: structural and functional linkage of DNA and histone methylation. *Biochemistry* 2010;**49**:2999–3008.
48. Fuks F, Hurd PJ, Deplus R, Kouzarides T. The DNA methyltransferases associate with HP1 and the SUV39H1 histone methyltransferase. *Nucleic Acids Res* 2003;**31**:2305–12.
49. Esteve PO, Chin HG, Smallwood A, Feehery GR, Gangisetty O, Karpf AR, et al. Direct interaction between DNMT1 and G9a coordinates DNA and histone methylation during replication. *Genes Dev* 2006;**20**:3089–103.
50. Rountree MR, Bachman KE, Baylin SB. DNMT1 binds HDAC2 and a new co-repressor, DMAP1, to form a complex at replication foci. *Nat Genet* 2000;**25**:269–77.
51. Kiskinis E, Hallberg M, Christian M, Olofsson M, Dilworth SM, White R, et al. RIP140 directs histone and DNA methylation to silence Ucp1 expression in white adipocytes. *EMBO J* 2007;**26**:4831–40.
52. Robertson KD, Jones PA. DNA methylation: past, present and future directions. *Carcinogenesis* 2000;**21**:461–7.
53. Di Croce L, Raker VA, Corsaro M, Fazi F, Fanelli M, Faretta M, et al. Methyltransferase recruitment and DNA hypermethylation of target promoters by an oncogenic transcription factor. *Science* 2002;**295**:1079–82.
54. Butler JS, Lee JH, Skalnik DG. CFP1 interacts with DNMT1 independently of association with the Setd1 Histone H3K4 methyltransferase complexes. *DNA Cell Biol* 2008;**27**:533–43.
55. Tatematsu KI, Yamazaki T, Ishikawa F. MBD2-MBD3 complex binds to hemi-methylated DNA and forms a complex containing DNMT1 at the replication foci in late S phase. *Genes Cells* 2000;**5**:677–88.

56. Vire E, Brenner C, Deplus R, Blanchon L, Fraga M, Didelot C, et al. The Polycomb group protein EZH2 directly controls DNA methylation. *Nature* 2006;**439**:871–4.
57. Reale A, Matteis GD, Galleazzi G, Zampieri M, Caiafa P. Modulation of DNMT1 activity by ADP-ribose polymers. *Oncogene* 2005;**24**:13–9.
58. Zardo G, Reale A, Passananti C, Pradhan S, Buontempo S, De Matteis G, et al. Inhibition of poly(ADP-ribosyl)ation induces DNA hypermethylation: a possible molecular mechanism. *FASEB J* 2002;**16**:1319–21.
59. Pradhan S, Kim GD. The retinoblastoma gene product interacts with maintenance human DNA (cytosine-5) methyltransferase and modulates its activity. *EMBO J* 2002;**21**:779–88.
60. Robertson KD, Ait-Si-Ali S, Yokochi T, Wade PA, Jones PL, Wolffe AP. DNMT1 forms a complex with Rb, E2F1 and HDAC1 and represses transcription from E2F-responsive promoters. *Nat Genet* 2000;**25**:338–42.
61. Esteve PO, Chin HG, Pradhan S. Human maintenance DNA (cytosine-5)-methyltransferase and p53 modulate expression of p53-repressed promoters. *Proc Natl Acad Sci USA* 2005;**102**:1000–5.
62. Liu Z, Fisher RA. RGS6 interacts with DMAP1 and DNMT1 and inhibits DMAP1 transcriptional repressor activity. *J Biol Chem* 2004;**279**:14120–8.
63. Lee GE, Kim JH, Taylor M, Muller MT. DNA methyltransferase 1-associated protein (DMAP1) is a co-repressor that stimulates DNA methylation globally and locally at sites of double strand break repair. *J Biol Chem* 2010;**285**:37630–40.
64. Ohsawa K, Imai Y, Ito D, Kohsaka S. Molecular cloning and characterization of annexin V-binding proteins with highly hydrophilic peptide structure. *J Neurochem* 1996;**67**:89–97.
65. Zhou Q, Agoston AT, Atadja P, Nelson WG, Davidson NE. Inhibition of histone deacetylases promotes ubiquitin-dependent proteasomal degradation of DNA methyltransferase 1 in human breast cancer cells. *Mol Cancer Res* 2008;**6**:873–83.
66. Zhang X, Verdine GL. Mammalian DNA cytosine-5 methyltransferase interacts with p23 protein. *FEBS Lett* 1996;**392**:179–83.
67. Fuks F, Hurd PJ, Wolf D, Nan X, Bird AP, Kouzarides T. The methyl-CpG-binding protein MeCP2 links DNA methylation to histone methylation. *J Biol Chem* 2003;**278**:4035–40.
68. Amir RE, Van den Veyver IB, Wan M, Tran CQ, Francke U, Zoghbi HY. Rett syndrome is caused by mutations in X-linked MECP2, encoding methyl-CpG-binding protein 2. *Nat Genet* 1999;**23**:185–8.
69. Carty SM, Greenleaf AL. Hyperphosphorylated C-terminal repeat domain-associating proteins in the nuclear proteome link transcription to DNA/chromatin modification and RNA processing. *Mol Cell Proteomics* 2002;**1**:598–610.
70. Tao Y, Xi S, Briones V, Muegge K. Lsh mediated RNA polymerase II stalling at HoxC6 and HoxC8 involves DNA methylation. *PLoS One* 2010;**5**:e9163.
71. Takeshima H, Yamashita S, Shimazu T, Niwa T, Ushijima T. The presence of RNA polymerase II, active or stalled, predicts epigenetic fate of promoter CpG islands. *Genome Res* 2009;**19**:1974–82.
72. Spada F, Rothbauer U, Zolghadr K, Schermelleh L, Leonhardt H. Regulation of DNA methyltransferase 1. *Adv Enzyme Regul* 2006;**46**:224–34.
73. Flynn J, Reich N. Murine DNA (cytosine-5-)-methyltransferase: steady-state and substrate trapping analyses of the kinetic mechanism. *Biochemistry* 1998;**37**:15162–9.
74. Svedruzic ZM, Reich NO. Mechanism of allosteric regulation of Dnmt1's processivity. *Biochemistry* 2005;**44**:14977–88.
75. Grace Goll M, Bestor TH. Eukaryotic cytosine methyltransferases. *Annu Rev Biochem* 2005;**74**:481–514.
76. Jeltsch A. Beyond Watson and Crick: DNA methylation and molecular enzymology of DNA methyltransferases. *Chembiochem* 2002;**3**:274–93.

77. Hermann A, Gowher H, Jeltsch A. Biochemistry and biology of mammalian DNA methyltransferases. *Cell Mol Life Sci* 2004;**61**:2571–87.
78. Kim JK, Samaranayake M, Pradhan S. Epigenetic mechanisms in mammals. *Cell Mol Life Sci* 2009;**66**:596–612.
79. Van Emburgh BO, Robertson KD. DNA methyltransferases and Methyl CpG binding proteins as multifunctional regulators of chromatin structure and development in mammalian cells. In: Tost J, editor. *Epigenetics*. Norfolk, UK: Caister Academic Press; 2008. p. 392.
80. Wu JC, Santi DV. Kinetic and catalytic mechanism of HhaI methyltransferase. *J Biol Chem* 1987;**262**:4778–86.
81. Esteller M. Epigenetics provides a new generation of oncogenes and tumour-suppressor genes. *Br J Cancer* 2006;**94**:179–83.
82. Robertson KD. DNA methylation and human disease. *Nat Rev Genet* 2005;**6**:597–610.
83. Ooi SK, Bestor TH. Cytosine methylation: remaining faithful. *Curr Biol* 2008;**18**:R174–6.
84. Jones PA, Liang G. Rethinking how DNA methylation patterns are maintained. *Nat Rev Genet* 2009;**10**:805–11.
85. Purdy MM, Holz-Schietinger C, Reich NO. Identification of a second DNA binding site in human DNA methyltransferase 3A by substrate inhibition and domain deletion. *Arch Biochem Biophys* 2010;**498**:13–22.
86. Yokochi T, Robertson KD. Preferential methylation of unmethylated DNA by mammalian de novo DNA methyltransferase Dnmt3a. *J Biol Chem* 2002;**277**:11735–45.
87. Flynn J, Fang JY, Mikovits JA, Reich NO. A potent cell-active allosteric inhibitor of murine DNA cytosine C5 methyltransferase. *J Biol Chem* 2003;**278**:8238–43.
88. Jeltsch A. On the enzymatic properties of Dnmt1: specificity, processivity, mechanism of linear diffusion and allosteric regulation of the enzyme. *Epigenetics* 2006;**1**:63–6.
89. Zhou L, Cheng X, Connolly BA, Dickman MJ, Hurd PJ, Hornby DP. Zebularine: a novel DNA methylation inhibitor that forms a covalent complex with DNA methyltransferases. *J Mol Biol* 2002;**321**:591–9.
90. Isakovic L, Saavedra OM, Llewellyn DB, Claridge S, Zhan L, Bernstein N, et al. Constrained (l-)-S-adenosyl-l-homocysteine (SAH) analogues as DNA methyltransferase inhibitors. *Bioorg Med Chem Lett* 2009;**19**:2742–6.
91. Uetrecht J. Current trends in drug-induced autoimmunity. *Autoimmun Rev* 2005;**4**:309–14.
92. Brueckner B, Garcia Boy R, Siedlecki P, Musch T, Kliem HC, Zielenkiewicz P, et al. Epigenetic reactivation of tumor suppressor genes by a novel small-molecule inhibitor of human DNA methyltransferases. *Cancer Res* 2005;**65**:6305–11.
93. Stresemann C, Brueckner B, Musch T, Stopper H, Lyko F. Functional diversity of DNA methyltransferase inhibitors in human cancer cell lines. *Cancer Res* 2006;**66**:2794–800.
94. Lee BH, Yegnasubramanian S, Lin X, Nelson WG. Procainamide is a specific inhibitor of DNA methyltransferase 1. *J Biol Chem* 2005;**280**:40749–56.
95. Castellano S, Kuck D, Sala M, Novellino E, Lyko F, Sbardella G. Constrained analogues of procaine as novel small molecule inhibitors of DNA methyltransferase-1. *J Med Chem* 2008;**51**:2321–5.
96. Stewart DJ, Donehower RC, Eisenhauer EA, Wainman N, Shah AK, Bonfils C, et al. A phase I pharmacokinetic and pharmacodynamic study of the DNA methyltransferase 1 inhibitor MG98 administered twice weekly. *Ann Oncol* 2003;**14**:766–74.
97. Lin J, Haffner MC, Zhang Y, Lee BH, Brennen WN, Britton J, et al. Disulfiram is a DNA demethylating agent and inhibits prostate cancer cell growth. *Prostate* 2011;**71**(4):333–43.
98. Medina-Franco JL, Lopez-Vallejo F, Kuck D, Lyko F. Natural products as DNA methyltransferase inhibitors: a computer-aided discovery approach. *Mol Divers* 2010;**1**: 1–12.
99. Christman JK. 5-Azacytidine and 5-aza-2′-deoxycytidine as inhibitors of DNA methylation: mechanistic studies and their implications for cancer therapy. *Oncogene* 2002;**21**:5483–95.

100. Bestor TH. Activation of mammalian DNA methyltransferase by cleavage of a Zn binding regulatory domain. *EMBO J* 1992;**11**:2611–7.
101. Posfai J, Bhagwat AS, Posfai G, Roberts RJ. Predictive motifs derived from cytosine methyltransferases. *Nucleic Acids Res* 1989;**17**:2421–35.
102. Margot JB, Ehrenhofer-Murray AE, Leonhardt H. Interactions within the mammalian DNA methyltransferase family. *BMC Mol Biol* 2003;**4**:7.
103. Youngblood B, Shieh FK, Buller F, Bullock T, Reich NO. S-adenosyl-L-methionine-dependent methyl transfer: observable precatalytic intermediates during DNA cytosine methylation. *Biochemistry* 2007;**46**:8766–75.
104. Shieh FK, Reich NO. AdoMet-dependent methyl-transfer: Glu119 is essential for DNA C5-cytosine methyltransferase M.HhaI. *J Mol Biol* 2007;**373**:1157–68.
105. Shieh FK, Youngblood B, Reich NO. The role of Arg165 towards base flipping, base stabilization and catalysis in M.HhaI. *J Mol Biol* 2006;**362**:516–27.
106. Zhou H, Purdy MM, Dahlquist FW, Reich NO. The recognition pathway for the DNA cytosine methyltransferase M.HhaI. *Biochemistry* 2009;**48**:7807–16.
107. Cheng X. Structure and function of DNA methyltransferases. *Annu Rev Biophys Biomol Struct* 1995;**24**:293–318.
108. O'Gara M, Roberts RJ, Cheng X. A structural basis for the preferential binding of hemimethylated DNA by HhaI DNA methyltransferase. *J Mol Biol* 1996;**263**:597–606.
109. Cheng X, Kumar S, Posfai J, Pflugrath JW, Roberts RJ. Crystal structure of the HhaI DNA methyltransferase complexed with S- adenosyl-L-methionine. *Cell* 1993;**74**:299–307.
110. O'Gara M, Klimasauskas S, Roberts RJ, Cheng X. Enzymatic C5-cytosine methylation of DNA: mechanistic implications of new crystal structures for HhaL methyltransferase-DNA-AdoHcy complexes. *J Mol Biol* 1996;**261**:634–45.
111. Schluckebier G, O'Gara M, Saenger W, Cheng X. Universal catalytic domain structure of AdoMet-dependent methyltransferases. *J Mol Biol* 1995;**247**:16–20.
112. O'Gara M, Zhang X, Roberts RJ, Cheng X. Structure of a binary complex of HhaI methyltransferase with S-adenosyl-L-methionine formed in the presence of a short non-specific DNA oligonucleotide. *J Mol Biol* 1999;**287**:201–9.
113. Dong A, Zhou L, Zhang X, Stickel S, Roberts RJ, Cheng X. Structure of the Q237W mutant of HhaI DNA methyltransferase: an insight into protein-protein interactions. *Biol Chem* 2004;**385**:373–9.
114. Svedruzic ZM, Reich NO. The mechanism of target base attack in DNA cytosine carbon 5 methylation. *Biochemistry* 2004;**43**:11460–73.
115. Lindstrom Jr. WM, Flynn J, Reich NO. Reconciling structure and function in HhaI DNA cytosine-C-5 methyltransferase. *J Biol Chem* 2000;**275**:4912–9.
116. Vilkaitis G, Merkiene E, Serva S, Weinhold E, Klimasauskas S. The mechanism of DNA cytosine-5 methylation. Kinetic and mutational dissection of Hhai methyltransferase. *J Biol Chem* 2001;**276**:20924–34.
117. Zingg JM, Shen JC, Jones PA. Enzyme-mediated cytosine deamination by the bacterial methyltransferase M.MspI. *Biochem J* 1998;**332**:223–30.
118. Zingg JM, Shen JC, Yang AS, Rapoport H, Jones PA. Methylation inhibitors can increase the rate of cytosine deamination by (cytosine-5)-DNA methyltransferase. *Nucleic Acids Res* 1996;**24**:3267–75.
119. Shen JC, Rideout III WM, Jones PA. High frequency mutagenesis by a DNA methyltransferase. *Cell* 1992;**71**:1073–80.
120. Rideout WMI, Coetzee GA, Olumi AF, Jones PA. 5-Methylcytosine as an endogenous mutagen in the human LDL receptor and p53 genes. *Science* 1990;**249**:1288–90.
121. Friedberg EC, Walker GC, Siede W, Wood RD, Schultz RA, Ellenberger T. *DNA repair and mutagenesis*. 2nd ed. Washington, DC: ASM Press; 2005.

122. D'Aiuto L, Marzulli M, Mohan KN, Borowczyk E, Saporiti F, Vandemark A, et al. Dissection of structure and function of the N-terminal domain of mouse DNMT1 using regional frameshift mutagenesis. *PLoS One* 2010;**5**:e9831.
123. Jeltsch A. Reading and writing DNA methylation. *Nat Struct Mol Biol* 2008;**15**:1003–4.
124. Fersht A. *Structure and mechanism in protein science: a guide to enzyme catalysis and protein folding (Hardcover)*. 1st ed. New York, United Media Business; 1998.
125. Motulsky H, Christopoulos A. *Fitting models to biological data using linear and nonlinear regression: a practical guide to curve fitting*. 1st ed. USA: Oxford University Press; 2004.
126. Tipton KF. Enzyme assays. In: Eisenthal R, Danson MJ, editors. *Practical approach*. Oxford, UK: University Press; 2002. p. 282.
127. Luo W, Karpf AR, Deeb KK, Muindi JR, Morrison CD, Johnson CS, et al. Epigenetic regulation of vitamin D 24-hydroxylase/CYP24A1 in human prostate cancer. *Cancer Res* 2010;**70**:5953–62.
128. Yang Q, Tian Y, Ostler KR, Chlenski A, Guerrero LJ, Salwen HR, et al. Epigenetic alterations differ in phenotypically distinct human neuroblastoma cell lines. *BMC Cancer* 2010;**10**:286.
129. Walsh C, Cromartie T, Marcotte P, Spencer R. Suicide substrates for flavoprotein enzymes. *Methods Enzymol* 1978;**53**:437–48.
130. Wang ZX. Kinetics of suicide substrates. *J Theor Biol* 1990;**147**:497–508.
131. Waley SG. Kinetics of suicide substrates. Practical procedures for determining parameters. *Biochem J* 1985;**227**:843–9.
132. Vilkaitis G, Dong A, Weinhold E, Cheng X, Klimasauskas S. Functional roles of the conserved threonine 250 in the target recognition domain of HhaI DNA methyltransferase. *J Biol Chem* 2000;**275**:38722–30.
133. Parker JB, Bianchet MA, Krosky DJ, Friedman JI, Amzel LM, Stivers JT. Enzymatic capture of an extrahelical thymine in the search for uracil in DNA. *Nature* 2007;**449**:433–7.
134. Qi Y, Spong MC, Nam K, Banerjee A, Jiralerspong S, Karplus M, et al. Encounter and extrusion of an intrahelical lesion by a DNA repair enzyme. *Nature* 2009;**462**:762–6.
135. Krosky DJ, Bianchet MA, Seiple L, Chung S, Amzel LM, Stivers JT. Mimicking damaged DNA with a small molecule inhibitor of human UNG2. *Nucleic Acids Res* 2006;**34**:5872–9.
136. Meeker JD, Sathyanarayana S, Swan SH. Phthalates and other additives in plastics: human exposure and associated health outcomes. *Philos Trans R Soc Lond B Biol Sci* 2009;**364**:2097–113.
137. Margot JB, Aguirre-Arteta AM, Di Giacco BV, Pradhan S, Roberts RJ, Cardoso MC, et al. Structure and function of the mouse DNA methyltransferase gene: Dnmt1 shows a tripartite structure. *J Mol Biol* 2000;**297**:293–300.
138. Klotz IM. *Ligand-receptor energetics: a guide for the perplexed*. New York, Wiley; 1997.
139. Haynie DT. *Biological thermodynamics*. 2nd ed. Cambridge, UK: Cambridge University Press; 2008.
140. Cantor C, Schimmel P. *Biophysical chemistry*. New York, United Media Business; 1980.
141. Shearman MS, Beher D, Clarke EE, Lewis HD, Harrison T, Hunt P, et al. L-685, 458, an aspartyl protease transition state mimic, is a potent inhibitor of amyloid beta-protein precursor gamma-secretase activity. *Biochemistry* 2000;**39**:8698–704.
142. Walker JR, Avvakumov GV, Xue S, Li Y, Bountra C, Weigelt J, et al. Structure of the replication foci-targeting sequence of human dna cytosine methyltransferase Dnmt1. Structural Genomics Consortium (Sgc), 2008/9/30: PDB 3EPZ.
143. Cheng X, Blumenthal RM. Mammalian DNA methyltransferases: a structural perspective. *Structure* 2008;**16**:341–50.
144. Kameshita I, Sekiguchi M, Hamasaki D, Sugiyama Y, Hatano N, Suetake I, et al. Cyclin-dependent kinase-like 5 binds and phosphorylates DNA methyltransferase 1. *Biochem Biophys Res Commun* 2008;**377**:1162–7.

145. Glickman JF, Pavlovich JG, Reich NO. Peptide mapping of the murine DNA methyltransferase reveals a major phosphorylation site and the start of translation. *J Biol Chem* 1997;**272**:17851–7.
146. Gale JM, Nissen KA, Smerdon MJ. UV-induced formation of pyrimidine dimers in nucleosome core DNA is strongly modulated with a period of 10.3 bases. *Proc Natl Acad Sci USA* 1987;**84**:6644–8.
147. Gale JM, Smerdon MJ. Photofootprint of nucleosome core DNA in intact chromatin having different structural states. *J Mol Biol* 1988;**204**:949–58.
148. Schermelleh L, Haemmer A, Spada F, Rosing N, Meilinger D, Rothbauer U, et al. Dynamics of Dnmt1 interaction with the replication machinery and its role in postreplicative maintenance of DNA methylation. *Nucleic Acids Res* 2007;**35**:4301–12.
149. Bronner C, Chataigneau T, Schini-Kerth VB, Landry Y. The "Epigenetic Code Replication Machinery", ECREM: a promising drugable target of the epigenetic cell memory. *Curr Med Chem* 2007;**14**:2629–41.
150. Bronner C, Fuhrmann G, Chédin F, Macaluso M, Dhe-Paganon S. UHRF1 links the histone code and DNA methylation to ensure faithful epigenetic memory inheritance. *Genet Epigenet* 2009;**2**:29–36.
151. Fellinger K, Rothbauer U, Felle M, Langst G, Leonhardt H. Dimerization of DNA methyltransferase 1 is mediated by its regulatory domain. *J Cell Biochem* 2009;**106**:521–8.
152. Chodavarapu RK, Feng S, Bernatavichute YV, Chen PY, Stroud H, Yu Y, et al. Relationship between nucleosome positioning and DNA methylation. *Nature* 2010;**466**:388–92.
153. Gowher H, Stockdale CJ, Goyal R, Ferreira H, Owen-Hughes T, Jeltsch A. De novo methylation of nucleosomal DNA by the mammalian Dnmt1 and Dnmt3A DNA methyltransferases. *Biochemistry* 2005;**44**:9899–904.
154. Okuwaki M, Verreault A. Maintenance DNA methylation of nucleosome core particles. *J Biol Chem* 2004;**279**:2904–12.
155. Robertson AK, Geiman TM, Sankpal UT, Hager GL, Robertson KD. Effects of chromatin structure on the enzymatic and DNA binding functions of DNA methyltransferases DNMT1 and Dnmt3a in vitro. *Biochem Biophys Res Commun* 2004;**322**:110–8.
156. Choy JS, Wei S, Lee JY, Tan S, Chu S, Lee TH. DNA methylation increases nucleosome compaction and rigidity. *J Am Chem Soc* 2010;**132**:1782–3.
157. Buttinelli M, Minnock A, Panetta G, Waring M, Travers A. The exocyclic groups of DNA modulate the affinity and positioning of the histone octamer. *Proc Natl Acad Sci USA* 1998;**95**:8544–9.
158. Segal E, Widom J. What controls nucleosome positions? *Trends Genet* 2009;**25**:335–43.
159. Kaplan N, Moore IK, Fondufe-Mittendorf Y, Gossett AJ, Tillo D, Field Y, et al. The DNA-encoded nucleosome organization of a eukaryotic genome. *Nature* 2009;**458**:362–6.
160. Segal E, Fondufe-Mittendorf Y, Chen L, Thastrom A, Field Y, Moore IK, et al. A genomic code for nucleosome positioning. *Nature* 2006;**442**:772–8.
161. Li G, Widom J. Nucleosomes facilitate their own invasion. *Nat Struct Mol Biol* 2004;**11**:763–9.
162. Hinz JM, Rodriguez Y, Smerdon MJ. Rotational dynamics of DNA on the nucleosome surface markedly impact accessibility to a DNA repair enzyme. *Proc Natl Acad Sci USA* 2010;**107**:4646–51.
163. Li Q, Wrange O. Translational positioning of a nucleosomal glucocorticoid response element modulates glucocorticoid receptor affinity. *Genes Dev* 1993;**7**:2471–82.
164. Li Q, Wrange O. Accessibility of a glucocorticoid response element in a nucleosome depends on its rotational positioning. *Mol Cell Biol* 1995;**15**:4375–84.
165. Polach KJ, Widom J. Mechanism of protein access to specific DNA sequences in chromatin: a dynamic equilibrium model for gene regulation. *J Mol Biol* 1995;**254**:130–49.

166. Svedruzic ZM, Wang C, Kosmoski JV, Smerdon MJ. Accommodation and repair of a UV photoproduct in DNA at different rotational settings on the nucleosome surface. *J Biol Chem* 2005;**280**:40051–7.
167. Mer G, Bochkarev A, Gupta R, Bochkareva E, Frappier L, Ingles CJ, et al. Structural basis for the recognition of DNA repair proteins UNG2, XPA, and RAD52 by replication factor RPA. *Cell* 2000;**103**:449–56.
168. Fatemi M, Hermann A, Gowher H, Jeltsch A. Dnmt3a and Dnmt1 functionally cooperate during de novo methylation of DNA. *Eur J Biochem* 2002;**269**:4981–4.
169. Avvakumov GV, Walker JR, Xue S, Li Y, Duan S, Bronner C, et al. Structural basis for recognition of hemi-methylated DNA by the SRA domain of human UHRF1. *Nature* 2008;**455**:822–5.
170. Kim JK, Esteve PO, Jacobsen SE, Pradhan S. UHRF1 binds G9a and participates in p21 transcriptional regulation in mammalian cells. *Nucleic Acids Res* 2009;**37**:493–505.
171. Flynn J, Azzam R, Reich N. DNA binding discrimination of the murine DNA cytosine-C5 methyltransferase. *J Mol Biol* 1998;**279**:101–16.
172. Shukla V, Coumoul X, Lahusen T, Wang RH, Xu X, Vassilopoulos A, et al. BRCA1 affects global DNA methylation through regulation of DNMT1. *Cell Res* 2010;**20**:1201–15.
173. Tan HH, Porter AG. p21(WAF1) negatively regulates DNMT1 expression in mammalian cells. *Biochem Biophys Res Commun* 2009;**382**:171–6.
174. Le Gac G, Esteve PO, Ferec C, Pradhan S. DNA damage-induced down-regulation of human Cdc25C and Cdc2 is mediated by cooperation between p53 and maintenance DNA (cytosine-5) methyltransferase 1. *J Biol Chem* 2006;**281**:24161–70.
175. Matzke MA, Birchler JA. RNAi-mediated pathways in the nucleus. *Nat Rev Genet* 2005;**6**:24–35.
176. Aravin AA, Sachidanandam R, Bourc'his D, Schaefer C, Pezic D, Toth KF, et al. A piRNA pathway primed by individual transposons is linked to de novo DNA methylation in mice. *Mol Cell* 2008;**31**:785–99.
177. Kanno T, Bucher E, Daxinger L, Huettel B, Bohmdorfer G, Gregor W, et al. A structural-maintenance-of-chromosomes hinge domain-containing protein is required for RNA-directed DNA methylation. *Nat Genet* 2008;**40**:670–5.
178. Kuramochi-Miyagawa S, Watanabe T, Gotoh K, Totoki Y, Toyoda A, Ikawa M, et al. DNA methylation of retrotransposon genes is regulated by Piwi family members MILI and MIWI2 in murine fetal testes. *Genes Dev* 2008;**22**:908–17.
179. Tam OH, Aravin AA, Stein P, Girard A, Murchison EP, Cheloufi S, et al. Pseudogene-derived small interfering RNAs regulate gene expression in mouse oocytes. *Nature* 2008;**453**:534–8.
180. Watanabe T, Totoki Y, Toyoda A, Kaneda M, Kuramochi-Miyagawa S, Obata Y, et al. Endogenous siRNAs from naturally formed dsRNAs regulate transcripts in mouse oocytes. *Nature* 2008;**453**:539–43.
181. Zaratiegui M, Irvine DV, Martienssen RA. Noncoding RNAs and gene silencing. *Cell* 2007;**128**:763–76.
182. Kimura H, Shiota K. Methyl-CpG-binding protein, MeCP2, is a target molecule for maintenance DNA methyltransferase, Dnmt1. *J Biol Chem* 2003;**278**:4806–12.
183. Jeffery L, Nakielny S. Components of the DNA methylation system of chromatin control are RNA-binding proteins. *J Biol Chem* 2004;**279**:49479–87.
184. Nan X, Campoy FJ, Bird A. MeCP2 is a transcriptional repressor with abundant binding sites in genomic chromatin. *Cell* 1997;**88**:471–81.
185. Cross SH, Meehan RR, Nan X, Bird A. A component of the transcriptional repressor MeCP1 shares a motif with DNA methyltransferase and HRX proteins. *Nat Genet* 1997;**16**:256–9.

186. Weinberg MS, Villeneuve LM, Ehsani A, Amarzguioui M, Aagaard L, Chen ZX, et al. The antisense strand of small interfering RNAs directs histone methylation and transcriptional gene silencing in human cells. *RNA* 2006;**12**:256–62.
187. Goll MG, Kirpekar F, Maggert KA, Yoder JA, Hsieh CL, Zhang X, et al. Methylation of tRNAAsp by the DNA methyltransferase homolog Dnmt2. *Science* 2006;**311**:395–8.
188. Jeltsch A, Nellen W, Lyko F. Two substrates are better than one: dual specificities for Dnmt2 methyltransferases. *Trends Biochem Sci* 2006;**31**:306–8.
189. Song J, Rechkoblit O, Bestor TH, Patel DJ. Structure of DNMT1-DNA complex reveals a role for autoinhibition in maintenance DNA methylation. *Science* 2001; **331**(6020): 1036–40.
190. Iakoucheva LM, Radivojac P, Brown CJ, O'Connor TR, Sikes JG, Obradovic Z, et al. The importance of intrinsic disorder for protein phosphorylation. *Nucleic Acids Res* 2004;**32**:1037–49.
191. Dunker AK, Brown CJ, Lawson JD, Iakoucheva LM, Obradovic Z. Intrinsic disorder and protein function. *Biochemistry* 2002;**41**:6573–82.
192. Luger K, Mader AW, Richmond RK, Sargent DF, Richmond TJ. Crystal structure of the nucleosome core particle at 2.8 Å resolution. *Nature* 1997;**389**:251–60.

The DNMT3 Family of Mammalian *De Novo* DNA Methyltransferases

Frédéric Chédin

Department of Molecular and Cellular Biology, University of California, Davis, California, USA

The deposition of DNA methylation at promoters of transposons, X-linked genes, imprinted genes, and other lineage-specific genes is clearly associated with long-term transcriptional silencing. Thus, DNA methylation represents a key layer of epigenetic information in mammals that is required for embryonic development, germline differentiation, and, as shown more recently, for the function and maturation of neuronal tissues. The DNMT3A, DNMT3B, and DNMT3L proteins are primarily responsible for the establishment of genomic DNA methylation patterns and, as such, play an important role in human

Progress in Molecular Biology
and Translational Science, Vol. 101
DOI: 10.1016/B978-0-12-387685-0.00007-X

developmental, reproductive, and mental health. Progress in our understanding of this important protein family has been rapid in recent years and has been accompanied by stunning developments in the analysis of the human DNA methylome in multiple cell types. This review focuses on recent developments in the characterization of the DNMT3 family of DNA methyltransferases at the biochemical, structural, and functional levels. Interconnections between the DNA-based and histone-based layers of epigenetic information are particularly highlighted, as it is now clear that *de novo* methylation occurs chiefly in the context of nucleosomal templates.

I. Introduction

In 1975, Riggs, Holliday, and Pugh proposed the groundbreaking hypothesis that cytosine DNA methylation could be used to regulate transcriptional activity, and function as a form of cellular memory.[1,2] Over three decades later, it is well established that DNA methylation indeed provides epigenetic information that plays critical roles in processes such as X chromosome inactivation, repeat silencing, and gene regulation during differentiation. Not surprisingly, the enzymes that establish and maintain genomic DNA methylation patterns have been the object of intense study. Cytosine DNA methyltransferases, and their many associated factors, are at the center of an ever expanding group of proteins involved in writing, maintaining, reading, and perhaps erasing DNA methylation.

In mammals, with which this chapter is primarily concerned, DNA methylation is almost exclusively observed at symmetrical CpG dinucleotides. Analyses of the mammalian methylome revealed that CpG methylation is very prevalent in these organisms.[3–5] Given that methylated cytosines undergo spontaneous deamination at a much higher rate than their unmethylated counterparts,[6] the frequency of CpG sites in the genomes of organisms that extensively use this epigenetic mark is very low. Today, the only genomic regions that exhibit normal or even enriched CpG densities correspond to so-called "CpG islands," which are maintained due to the fact that they remain largely free of DNA methylation.[7] Readers are referred to recent breakthrough articles[5,8,9] and reviews[10–12] describing how genomic methylation patterns are distributed in mammals, as well as to Chapter by Eleanor Wong and Chia-Lin Wei. This chapter instead will focus on a small, but critical, piece of the DNA methylation life cycle: the family of proteins that establish DNA methylation patterns *de novo*. Numerous other excellent reviews have been published on the topic in recent years and readers are encouraged to refer to them for complementary information,[13–16] and see also Chapter by Željko M. Svedružić.

II. Discovery, Primary Structure, and Expression of Mammalian DNMT3A and DNMT3B

For about a decade, the only known DNA methyltransferase in mammals was DNMT1, a large protein carrying a C-terminal domain with clear homology to bacterial cytosine C5-methyltransferases.[17] Even though DNMT1 shows robust *de novo* methylation activity *in vitro*, its preference for hemimethylated DNA substrates at least under some conditions[18] suggested early on that it might be preferentially involved in maintaining existing DNA methylation patterns (but see Chapter by Željko M. Svedružić for a more detailed discussion). The observations that mouse embryonic stem cells (ESCs) deficient for *DNMT1* still contain low but detectable levels of 5-methylcytosine in their genomes and are capable of carrying out the methylation of retroviral constructs *de novo*,[19] support this model. This, in turn, suggested that mammalian cells carry at least one additional functional DNA methyltransferase. Analysis of expressed sequence tags (ESTs) enabled En Li and colleagues to identify and clone two novel genes, each carrying sequence motifs with high similarity to conserved motifs in bacterial cytosine C5-methyltransferases.[20] These two genes, *DNMT3A* and *DNMT3B*, map respectively to 2p23.3 and 20q11.21[21] and specify two proteins with similar primary structures and domain organization (Fig. 1). Both proteins contain a variable N-terminal domain followed by a moderately conserved PWWP (Pro-Trp-Trp-Pro) domain, a member of the "Royal" family of putative histone-binding modules.[22] This is followed by a cysteine-rich zinc-finger binding domain related to the plant homeodomain (PHD) family and termed ADD domain (ATRX-DNMT3-DNMT3L). This domain is strongly conserved between DNMT3A and DNMT3B, and is implicated in histone binding (see below for details concerning the interaction between DNMT3 family members and chromatin). Finally, a C-terminal domain carries six highly conserved cytosine C5-DNA methyltransferase motifs as defined earlier from the studies of bacterial enzymes.[23] This structural organization is clearly distinct from those observed for DNMT1 and DNMT2, thus leading to the founding of the DNMT3 family of cytosine DNA methyltransferases. This family of proteins is widely distributed in the animal kingdom, with clear DNMT3 homologs observed in cnidarians (*Nematostella vectensis*), insects (*Apis mellifera*), tunicates (*Ciona intestinalis*), and, most particularly, vertebrates[24] (see Part 2 of this volume).

Expression studies in human and mouse show that *DNMT3A* and *DNMT3B* are highly expressed around the blastocyst stage and during germ cell development, two developmental stages during which new DNA methylation patterns are being laid genome wide.[25] As development progresses, both genes undergo tissue-specific repression such that their expression becomes mostly undetectable past midgestation, with the exception of spleen, thymus,

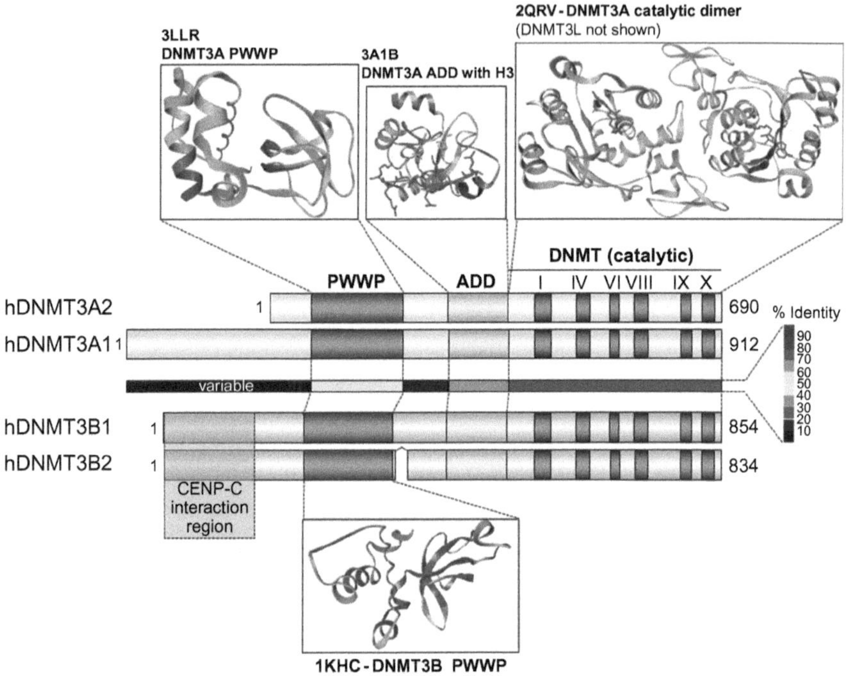

FIG. 1. Schematic representation of human DNMT3A and DNMT3B proteins. Both full-length and shorter active isoforms (DNMT3A2 and DNMT3B2) are depicted here. Conserved domains are indicated and the extent of conservation (% identity) between DNMT3A and DNMT3B is represented by a heat map. Available structural information is depicted above or below each protein in a simple ribbon format. In each case, the PDB code for the corresponding structure is indicated.

and brain.[26] In the case of *DNMT3A*, transcription in ESCs is initiated from an internal promoter and leads to the production of a shorter protein isoform named DNMT3A2 lacking most of the variable N-terminal region[27] (Fig. 1). Interestingly, DNMT3A2 expression gradually decreases upon ESC differentiation and switches to the full-length DNMT3A1 isoform which remains expressed at low levels in most somatic tissues.[27,28] *DNMT3B* expression in ESCs gives rise primarily to the full-length DNMT3B1 isoform.[20,29] In humans, DNMT3B1 is by far the dominant *de novo* DNMT in early peri-implantation development.[9,30] Upon differentiation, however, *DNMT3B* expression switches to produce alternatively spliced isoforms such as DNMT3B2, which lacks exons 10 and 11 in the N-terminus of the protein and DNMT3B3, which in addition to exons 10 and 11 also lacks exons 21 and 22 from the catalytic domain.[20,29] *DNMT3B* expression is low in most somatic tissues, and is associated with the production of DNMT3B2 and DNMT3B3, among several other isoforms.[21,27,31] Interestingly, the patterns of alternative splicing and the temporal regulation of isoform expression through development appears largely conserved between human and mouse. However, the biological significance of expressing specific variants, particularly in the case of *DNMT3B*, remains mostly unclear. Apart from pluripotent ESCs, *DNMT3A* and *DNMT3B* are also highly expressed in germ cells. Careful expression analyses revealed that expression of *DNMT3A* is under strict temporal control, such that the peak of *DNMT3A* expression coincides with the deposition of genome-wide *de novo* methylation patterns and the acquisition of genomic methylation imprints. In male germ cells, expression is highest in resting prenatal prospermatogonia prior to entry into meiosis.[32,33] By contrast, expression of *DNMT3A* is highest in postnatal growing oocytes.[34] *DNMT3B* also undergoes dynamic and regulated variation in expression in germ cells.[32–34] Together, these studies indicate that the peak of DNMT3A and DNMT3B expression coincides with active *de novo* methylation, and suggest that DNMT3A may play a particular role in setting DNA methylation in germ cells, while DNMT3B appears to play a dominant role in early development.

III. DNMT3A and DNMT3B Drive the Establishment of DNA Methylation Patterns *De Novo*

A. DNMT3A and DNMT3B Are Active DNA Methyltransferases

Shortly after their discovery, the murine DNMT3A and DNMT3B proteins were overexpressed using a baculovirus system, and shown to be catalytically active as DNA methyltransferases *in vitro*.[20] Further, it appeared that their main target corresponded to CpG sites, as DNA substrates in which all CpG sites (but not other cytosines) were premethylated did not support activity.[20] In contrast to DNMT1, it was noted from the onset that DNMT3A and DNMT3B do not show any preference for hemimethylated DNA substrates under the conditions tested. Thus, while DNMT1 showed the characteristics expected of a maintenance DNA methyltransferase, DNMT3A and 3B appear more adapted toward the establishment of DNA methylation patterns *de novo*. Subsequent biochemical studies using highly purified preparations have confirmed that both enzymes, either in their full-length forms or purified as isolated catalytic domains, are enzymatically active and show little preference for hemi-versus unmethylated DNA substrates.[35–37] An episomal cell culture assay also confirmed that the two proteins were active *in vivo*.[38] Analysis of *in vitro* methylated DNA using bisulfite sequencing revealed that both

DNMT3A and DNMT3B mainly target CpG sites, as originally described, but that they also target CpA and CpT sites to a lesser degree.[35,36,39] Interestingly, human and mouse ESCs, which are characterized by high levels of *de novo* DNA methylation, show significant non-CpG methylation in their genomes.[5,9,40] The function of such non-CpG methylation, if any, is unknown, but it is interesting that non-CpG methylation appears particularly enriched in the gene bodies of a subset of genes involved in RNA processing, RNA splicing, and RNA metabolic processes.[5] The biological significance of this observation is unclear. DNA methylation in somatic cells, by contrast, is focused almost exclusively at CpG sites due to the fact that non-CpG methylation is not maintained by DNMT1, and that somatic cells exhibit only low levels of ongoing *de novo* methylation.

B. DNMT3A and DNMT3B Drive DNA Methylation *De Novo*

A key step in the understanding of DNMT3A and DNMT3B function came with the development of targeted gene replacements in mouse. *DNMT3A*-deficient mice survive to term but are runted and die within 4 weeks, showing signs of gastrointestinal dysfunction, gait ataxia, and reduced germ cell counts in males. *DNMT3B*-deficient mice die during development around midgestation due to multiple developmental defects including rostral tube defects, liver hypotrophy, ventricular septal defects, and hemorrhaging.[26,41] However ($DNMT3A^{-/-}$, $DNMT3B^{-/-}$), double knockout embryos show profound embryonic defects, lacking somites and failing to undergo embryonic turning, and die prior to E11.5. Thus, DNMT3A and DNMT3B have critical and partially overlapping functions during early embryogenesis. In terms of DNA methylation, DNMT3B-deficient embryos showed demethylation of minor satellite repeats. Similar demethylation was observed in ($DNMT3A^{-/-}$, $DNMT3B^{-/-}$) embryos, but not in single $DNMT3A^{-/-}$ animals, indicating that minor satellite repeats represent a specific target for DNMT3B. Other dispersed repetitive elements such as C-type retroviruses and IAP elements showed demethylation in double knockout embryos, although the extent of demethylation was significantly lower than was observed in $DNMT1^{-/-}$ embryos.

Single and double mutant ESC lines derived from the knockout animals were particularly useful in assigning functional roles to DNMT3A and DNMT3B relative to DNMT1. Okano and colleagues showed that ($DNMT3A^{-/-}$, $DNMT3B^{-/-}$) double mutant ESCs fail to catalyze the methylation of proviral DNA *de novo*, indicating that DNMT3A and DNMT3B are responsible for establishing DNA methylation patterns at least at this target.[26] More globally, reintroduction of DNMT3A, DNMT3A2, or DNMT3B1 in highly demethylated double mutant ESCs restored DNA methylation patterns to both tandem and dispersed repeats as well as to several single-copy genes across the genome. However, reintroduction of DNMT1 failed to do so.[42] These observations clearly establish that DNMT3A and DNMT3B are the main engines of *de novo* DNA methylation in mammals. In support of this notion, expression of DNMT3A in the heterologous host *Drosophila melanogaster* resulted in the deposition of DNA methylation patterns *de novo*, while expression of DNMT1 did not.[43] Likewise, expression of DNMT3A or DNMT3B can drive the deposition of DNA methylation patterns on unmethylated episomal constructs in human cells in culture,[38] but expression of DNMT1 does not (F. Chédin, unpublished observations). These studies support a model whereby DNMT3A and DNMT3B establish DNA methylation patterns *de novo* and DNMT1—in concert with its partner UHRF1[44,45]—functions in maintaining these patterns upon replication owing to its preference for hemimethylated substrates and association with the replication fork complex.[46]

It should be noted, however, that the division of labor between DNMT3 proteins and DNMT1 in performing *de novo* versus maintenance methylation is not strict (see also Chapter by Željko M. Svedružić). Multiple lines of investigation indicate that DNMT3A and DNMT3B may also contribute to maintenance methylation. For instance, while double mutant ESC lines only show modest demethylation at low-passage, prolonged culturing of these cells causes a progressive and global loss of methylation at various repeated DNA elements and single-copy genes.[42] At high passages, the extent of demethylation is close to that observed for DNMT1-deficient animals (although it should be noted that the absence of DNMT1 leads to severe and immediate loss of methylation[47,48]). This suggests that DNMT3A and DNMT3B are required for the long-term stability of DNA methylation patterns at multiple regions across the genome. Evidence for the presence of hemimethylated DNA in (*DNMT3A* $^{-/-}$, *DNMT3B* $^{-/-}$) mutant ESCs[49] suggests a model whereby ongoing *de novo* DNA methylation by DNMT3A and DNMT3B serves to compensate for inefficient maintenance methylation by DNMT1.[50,51]

C. DNMT3A Plays a Specialized Role in Germ Cells and Neurons

Conditional knockout technology was used to study the roles of *DNMT3A* and *DNMT3B* in germ cell methylation.[52] Importantly, *DNMT3A* ablation resulted in clear but distinct phenotypes in male and female animals. Males displayed impaired spermatogenesis leading to azoospermia, and loss of DNA methylation at several paternally methylated imprinting control regions. In contrast, females produced mature oocytes, but their fertilization by wild-type sperm never resulted in live birth. Instead, the embryos derived from these crosses died around E10.5 and showed multiple developmental defects, such as open neural tube and lack of branchial arches.[52] DNMT3A-deficient oocytes showed loss of methylation at dispersed IAP repeats, indicating that DNMT3A plays an important role in silencing this class of repeated elements. In embryos, loss of methylation was clearly observed at multiple maternally methylated imprinting centers but not IAP elements, indicating that DNMT3A plays a key role in imprinting and that loss of methylation at IAP elements in oocytes can be rescued during early development while defects at imprinted loci cannot.[52] Strikingly, conditional knockouts of *DNMT3B* did not lead to any phenotype, suggesting that DNMT3A plays the dominant role in establishing DNA methylation patterns in mouse germ cells.

In mouse, *Dnmt3a* is expressed in multiple neuronal tissues during postnatal development and early adulthood, suggesting that it may be important for the maturation of the central nervous system (CNS).[53,54] In humans, *DNMT3A* is also expressed in a subset of cortical neurons, and evidence indicates that DNA methylation is dynamically regulated in the cerebral cortex throughout life.[55] Mice lacking *Dnmt3a* in the entire CNS, while born healthy, showed reduced life span and displayed hypoactivity, gait abnormalities, and lower neuromuscular function.[56] This suggested that Dnmt3a is implicated in the neuromuscular control of motor movement.[56] More recently, *Dnmt3a*, together with *Dnmt1*, was shown to regulate synaptic plasticity, learning, and memory in forebrain excitatory neurons.[57] Likewise, *Dnmt3a* was implicated in regulating cellular and behavioral plasticity in response to emotional stimuli.[58] While the manner by which DNMT3A affects these complex processes is not entirely clear, recent evidence suggests that DNMT3A may paradoxically favor expression of neurogenic genes.[54] DNMT3A was indeed reported to bind to and methylate DNA sequences flanking the promoters of a subset of highly expressed neurogenesis genes in neural stem cells. Such nonpromoter methylation was proposed to facilitate transcription by preventing Polycomb group complexes from depositing silencing chromatin marks.[54] How DNMT3A is recruited to target genes and how it antagonizes Polycomb group complexes are important questions that remain to be answered.

It is worth noting that in addition to a role in neuronal function, recent evidence establishes that human *DNMT3A* is frequently mutated in acute myeloid leukemia (AML).[59]

D. DNMT3B is Mutated in ICF Syndrome

Immunodeficiency, centromeric instability, and facial anomalies syndrome (ICF; OMIM: 242860) is a rare autosomal recessive disorder characterized by immunoglobulin deficiency and facial dysmorphism (see Chapter by Jon F. Wilkins and Francisco Úbeda). Most patients affected by ICF syndrome succumb to opportunistic infections early in life. At the molecular level, ICF patients exhibit severe DNA hypomethylation of classical satellite tandem repeats on chromosomes 1, 9, and 16.[60] This defect leads to the formation of abnormal cytogenetic structures such as branched multiradial structures in which chromosomes 1, 9, and 16 appear fused at their centromeric regions.[61] Such chromosomal aberrations are most apparent in T lymphocytes that have undergone stimulation, despite the fact that demethylation is observed in all tissues. Several groups traced these defects to a series of compound heterozygous mutations in the *DNMT3B* gene.[26,31,61] In nearly all cases, these mutations map to the DNA methyltransferase catalytic domain residing in the C-terminal portion of DNMT3B. These mutations appear to reduce DNMT3B catalytic activity.[62,63] Significantly, ICF mutations must not fully inactivate DNMT3B since such a drastic change leads to embryonic lethality in mouse while ICF individuals survive to term.[41] At the molecular level, analysis of lymphoblastoid cell lines derived from ICF patients has revealed that ICF mutations are associated with changes in expression of a subset of genes associated with immune function, development, and neurogenesis.[64] Interestingly, ICF mutations were only associated with subtle changes in DNA methylation patterns at a subset of upregulated genes, with some genes showing no significant change over the regions analyzed.[64] This suggests that the impact of DNMT3B deficiency may not be felt at the level of DNA methylation for most genes. However, a significant subset of upregulated genes in ICF cell lines showed a loss of the repressive histone H3 lysine 27 trimethylation mark, and associated gain of activating marks.[64] This suggests that DNMT3B function is integrated with chromatin-based modifications, particularly those mediated by Polycomb group proteins.

A recent report, that two ICF mutations may not affect DNMT3B catalytic activity but rather its ability to interact with and be stimulated by DNMT3L,[63] is in doubt since these mutations are located far from the DNMT3L:DNMT3B interaction surface.[65] Our unpublished observations indicate that these two mutations do affect DNMT3B catalytic activity rather than its functional interaction with DNMT3L (Amir H. Moarefi and F. Chédin, unpublished). This suggests that DNMT3L is not involved in the pathology of ICF syndrome. Interestingly, however, a significant portion of ICF cases with satellite DNA hypomethylation is not associated with mutations in *DNMT3B*.[66] This suggests that additional genes controlling the deposition or maintenance of DNA methylation patterns may exist.

E. Functional Characterization of DNMT3A and DNMT3B

1. Catalytic Mechanism

The catalytic mechanism of bacterial C5-cytosine methyltransferases has been well defined.[67,68] In brief, the sulfhydryl group of a catalytic cysteine residue (motif IV) mediates a nucleophilic attack on carbon 6 of the target cytosine, thus resulting in a covalent DNMT–cytosine intermediate. This event is facilitated by the protonation of the N3 position by a conserved glutamate residue (motif VI) and involves the eversion of the target cytosine from the DNA double helix into the enzyme's active site pocket.[69] The deprotonation of N3 then activates the C5 position, leading to a second nucleophilic attack, this time on the methyl group carried by S-adenosyl-L-methionine (AdoMet, the universal methyl group donor). This complex is finally resolved upon deprotonation of the C5 position which liberates the enzyme, methylated DNA, and S-adenosyl-homocysteine as reaction by-products. Inspection of protein sequence alignments indicate that the catalytic motifs of DNMT3A and DNMT3B have evolved from a common ancestor with bacterial C5-cytosine methyltransferases.[16,70] In particular, key residues such as the catalytic cysteine in motif IV and the glutamate in motif VI are strictly conserved. Recent evidence further shows that the structure of the DNMT3A catalytic domain in complex with the DNMT3L cofactor is very similar to the bacterial M.HhaI enzyme.[65] Computational simulations in which a flipped-out cytosine target is docked onto the DNMT3A active site in the presence of AdoMet demonstrates that the geometry expected for the canonical C6 nucleophilic attack is realized.[15] Accordingly, all expectations are that the mammalian DNMT3A and DNMT3B enzymes use the same catalytic mechanism as the bacterial enzymes. In support of this, multiple independent groups have reported that mutating the conserved catalytic cysteine abolishes function *in vivo* and *in vitro*[27,38,42,71](Amir H. Moarefi and F. Chédin, unpublished observations). The existence of an alternative catalytic mechanism independent of a covalent intermediate[72] seems highly unlikely.

2. Sequence Specificity

Most bacterial DNA methyltransferases are sequence-specific modification enzymes. This opens the possibility that DNMT3A and DNMT3B might also possess a preference for particular sequences flanking their target CpG sites. Analysis of the complete human methylome did not reveal any flanking preference for CpG methylation,[5] which is not surprising given the prevalence of methylation across millions of CpG sites. Indeed, when focusing on less frequent non-CpG methylation events, Lister *et al.*[5] report observing significant compositional biases at the two positions flanking the target cytosines. Deciphering whether these biases result from DNMT3A or DNMT3B action requires a more defined system. Using a full-length GST-tagged mouse DNMT3A1 protein purified from human cells, Lin *et al.*[71] reported that DNMT3A1 shows a modest but significant flanking sequence preference. In this system, pyrimidines were preferred over purines at positions -2 and $+1$ flanking the target CpG site.[71] A subsequent study reported that, in contrast, the murine catalytic domains of DNMT3A and DNMT3B show a profound and shared preference for certain flanks.[73] Here, the discrimination between good and bad flanks was mediated by the composition at the -1 and $+1$ positions (purines at -1 and pyrimidines at $+1$ constituting a good flank) with additional influence for sequences located at the ± 2, 3, and 4 positions. This preference reportedly drove a highly unlikely >500-fold difference in methylation rates between good and bad flanks,[73] although this number was later revised sharply down, to three- to fivefold.[74]

Using an episomal assay for *de novo* methylation, we recently showed that DNMT3A and DNMT3B possess significant and distinct sequence preferences *in vivo*.[75] In close agreement with Lin *et al.*,[71] we reported that DNMT3A discriminates between good and bad flanks by the composition at the -2 and $+2$ positions around the target CpG site (*in vitro*, the preference shifted to -2 and $+1$, as described by Lin *et al.*[71]). In contrast, DNMT3B selects target CpG sites through the -1 and $+1$ positions. The selectivity between high and low sites under this experimental system ranged from five- to eightfold with DNMT3B appearing more specific than DNMT3A. These preferences are modest enough that they are compatible with the methylation of a wide variety of CpG sites, particularly since we also showed that DNMT3L attenuates the effects of sequence preference.[75] At the same time, these preferences are strong enough that they influence DNA methylation patterns, and may have contributed to the evolution of human CpG island promoters. Indeed, we reported that these regions appear enriched for DNMT3B coldspots and depleted for DNMT3B hotspots, which might favor their maintenance as unmethylated genomic islands. The observation that DNMT3A and DNMT3B show distinct preferences is compatible with the fact that these two proteins show the most divergence in their catalytic domains for a region located between motifs VIII and IX which is thought to correspond to the target recognition domain (TRD).[70]

3. *De Novo* Activity Leads to Hemimethylation

As discussed above, *de novo* methylation by DNMT3A and DNMT3B shows modest but significant intrinsic sequence preferences. Given that the preferred sites are not palindromic, one expects that DNMT3A/B activity may lead to the establishment of strand-specific hemimethylation patterns. In keeping with this, prior studies have reported the formation of strand-specific methylation patterns. Hence, analysis of *de novo* methylation of viral sequences in DNMT1-deficient ESCs revealed clear strand specificities in methylation patterns.[76] More globally, DNMT3A and DNMT3B function in DNMT1-deficient ESCs leads to a high load of hemimethylated sites, particularly at pericentric regions, as seen by strong UHRF1 staining (UHRF1 specifically recognizes hemimethylated CpG sites through its specialized SRA domain[45,77]). Finally, biochemical assays using purified murine DNMT3A clearly led to strand-specific patterns reflective of hemimethylation.[71] The deposition of hemimethylation patterns in the context of early development, during which cells rapidly divide, offers a challenge in terms of establishing uniform methylation patterns. Indeed, if maintenance DNA methylation is exclusively associated with DNA replication—as most models for maintenance methylation suggest—then daughter cells will inherit variable DNA methylation patterns (Fig. 2A). Since DNA methylation plays a critical role in silencing transposable elements [48,78] and contributes to gene regulation during development (Chapters by Anton Wutz; Pierre-Antoine Defossez and Irina Stancheva),[8] the formation of variable methylation patterns is a dangerous proposition. Instead, we suggest that maintenance methylation can occur independently of DNA replication so as to rapidly convert hemimethylation into full methylation prior to the passage of the replication fork (Fig. 2B). The combination of replication-associated and replication-independent maintenance methylation would ensure the inheritance of uniform, highly methylated patterns. Recent evidence shows that DNMT1 can function at least partially outside the replication fork (Chapter by Željko M. Svedružić).[79] Likewise, the critical methylation maintenance factor UHRF1 (Np95) was shown to physically interact with DNMT3A (Chapter by Jafar Sharif and Haruhiko Koseki) and DNMT3B,[80] thus suggesting that it could recruit the DNA methylation machinery at hemimethylated sites in order to ensure full methylation. However, rapid and efficient maintenance methylation outside the replication fork remains to be directly documented.

4. CpG Spacing

Structural analysis of the human DNMT3A catalytic domain, in complex with DNMT3L, revealed that DNMT3A exists in a dimeric arrangement in the complex (Fig. 1) with the active sites of the monomers spaced by 40 Å.[65] While the exact configuration of the DNA in the cocrystals could not be solved, modeling studies suggested that a single DNA molecule might run across both active sites. Under this proposal, methylation would occur in concert at two CpG sites, located across both strands and separated by roughly one helical turn of DNA.[15,65] This model suggests that CpG spacing might be an important factor in setting DNA methylation patterns. Early support for this notion came from the analysis of the sequences of a limited set of maternally methylated imprinting regions which are known DNMT3L targets. Indeed, these regions showed a modest enrichment for CpG sites spaced by 8–10 bp.[65] However, a more comprehensive analysis of CpG periodicity in the human and mouse genomes concluded that the role played by CpG spacing, if any, cannot explain many aspects of CpG methylation patterning.[81] For instance, CpG-dense regions, which account for the majority of properly spaced CpG dyads, are rarely methylated. Likewise, L1 LINE elements, which are highly methylated and respond to DNMT3L deficiency,[82] do not show enrichment for a 8- to 10-bp CpG periodicity.[81] While

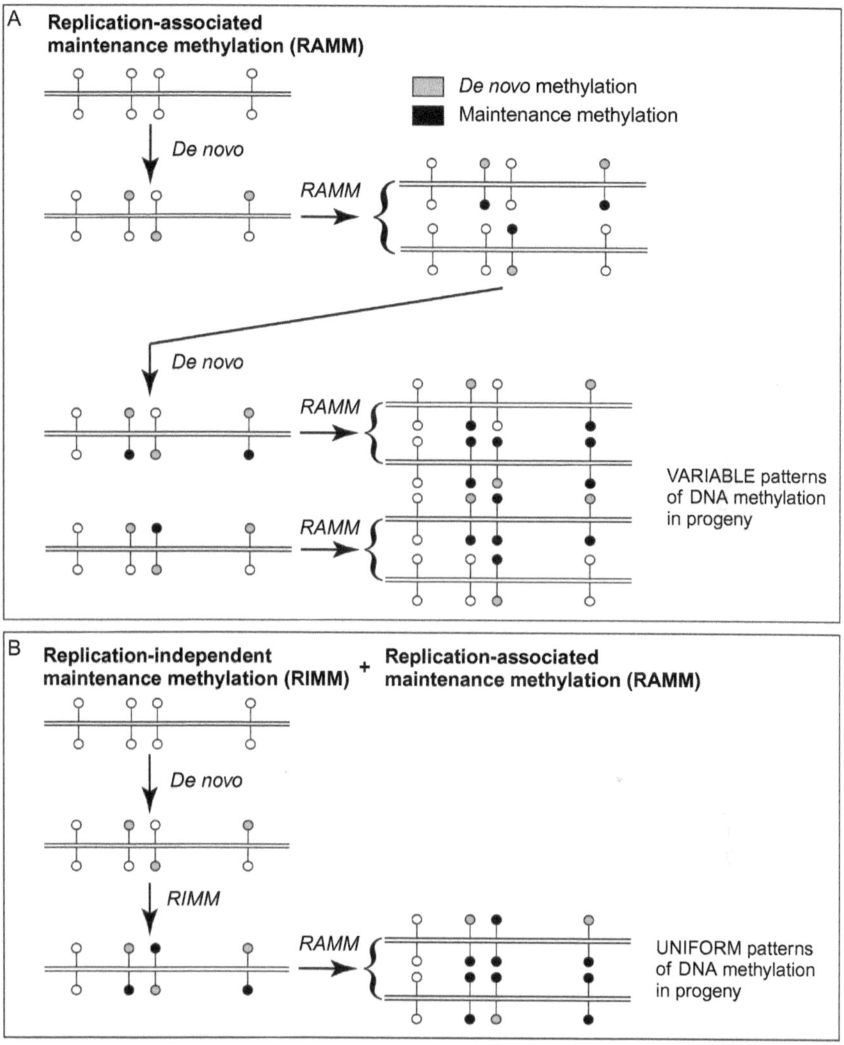

FIG. 2. The deposition of hemimethylated patterns suggests the existence of an efficient replication-independent maintenance methylation system. CpG sites are represented as lollipops. Open symbols correspond to unmethylated sites, while filled-in symbols represent methylated CpGs. Gray color is used to represent the sites that were methylated *de novo* while black is used to indicate maintenance methylation. Panel (A) describes the classic postreplicative maintenance methylation system and shows how it is expected to generate variable genomic methylation patterns in progeny. Panel (B) shows how a proposed replication-independent maintenance methylation system might help to correct the variability and result in the inheritance of uniform patterns in progeny. See text for details.

weak biochemical evidence in support of an influence of spacing has been presented,[65,83] other recent studies of the activity of DNMT3A and DNMT3B (in the presence of DNMT3L or not) have failed to detect an influence of CpG spacing[75,84] (see also Chapter by Eleanor Wong and Chia-Lin Wei).

Interestingly, analysis of the complete human methylome by autocorrelation analysis has suggested that methylated cytosines in the non-CpG context (CHG and CHH, where H = A, C, or T) are preferentially separated by 8–10 bp or multiple units thereof.[5] More recently, a similar 10 bp periodicity has been reported for CpG methylation of nucleosomal DNA in the human genome.[8] A similar patterning was also observed for methylation of cytosines in the CHH context in the model plant, *Arabidopsis thaliana*,[85] suggesting deep biological conservation. Analysis of the origin of this periodicity in the context of nucleosomal DNA—the main target for *de novo* methylation in human and plants—suggests that it likely results from constraints placed onto DNA sequence by the necessities of wrapping DNA around nucleosomes.[8] According to this proposal, the well-known 10 bp periodicities in the distribution of SS (S = A or T) and WW (W = G or C) dinucleotides around nucleosomes[86] might serve to regulate access of DNMTs to methylatable CpG sites and thus determine the registry of methylation. It is therefore possible that the methylation periodicities observed *in vivo* are independent from the structural architecture of the DNMT3A/B active sites. Addressing how DNA is bound to the active site of DNMT3A and DNMT3B, through biochemical and structural assays, remains a high priority in order to improve our understanding of the function of these enzymes. Likewise, studies are needed to address the contribution of DNMT3A and DNMT3B sequence-nonspecific PWWP domains[87,88] to the regulation of catalytic efficiency and/or specificity.

5. A Role for Oligomerization?

As discussed above, DNMT3A forms a catalytically active dimer in the DNMT3A:DNMT3L complex. In addition, DNMT3A and DNMT3B are capable of further homo-oligomerization in the absence of DNMT3L, forming large and heterogeneous complexes in solution[65,87,89] (Amir H. Moarefi, Catherine Gordon, and F. Chédin, unpublished observations). *In vivo* studies indicate that DNMT3A and DNMT3B can form mixed complexes.[27,41,90] The formation of such complexes is unusual among C5-DNA methyltransferases, and the significance of these complexes remains unclear. Mutational studies indicate that dimerization is key to the catalytic function of DNMT3A and DNMT3B[65,83] (Amir H. Moarefi and F. Chédin, unpublished observations), indicating that the minimal unit capable of catalytic activity corresponds to a dimer. The function of higher order oligomers, however, remains elusive. One possibility is that large oligomers allow the binding of multiple "subunits" to a single DNA molecule, thereby enabling high efficiency methylation of adjacent CpG sites. This hypothesis is consistent with the fact that most methylated DNA regions in the human genome are highly methylated and that very few regions show intermediate methylation levels.[5] This view is also consistent with recent evidence that DNMT3A, and probably DNMT3B, acts processively on DNA.[91]

IV. The DNMT3L Protein, an Essential Accessory Factor for *De Novo* Methylation

A. Discovery, Structure, and Expression of DNMT3L

The DNMT3-like protein DNMT3L was discovered on the basis of the similarity of its N-terminal cysteine-rich domain to the corresponding domains of DNMT3A and DNMT3B[92,93] (Fig. 3A). The C-terminal domain of DNMT3L also resembles the catalytic domain of DNMT3A and DNMT3B, but is truncated and is missing key residues mapping to critical catalytic motifs. Not surprisingly, DNMT3L on its own is inactive as a DNA methyltransferase. Expression studies in the mouse, however, revealed that *DNMT3L* is expressed precisely at key stages of male and female germ cell development, characterized by high rates of *de novo* DNA methylation.[28,32,34,82,94–96] *DNMT3L* is also highly expressed in mouse and human ESCs,[29] which are characterized by high *de novo* methylation activity.[19,97] This suggested that DNMT3L may play a role in methylation despite its apparent lack of activity. In somatic cells, *DNMT3L* is either not expressed or expressed at low levels, with evidence suggesting that *DNMT3L* is regulated primarily at the transcriptional level and that repression is accompanied by methylation of its own promoter region.[28,98,99]

Structural characterization of the human DNMT3L protein by X-ray crystallography revealed that its C-terminal region adopts a methyltransferase fold characteristic of C5-DNMTs (Fig. 3B). It was noted, however, that its presumptive DNA-binding region was largely lacking.[100] Biochemical studies confirmed that DNMT3L binds to DNA only weakly and does not bind AdoMet.[89] Its conserved cysteine-rich N-terminal domain defines a distinct globular domain linked to the C-terminal by a small flexible linker. This N-terminal region contains a zinc-finger domain that coordinates three Zn atoms through a series of six CxxC motifs.[100] Conservation and structural arguments indicate that this domain resembles the broadly distributed class of PHDs, which serve as chromatin-interacting zinc-finger domains. However, due to the particularly high conservation of this region within the DNMT3 family and the ATRX SWI/SNF chromatin remodeling factor, it was termed an ADD domain (ATRX-DNMT3-DNMT3L).[101,102] The ADD domain directly interacts with histone H3 and links DNA methylation with chromatin,[100] as described further below.

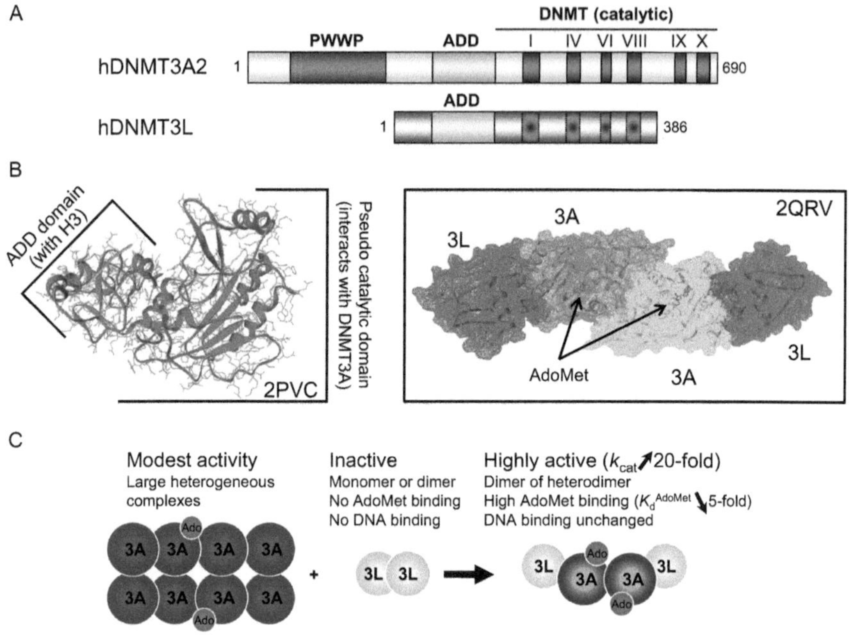

FIG. 3. Panel (A) shows a schematic representation of the human DNMT3A and DNMT3L proteins. Conserved domains are indicated as in Fig. 1. Panel (B) summarizes available structural information available for the DNMT3L protein (shown at left bound to a histone H3 peptide[100]) and for the DNMT3A:DNMT3L complex (shown at right as a ribbon diagram with a calculated 3D surface[65]). Panel (C) depicts a model for DNMT3L function. See text for details.

B. Conservation of DNMT3L Through Evolution—Coevolution with Imprinting

While DNMT3A and DNMT3B orthologs are distributed widely through evolutionarily distant species,[24] clear DNMT3L orthologs appear to be more restricted. For instance, while DNMT3A/B orthologs exist in fish and chicken, *DNMT3L*-type genes have been found only in eutherian mammals (from primates to rodents) and some marsupials (opossum)[103] (see Chapter by Lakshminarayan M. Iyer *et al.*). The platypus, a member of the monotremes (egg-laying mammals), carries a DNMT3A/B ortholog but no DNMT3L equivalent. Interestingly, the distribution of DNMT3L orthologs among species was noted to mirror the one of genomic imprinting, the process by which a specific subset of genes acquire parent-of-origin-specific epigenetic marks which ensure monoallelic maternal or paternal expression.[103] This led to the suggestions that DNMT3L may have coevolved with genomic imprinting, perhaps upon the emergence of placentation and *in utero* fetal growth. Interestingly too, DNMT3L homologs seem to evolve much faster than their DNMT3A counterparts, even when only the noncatalytic portions of DNMT3A are taken into account (F. Chédin, unpublished observations). Given the role of DNMT3L in ensuring the methylation of a diverse and fast-evolving class of transposable elements and other repeats (see below), it may not be surprising that DNMT3L is evolving fast too. However, as noted above, biochemical studies suggest that DNMT3L does not bind DNA, which makes the rapid evolution of DNMT3L members more difficult to grasp. This opens up the possibility that DNMT3L may possess an as-yet uncharacterized function in preventing transposon spreading prior to their integration in the genome. Interestingly, the *MIWI2* and *MILI* genes, which are involved in the piRNA-mediated transposon defense pathway, are expressed in male prospermatogonia during the wave of *de novo* methylation.[104] Moreover, knockouts of these genes result in a failure to methylate Line-1 and IAP retrotransposons and a global reactivation of the expression of these elements.[105,106] Thus, MILI and MIWI2 deficiency strikingly resembles DNMT3L deficiency, suggesting that DNMT3L may interface with the piRNA pathway of transposon suppression.[104] How these pathways interconnect at the molecular level is currently unclear but will likely reveal important insights into how parasitic DNA elements are managed by developing germ cells (see also Chapter by Mary G. Goll and Marnie E. Halpern).

C. DNMT3L Plays an Essential Role in *De Novo* DNA Methylation

Gene knockout studies clearly established the importance of DNMT3L in *de novo* DNA methylation. *DNMT3L*$^{-/-}$ animals are viable and largely asymptomatic, indicating that DNMT3L does not play an important postzygotic role in mouse. However, most offspring from crosses between *DNMT3L*$^{-/-}$ females and wild-type males die during development.[94,95,107] These embryos showed delayed growth, neural tube defects, and abnormal placental structures, likely due to improper expression of multiple imprinted genes including *SNRPN*, *NDN*, *p57Kip2*, *Airn*, *IGF2R*, *ZFP127*, *PEG1*, *PEG3*, *IPL*, and *CDKN1*.[94,95,108] Oocytes from *DNMT3L*$^{-/-}$ females show lack of DNA methylation at maternally methylated imprinting centers such as *SNRPN*, *Airn*, and *PEG3*.[94,109] This deficiency results in either biallelic or nonexpression of downstream imprinted genes, indicating that DNMT3L is critical for the establishment of maternal methylation imprints during murine oogenesis. Interestingly, *DNMT3L* expression has so far not been detected in human oocytes while it is clearly expressed in early development.[9,29] This suggests that imprints may be acquired postzygotically in human, as demonstrated by some studies,[110,111] though evidence that imprints are acquired during oogenesis has also been presented[112] (see Chapter by Jon F. Wilkins and

Francisco Úbeda). A recent study is consistent with DNMT3L possessing a somatic role in humans, showing that a rare DNMT3L mutation that perturbs interaction with DNMT3A is associated with hypomethylation at subtelomeric regions in DNA from adult blood cells.[113]

DNMT3L$^{-/-}$ males, in contrast to females, are sterile; they show reduced testis size and complete azoospermia.[82,95,114] This extreme phenotype results from the inability of spermatogonial cells to complete meiosis, as evidenced by a global failure of chromosomal synapsis during the leptotene and zygotene phases. These defects, in turn, trigger a wave of apoptosis which progressively ablates germ cell populations.[82,114] Interestingly, DNMT3L in males is only expressed during a brief developmental window in fetal nondividing prospermatogonial cells prior to meiosis, suggesting that the meiotic defect observed in adult spermatocytes was initiated long before in a population of precursor cells.[82] In contrast, the defects observed in oocytes from *DNMT3L*$^{-/-}$ females occur during the dictyate stage of meiosis at which point oocytes enter a quiescent growth phase. Thus, DNMT3L function shows profound sexual dimorphism.[115] Not surprisingly, DNA methylation in *DNMT3L*-deficient spermatocytes is profoundly deficient: in addition to reduced DNA methylation at some paternally methylated imprinted genes, dispersed repeated DNA elements such as LINE-1 and IAP elements show dramatic loss of DNA methylation.[82,114] The hypomethylation of these repeats leads to their transcriptional reactivation and mobilization, highlighting the role played by DNMT3L in suppressing parasitic DNA elements[82,114] (see Chapter by Mary G. Goll and Marnie E. Halpern). Further analysis of DNA methylation defects in DNMT3L-deficient spermatocytes revealed that DNMT3L also plays a role in methylating a large class of nonrepetitive sequences dispersed throughout the genome[116] and that it is also involved in methylating centromeric satellite repeats.[117] While similar defects in repeat methylation were initially not observed in *DNMT3L*-deficient oocytes, recent results convincingly show that DNMT3L plays a similar role in female germ cells.[109] Likewise, global DNA hypomethylation at repeated sequences was observed in DNMT3L$^{-/-}$ ESCs over multiple passages.[118] DNMT3L is therefore critically and broadly required for setting DNA methylation patterns genome wide in germ cells and for maintaining them in early development.

D. DNMT3L Physically Interacts with, Structurally Reorganizes, and Functionally Stimulates DNMT3A and DNMT3B

Coimmunoprecipitation experiments indicated that DNMT3L interacts with DNMT3A and DNMT3B, leading to the suggestion that DNMT3 family members cooperate in the deposition of DNA methylation patterns.[95] Further experiments confirmed that the DNMT3L pseudocatalytic domain mediates physical interactions with the catalytic domain of DNMT3A and DNMT3B.[119–121] This is so far one of the only two known protein–protein interactions that involve the catalytic domain of DNMT3A and DNMT3B (the other one being the interaction with thymine-DNA glycosylase[122,123]). Using a cell culture assay, DNMT3L was reported to significantly stimulate the activity of DNMT3A and DNMT3B.[119,124] Initial biochemical analyses confirmed that while DNMT3L does not bind to DNA or AdoMet, it does physically interact with DNMT3A and stimulates its activity.[121,125] Further careful analysis showed that the full-length human DNMT3L protein stimulates the catalytic rate of the full-length human DNMT3A2 enzyme (up to 20-fold) and that the optimal stimulation is reached at a 1:1 molar stoichiometry of DNMT3L to DNMT3A.[89] This strongly implies that DNMT3L and DNMT3A form heterodimeric complexes. Further, while DNMT3L did not modify binding of DNMT3A to DNA, it significantly increased its affinity for AdoMet,[89] thereby rendering the DNMT3A:DNMT3L complexes better able to compete with numerous other methyltransferases for this universal cofactor.

Importantly, these studies also showed that DNMT3A and DNMT3L exist in different oligomeric states. Recombinant DNMT3A purified from *Escherichia coli* forms large and stable soluble aggregates.[65,87,89] These aggregates result from the ability of DNMT3A to interact with itself through at least two interfaces.[15,65] Whether or not they represent a physiologically relevant form of DNMT3A is hard to determine, though evidence from immunoprecipitation experiments suggests that DNMT3A and DNMT3B form homo-oligomers and even mixed DNMT3A:DNMT3B hetero-oligomers *in vivo*.[41,126] Whether these oligomers have a particular function is also hard to discern, though recent evidence suggests that purified DNMT3A oligomers are only partially active,[87] suggesting that oligomerization could be a way to regulate the activity of DNMT3 proteins. DNMT3L, in contrast, is mostly observed as a homodimer.[65] Importantly, DNMT3L has the unique ability to reorganize DNMT3A oligomers into distinct DNMT3A:DNMT3L heterodimeric complexes.[89] Crystallographic evidence, combined with gel filtration and sedimentation analyses, showed that these DNMT3A:DNMT3L complexes correspond to dimers of heterodimers (2:2)[65,83] (Fig. 3B). DNMT3L therefore plays a structural role in addition to its functional role, by reorganizing DNMT3A into specific complexes with enhanced DNA methyltransferase activity and increased AdoMet binding (Fig. 3C).

In vivo episomal methylation analysis suggests that DNMT3L does not change the flanking sequence preference of DNMT3A or DNMT3B,[75] consistent with the evidence that DNMT3L does not bind to DNA. However, inclusion of DNMT3L led to the deposition of more uniform methylation patterns, which is consistent with patterns observed *in vivo* at known DNMT3L targets.[82,94] The greater uniformity of DNA methylation patterns appears to be due to DNMT3L preferentially enhancing methylation of CpG sites that were intrinsically poorly methylated by DNMT3A or DNMT3B on their own (coldspots).[75] This may not reflect modulation of the sequence preference by DNMT3A and DNMT3B, but rather result from DNMT3L anchoring its partner proteins to nucleosomes, thus increasing the dwell time of the enzyme on its DNA substrate. A methylation coldspot is much more likely to benefit from increased residence time, compared to a hotspot that is normally rapidly and efficiently methylated. Interestingly, purified DNMT3L: DNMT3A complexes also appear to oligomerize further forming nucleoprotein filaments on naked DNA substrates.[83] Such filaments might account for long methylation patterns across multiple CpG sites.[11] Recently, DNMT3L was indeed suggested to increase the processivity of DNMT3A.[91] It remains to be investigated whether oligomerization and processivity are somehow coupled and how these processes may be relevant to the deposition of methylation patterns on nucleosomal templates (the physiologically relevant substrate for *de novo* methylation).

V. Interactions Between the *De Novo* DNA Methylation Machinery and Chromatin

DNA in mammalian cells exists in the context of nucleosomes, and multiple lines of evidence suggest that *de novo* methylation *in vivo* occurs in the context of nucleosomes. Careful analysis of the distribution of DNMT3A and DNMT3B within nuclei, via sedimentation analysis, revealed that these two proteins strongly associate with nucleosomes.[127,128] In contrast, DNMT1 only showed a weak and perhaps transient association with chromatin. Anchoring active DNMT3 proteins to chromatin required intact nucleosomes and did not depend on any of the tested secondary chromatin-binding proteins (EZH2, HP1, UHRF1, MeCP2, HDAC1).[127] This suggests that DNMT3A and DNMT3B directly bind to nucleosomes, though the relative contributions of DNA binding and histone binding remain to be clarified.

A. The Conserved ADD Domain Interacts with Histone H3 and Discriminates Against Trimethylation of Lysine 4

Immunoprecipitation of tagged DNMT3L in mouse ESCs brought down all four core histones in addition to DNMT3A and DNMT3B, suggesting that these enzymes physically interact with chromatin.[100] Binding studies clearly established that DNMT3L itself interacted with chromatin and that this interaction involved the N-terminal tail of histone H3. Further, the use of chemically modified peptides revealed that binding to H3 was severely compromised by methylation of the histone H3 lysine 4 (H3K4) residue (mono- > di- > trimethylation).[100,129] Crystallographic data established that the binding of DNMT3L to the H3 tail is mediated by the cysteine-rich ADD zinc-finger domain and that the H3K4 position in particular is coordinated by two conserved aspartic acid residues that cannot accommodate modification of H3K4 due to steric constraints.[100] As discussed above, the ADD domain is highly conserved between DNMT3A, DNMT3B, and DNMT3L. This suggests that the corresponding ADD domains of DNMT3A and DNMT3B may also bind to the tail of histone H3 and sense the modification state of H3K4. Clear biochemical and structural data support this view,[102] despite an earlier report that the DNMT3A ADD domain binds to the symmetrically dimethylated H4R3 residue.[130] These results therefore indicate that all three members of the DNMT3 family are capable of binding to the histone H3 N-terminal tail and are sensitive to the modification of H3K4. This, in turn, suggests that *de novo* DNA methylation should correlate with chromatin regions carrying an unmodified H3K4 residue[100] and that, conversely, nucleosomes methylated at H3K4 should be refractory to DNA methylation. Such an inverse relationship between H3K4 methylation and DNA methylation is clearly supported.[9,131–133] This anticorrelation is particularly relevant in the case of CpG islands, which coincide with clear peaks of H3K4 trimethylation,[134] and remain largely protected from DNA methylation despite their high CpG density.

The redundancy in ADD domain distribution among DNMT3A, DNMT3B, and DNMT3L raises interesting issues regarding the function and eventual hierarchy of each domain, particularly in the context of the DNMT3A:DNMT3L complexes. H3K4-binding-deficient DNMT3L variants show reduced *de novo* methylation activity *in vivo*,[118,132] arguing that nucleosome binding by DNMT3L is important. It will be key to determine the relative contribution of the DNMT3A and DNMT3B ADD domains to this process. Based on biochemical measurement of binding affinities,[102] the DNMT3A ADD domain is predicted to bind to the H3 tail with a significantly higher affinity than the DNMT3L ADD domain, suggesting that DNMT3A may play the primary role in engaging chromatin. Because *de novo* methylation in somatic cells is strongly anticorrelated with H3K4 methylation despite the absence of DNMT3L, DNMT3A and/or DNMT3B on their own probably bind to nucleosomes and discriminate against the presence of H3K4 modifications. Whether the two H3K4me0-binding domains present in DNMT3A: DNMT3L complexes show a binding hierarchy or enable the complex to either bridge several chromatin fibers or spread along chromatin are interesting possibilities that remain to be examined.

B. A Role for the PWWP Domain in Chromatin Interaction?

The conserved PWWP domain that is found in DNMT3A and DNMT3B (Fig. 1) is a member of the royal family of potential histone-binding domains.[22] Using cell culture assays, Chen *et al.*[135] showed that while murine DNMT3A and DNMT3B are preferentially located to pericentric heterochromatin, mutations in the PWWP domains abrogated this localization. Further, PWWP domain mutations abolished the methylation of major satellite repeats by DNMT3A and DNMT3B, suggesting that the PWWP domain is critical in targeting these enzymes to pericentric regions.[135] However, no evidence exists to date for direct binding of the PWWP domain to nucleosomes carrying modifications characteristic of pericentric heterochromatin (H3K9me3, H4K20me3). Recent data suggest that the DNMT3A (but not DNMT3B) PWWP domain binds to nucleosomes trimethylated at the H3K36 position.[136] Interestingly, H3K36me3 is a mark of transcriptional elongation, and this binding offers a possible mechanism to explain how gene bodies may become highly methylated, a common feature of the human methylome.[137] The exact mechanism(s) by which the PWWP domain may guide DNMT3A to both heterochromatic (major satellite) and euchromatic (transcribed genes) regions remain to be clarified. Likewise, it remains to be determined if the PWWP domain of DNMT3B can bind directly to nucleosomes and, if so, if it exhibits a preference for a specific chromatin modification(s).

C. Interaction Between DNMT3B and Components of the Pericentric Heterochromatin

As emphasized above, DNMT3B plays an important role in the methylation of centromeric and pericentric heterochromatin. A yeast two-hybrid screen identified a specific interaction between the N-terminal of DNMT3B and CENP-C, a constitutive centromere protein.[138] This interaction was confirmed by coimmunoprecipitation experiments using endogenous proteins in human cells, indicating that at least a portion of DNMT3B exists as a centromere-associated protein. The interaction between DNMT3B and CENP-C appears functionally relevant since knockdown of DNMT3B or CENP-C leads to a reduction of DNA methylation at alpha satellite sequence and pericentromeric satellite 2 repeats.[138] Interestingly, knockdown of DNMT3B or CENP-C also led to increased transcript levels at alpha satellite sequence and pericentromeric satellite 2 repeat, and led to an increase in misaligned chromosome and anaphase bridges in mitosis, suggesting that methylation and silencing of these regions are required for proper chromosome segregation.[138] There is so far no evidence that DNMT3A shows similar interactions, though it is interesting that while the full-length DNMT3A isoform is preferentially located to pericentric regions, the shorter DNMT3A2 isoform is located to euchromatin.[27] This argues for a role of the N-terminal portion of DNMT3A in specifying the pericentric location. Recent evidence suggest that incorporation of the linker histone, histone H1, to nucleosomal arrays favors association of DNMT3B to chromatin, further suggesting that DNMT3B may recognize several higher order features of heterochromatin.[128]

VI. Concluding Remarks and Outstanding Questions

Our understanding of the function of the DNMT3 family of proteins has increased considerably since their discovery about a decade ago. It is now clear that these proteins play key roles in setting DNA methylation patterns genome wide at specific developmental time points, particularly during germ cell differentiation and early development, or in specific tissues undergoing dynamic methylation such as neurons. Most recent studies consistently indicate that these proteins methylate DNA in the context of nucleosomes. Future biochemical and structural studies will thus need to focus on using appropriate nucleosomal templates, in order to assess the *in vivo* function of these proteins. Tackling the function of the DNMT3 proteins in the broader context of the various multiprotein complexes in which they participate will also be required. While these protein–protein interactions have not been specifically discussed here, they undoubtedly play important roles in DNMT3 function. In particular, understanding how DNMT3 proteins may be connected to mitotic chromosome condensation,[139] silencing during double-strand break repair,[140] and chromatin remodeling[141–143] are likely to bring important insights into epigenetic programs. More broadly, addressing the interconnections between the DNA- and the histone-based layer of epigenetics will be necessary to understand how DNA methylation is ultimately "read" and translated into biological output.

Acknowledgments

I thank members of my group, in particular, Paul Ginno for the critical reading of the chapter. Research in my laboratory is supported by a grant from the National Institutes of Health.

References

1. Holliday R, Pugh JE. DNA modification mechanisms and gene activity during development. *Science* 1975;**187**:226–32.
2. Riggs AD. X inactivation, differentiation, and DNA methylation. *Cytogenet Cell Genet* 1975;**14**:9–25.

3. Bird AP, Taggart MH. Variable patterns of total DNA and rDNA methylation in animals. *Nucleic Acids Res* 1980;**8**:1485–97.
4. Ehrlich M, Gama-Sosa MA, Huang LH, Midgett RM, Kuo KC, McCune RA, et al. Amount and distribution of 5-methylcytosine in human DNA from different types of tissues of cells. *Nucleic Acids Res* 1982;**10**:2709–21.
5. Lister R, Pelizzola M, Dowen RH, Hawkins RD, Hon G, Tonti-Filippini J, et al. Human DNA methylomes at base resolution show widespread epigenomic differences. *Nature* 2009;**462**:315–22.
6. Bird AP. DNA methylation and the frequency of CpG in animal DNA. *Nucleic Acids Res* 1980;**8**:1499–504.
7. Bird A, Taggart M, Frommer M, Miller OJ, Macleod D. A fraction of the mouse genome that is derived from islands of nonmethylated, CpG-rich DNA. *Cell* 1985;**40**:91–9.
8. Chodavarapu RK, Feng S, Bernatavichute YV, Chen PY, Stroud H, Yu Y, et al. Relationship between nucleosome positioning and DNA methylation. *Nature* 2010;**466**:388–92.
9. Laurent L, Wong E, Li G, Huynh T, Tsirigos A, Ong CT, et al. Dynamic changes in the human methylome during differentiation. *Genome Res* 2010;**20**:320–31.
10. Guibert S, Forne T, Weber M. Dynamic regulation of DNA methylation during mammalian development. *Epigenomics* 2009;**1**:81–98.
11. Law JA, Jacobsen SE. Establishing, maintaining and modifying DNA methylation patterns in plants and animals. *Nat Rev Genet* 2010;**11**:204–20.
12. Mohn F, Schubeler D. Genetics and epigenetics: stability and plasticity during cellular differentiation. *Trends Genet* 2009;**25**:129–36.
13. Bestor TH. The DNA methyltransferases of mammals. *Hum Mol Genet* 2000;**9**:2395–402.
14. Chen T, Li E. Structure and function of eukaryotic DNA methyltransferases. *Curr Top Dev Biol* 2004;**60**:55–89.
15. Cheng X, Blumenthal RM. Mammalian DNA methyltransferases: a structural perspective. *Structure* 2008;**16**:341–50.
16. Goll MG, Bestor TH. Eukaryotic cytosine methyltransferases. *Annu Rev Biochem* 2004;**74**:481–574.
17. Bestor T, Laudano A, Mattaliano R, Ingram V. Cloning and sequencing of a cDNA encoding DNA methyltransferase of mouse cells. The carboxyl-terminal domain of the mammalian enzymes is related to bacterial restriction methyltransferases. *J Mol Biol* 1988;**203**:971–83.
18. Pradhan S, Bacolla A, Wells RD, Roberts RJ. Recombinant human DNA (cytosine-5) methyltransferase. I. Expression, purification, and comparison of de novo and maintenance methylation. *J Biol Chem* 1999;**274**:33002–10.
19. Lei H, Oh SP, Okano M, Juttermann R, Goss KA, Jaenisch R, et al. De novo DNA cytosine methyltransferase activities in mouse embryonic stem cells. *Development* 1996;**122**:3195–205.
20. Okano M, Xie S, Li E. Cloning and characterization of a family of novel mammalian DNA (cytosine-5) methyltransferases. *Nat Genet* 1998;**19**:219–20.
21. Robertson KD, Uzvolgyi E, Liang G, Talmadge C, Sumegi J, Gonzales FA, et al. The human DNA methyltransferases (DNMTs) 1, 3a and 3b: coordinate mRNA expression in normal tissues and overexpression in tumors. *Nucleic Acids Res* 1999;**27**:2291–8.
22. Maurer-Stroh S, Dickens NJ, Hughes-Davies L, Kouzarides T, Eisenhaber F, Ponting CP. The Tudor domain 'Royal Family': Tudor, plant Agenet, Chromo, PWWP and MBT domains. *Trends Biochem Sci* 2003;**28**:69–74.
23. Posfai J, Bhagwat AS, Posfai G, Roberts RJ. Predictive motifs derived from cytosine methyltransferases. *Nucleic Acids Res* 1989;**17**:2421–35.
24. Zemach A, McDaniel IE, Silva P, Zilberman D. Genome-wide evolutionary analysis of eukaryotic DNA methylation. *Science* 2010;**328**:916–9.

25. Li E. Chromatin modification and epigenetic reprogramming in mammalian development. *Nat Rev Genet* 2002;**3**:662–73.
26. Okano M, Bell DW, Haber DA, Li E. DNA methyltransferases Dnmt3a and Dnmt3b are essential for de novo methylation and mammalian development. *Cell* 1999;**99**:247–57.
27. Chen T, Ueda Y, Xie S, Li E. A novel Dnmt3a isoform produced from an alternative promoter localizes to euchromatin and its expression correlates with active de novo methylation. *J Biol Chem* 2002;**277**:38746–54.
28. Lees-Murdock DJ, Shovlin TC, Gardiner T, De Felici M, Walsh CP. DNA methyltransferase expression in the mouse germ line during periods of de novo methylation. *Dev Dyn* 2005;**232**:992–1002.
29. Huntriss J, Hinkins M, Oliver B, Harris SE, Beazley JC, Rutherford AJ, et al. Expression of mRNAs for DNA methyltransferases and methyl-CpG-binding proteins in the human female germ line, preimplantation embryos, and embryonic stem cells. *Mol Reprod Dev* 2004;**67**:323–36.
30. Takahashi K, Tanabe K, Ohnuki M, Narita M, Ichisaka T, Tomoda K, et al. Induction of pluripotent stem cells from adult human fibroblasts by defined factors. *Cell* 2007;**131**:861–72.
31. Hansen RS, Wijmenga C, Luo P, Stanek AM, Canfield TK, Weemaes CM, et al. The DNMT3B DNA methyltransferase gene is mutated in the ICF immunodeficiency syndrome. *Proc Natl Acad Sci USA* 1999;**96**:14412–7.
32. La Salle S, Mertineit C, Taketo T, Moens PB, Bestor TH, Trasler JM. Windows for sex-specific methylation marked by DNA methyltransferase expression profiles in mouse germ cells. *Dev Biol* 2004;**268**:403–15.
33. La Salle S, Trasler JM. Dynamic expression of DNMT3a and DNMT3b isoforms during male germ cell development in the mouse. *Dev Biol* 2006;**296**:71–82.
34. Lucifero D, La Salle S, Bourc'his D, Martel J, Bestor TH, Trasler JM. Coordinate regulation of DNA methyltransferase expression during oogenesis. *BMC Dev Biol* 2007;**7**:36.
35. Gowher H, Jeltsch A. Enzymatic properties of recombinant Dnmt3a DNA methyltransferase from mouse: the enzyme modifies DNA in a non-processive manner and also methylates non-CpG [correction of non-CpA] sites. *J Mol Biol* 2001;**309**:1201–8.
36. Suetake I, Miyazaki J, Murakami C, Takeshima H, Tajima S. Distinct enzymatic properties of recombinant mouse DNA methyltransferases Dnmt3a and Dnmt3b. *J Biochem (Tokyo)* 2003;**133**:737–44.
37. Yokochi T, Robertson KD. Preferential methylation of unmethylated DNA by Mammalian de novo DNA methyltransferase Dnmt3a. *J Biol Chem* 2002;**277**:11735–45.
38. Hsieh CL. In vivo activity of murine de novo methyltransferases, Dnmt3a and Dnmt3b. *Mol Cell Biol* 1999;**19**:8211–8.
39. Aoki A, Suetake I, Miyagawa J, Fujio T, Chijiwa T, Sasaki H, et al. Enzymatic properties of de novo-type mouse DNA (cytosine-5) methyltransferases. *Nucleic Acids Res* 2001;**29**:3506–12.
40. Ramsahoye BH, Biniszkiewicz D, Lyko F, Clark V, Bird AP, Jaenisch R. Non-CpG methylation is prevalent in embryonic stem cells and may be mediated by DNA methyltransferase 3a. *Proc Natl Acad Sci USA* 2000;**97**:5237–42.
41. Ueda Y, Okano M, Williams C, Chen T, Georgopoulos K, Li E. Roles for Dnmt3b in mammalian development: a mouse model for the ICF syndrome. *Development* 2006;**133**:1183–92.
42. Chen T, Ueda Y, Dodge JE, Wang Z, Li E. Establishment and maintenance of genomic methylation patterns in mouse embryonic stem cells by Dnmt3a and Dnmt3b. *Mol Cell Biol* 2003;**23**:5594–605.
43. Lyko F, Ramsahoye BH, Kashevsky H, Tudor M, Mastrangelo MA, Orr-Weaver TL, et al. Mammalian (cytosine-5) methyltransferases cause genomic DNA methylation and lethality in Drosophila. *Nat Genet* 1999;**23**:363–6.

44. Bostick M, Kim JK, Esteve PO, Clark A, Pradhan S, Jacobsen SE. UHRF1 plays a role in maintaining DNA methylation in mammalian cells. *Science* 2007;**317**:1760–4.
45. Sharif J, Muto M, Takebayashi S, Suetake I, Iwamatsu A, Endo TA, et al. The SRA protein Np95 mediates epigenetic inheritance by recruiting Dnmt1 to methylated DNA. *Nature* 2007;**450**:908–12.
46. Leonhardt H, Page AW, Weier HU, Bestor TH. A targeting sequence directs DNA methyltransferase to sites of DNA replication in mammalian nuclei. *Cell* 1992;**71**:865–73.
47. Chen T, Hevi S, Gay F, Tsujimoto N, He T, Zhang B, et al. Complete inactivation of DNMT1 leads to mitotic catastrophe in human cancer cells. *Nat Genet* 2007;**39**:391–6.
48. Li E, Bestor TH, Jaenisch R. Targeted mutation of the DNA methyltransferase gene results in embryonic lethality. *Cell* 1992;**69**:915–26.
49. Liang G, Chan MF, Tomigahara Y, Tsai YC, Gonzales FA, Li E, et al. Cooperativity between DNA methyltransferases in the maintenance methylation of repetitive elements. *Mol Cell Biol* 2002;**22**:480–91.
50. Jones PA, Liang G. Rethinking how DNA methylation patterns are maintained. *Nat Rev Genet* 2009;**10**:805–11.
51. Riggs AD, Xiong Z. Methylation and epigenetic fidelity. *Proc Natl Acad Sci USA* 2004;**101**:4–5.
52. Kaneda M, Okano M, Hata K, Sado T, Tsujimoto N, Li E, et al. Essential role for de novo DNA methyltransferase Dnmt3a in paternal and maternal imprinting. *Nature* 2004;**429**:900–3.
53. Feng J, Chang H, Li E, Fan G. Dynamic expression of de novo DNA methyltransferases Dnmt3a and Dnmt3b in the central nervous system. *J Neurosci Res* 2005;**79**:734–46.
54. Wu H, Coskun V, Tao J, Xie W, Ge W, Yoshikawa K, et al. Dnmt3a-dependent nonpromoter DNA methylation facilitates transcription of neurogenic genes. *Science* 2010;**329**:444–8.
55. Siegmund KD, Connor CM, Campan M, Long TI, Weisenberger DJ, Biniszkiewicz D, et al. DNA methylation in the human cerebral cortex is dynamically regulated throughout the life span and involves differentiated neurons. *PLoS One* 2007;**2**:e895.
56. Nguyen S, Meletis K, Fu D, Jhaveri S, Jaenisch R. Ablation of de novo DNA methyltransferase Dnmt3a in the nervous system leads to neuromuscular defects and shortened lifespan. *Dev Dyn* 2007;**236**:1663–76.
57. Feng J, Zhou Y, Campbell SL, Le T, Li E, Sweatt JD, et al. Dnmt1 and Dnmt3a maintain DNA methylation and regulate synaptic function in adult forebrain neurons. *Nat Neurosci* 2010;**13**:423–30.
58. LaPlant Q, Vialou V, Covington 3rd HE, Dumitriu D, Feng J, Warren BL, et al. Dnmt3a regulates emotional behavior and spine plasticity in the nucleus accumbens. *Nat Neurosci* 2010;**13**:1137–43.
59. Ley TJ, Ding L, Walter MJ, McLellan MD, Lamprecht T, Larson DE, et al. DNMT3A mutations in acute myeloid leukemia. *N Engl J Med* 2010;**363**:2424–33.
60. Jeanpierre M, Turleau C, Aurias A, Prieur M, Ledeist F, Fischer A, et al. An embryonic-like methylation pattern of classical satellite DNA is observed in ICF syndrome. *Hum Mol Genet* 1993;**2**:731–5.
61. Xu GL, Bestor TH, Bourc'his D, Hsieh CL, Tommerup N, Bugge M, et al. Chromosome instability and immunodeficiency syndrome caused by mutations in a DNA methyltransferase gene. *Nature* 1999;**402**:187–91.
62. Gowher H, Jeltsch A. Molecular enzymology of the catalytic domains of the Dnmt3a and Dnmt3b DNA methyltransferases. *J Biol Chem* 2002;**277**:20409–14.
63. Xie ZH, Huang YN, Chen ZX, Riggs AD, Ding JP, Gowher H, et al. Mutations in DNA methyltransferase DNMT3B in ICF syndrome affect its regulation by DNMT3L. *Hum Mol Genet* 2006;**15**:1375–85.

64. Jin B, Tao Q, Peng J, Soo HM, Wu W, Ying J, et al. DNA methyltransferase 3B (DNMT3B) mutations in ICF syndrome lead to altered epigenetic modifications and aberrant expression of genes regulating development, neurogenesis and immune function. *Hum Mol Genet* 2008;**17**:690–709.
65. Jia D, Jurkowska RZ, Zhang X, Jeltsch A, Cheng X. Structure of Dnmt3a bound to Dnmt3L suggests a model for de novo DNA methylation. *Nature* 2007;**449**:248–51.
66. Ehrlich M, Sanchez C, Shao C, Nishiyama R, Kehrl J, Kuick R, et al. ICF, an immunodeficiency syndrome: DNA methyltransferase 3B involvement, chromosome anomalies, and gene dysregulation. *Autoimmunity* 2008;**41**:253–71.
67. Santi DV, Norment A, Garrett CE. Covalent bond formation between a DNA-cytosine methyltransferase and DNA containing 5-azacytosine. *Proc Natl Acad Sci USA* 1984;**81**:6993–7.
68. Wu JC, Santi DV. Kinetic and catalytic mechanism of HhaI methyltransferase. *J Biol Chem* 1987;**262**:4778–86.
69. Klimasauskas S, Kumar S, Roberts RJ, Cheng X. HhaI methyltransferase flips its target base out of the DNA helix. *Cell* 1994;**76**:357–69.
70. Kumar S, Cheng X, Klimasauskas S, Mi S, Posfai J, Roberts RJ, et al. The DNA (cytosine-5) methyltransferases. *Nucleic Acids Res* 1994;**22**:1–10.
71. Lin IG, Han L, Taghva A, O'Brien LE, Hsieh CL. Murine de novo methyltransferase Dnmt3a demonstrates strand asymmetry and site preference in the methylation of DNA in vitro. *Mol Cell Biol* 2002;**22**:704–23.
72. Reither S, Li F, Gowher H, Jeltsch A. Catalytic mechanism of DNA-(cytosine-C5)-methyltransferases revisited: covalent intermediate formation is not essential for methyl group transfer by the murine Dnmt3a enzyme. *J Mol Biol* 2003;**329**:675–84.
73. Handa V, Jeltsch A. Profound flanking sequence preference of Dnmt3a and Dnmt3b mammalian DNA methyltransferases shape the human epigenome. *J Mol Biol* 2005;**348**:1103–12.
74. Gowher H, Loutchanwoot P, Vorobjeva O, Handa V, Jurkowska RZ, Jurkowski TP, et al. Mutational analysis of the catalytic domain of the murine Dnmt3a DNA-(cytosine C5)-methyltransferase. *J Mol Biol* 2006;**357**:928–41.
75. Wienholz BL, Kareta MS, Moarefi AH, Gordon CA, Ginno PA, Chedin F. DNMT3L modulates significant and distinct flanking sequence preference for DNA methylation by DNMT3A and DNMT3B in vivo. *PLoS Genet* 2010;**6**:e1001106.
76. Dodge JE, Ramsahoye BH, Wo ZG, Okano M, Li E. De novo methylation of MMLV provirus in embryonic stem cells: CpG versus non-CpG methylation. *Gene* 2002;**289**:41–8.
77. Hashimoto H, Horton JR, Zhang X, Bostick M, Jacobsen SE, Cheng X. The SRA domain of UHRF1 flips 5-methylcytosine out of the DNA helix. *Nature* 2008;**455**:826–9.
78. Yoder JA, Walsh CP, Bestor TH. Cytosine methylation and the ecology of intragenomic parasites. *Trends Genet* 1997;**13**:335–40.
79. Egger G, Jeong S, Escobar SG, Cortez CC, Li TW, Saito Y, et al. Identification of DNMT1 (DNA methyltransferase 1) hypomorphs in somatic knockouts suggests an essential role for DNMT1 in cell survival. *Proc Natl Acad Sci USA* 2006;**103**:14080–5.
80. Meilinger D, Fellinger K, Bultmann S, Rothbauer U, Bonapace IM, Klinkert WE, et al. Np95 interacts with de novo DNA methyltransferases, Dnmt3a and Dnmt3b, and mediates epigenetic silencing of the viral CMV promoter in embryonic stem cells. *EMBO Rep* 2009;**10**:1259–64.
81. Glass JL, Fazzari MJ, Ferguson-Smith AC, Greally JM. CG dinucleotide periodicities recognized by the Dnmt3a-Dnmt3L complex are distinctive at retroelements and imprinted domains. *Mamm Genome* 2009;**20**:633–43.
82. Bourc'his D, Bestor TH. Meiotic catastrophe and retrotransposon reactivation in male germ cells lacking Dnmt3L. *Nature* 2004;**431**:96–9.

83. Jurkowska RZ, Anspach N, Urbanke C, Jia D, Reinhardt R, Nellen W, et al. Formation of nucleoprotein filaments by mammalian DNA methyltransferase Dnmt3a in complex with regulator Dnmt3L. *Nucleic Acids Res* 2008;**36**:6656–63.
84. Ross JP, Suetake I, Tajima S, Molloy PL. Recombinant mammalian DNA methyltransferase activity on model transcriptional gene silencing short RNA:DNA heteroduplex substrates. *Biochem J* 2010;**432**:323–32.
85. Cokus SJ, Feng S, Zhang X, Chen Z, Merriman B, Haudenschild CD, et al. Shotgun bisulphite sequencing of the Arabidopsis genome reveals DNA methylation patterning. *Nature* 2008;**452**:215–9.
86. Segal E, Fondufe-Mittendorf Y, Chen L, Thastrom A, Field Y, Moore IK, et al. A genomic code for nucleosome positioning. *Nature* 2006;**442**:772–8.
87. Purdy MM, Holz-Schietinger C, Reich NO. Identification of a second DNA binding site in human DNA methyltransferase 3A by substrate inhibition and domain deletion. *Arch Biochem Biophys* 2010;**498**:13–22.
88. Qiu C, Sawada K, Zhang X, Cheng X. The PWWP domain of mammalian DNA methyltransferase Dnmt3b defines a new family of DNA-binding folds. *Nat Struct Biol* 2002;**9**:217–24.
89. Kareta MS, Botello ZM, Ennis JJ, Chou C, Chedin F. Reconstitution and mechanism of the stimulation of de novo methylation by human DNMT3L. *J Biol Chem* 2006;**281**:25893–902.
90. Li JY, Pu MT, Hirasawa R, Li BZ, Huang YN, Zeng R, et al. Synergistic function of DNA methyltransferases Dnmt3a and Dnmt3b in the methylation of Oct4 and Nanog. *Mol Cell Biol* 2007;**27**:8748–59.
91. Holz-Schietinger C, Reich NO. The inherent processivity of the human de novo methyltransferase 3A (DNMT3A) is enhanced by DNMT3L. *J Biol Chem* 2010;**285**:29091–100.
92. Aapola U, Kawasaki K, Scott HS, Ollila J, Vihinen M, Heino M, et al. Isolation and initial characterization of a novel zinc finger gene, DNMT3L, on 21q22.3, related to the cytosine-5-methyltransferase 3 gene family. *Genomics* 2000;**65**:293–8.
93. Aapola U, Lyle R, Krohn K, Antonarakis SE, Peterson P. Isolation and initial characterization of the mouse Dnmt3l gene. *Cytogenet Cell Genet* 2001;**92**:122–6.
94. Bourc'his D, Xu GL, Lin CS, Bollman B, Bestor TH. Dnmt3L and the establishment of maternal genomic imprints. *Science* 2001;**294**:2536–9.
95. Hata K, Okano M, Lei H, Li E. Dnmt3L cooperates with the Dnmt3 family of de novo DNA methyltransferases to establish maternal imprints in mice. *Development* 2002;**129**:1983–93.
96. Sakai Y, Suetake I, Shinozaki F, Yamashina S, Tajima S. Co-expression of de novo DNA methyltransferases Dnmt3a2 and Dnmt3L in gonocytes of mouse embryos. *Gene Expr Patterns* 2004;**5**:231–7.
97. Kameda T, Smuga-Otto K, Thomson JA. A severe de novo methylation of episomal vectors by human ES cells. *Biochem Biophys Res Commun* 2006;**349**:1269–77.
98. Aapola U, Maenpaa K, Kaipia A, Peterson P. Epigenetic modifications affect Dnmt3L expression. *Biochem J* 2004;**380**:705–13.
99. Hu YG, Hirasawa R, Hu JL, Hata K, Li CL, Jin Y, et al. Regulation of DNA methylation activity through Dnmt3L promoter methylation by Dnmt3 enzymes in embryonic development. *Hum Mol Genet* 2008;**17**:2654–64.
100. Ooi SK, Qiu C, Bernstein E, Li K, Jia D, Yang Z, et al. DNMT3L connects unmethylated lysine 4 of histone H3 to de novo methylation of DNA. *Nature* 2007;**448**:714–7.
101. Argentaro A, Yang JC, Chapman L, Kowalczyk MS, Gibbons RJ, Higgs DR, et al. Structural consequences of disease-causing mutations in the ATRX-DNMT3-DNMT3L (ADD) domain of the chromatin-associated protein ATRX. *Proc Natl Acad Sci USA* 2007;**104**:11939–44.
102. Otani J, Nankumo T, Arita K, Inamoto S, Ariyoshi M, Shirakawa M. Structural basis for recognition of H3K4 methylation status by the DNA methyltransferase 3A ATRX-DNMT3-DNMT3L domain. *EMBO Rep* 2009;**10**:1235–41.

103. Yokomine T, Hata K, Tsudzuki M, Sasaki H. Evolution of the vertebrate DNMT3 gene family: a possible link between existence of DNMT3L and genomic imprinting. *Cytogenet Genome Res* 2006;**113**:75–80.
104. Aravin AA, Sachidanandam R, Bourc'his D, Schaefer C, Pezic D, Toth KF, et al. A piRNA pathway primed by individual transposons is linked to de novo DNA methylation in mice. *Mol Cell* 2008;**31**:785–99.
105. Aravin AA, Sachidanandam R, Girard A, Fejes-Toth K, Hannon GJ. Developmentally regulated piRNA clusters implicate MILI in transposon control. *Science* 2007;**316**:744–7.
106. Kuramochi-Miyagawa S, Watanabe T, Gotoh K, Totoki Y, Toyoda A, Ikawa M, et al. DNA methylation of retrotransposon genes is regulated by Piwi family members MILI and MIWI2 in murine fetal testes. *Genes Dev* 2008;**22**:908–17.
107. Arnaud P, Hata K, Kaneda M, Li E, Sasaki H, Feil R, et al. Stochastic imprinting in the progeny of Dnmt3L−/− females. *Hum Mol Genet* 2006;**15**:589–98.
108. Arima T, Hata K, Tanaka S, Kusumi M, Li E, Kato K, et al. Loss of the maternal imprint in Dnmt3L(mat−/−) mice leads to a differentiation defect in the extraembryonic tissue. *Dev Biol* 2006;**297**:361–73.
109. Kaneda M, Hirasawa R, Chiba H, Okano M, Li E, Sasaki H. Genetic evidence for Dnmt3a-dependent imprinting during oocyte growth obtained by conditional knockout with Zp3-Cre and complete exclusion of Dnmt3b by chimera formation. *Genes Cells* 2010;**15**:169–79.
110. Bielinska B, Blaydes SM, Buiting K, Yang T, Krajewska-Walasek M, Horsthemke B, et al. De novo deletions of SNRPN exon 1 in early human and mouse embryos result in a paternal to maternal imprint switch. *Nat Genet* 2000;**25**:74–8.
111. El-Maarri O, Buiting K, Peery EG, Kroisel PM, Balaban B, Wagner K, et al. Maternal methylation imprints on human chromosome 15 are established during or after fertilization. *Nat Genet* 2001;**27**:341–4.
112. Geuns E, De Rycke M, Van Steirteghem A, Liebaers I. Methylation imprints of the imprint control region of the SNRPN-gene in human gametes and preimplantation embryos. *Hum Mol Genet* 2003;**12**:2873–9.
113. El-Maarri O, Kareta MS, Mikeska T, Becker T, Diaz-Lacava A, Junen J, et al. A systematic search for DNA methyltransferase polymorphisms reveals a rare DNMT3L variant associated with subtelomeric hypomethylation. *Hum Mol Genet* 2009;**18**:1755–68.
114. Webster KE, O'Bryan MK, Fletcher S, Crewther PE, Aapola U, Craig J, et al. Meiotic and epigenetic defects in Dnmt3L-knockout mouse spermatogenesis. *Proc Natl Acad Sci USA* 2005;**102**:4068–73.
115. Schaefer CB, Ooi SK, Bestor TH, Bourc'his D. Epigenetic decisions in mammalian germ cells. *Science* 2007;**316**:398–9.
116. La Salle S, Oakes CC, Neaga OR, Bourc'his D, Bestor TH, Trasler JM. Loss of spermatogonia and wide-spread DNA methylation defects in newborn male mice deficient in DNMT3L. *BMC Dev Biol* 2007;**7**:104.
117. Kato Y, Kaneda M, Hata K, Kumaki K, Hisano M, Kohara Y, et al. Role of the Dnmt3 family in de novo methylation of imprinted and repetitive sequences during male germ cell development in the mouse. *Hum Mol Genet* 2007;**16**:2272–80.
118. Ooi SK, Wolf D, Hartung O, Agarwal S, Daley GQ, Goff SP, et al. Dynamic instability of genomic methylation patterns in pluripotent stem cells. *Epigenet. Chromatin* 2010;**3**:17.
119. Chen ZX, Mann JR, Hsieh CL, Riggs AD, Chedin F. Physical and functional interactions between the human DNMT3L protein and members of the de novo methyltransferase family. *J Cell Biochem* 2005;**95**:902–17.
120. Margot JB, Ehrenhofer-Murray AE, Leonhardt H. Interactions within the mammalian DNA methyltransferase family. *BMC Mol Biol* 2003;**4**:7.

121. Suetake I, Shinozaki F, Miyagawa J, Takeshima H, Tajima S. DNMT3L stimulates the DNA methylation activity of Dnmt3a and Dnmt3b through a direct interaction. *J Biol Chem* 2004;**279**:27816–23.
122. Boland MJ, Christman JK. Characterization of Dnmt3b:thymine-DNA glycosylase interaction and stimulation of thymine glycosylase-mediated repair by DNA methyltransferase(s) and RNA. *J Mol Biol* 2008;**379**:492–504.
123. Li YQ, Zhou PZ, Zheng XD, Walsh CP, Xu GL. Association of Dnmt3a and thymine DNA glycosylase links DNA methylation with base-excision repair. *Nucleic Acids Res* 2007;**35**:390–400.
124. Chedin F, Lieber MR, Hsieh CL. The DNA methyltransferase-like protein DNMT3L stimulates de novo methylation by Dnmt3a. *Proc Natl Acad Sci USA* 2002;**99**:16916–21.
125. Gowher H, Liebert K, Hermann A, Xu G, Jeltsch A. Mechanism of stimulation of catalytic activity of Dnmt3A and Dnmt3B DNA-(cytosine-C5)-methyltransferases by Dnmt3L. *J Biol Chem* 2005;**280**:13341–8.
126. Kim GD, Ni J, Kelesoglu N, Roberts RJ, Pradhan S. Co-operation and communication between the human maintenance and de novo DNA (cytosine-5) methyltransferases. *EMBO J* 2002;**21**:4183–95.
127. Jeong S, Liang G, Sharma S, Lin JC, Choi SH, Han H, et al. Selective anchoring of DNA methyltransferases 3A and 3B to nucleosomes containing methylated DNA. *Mol Cell Biol* 2009;**29**:5366–76.
128. Kashiwagi K, Nimura K, Ura K, Kaneda Y. DNA methyltransferase 3b preferentially associates with condensed chromatin. *Nucleic Acids Res* 2010;**39**:874–88.
129. Nady N, Min J, Kareta MS, Chedin F, Arrowsmith CH. A SPOT on the chromatin landscape? Histone peptide arrays as a tool for epigenetic research. *Trends Biochem Sci* 2008;**33**:305–13.
130. Zhao Q, Rank G, Tan YT, Li H, Moritz RL, Simpson RJ, et al. PRMT5-mediated methylation of histone H4R3 recruits DNMT3A, coupling histone and DNA methylation in gene silencing. *Nat Struct Mol Biol* 2009;**16**:304–11.
131. Edwards JR, O'Donnell AH, Rollins RA, Peckham HE, Lee C, Milekic MH, et al. Chromatin and sequence features that define the fine and gross structure of genomic methylation patterns. *Genome Res* 2010;**20**:972–80.
132. Hu JL, Zhou BO, Zhang RR, Zhang KL, Zhou JQ, Xu GL. The N-terminus of histone H3 is required for de novo DNA methylation in chromatin. *Proc Natl Acad Sci USA* 2009;**106**:22187–92.
133. Zhang Y, Jurkowska R, Soeroes S, Rajavelu A, Dhayalan A, Bock I, et al. Chromatin methylation activity of Dnmt3a and Dnmt3a/3L is guided by interaction of the ADD domain with the histone H3 tail. *Nucleic Acids Res* 2010;**38**:4246–53.
134. Barski A, Cuddapah S, Cui K, Roh TY, Schones DE, Wang Z, et al. High-resolution profiling of histone methylations in the human genome. *Cell* 2007;**129**:823–37.
135. Chen T, Tsujimoto N, Li E. The PWWP domain of Dnmt3a and Dnmt3b is required for directing DNA methylation to the major satellite repeats at pericentric heterochromatin. *Mol Cell Biol* 2004;**24**:9048–58.
136. Dhayalan A, Rajavelu A, Rathert P, Tamas R, Jurkowska RZ, Ragozin S, et al. The Dnmt3a PWWP domain reads histone 3 lysine 36 trimethylation and guides DNA methylation. *J Biol Chem* 2010;**285**:26114–20.
137. Feng S, Cokus SJ, Zhang X, Chen PY, Bostick M, Goll MG, et al. Conservation and divergence of methylation patterning in plants and animals. *Proc Natl Acad Sci USA* 2010;**107**:8689–94.
138. Gopalakrishnan S, Sullivan BA, Trazzi S, Della Valle G, Robertson KD. DNMT3B interacts with constitutive centromere protein CENP-C to modulate DNA methylation and the histone code at centromeric regions. *Hum Mol Genet* 2009;**18**:3178–93.

139. Geiman TM, Sankpal UT, Robertson AK, Chen Y, Mazumdar M, Heale JT, et al. Isolation and characterization of a novel DNA methyltransferase complex linking DNMT3B with components of the mitotic chromosome condensation machinery. *Nucleic Acids Res* 2004;**32**:2716–29.
140. O'Hagan HM, Mohammad HP, Baylin SB. Double strand breaks can initiate gene silencing and SIRT1-dependent onset of DNA methylation in an exogenous promoter CpG island. *PLoS Genet* 2008;**4**:e1000155.
141. Dennis K, Fan T, Geiman T, Yan Q, Muegge K. Lsh, a member of the SNF2 family, is required for genome-wide methylation. *Genes Dev* 2001;**15**:2940–4.
142. Yan Q, Cho E, Lockett S, Muegge K. Association of Lsh, a regulator of DNA methylation, with pericentromeric heterochromatin is dependent on intact heterochromatin. *Mol Cell Biol* 2003;**23**:8416–28.
143. Zhu H, Geiman TM, Xi S, Jiang Q, Schmidtmann A, Chen T, et al. Lsh is involved in de novo methylation of DNA. *EMBO J* 2006;**25**:335–45.

Section III

DNA Methylation and Demethylation

Recruitment of Dnmt1: Roles of the SRA Protein Np95 (Uhrf1) and Other Factors

JAFAR SHARIF AND HARUHIKO KOSEKI

RIKEN Research Center for Allergy & Immunology (RCAI), Tsurumi-ku, Yokohama, Kanagawa, Japan

Dnmt1, the major DNA methyltransferase enzyme in mammals, mediates inheritance of DNA methylation from parent cell to daughter cells. However, the mechanism by which Dnmt1 is recruited into the replication fork (RF)

Progress in Molecular Biology and Translational Science, Vol. 101
DOI: 10.1016/B978-0-12-387685-0.00008-1

during cell division was not fully understood. Recently, the mammalian SRA (Set and RING-finger-associated domain) protein Np95 (nuclear protein 95 kDa) (also known as Uhrf1) was found to help recruit Dnmt1 to hemimethylated CpGs generated at the RF during the S-phase. Here, we will discuss in detail how Np95 recognizes hemimethylated DNA through its SRA domain and recruits Dnmt1 to these sites. In addition, we will examine the functions of other recruiter/interactor molecules that are involved in Dnmt1 localization both locus specifically and globally. Elucidation of the pathways by which Dnmt1 is recruited and identification of the molecules that are involved in this process will provide key information about the mechanisms of genome reprogramming and the epigenetic differences between normal and transformed cells.

I. Roles for DNA Methylation in Transcriptional Regulation and Development

DNA (5-cytosine) methylation is a major epigenetic modification in plants and animals, including mammals (see Part II of this volume). DNA methylation occurs predominantly within the CpG dinucleotides in vertebrates. Indeed, in both humans and mice, the overwhelming majority (70–80%) of the CpG sites in the genome are methylated,[1] with the exception of the CpG-dense CpG islands (CGIs) that remain unmethylated and are found near the transcription start sites of genes[2] (see Chapter by Eleanor Wong and Chia Lin Wei). This pervasive (genome-wide) methylation pattern is a unique signature of the vertebrates[3] because, although DNA methylation is present in other organisms such as plants and invertebrates, it is imparted in a mosaic fashion in the genome with interspersed methylated and unmethylated regions.[4]

The pervasive DNA methylation pattern could have emerged as a primary defense mechanism to silence genomic retrotransposons that are abundant in the vertebrate genome (40–50% in case of mammals).[5] Consistent with this notion, loss of DNA methylation induces activation of retrotransposons in the genome.[6,7] Apart from repressing retrotransposons, DNA methylation is also required to silence single-copy genes such as the X-linked *Rhox* cluster in mouse.[8] In fact, DNA methylation and gene expression patterns could be connected in a more general manner. The vertebrate (especially the warm-blooded class) gene promoters exhibit a clear divergence in terms of their susceptibility/refractivity to DNA methylation.[9] Promoters with classical long CGIs (with high CpG density) are generally unmethylated and associate with frequently and robustly transcribed genes. In contrast, promoters that do not possess CGIs (with low CpG density) are usually enriched in DNA methylation and link with genes that are strictly regulated and transcribed in a tissue-specific manner.[9,10]

In addition, DNA methylation plays a crucial role in key developmental processes that are unique for the mammals such as genome imprinting (Chapter by Jon F. Wilkins and Francisco Úbeda) and X-chromosome inactivation.[11,12] Fetuses that are impaired for DNA methylation die during mid-gestation, which demonstrates an essential role for this epigenetic mark in embryogenesis and development.[7,13] DNA methylation is also important for cell differentiation. DNA methylation-deficient murine embryonic stem (ES) cells have apparently normal self-renewal ability, but they cannot differentiate.[14,15] In the case of adult stem cells, disruption of the DNA methylation machinery could have even more severe effects. DNA methylation-impaired hematopoietic stem cells (HSCs) are defective in self-renewal, niche retention, and their ability to give rise to multilineage hematopoiesis.[16]

In summary, DNA methylation is a major epigenetic modification of the DNA that has crucial roles for silencing of retrotransposons and single-copy genes and is essential for embryonic development and cellular differentiation.

II. Maintenance and *De Novo* DNA Methyltransferases: Two Modes of Action

Enzymatic transfer of the methyl group into the 5-carbon position of the cytosine molecule is mediated by DNA methyltransferase (Dnmt) enzymes. In mammals, three active Dnmts have been identified so far, namely Dnmt1, Dnmt3a, and Dnmt3b.[17,18] There is another molecule known as Dnmt3L that is closely related to Dnmt3a and Dnmt3b in structure. Dnmt3L is essential for the sequence preference and activity of Dnmt3a and Dnmt3b but is not enzymatically active itself.[19]

DNA methylation requires to be faithfully copied from parent to daughter cells. Classically, this activity has been associated with the maintenance Dnmt1 (see Chapter by Željko M. Svedružić). During replication of DNA, CpG methylation marks reside only in the original (parent) strand but not in the nascent (daughter) strand. As a result, newly formed "hemimethylated" (DNA methylation in only one strand) DNA sites are generated at the replication fork (RF). Dnmt1 has a high affinity for such hemimethylated sites (40-fold preference over unmethylated DNA) and is recruited into the RF during the S-phase, where it converts these hemimethylated CpGs into full methylated ones in a processive (moving on the DNA in a random walk) method.[20,21]

In contrast, the *de novo* enzymes are linked with establishing new (not preexisting) DNA methylation patterns using unmethylated CpGs as substrates. Dnmt3a and Dnmt3b have *de novo* activities (see Chapter by Frédéric Chédin), while Dnmt3L functions as a cofactor. *De novo* enzymes are thought to work on

DNA templates in a distributive (bind and release) method. However, as our main interest will be the maintenance methyltransferase Dnmt1 in this study, we will not further discuss the roles of the *de novo* enzymes.

III. Localization of Dnmt1 into Hemimethylated DNA: Search for Recruiter Molecule(s)

Although *in vitro* analyses revealed that Dnmt1 has a strong preference for hemimethylated substrates, these studies also pointed out a critical problem: that this affinity alone would not be sufficient to ensure faithful transmission of DNA methylation upon unlimited cell division. Indeed, it has been suggested that as much as 3–20% of DNA methylation could be lost if cells had to depend solely on the activity of Dnmt1 for converting hemimethylated DNA.[22] Further, apart from the classical "maintenance" function, Dnmt1 appears to possess considerable "*de novo*" activity. In fact, this activity could be more robust than that of the so-called *de novo* enzymes Dnmt3a and Dnmt3b.[23] This implies that if the *de novo* activity of Dnmt1 is not rigorously suppressed by some inhibitory mechanism(s), this could lead to aberrantly methylated CpGs in daughter cells.

The above insights suggest that additional factors might be needed to recruit (and regulate) Dnmt1 to the RF, to increase its preference for hemimethylated CpGs for maintenance methylation functions and/or suppress its promiscuous loading into unmethylated DNA, to prevent aberrant *de novo* methylation activity. As a result, the putative Dnmt1 recruiter(s) should possess (at least) three distinct features. First, these molecules should be able to physically interact with/recruit Dnmt1. Second, these should localize to the RF during S-phase. Third, these factors should possess a specific affinity for hemimethylated CpGs.

An attractive candidate for such a role was the Pcna (proliferating cell nuclear antigen) protein because it functions as an integral component of the replication machinery and also interacts with the N-terminal of the Dnmt1 protein during replication.[24] However, expression of a truncated DNMT1 lacking the PCNA-interacting domain in cultured cells, in which the endogenous *DNMT1* gene had been inactivated, did not result in any major reduction in the overall DNA methylation level.[25] Besides, Pcna does not possess specificity for hemimethylated DNA. Taken together, these results suggest that Pcna is not the putative Dnmt1 recruiter.

Another class of candidate molecules is the MBD (methyl DNA-binding domain) proteins (see Chapter by Pierre-Antoine Defossez and Irina Stancheva). Several MBD factors, namely, Mbd1-4 and MeCP2, have been identified as methylated DNA-interacting molecules. However, MBD proteins do not show specificity for hemimethylated DNA. Further, mutant mice lacking any of these genes do not exhibit the drastic hypomethylated phenotype that is characteristic for *Dnmt1*$^{-/-}$ embryos.[13,26,27] These results indicate that MBD proteins are not major players for recruiting Dnmt1.

At this stage, a significant breakthrough came from a series of studies performed on plants. Several groups identified a family of plant SRA-containing proteins and showed that these factors were linked with DNA methylation.[28–30] Two SRA proteins Np95/Uhrf1 (human ortholog NP95/UHRF1/ICBP90) and Np97 (human ortholog NP97/UHRF2) occur in mammals.[31,32] Prompted by the plant studies, investigations sought a possible link between the SRA molecule Np95 and DNA methylation in mammalian cells, taking advantage of an already established *Np95*$^{-/-}$ murine ES cell line.[33]

We and others showed that Np95 was indeed the molecule that was required for recruiting Dnmt1 to RFs during S-phase.[9,34] The *Np95*$^{-/-}$ genotype was embryonically lethal at 10.5 dpc, a phenotype that was strikingly similar to *Dnmt1*$^{-/-}$ mutants.[13] In addition, *Np95*-null embryos exhibited global loss of DNA methylation, derepression of retrotransposons, and impairment of genomic imprinting, which are the hallmarks for *Dnmt1*-knockout embryos.[13]

Consistent with these observations, *in vitro* analyses revealed that the SRA domain of NP95 had an affinity for hemimethylated DNA.[34] X-ray crystallographic studies resolved the SRA domain–hemimethylated DNA interaction to a much more detailed level, revealing that the SRA domain of human NP95 contacted hemimethylated DNA from both major and minor grooves and flipped out the hemimethylated cytosine molecule from the DNA helix.[35–38] Collectively, these studies identified Np95 as the first molecule to have specificity for hemimethylated DNA.

In the following sections of this chapter, we will discuss how Np95 localizes to the RF during S-phase and specifically recognizes hemimethylated CpGs. We will also examine the mechanisms by which Np95 recruits Dnmt1 to hemimethylated DNA for maintenance of epigenetic inheritance through successive cell divisions.

IV. Np95 Is a Cell Cycle-Associated Protein That Localizes to the RF During S-Phase

Although the roles of Np95 in mammalian epigenetics had been under intense investigation recently, the Np95 molecule itself was first identified 15 years ago.[31] Muto and colleagues derived a monoclonal antibody, Th-10a, by immunizing Wistar rats with a murine thymic lymphoma (TIGN). Interestingly, this antibody interacted with a nuclear protein that appeared during the S-phase in normal thymocytes. This factor was 95 kDa in size and was therefore named Np95. Fujimori and colleagues isolated the cDNA encoding Np95 from a λgt-11 cDNA expression library and showed that the Np95 cDNA encoded a 782 amino acid nuclear protein that contained a zinc finger motif, a potential ATP/GTP-binding site, a putative cyclin A/E-cdk2 phosphorylation site, and the retinoblastoma (Rb) protein-binding motif "IXCXE".[39]

Uemura and colleagues investigated the spatiotemporal localization of Np95 in mitotic and meiotic cells and showed that Np95 colocalized with Pcna in RFs during the S-phase but not in G1 and G2/M phases during mitosis.[40] For meiotic cells, Np95 was present not only in proliferating spermatogonia but also in nonproliferating spermatocytes and differentiating spermatids.[40] Several studies addressed the question whether Np95 had a direct role in cell cycle progression. It was shown that Np95 regulated cell cycle by modulating the expression of *topoisomerase IIα*.[41,42] Another report revealed that the human NP95 ortholog ICBP90 promoted entry to S-phase in culture cells by repressing the tumor suppressor *Rb* gene.[43]

The above insights established two key functions for Np95. First, Np95 localizes to the RF during the S-phase. Second, Np95 possesses a direct role in promoting cell cycle, by activating cell cycle inducer genes such as *topoisomerase IIα*, and blocking cell cycle inhibitory factors like *Rb*.

V. Np95 Protein Domains: SRA and More

Np95 has several interesting motifs (Fig. 1A). But the most striking feature of Np95 is probably the SRA domain. As we have mentioned earlier, only two SRA proteins have been identified in mammals so far, namely, Np95/Uhrf1 and Np97/Uhrf2. The plant SRA homologs are much more numerous. In fact, the initial insights about the link between SRA proteins and DNA methylation came from the studies performed in plants. Plant SRA molecules include the VIMs and the histone methyltransferases called SUVHs. The VIM proteins (VIM1/ORTH2, VIM2/ORTH5, VIM3/ORTH1, VIM4/ORTH4, and VIM5/ORTH3) contain an N-terminal PHD domain, followed by RING, SRA, and C-terminal RING motifs, and have more similarity with their mammalian counterparts in terms of domain structure.[44] The SUVH plant histone methyltransferases possess SRA and SET domains but not PHD and RING domains (Fig. 1A). We will discuss the functions of the SRA domain in details in the ensuing sections of this chapter.

Np95 contains an N-terminal UBL domain (1–74 amino acids) that has similarities with small proteins such as ubiquitin (Ubb, Ubd), ubiquitin-like (Ubl3, Ubl4, Ubl4b, Ubl7), and sumo (Sumo1, Sumo2, Sumo3). This domain is

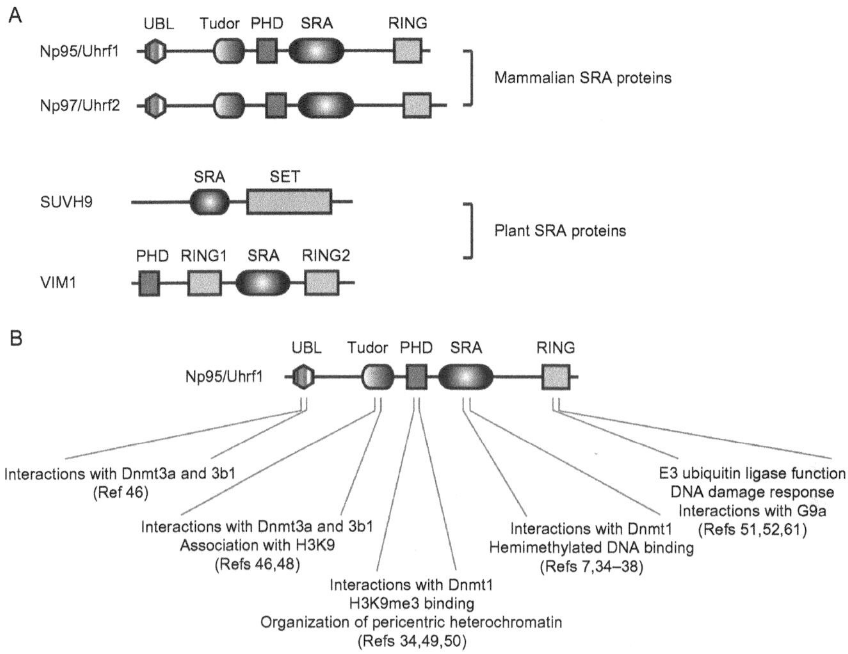

FIG. 1. Structure and functions of Np95 protein domains. (A) Protein domain structures of the mammalian SRA molecules Np95/Uhrf1 and Np97/Uhrf2 are shown (top panel). Both Np95 and Np97 have very similar domain organizations. In contrast, the plant SRA factors exhibit remarkable variation in their protein domain arrangements (bottom panel). The SUVH molecules have an SRA domain, followed by a SET motif (SUVH9 is shown as an example). The VIM proteins demonstrate a closer resemblance to their mammalian counterparts, possessing PHD, RING (1 and 2), and SRA domains (VIM1 structure is shown as a representative). (B) Several functions have already been reported for the Np95 protein domains. These include the recruitment of the maintenance methyltransferase Dnmt1 and the *de novo* methyltransferases Dnmt3a and Dnmt3b1, organization of centromeric heterochromatin, and DNA damage response. The schematic figure summarizes the functions for each domain (references are given in parentheses).

also present in the UV-excision repair-associated Rad23a and Rad23b factors. The Np95-UBL domain has been implicated in *de novo* DNA methylation for its interactions with Dnmt3a and Dnmt3b1 enzymes.[45]

There is a tudor domain next to the UBL motif. Although the functions of the tudor domain are presently not well understood, recent work suggests that it could be involved in interactions with Dnmt3a and Dnmt3b1.[45] The structure of the human NP95 tudor domain has been solved via X-ray crystallography and suggested that this domain might have a role in methylated H3K9 interactions.[46,47]

A number of functions have been assigned to the Np95 PHD domain (322–369 amino acids). This domain is required for interactions with Dnmt1 *in vitro*.[34] It is also necessary for large-scale organization of the pericentric heterochromatin.[48] Consistent with this observation, the PHD domain of human NP95 was shown to be involved in H3K9me3 binding.[49]

Presence of both PHD and RING domains is a hallmark of E3 ubiquitin ligases. Indeed, the RING domain of NP95 functions as an E3 ubiquitin ligase and promoted cell survival under genotoxic conditions in cultured human cells.[50] Supporting this notion, the RING domain is essential for recruitment of Eme1, a protein associated with genomic stability and response to DNA crosslink damages.[51] A summary of already reported functions for each Np95 protein domain is given in Fig. 1B.

VI. Np95 Recruits Dnmt1 into Hemimethylated Sites During DNA Replication

A. Np95 Colocalizes with Dnmt1 and Pcna at RFs During S-Phase

Previous studies suggested that Np95 localizes to replicating DNA during the S-phase.[40] This was recently investigated in detail in murine ES cells.[7] ES cells were sorted by respective cell cycle stages and then stained with antibodies for Np95, Dnmt1, and Pcna. Np95 robustly accumulated into the DAPI-dense heterochromatin loci during mid-to-late S-phase but not in the G1 and G2/M phases. Interestingly, during S-phase, Np95 showed strong colocalization with both Pcna and Dnmt1, raising the possibility that Np95 might form complexes with Dnmt1. To examine this question, a biotin-taggable human NP95 protein was coexpressed with the *Escherichia coli* biotin ligase BirA in HeLa cells, and the NP95 complexes were purified on streptavidin beads. Immunoblotting showed that both DNMT1 and PCNA were indeed present in the NP95 protein complex in significant quantities. This evidence indicates that NP95 forms complexes with DNMT1 and PCNA at the replicating heterochromatin.

B. Recruitment of Dnmt1 into RF Is Dependent on Np95

Colocalization of Np95 with Dnmt1 at RF during the S-phase suggested the possibility that Np95 might have a direct role in recruiting Dnmt1. This question was addressed by taking advantage of $Np95^{-/-}$ murine ES cells.[7] In these cells, Dnmt1 showed a diffused pattern and no significant enrichment in the replicating heterochromatin during the S-phase. Expressing a Myc-tagged

Np95 protein in *Np95*$^{-/-}$ cells rescued the Dnmt1 localization pattern. These results imply that Np95 is required for recruitment of Dnmt1 into replication foci during mitosis.

C. Np95 Localization into Replicating DNA Depends on Methylated DNA

Np95 recruits Dnmt1 into RFs during the S-phase. However, to qualify as a canonical "recruiter" of Dnmt1, Np95 must also possess the function of (specifically) binding hemimethylated DNA sites. To investigate this point, studies were performed in various DNA methylation mutant cells to check whether Np95 localization to RF depends on the presence of methylated DNA.[7] Np95 was unable to localize into newly replicating DNA sites in Dnmt triple knockout (*TKO: Dnmt1*$^{-/-}$, *Dnmt3a*$^{-/-}$, *Dnmt3b*$^{-/-}$) ES cells, in which DNA methylation is absent (or extremely low). In contrast, Np95 localization to RFs was observed in *Dnmt1*$^{-/-}$ cells that exhibit severely reduced but not completely absent DNA methylation levels. Similarly, in *Dnmt3a*$^{-/-}$,*3b*$^{-/-}$ *DKO* cells, in which the *de novo* Dnmts are knocked out, Np95 showed apparently normal localization patterns. Further, expressing either Dnmt3a or Dnmt3b in *TKO* cells could rescue the binding pattern of Np95 at replicating heterochromatin, indicating that Np95 localization to RF was dependent on methylated DNA.

Next, to examine the question of whether Np95 specifically binds hemimethylated DNA, replication labeling with 5-methyl-dCTP incorporation was performed in unmethylated TKO cells. Strikingly, Np95 prominently accumulated into replicating heterochromatin where 5-methyl-dCTP was incorporated. In summary, these observations revealed that Np95 localization into replicating DNA was dependent on the presence of hemimethylated DNA.

D. The SRA Domain Is Involved in Hemimethylated DNA Recognition

Biochemical studies revealed that the SRA domain of Np95 contains the recognition motif for hemimethylated DNA. The human NP95-SRA domain has a sevenfold higher affinity for hemimethylated DNA compared to full methylated DNA but, notably, did not have substantial affinity for unmethylated DNA.[34] A series of studies were carried out to elucidate the X-ray crystal structure of the NP95 SRA domain bound to hemimethylated oligonucleotides.[35–38] These investigations suggested an intriguing mode of action: that the SRA domain contacts DNA from both the major and the minor grooves by two loops that penetrate into the middle of the DNA double helix. One loop recognizes the hemimethylated DNA, while the other flips out the hemimethylated cytosine molecule from the DNA helix, presumably to accommodate Dnmt1 so that it can methylate the cytosine molecule on the other (nascent)

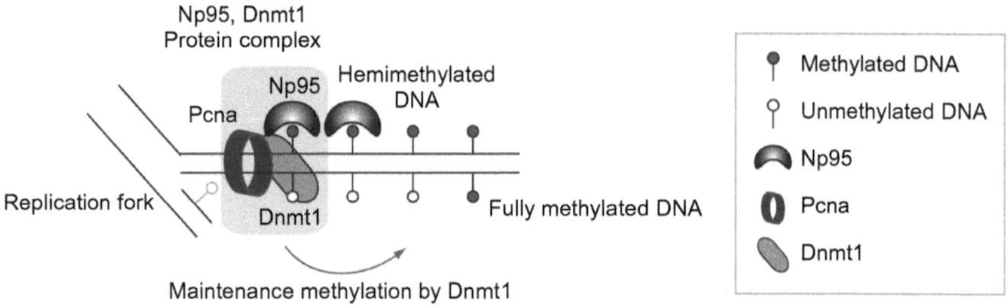

FIG. 2. The role of Np95 for inheritance of DNA methylation. Np95 specifically recognizes hemimethylated CpGs by SRA domain and recruits Dnmt1 into these sites to convert hemimethylated DNA into fully methylated ones. Other cofactors such as Pcna are also involved in this process.

strand. Moreover, it was shown that two key residues in each loop, R496 on the CpG recognition loop and V451 on the base flipping loop, contacted each other through direct van der Waals contacts.[37]

NP95 is the first mammalian protein shown to have preferential affinity for hemimethylated DNA, thus establishing NP95 as a major player in the maintenance DNA methylation mechanism in mammals (Fig. 2).

VII. Np95 Is Required for Global DNA Methylation

Above, we have discussed that Np95 has a role as an essential "recruiter" for the localization of Dnmt1 to hemimethylated sites during replication. To address whether inactivation of *Np95* has an impact on global DNA methylation, investigations were performed in *Np95*$^{-/-}$ ES cells and embryos.[7] As expected, both *Np95*$^{-/-}$ ES cells and embryos exhibited severe hypomethylation of the genome, which was reminiscent of the *Dnmt1* mutants. Further, *Np95*$^{-/-}$ fetuses were embryonic lethal during mid-gestation, a phenotype that is also remarkably similar to *Dnmt1-KO* embryos.[7,13]

DNA methylation is necessary to repress endogenous retrotransposons (Chapter by Mary G. Goll and Marnie E. Halpern) and regulation of genome imprinting (Chapter by Jon F. Wilkins and Francisco Úbeda).[6,11] In *Np95*$^{-/-}$ embryos, endogenous retrotransposons were derepressed, both autonomous (IAP: intercisternal A particles and LINE: long interspersed nuclear elements) and nonautonomous (SINE: short interspersed nuclear elements).[7] Consistent with this observation, bisulfite DNA sequencing analyses revealed that IAP and LINE retrotransposons were severely demethylated in *Np95*$^{-/-}$ ES cells. In addition, inactivation of *Np95* had impacts on genome imprinting. In *Np95*-KO embryos, the maternally expressed *H19* and paternally expressed *Lit1/Kcnq1ot1* noncoding RNAs were biallelically transcribed, while the maternally expressed *Cdkn1c* and paternally transcribed *Igf2* genes were silenced.[7] Genome imprinting is regulated by allele-specific DNA methylation patterns. Indeed, drastic loss of DNA methylation in the *Igf2-H19*, *Kcnq1ot1*, and *Dlk1-Gtl2* imprinting control regions (ICRs) was found in *Np95*$^{-/-}$ ES cells. These results confirmed that Np95 plays a crucial role for maintaining genome-wide DNA methylation by cooperating with Dnmt1.

VIII. Other Factors That Recruit/Interact with Dnmt1

Above, we have discussed the role of the SRA protein Np95 in recruiting Dnmt1 into replication sites during mitosis. However, there are other molecules (and noncoding RNAs) that have been implicated in Dnmt1 recruitment/interaction in both locus-specific and global manners. None of these factors (other than Np95) exhibit the property of preferentially binding hemimethylated DNA and, therefore, do not rigorously qualify as canonical recruiters of Dnmt1. Nevertheless, they appear to have additional and important roles, such as heritable transmission of repressive chromatin by cooperating with Dnmt1 (CFP1, G9a, histone deacetylases 1/2 [HDAC1/2], heterochromatin protein 1 [HP1]), locus-specific gene silencing in normal and transformed cells (Daxx, hNaa10p, STAT3, *Kcnq1ot1* noncoding RNA), and stabilization of DNMT1 protein (AKT1, CFP1, LSD1) and DNMT1 mRNA transcription (CFP1, STAT3). In the following sections, we will analyze the roles of these factors in detail (Fig. 3 and Table I for summary).

A. AKT1 (v-akt Murine Thymoma Viral Oncogene Homolog 1)

The serine/threonine protein kinase AKT1 modulates DNMT1 protein stability.[52] AKT1 phosphorylates Ser143 of DNMT1, which prevents its degradation. Further, phosphorylation of Ser143 interferes with the methylation of the adjacent Lys142 residue, a modification that destabilizes DNMT1 (discussed later in this chapter). *In vitro* analyses show that the kinase domain of AKT1 physically interacts with the N-terminus (residues 1–446) of DNMT1.[52] Importantly, blocking the activity of AKT1 with an inhibitor, LY294002, decreases global DNA methylation. These findings indicate that AKT1 is an important regulator of DNMT1.

B. CFP1/CXXC1 (CXXC Finger Protein 1)

The CFP1/CXXC1 molecule contains a CXXC finger that specifically binds unmethylated CpGs.[53] However, knockout of *CFP1* paradoxically reduces genome-wide DNA methylation.[54] CFP1 directly interacts with DNMT1 and

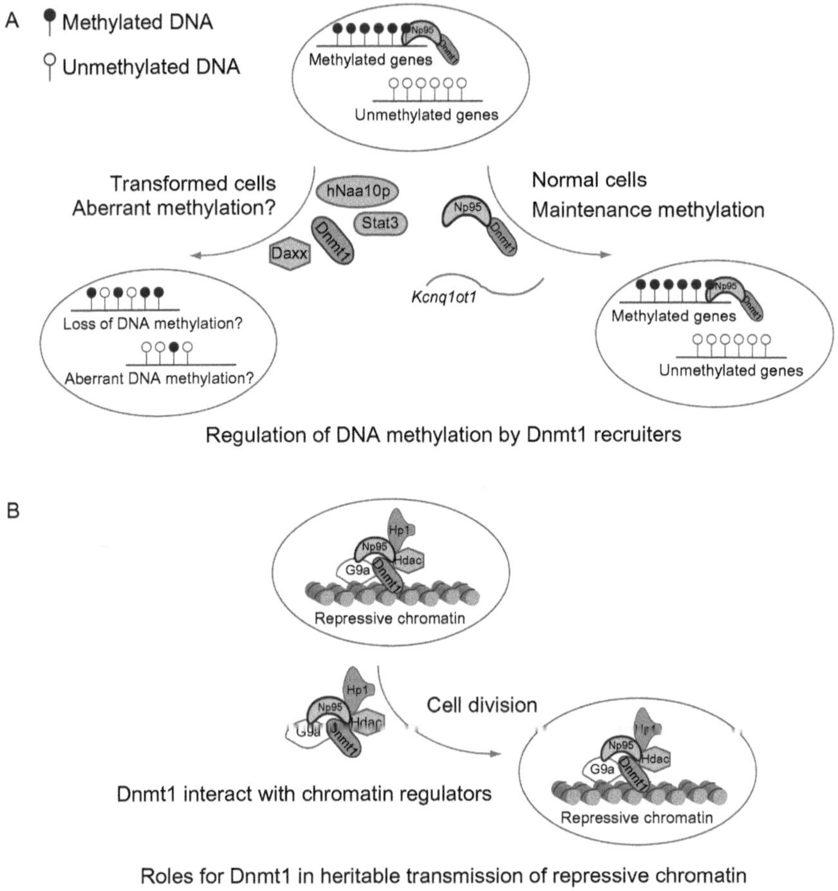

FIG. 3. Inheritable transmission of DNA methylation and repressive chromatin marks is mediated by the interactions of various cofactors with Dnmt1. (A) In normal cells, Np95 recruits Dnmt1 for global maintenance of DNA methylation. However, in transformed cells, Dnmt1 might generate aberrant DNA methylation patterns by associating with oncogenic factors such as Daxx and hNaa10p. (B) Important roles for Dnmt1 have been implicated in heritable transmission of repressive chromatin. Dnmt1 interacts with histone-modifying factors such as G9a, Hdac1/2, and Hp1 (and also Np95) to carry out these functions.

appears to stabilize it, which may lead to the observed hypomethylated phenotype in *CFP1* inactivated cells and embryos.[54,55] Specificity of CFP1 for unmethylated CpG sites suggests that CFP1 could preferentially localize to CGIs because unmethylated CpGs are mostly found in CGIs.[1,2] CGIs are generally present in the euchromatin, which is nucleosome poor, and enriched in permissive histone marks.[9] These characteristics enable CGIs to remain in a

TABLE I
FACTORS THAT RECRUIT/INTERACT WITH DNMT1

Factor	Type	Mode of association	Involved protein domain(s)	Reference(s)
Np95/Uhrf1	Recruiter and interactor	Recruits Dnmt1 into hemimethylated DNA during replication (global)	SRA, PHD domains of Np95	7,34–38
AKT1	Interactor	Increases DNMT1 stability by phosphorylating Ser143	The N-terminal (1–446) of DNMT1	52
CFP1/CXXC1	Interactor	Increases DNMT1 protein stability and mRNA transcription	Amino acids (169–493) and (970–1617) of DNMT1; (1–123), (103–367), and (361–656) of CFP1	53–55
DAXX	Recruiter and interactor	Recruits DNMT1 into target genes (locus specific)	N-terminal of DNMT1 (amino acids 1–126)	56–58
G9a	Interactor	Interacts with DNMT1, NP95, and HDAC1 to silence genes such as *p21* (locus specific)	N-terminal of G9a (amino acids 1–463) and N-terminal of Dnmt1 (1–446aa)	59,60
HDAC1/2	Interactor	Silences E2F target genes (locus specific) and interacts with DNMT1 for heritable maintenance of heterochromatin (global)	Amino acids (686–812) of DNMT1 and (150–242) of HDAC1 for DNMT1–HDAC1 interactions; N-terminal domain of DNMT1 for DNMT1–HDAC2 interactions	48,61–64
hNaa10p	Recruiter and interactor	Recruits DNMT1 into target genes such as *E-cadherin* (locus specific)	Amino acids (291–570) of DNMT1 and (102–122) of hNaa10p	65–67
HP1	Interactor	Involved in heritable transmission of repressive chromatin by cooperating with Dnmt1 (global), and repression of the *Survivin* gene by interacting with G9a, Dnmt1 (locus specific)	The chromodomain of Hp1α,β,γ	68

(*Continues*)

TABLE I (*Continued*)

Factor	Type	Mode of association	Involved protein domain(s)	Reference(s)
LSD1	Interactor	Demethylates and stabilizes DNMT1 protein	The C-terminal domain of DNMT1 (amino acids 956–1619)	69
SET7	Interactor	Decreases DNMT1 stability by methylating Lys142	Both the N-terminal (1–446) and C-terminal (1081–1408) of DNMT1	70
STAT3	Recruiter and activator	Phosphorylated STAT3 binds to DNMT1 promoter to activate its expression and recruits DNMT1 into target genes like *SHP-1* to silence transcription (locus specific)	DNMT1 recruiter domain not identified yet	71–73
Kcnq1ot1 noncoding RNA	Recruiter and interactor	Directly interacts and recruits Dnmt1 into an imprinted loci (locus specific)	A 890-bp domain located downstream of the *Kcnq1ot1* promoter	74,75

transcriptionally active environment. To silence such euchromatinic genes that possess CGIs in their promoters, CFP1 might recruit DNMT1 (and other cofactors) to enforce a repressive chromatin state. Indeed, a role has been suggested for Dnmt1 in regulating repressive chromatin in cooperation with other chromatin modifiers such as HP1 and G9a.[68]

C. DAXX (Fas Death Domain-Associated Protein)

Daxx, an apoptosis-associated nuclear protein, was previously identified as a Dnmt1-interacting factor by yeast two-hybrid experiments.[56] This molecule has been implicated in binding and suppression of the nuclear factor Relb. Daxx recruits Dnmt1 into the Relb targets *dapk1*, *dapk3*, *c-flip*, and *birc3* to silence them through DNA methylation.[57] It appears that the N-terminal domain (amino acids 1–126) of DNMT1 is responsible for association with Daxx, by interacting with DMAP1 (DNMT1-associated protein).[58]

D. G9a/Euchromatic Histone Lysine *N*-Methyltransferase 2

The histone methyltransferase G9a colocalizes with DNMT1 at H3K9me2 regions during DNA replication, indicating that DNMT1 plays a role in the epigenetic inheritance of heterochromatin.[59] DNMT1 appears to be the primary "loading molecule" in this interaction because siRNA-mediated knockdown of Dnmt1 causes loss of DNA methylation, G9a recruitment, and H3K9me2 modifications.[59] Interestingly, G9a is present in a protein complex with NP95, the canonical recruiter of DNMT1.[7] NP95 facilitates the accumulation of HDAC1, DNMT1, and G9a to silence transcription at the tumor suppressor p21 gene promoter.[60] This suggests that the NP95 and DNMT1 might also be involved in organizing repressive chromatin by cooperating with other chromatin-associated factors such as G9a and HDACs.

E. Histone Deacetylases 1/2

Synergistic functions for Hdacs and Dnmt1 in gene repression have been suggested by a number of studies. Human DNMT1 copurifies with the Rb protein, E2F1, and HDAC1 (*note:* murine Hdacs are shown in lower case, human ones in uppercase), and cooperates with Rb to silence genes that have E2F-binding sites in their promoters.[61] Further, in murine cells, Dnmt1 affects the histone deacetylase activity of Hdac1, suggesting that DNA methylation might depend on or generate an altered chromatin state via histone deacetylase activity.[62] A function for Dnmt1 in heritable maintenance of heterochromatin by interacting with Hdacs has been indicated in other studies as well. Human DNMT1 forms a complex with DMAP1 during S-phase and is joined by HDCA2 in late S-phase, an interaction that might impact transmission of transcriptionally silent heterochromatin to daughter cells.[63] Interestingly, Np95 has been implicated in replication of pericentric heterochromatin by associating with Hdac1, suggesting that Np95, Dnmt1, and Hdac1/2 might work together to ensure proper inheritance of repressive heterochromatin through successive cells divisions.[48,64] However, it is not clear whether this activity of Dnmt1, Np95, and Hdac1/2 for maintenance of heterochromatin is dependent on DNA methylation or Dnmt1 binding to hemimethylated DNA.

F. hNaa10p/hARD1 (Human *N*-α-Acetyltransferase 10, NatA Catalytic Subunit)

A yeast two-hybrid screen identified DNMT1 as an interacting molecule for the putative oncogene hNaa10p. hNaa10p has a role in recruiting DNMT1 to the promoters of tumor suppressor genes such as *E-cadherin*.[65] It was reported that the N-terminal 291–570 region of DNMT1 and 21-residue fragment (102–122) of hNaa10p are required for this interaction.[65] Aberrant methylation/hypermethylation of tumor suppressor genes by Dnmt1 in transformed cells has been described previously, and molecules like hNaa10p might be involved in such pathways.[66,67]

G. Heterochromatin Protein 1

The heterochromatin-binding protein HP1 functions as a transcriptional repressor by recognizing methylated H3K9 marks. Recently, it has been reported that HP1 is involved in transcriptional silencing by cooperating with DNMT1 *in vitro* and *in vivo*, in a process that is independent of the maintenance methyltransferase activity of DNMT1.[68] The binding of a GAL4-HP1 hybrid to a reporter plasmid was sufficient to induce DNA methylation by recruiting DNMT1. In addition, the combined functions of HP1, G9a, and DNMT1 were required to repress the *Survivin* gene *in vivo*.

H. LSD1/KDM1A (Lysine (K)-Specific Demethylase 1A)

The histone demethylase LSD1 mediates the demethylation of H3K4 and H3K9. Interestingly, LSD1 also demethylates DNMT1.[69] DNMT1 protein could be methylated *in vivo*, a modification that could lead to its destabilization and reduced activity. LSD1 demethylates and stabilizes DNMT1 and, as a result, is required for global DNA methylation (see Chapter by Shannon R. Morey Kinney and Sriharsa Pradhan).

I. SET Domain Containing Lysine Methyltransferase 7

The lysine methyltransferase SET7 regulates DNMT1 protein stability by methylating Lys142.[70] SET7 and DNMT1 colocalize during the cell cycle, and the extent of Lys142 methylation increases during the progression from early S to G2 phase. The amino terminus of SET7 interacts with both amino (residues 1–446) and carboxyl (residues 1081–1408) termini of DNMT1. Interestingly, methylation of Lys142 promotes proteasome-mediated degradation of the DNMT1 protein. Consistent with this observation, overexpression of SET7 causes approximately 10% reduction of global DNA methylation. We have mentioned earlier that the adjacent Ser143 residue is phosphorylated by AKT1,[52] which promotes DNMT1 protein stability and interferes with methylation of Lys142. Methylation and phosphorylation of the Lys142 and Ser143 residues are reminiscent of the modifications observed in the histone 3 Lys9 and Ser10 cassette. These insights suggest that the methylation and phosphorylation switch may be regulated by cell cycle-specific communication between AKT1, SET7, and DNMT1.[52]

J. Signal Transducer and Activator of Transcription 3

The transcription factor STAT3 has been associated with recruiting DNMT1 to the *SHP-1* gene, to silence its expression by DNA methylation in malignant T cells.[71] In addition, STAT3 itself binds *DNMT1* promoter to activate *DNMT1* expression.[72] Previous reports reveal that in EBV-associated gastric carcinoma, the LMP2A (viral latent membrane protein 2A) protein phosphorylates STAT3 and induces *DNMT1* overexpression, which leads to methylation and silencing of the tumor suppressor *PTEN* gene.[73]

K. *Kcnq1ot1* Noncoding RNA

Dnmt1 can be recruited not only by protein–protein interactions but also by noncoding RNAs.[74] The *Kcnq1ot1* long noncoding RNA plays an essential role in genome imprinting in humans and mice.[75] *Kcnq1ot1* regulates imprinting by maintaining the DNA methylation pattern through recruitment of Dnmt1, and that an 890-bp silencing domain located at the downstream of *Kcnq1ot1* promoter is required for this function.[74] Moreover, it appears that the *Kcnq1ot1* RNA directly interacts with Dnmt1.

IX. Concluding Remarks and Future Directions

Although the classical view recognizes Dnmt1 as the maintenance methyltransferase, it also appears to have considerable *de novo* activity, which could in fact be greater than the activity of the traditional *de novo* methyltransferases Dnmt3a and Dnmt3b[23] (see Chapter by Željko M. Svedružić). It has been suggested that Dnmt1 has two modes of action as a methyltransferase.[21] One is the classical maintenance activity that is imparted in a "processive" (moving on the DNA in a random quick walk) fashion on the DNA, and the other is the putative *de novo* activity that is mediated by a "distributive" (bind and slow release) manner. In normal cells, the *de novo* activity of Dnmt1 is presumably strictly repressed because such a function of Dnmt1 would lead to establishment of aberrant methylation patterns, which could lead to carcinogenesis. Indeed, these kinds of abnormal methylation patterns found in the promoters/regulatory regions of tumor suppressor genes are a hallmark of transformed cells (see Chapter by Minoru Toyota and Eiichiro Yamamoto). This suggests that, in normal cells, there may be some "repressor" or "checkpoint" mechanisms to prevent the *de novo* activity of Dnmt1, which somehow are impaired in cancer cells.

Some of the Dnmt1 interactors/recruiters that we have discussed here have been connected to carcinogenesis already, such as Daxx and hNaa10p. In addition, factors like STAT3 that do not interact with Dnmt1 in normal cells might behave abnormally in transformed cells to recruit Dnmt1 for silencing tumor suppressor genes (Fig. 3). Elucidation of the detailed mechanisms behind the aberrant recruitment pattern of Dnmt1 in transformed cells and the factors that are linked with this process will be crucial for designing novel therapeutic strategies for cancer.

We have described that Np95 functions as a recruiter molecule of Dnmt1 by virtue of its unique SRA domain. It is worthy of note that Np95/Uhrf1 is not the only SRA protein present in the mammalian genome. Another factor identified as Np97/Uhrf2 also possesses the SRA motif and is highly similar to Np95 in terms of protein domain structures (Fig. 1A). Akin to Np95, Np97 has been implicated in cell cycle progression and tumorigenesis.[32] It appears that Np97 also possesses E3 ubiquitin ligase activity, which is also similar to Np95.[50,76] Further, Np97 exerts a negative effect on the transcription of the *p21* tumor suppressor gene, a function that has been previously attributed to Np95.[50,77]

The similar protein domain structure and functions suggest that Np97 might also recruit Dnmt1. However, the knockout mice for *Np97* are viable and are apparently normal in terms of development (Sharif and Koseki, unpublished data). This raises two possibilities. First, the function of Np97 could be backed up by Np95 in a redundant manner (but not *vice versa*). Second, Np97 might have different functions from Np95 in terms of Dnmt1 recruitment. Elucidation of the roles of Np97 in DNA methylation and Dnmt1 interactions should be a rewarding field for future research.

Given the fact that Dnmt1 is the principal Dnmt in mammals, its functions in maintenance DNA methylation has been the major focus of previous research. However, recent results indicate that Dnmt1 forms complexes with various histone-modifying factors (such as HDAC1/2, G9a, and HP1) and could be involved in the transmission of repressive chromatin upon cell division (Fig. 3B). Further, the histone demethylase LSD1 stabilizes and increases the activity of DNMT1, showing that an interlinked epigenetic network exists between the histone modification and DNA methylation pathways (see Chapters by Xiaodong Cheng and Robert M. Blumenthal; and Taiping Chen). Involvement of Dnmt1 in epigenetic regulations that could be independent of DNA methylation must be addressed in future investigations.

References

1. Bird AP. DNA methylation and the frequency of CpG in animal DNA. *Nucleic Acids Res* 1980;**8**(7):1499–504.
2. Bird A, Taggart M, Frommer M, Miller OJ, Macleod D. A fraction of the mouse genome that is derived from islands of nonmethylated, CpG-rich DNA. *Cell* 1985;**40**(1):91–9.
3. Feng S, Cokus SJ, Zhang X, Chen PY, Bostick M, Goll MG, et al. Conservation and divergence of methylation patterning in plants and animals. *Proc Natl Acad Sci USA* 2010;**107**(19):8689–94.
4. Suzuki MM, Bird A. DNA methylation landscapes: provocative insights from epigenomics. *Nat Rev Genet* 2008;**9**:465–76.
5. Kidwell MG. Transposable elements and the evolution of genome size in eukaryotes. *Genetica* 2002;**115**(1):49–63.
6. Walsh CP, Chaillet JR, Bestor TH. Transcription of IAP endogenous retroviruses is constrained by cytosine methylation. *Nat Genet* 1998;**20**(2):116–7.
7. Sharif J, Muto M, Takebayashi S, Suetake I, Iwamatsu A, Endo TA, et al. The SRA protein Np95 mediates epigenetic inheritance by recruiting Dnmt1 to methylated DNA. *Nature* 2007;**450**(7171):908–12.
8. Oda M, Yamagiwa A, Yamamoto S, Nakayama T, Tsumura A, Sasaki H, et al. DNA methylation regulates long-range gene silencing of an X-linked homeobox gene cluster in a lineage-specific manner. *Genes Dev* 2006;**20**(24):3382–94.
9. Sharif J, Endo TA, Toyoda T, Koseki H. Divergence of CpG island promoters: a consequence or cause of evolution? *Dev Growth Differ* 2010;**52**(6):545–54.
10. Carninci P, Sandelin A, Lenhard B, Katayama S, Shimokawa K, Ponjavic J, et al. Genome-wide analysis of mammalian promoter architecture and evolution. *Nat Genet* 2006;**38**(6):626–35.
11. Li E, Beard C, Jaenisch R. Role for DNA methylation in genomic imprinting. *Nature* 1993;**366**:362–5.
12. Panning B, Jaenisch R. DNA hypomethylation can activate Xist expression and silence X-linked genes. *Genes Dev* 1996;**10**(16):1991–2002.
13. Li E, Bestor TH, Jaenisch R. Targeted mutation of the DNA methyltransferase gene results in embryonic lethality. *Cell* 1992;**69**:915–26.
14. Tsumura A, Hayakawa T, Kumaki Y, Takebayashi S, Sakaue M, Matsuoka C, et al. Maintenance of self-renewal ability of mouse embryonic stem cells in the absence of DNA methyltransferases Dnmt1, Dnmt3a and Dnmt3b. *Genes Cells* 2006;**11**(7):805–14.
15. Lei H, Oh SP, Okano M, Jüttermann R, Goss KA, Jaenisch R, et al. De novo DNA cytosine methyltransferase activities in mouse embryonic stem cells. *Development* 1996;**122**(10):3195–205.
16. Trowbridge JJ, Snow JW, Kim J, Orkin SH. DNA methyltransferase 1 is essential for and uniquely regulates hematopoietic stem and progenitor cells. *Cell Stem Cell* 2009;**5**(4):442–9.
17. Bestor TH, Ingram VM. Two DNA methyltransferases from murine erythroleukemia cells: purification, sequence specificity, and mode of interaction with DNA. *Proc Natl Acad Sci USA* 1983;**80**(18):5559–63.
18. Okano M, Bell DW, Haber DA, Li E. DNA methyltransferases Dnmt3a and Dnmt3b are essential for de novo methylation and mammalian development. *Cell* 1999;**99**(3):247–57.
19. Hata K, Okano M, Lei H, Li E. Dnmt3L cooperates with the Dnmt3 family of de novo DNA methyltransferases to establish maternal imprints in mice. *Development* 2002;**129**(8):1983–93.
20. Fatemi M, Hermann A, Pradhan S, Jeltsch A. The activity of the murine DNA methyltransferase Dnmt1 is controlled by interaction of the catalytic domain with the N-terminal part of the enzyme leading to an allosteric activation of the enzyme after binding to methylated DNA. *J Mol Biol* 2001;**309**(5):1189–99.

21. Hermann A, Goyal R, Jeltsch A. The Dnmt1 DNA-(cytosine-C5)-methyltransferase methylates DNA processively with high preference for hemimethylated target sites. *J Biol Chem* 2004;**279** (46):48350–9.
22. Ooi SK, Bestor TH. Cytosine methylation: remaining faithful. *Curr Biol* 2008;**18**(4):R174–6.
23. Okano M, Xie S, Li E. Cloning and characterization of a family of novel mammalian DNA (cytosine-5) methyltransferases. *Nat Genet* 1998;**19**:219–20.
24. Schermelleh L, Haemmer A, Spada F, Rösing N, Meilinger D, Rothbauer U, et al. Dynamics of Dnmt1 interaction with the replication machinery and its role in postreplicative maintenance of DNA methylation. *Nucleic Acids Res* 2007;**35**(13):4301–12.
25. Spada F, Haemmer A, Kuch D, Rothbauer U, Schermelleh L, Kremmer E, et al. DNMT1 but not its interaction with the replication machinery is required for maintenance of DNA methylation in human cells. *J Cell Biol* 2007;**176**(5):565–71.
26. Zhao X, Ueba T, Christie BR, Barkho B, McConnell MJ, Nakashima K, et al. Mice lacking methyl-CpG binding protein 1 have deficits in adult neurogenesis and hippocampal function. *Proc Natl Acad Sci USA* 2003;**100**(11):6777–82.
27. Guy J, Hendrich B, Holmes M, Martin JE, Bird A. A mouse Mecp2-null mutation causes neurological symptoms that mimic Rett syndrome. *Nat Genet* 2001;**27**(3):322–6.
28. Johnson LM, Bostick M, Zhang X, Kraft E, Henderson I, Callis J, et al. The SRA methyl-cytosine-binding domain links DNA and histone methylation. *Curr Biol* 2007;**17**(4):379–84.
29. Woo HR, Pontes O, Pikaard CS, Richards EJ. VIM1, a methylcytosine-binding protein required for centromeric heterochromatinization. *Genes Dev* 2007;**21**(3):267–77.
30. Liu S, Yu Y, Ruan Y, Meyer D, Wolff M, Xu L, et al. Plant SET- and RING-associated domain proteins in heterochromatinization. *Plant J* 2007;**52**(5):914–26.
31. Muto M, Utsuyama M, Horiguchi T, Kubo E, Sado T, Hirokawa K. The characterization of the monoclonal antibody Th-10a, specific for a nuclear protein appearing in the S phase of the cell cycle in normal thymocytes and its unregulated expression in lymphoma cell lines. *Cell Prolif* 1995;**28**(12):645–57.
32. Mori T, Li Y, Hata H, Ono K, Kochi H. NIRF, a novel RING finger protein, is involved in cell-cycle regulation. *Biochem Biophys Res Commun* 2002;**296**(3):530–6.
33. Muto M, Kanari Y, Kubo E, Takabe T, Kurihara T, Fujimori A, et al. Targeted disruption of Np95 gene renders murine embryonic stem cells hypersensitive to DNA damaging agents and DNA replication blocks. *J Biol Chem* 2002;**277**(37):34549–55.
34. Bostick M, Kim JK, Estève PO, Clark A, Pradhan S, Jacobsen SE. UHRF1 plays a role in maintaining DNA methylation in mammalian cells. *Science* 2007;**317**(5845):1760–4.
35. Arita K, Ariyoshi M, Tochio H, Nakamura Y, Shirakawa M. Recognition of hemi-methylated DNA by the SRA protein UHRF1 by a base-flipping mechanism. *Nature* 2008;**455** (7214):818–21.
36. Avvakumov GV, Walker JR, Xue S, Li Y, Duan S, Bronner C, et al. Structural basis for recognition of hemi-methylated DNA by the SRA domain of human UHRF1. *Nature* 2008;**455**(7214):822–5.
37. Hashimoto H, Horton JR, Zhang X, Bostick M, Jacobsen SE, Cheng X. The SRA domain of UHRF1 flips 5-methylcytosine out of the DNA helix. *Nature* 2008;**455**(7214):826–9.
38. Qian C, Li S, Jakoncic J, Zeng L, Walsh MJ, Zhou MM. Structure and hemimethylated CpG binding of the SRA domain from human UHRF1. *J Biol Chem* 2008;**283**(50):34490–4.
39. Fujimori A, Matsuda Y, Takemoto Y, Hashimoto Y, Kubo E, Araki R, et al. Cloning and mapping of Np95 gene which encodes a novel nuclear protein associated with cell proliferation. *Mamm Genome* 1998;**9**(12):1032–5.
40. Uemura T, Kubo E, Kanari Y, Ikemura T, Tatsumi K, Muto M. Temporal and spatial localization of novel nuclear protein NP95 in mitotic and meiotic cells. *Cell Struct Funct* 2000;**25** (3):149–59.

41. Hopfner R, Mousli M, Jeltsch JM, Voulgaris A, Lutz Y, Marin C, et al. ICBP90, a novel human CCAAT binding protein, involved in the regulation of topoisomerase IIalpha expression. *Cancer Res* 2000;**60**(1):121–8.
42. Hopfner R, Mousli M, Garnier JM, Redon R, du Manoir S, Chatton B, et al. Genomic structure and chromosomal mapping of the gene coding for ICBP90, a protein involved in the regulation of the topoisomerase II alpha gene expression. *Gene* 2001;**266**(1–2):15–23.
43. Jeanblanc M, Mousli M, Hopfner R, Bathami K, Martinet N, Abbady AQ, et al. The retinoblastoma gene and its product are targeted by ICBP90: a key mechanism in the G1/S transition during the cell cycle. *Oncogene* 2005;**24**(49):7337–45.
44. Kraft E, Bostick M, Jacobsen SE, Callis J. ORTH/VIM proteins that regulate DNA methylation are functional ubiquitin E3 ligases. *Plant J* 2008;**56**(5):704–15.
45. Meilinger D, Fellinger K, Bultmann S, Rothbauer U, Bonapace IM, Klinkert WE, et al. Np95 interacts with de novo DNA methyltransferases, Dnmt3a and Dnmt3b, and mediates epigenetic silencing of the viral CMV promoter in embryonic stem cells. *EMBO Rep* 2009;**10**(11):1259–64.
46. Walker JR, Wybenga-Groot L, Doherty RS, Finerty Jr. PJ, Newman E, Mackenzie FM, et al. *Ubiquitin-Like Domain of Human Nuclear Zinc Finger Protein NP95.*
47. Hashimoto H, Horton JR, Zhang X, Cheng X. UHRF1, a modular multi-domain protein, regulates replication-coupled crosstalk between DNA methylation and histone modifications. *Epigenetics* 2009;**4**(1):8–14.
48. Papait R, Pistore C, Grazini U, Babbio F, Cogliati S, Pecoraro D, et al. The PHD domain of Np95 (mUHRF1) is involved in large-scale reorganization of pericentromeric heterochromatin. *Mol Biol Cell* 2008;**19**(8):3554–63.
49. Karagianni P, Amazit L, Qin J, Wong J. ICBP90, a novel methyl K9 H3 binding protein linking protein ubiquitination with heterochromatin formation. *Mol Cell Biol* 2008;**28**(2):705–17.
50. Jenkins Y, Markovtsov V, Lang W, Sharma P, Pearsall D, Warner J, et al. Critical role of the ubiquitin ligase activity of UHRF1, a nuclear RING finger protein, in tumor cell growth. *Mol Biol Cell* 2005;**16**(12):5621–9.
51. Mistry H, Gibson L, Yun JW, Sarras H, Tamblyn L, McPherson JP. Interplay between Np95 and Eme1 in the DNA damage response. *Biochem Biophys Res Commun* 2008;**375**(3):321–5.
52. Estève PO, Chang Y, Samaranayake M, Upadhyay AK, Horton JR, Feehery GR, et al. A methylation and phosphorylation switch between an adjacent lysine and serine determines human DNMT1 stability. *Nat Struct Mol Biol* 2010; in press.
53. Voo KS, Carlone DL, Jacobsen BM, Flodin A, Skalnik DG. Cloning of a mammalian transcriptional activator that binds unmethylated CpG motifs and shares a CXXC domain with DNA methyltransferase, human trithorax, and methyl-CpG binding domain protein 1. *Mol Cell Biol* 2000;**20**(6):2108–21.
54. Carlone DL, Lee JH, Young SR, Dobrota E, Butler JS, Ruiz J, et al. Reduced genomic cytosine methylation and defective cellular differentiation in embryonic stem cells lacking CpG binding protein. *Mol Cell Biol* 2005;**25**(12):4881–91.
55. Butler JS, Lee JH, Skalnik DG. CFP1 interacts with DNMT1 independently of association with the Setd1 Histone H3K4 methyltransferase complexes. *DNA Cell Biol* 2008;**27**(10):533–43.
56. Michaelson JS, Bader D, Kuo F, Kozak C, Leder P. Loss of Daxx, a promiscuously interacting protein, results in extensive apoptosis in early mouse development. *Genes Dev* 1999;**13**(15):1918–23.
57. Puto LA, Reed JC. Daxx represses RelB target promoters via DNA methyltransferase recruitment and DNA hypermethylation. *Genes Dev* 2008;**22**(8):998–1010.
58. Muromoto R, Sugiyama K, Takachi A, Imoto S, Sato N, Yamamoto T, et al. Physical and functional interactions between Daxx and DNA methyltransferase 1-associated protein, DMAP1. *J Immunol* 2004;**172**(5):2985–93.

59. Estève PO, Chin HG, Smallwood A, Feehery GR, Gangisetty O, Karpf AR, et al. Direct interaction between DNMT1 and G9a coordinates DNA and histone methylation during replication. *Genes Dev* 2006;**20**(22):3089–103.
60. Kim JK, Estève PO, Jacobsen SE, Pradhan S. UHRF1 binds G9a and participates in p21 transcriptional regulation in mammalian cells. *Nucleic Acids Res* 2009;**37**(2):493–505.
61. Robertson KD, Ait-Si-Ali S, Yokochi T, Wade PA, Jones PL, Wolffe AP. DNMT1 forms a complex with Rb, E2F1 and HDAC1 and represses transcription from E2F-responsive promoters. *Nat Genet* 2000;**25**(3):338–42.
62. Fuks F, Burgers WA, Brehm A, Hughes-Davies L, Kouzarides T. DNA methyltransferase Dnmt1 associates with histone deacetylase activity. *Nat Genet* 2000;**24**(1):88–91.
63. Rountree MR, Bachman KE, Baylin SB. DNMT1 binds HDAC2 and a new co-repressor, DMAP1, to form a complex at replication foci. *Nat Genet* 2000;**25**(3):269–77.
64. Papait R, Pistore C, Negri D, Pecoraro D, Cantarini L, Bonapace IM. Np95 is implicated in pericentromeric heterochromatin replication and in major satellite silencing. *Mol Biol Cell* 2007;**18**(3):1098–106.
65. Lee CF, Ou DS, Lee SB, Chang LH, Lin RK, Li YS, et al. hNaa10p contributes to tumorigenesis by facilitating DNMT1-mediated tumor suppressor gene silencing. *J Clin Invest* 2010;**120**(8):2920–30. doi:10.1172/JCI42275.
66. Jair KW, Bachman KE, Suzuki H, Ting AH, Rhee I, Yen RW, et al. De novo CpG island methylation in human cancer cells. *Cancer Res* 2006;**66**(2):682–92.
67. Ting AH, Jair KW, Schuebel KE, Baylin SB. Differential requirement for DNA methyltransferase 1 in maintaining human cancer cell gene promoter hypermethylation. *Cancer Res* 2006;**66**(2):729–35.
68. Smallwood A, Estève PO, Pradhan S, Carey M. Functional cooperation between HP1 and DNMT1 mediates gene silencing. *Genes Dev* 2007;**21**(10):1169–78.
69. Wang J, Hevi S, Kurash JK, Lei H, Gay F, Bajko J, et al. The lysine demethylase LSD1 (KDM1) is required for maintenance of global DNA methylation. *Nat Genet* 2009;**41**(1):125–9.
70. Estève PO, Chin HG, Benner J, Feehery GR, Samaranayake M, Horwitz GA, et al. Regulation of DNMT1 stability through SET7-mediated lysine methylation in mammalian cells. *Proc Natl Acad Sci USA* 2009;**106**(13):5076–81.
71. Zhang Q, Wang HY, Marzec M, Raghunath PN, Nagasawa T, Wasik MA. STAT3- and DNA methyltransferase 1-mediated epigenetic silencing of SHP-1 tyrosine phosphatase tumor suppressor gene in malignant T lymphocytes. *Proc Natl Acad Sci USA* 2005;**102**(19):6948–53.
72. Zhang Q, Wang HY, Woetmann A, Raghunath PN, Odum N, Wasik MA. STAT3 induces transcription of the DNA methyltransferase 1 gene (DNMT1) in malignant T lymphocytes. *Blood* 2006;**108**(3):1058–64.
73. Hino R, Uozaki H, Murakami N, Ushiku T, Shinozaki A, Ishikawa S, et al. Activation of DNA methyltransferase 1 by EBV latent membrane protein 2A leads to promoter hypermethylation of PTEN gene in gastric carcinoma. *Cancer Res* 2009;**69**(7):2766–74.
74. Mohammad F, Mondal T, Guseva N, Pandey GK, Kanduri C. Kcnq1ot1 noncoding RNA mediates transcriptional gene silencing by interacting with Dnmt1. *Development* 2010;**137**(15):2493–9.
75. Smilinich NJ, Day CD, Fitzpatrick GV, Caldwell GM, Lossie AC, Cooper PR, et al. A maternally methylated CpG island in KvLQT1 is associated with an antisense paternal transcript and loss of imprinting in Beckwith-Wiedemann syndrome. *Proc Natl Acad Sci USA* 1999;**96**(14):8064–9.
76. Mori T, Li Y, Hata H, Kochi H. NIRF is a ubiquitin ligase that is capable of ubiquitinating PCNP, a PEST-containing nuclear protein. *FEBS Lett* 2004;**557**(1–3):209–14.
77. He X, Duan C, Chen J, Ou-Yang X, Zhang Z, Li C, et al. Let-7a elevates p21(WAF1) levels by targeting of NIRF and suppresses the growth of A549 lung cancer cells. *FEBS Lett* 2009;**583**(21):3501–7.

Regulation of Expression and Activity of DNA (Cytosine-5) Methyltransferases in Mammalian Cells

Shannon R. Morey Kinney and Sriharsa Pradhan

New England Biolabs, Ipswich, Massachusetts, USA

Three active DNA (cytosine-5) methyltransferases (DNMTs) have been identified in mammalian cells, Dnmt1, Dnmt3a, and Dnmt3b. DNMT1 is primarily a maintenance methyltransferase, as it prefers to methylate hemimethylated DNA during DNA replication and *in vitro*. DNMT3A and DNMT3B are *de novo* methyltransferases and show similar activity on unmethylated and hemimethylated DNA. DNMT3L, which lacks the catalytic domain, binds to DNMT3A and DNMT3B variants and facilitates their chromatin targeting, presumably for *de novo* methylation. There are several mechanisms by which mammalian cells regulate DNMT levels, including varied transcriptional activation of the respective genes and posttranslational modifications of the enzymes that can affect catalytic activity, targeting, and enzyme degradation. In addition, binding of miRNAs or RNA-binding proteins can also alter the expression of *DNMTs*. These regulatory processes can be

Progress in Molecular Biology
and Translational Science, Vol. 101
DOI: 10.1016/B978-0-12-387685-0.00009-3

disrupted in disease or by environmental factors, resulting in altered DNMT expression and aberrant DNA methylation patterns.

I. Transcriptional and Posttranscriptional Regulation of DNMTs

Progression through the cell cycle requires carefully controlled gene expression for the proper initiation and cessation of cell division, as well as for the correct inheritance of DNA methylation patterns. The process of shifting from an omnipotent or pluripotent cell (actively dividing) to a differentiated cell (normally arrested) also requires the coordinated regulation of many genes. Epigenetic factors are especially important for this process because differentiation does not coincide with changes in the genomic DNA sequence but does result in altered gene expression. Transcription is an early step of gene expression, and mammalian cells have developed complex and integrated mechanisms to regulate this process. Further, mRNA levels can be controlled once transcription has occurred by changes in mRNA stability. Regulation of mRNA encoding DNA methyltransferases (DNMTs) is not yet fully understood, but current knowledge indicates that it is an intricate system with multiple pathways converging to regulate transcription of these genes in various biological settings.

A. DNMT Expression During the Cell Cycle

The DNMT family displays cell cycle-specific expression, such that DNMT1 and DNMT3B mRNAs are expressed primarily during S-phase, whereas the DNMT3A transcript is expressed equally throughout cell division.[1,2] It is believed that increased expression of DNMT1 during S-phase correlates to its maintenance of DNA methylation patterns during replication. Because DNMT3A and DNMT3B are also expressed in S-phase, it is thought that they may contribute to maintenance methylation in addition to *de novo* methylation. This is supported by the fact that all DNMTs (DNMT1, DNMT3A, and DNMT3B) are localized in the nucleoplasm during S-phase, although methylation by DNMT3A and DNMT3B may not be occurring at the replication fork.[3] Little is known about the expression and cellular distribution of mammalian DNMT3L during cell division.

DNMT1 transcriptional activation was first identified at four regions in the *DNMT1* enhancer and promoter that include several putative AP-I and E2F binding sites.[4] Based on this observation, it was suggested that *DNMT1* transcription might be repressed by retinoblastoma (Rb) protein interaction with the E2F transcription factor (Fig. 1). Several years later, it was reported that *DNMT1* is indeed transcriptionally activated by E2F, and that Rb represses this activation.[5] In fact, loss of Rb resulted in deregulated cell cycle-specific

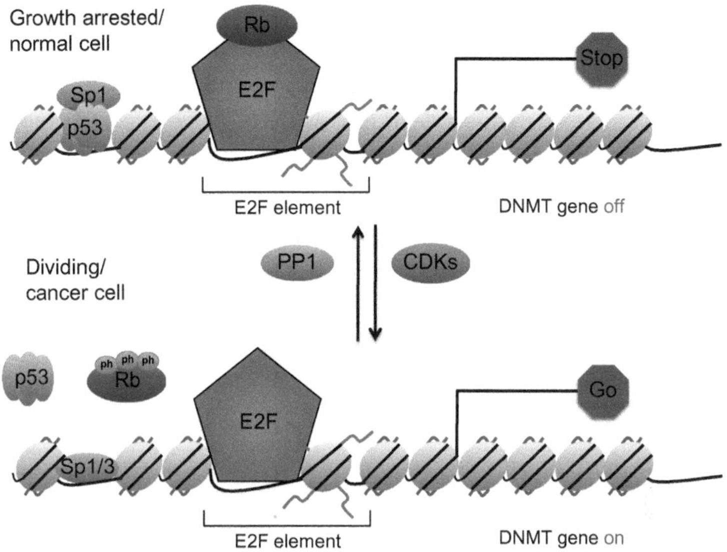

FIG. 1. *Transcriptional regulation of DNMTs.* This model represents the transcriptional regulation of DNMTs, especially DNMT1, by the Rb/E2F and p53/Sp1 pathways. The top section represents events during a growth arrested and/or normal state and the bottom outlines transcriptional activation in a dividing and/or tumor state. In an arrested cell, Rb is dephosphorylated by protein phosphatase 1 (PP1) or other phosphatases resulting in an unphosphorylated species, which then binds to E2F and prevents transcriptional activation. Sp1 and p53 also form a binary complex to repress *DNMT1* transcription in arrested or normal cells. As cells begin cycling, Rb is phosphorylated by cyclin-dependent kinases (CDKs) and can no longer bind to E2F. Rb may be deleted or mutated in cancer and therefore cannot repress E2F activity. The p53 protein is not normally expressed in an actively dividing cell and is commonly deleted or mutated in tumor cells. Thus, p53 cannot complete its normal functions, thereby allowing Sp1-mediated DNMT transcription.

transcription of DNMT1 mRNA.[5] The Rb/E2F pathway is well known for the role it plays in cell cycle regulation. When a normal cell is in the G_0 or G_1 phases of the cell cycle, Rb is hypophosphorylated and thus can bind to E2F proteins. This binding inhibits transcriptional activation of cell cycle proteins, such as cyclins, PCNA, and DNMT1.[6] As the cell begins dividing, cyclin-dependent kinases (CDKs) and cyclins hyperphosphorylate each Rb molecule, disrupting their binding to E2F. During progression through S-phase, RB becomes progressively dephosphorylated, allowing it to again bind to E2F, thereby inhibiting expression of E2F target genes.

DNMT1 mRNA expression is also regulated by transcriptional repression and mRNA destabilization. For example, p53 represses DNMT1 transcription, and this repression is abrogated by the specificity protein 1 (Sp1) transcription factor. Thus the p53/Sp1 ratio can determine whether *DNMT1* is activated or inactivated (Fig. 1).[7,8] Specifically, at low cellular concentrations, Sp1 acts as a corepressor forming a complex with p53 and binding to the DNMT1 promoter, whereas at high levels, it activates *DNMT1* transcription.[7] The p53 protein causes G_1/S arrest of the cell cycle in response to many types of cellular stress, including DNA damage.[9] In these situations, it would be necessary to stop DNA replication, eliminating the need for ancillary cellular processes such as maintenance methylation. In an additional mechanism, the AU-rich element RNA-binding protein 1 (AUF1), which is regulated in a cell cycle-specific manner, binds to an AU-rich sequence at the 3′ untranslated region (UTR) of DNMT1 mRNA accelerating degradation by the exosome.[10]

Less is known about the transcriptional regulation of the *DNMT3*. However, there is evidence that, despite DNMT3B displaying a different expression pattern than DNMT3A during the cell cycle, they are both regulated through similar transcriptional pathways.[11,12] The Sp family of transcription factors is ubiquitously expressed and interacts with many types of enzymes, including several cell cycle-specific proteins.[7,13] Several studies indicate that *DNMT3A* and *DNMT3B* are transcriptionally activated by Sp1 and Sp3 in a similar manner to DNMT1 (Fig. 1).[11,12]

Additional mechanisms have been identified for upregulation of DNMT3B mRNA levels other than Sp protein family-mediated activation. For example, knockout of *Vezf1*, a zinc finger DNA binding protein, results in decreased dnmt3b mRNA levels leading to DNA hypomethylation, suggesting that Vezf1 plays a role (direct or indirect) in the transcriptional activation of dnmt3b.[14] In a manner opposite to AUF1-mediated degradation of *DNMT1* transcripts, the HuR AUF stabilizes DNMT3B mRNA.[15] HuR is also known to stabilize cell cycle proteins Cyclin A and Cyclin B1 and thus plays a role in regulation of gene expression during cell division.[16]

The studies cited above demonstrate that multiple pathways are required for proper control of *DNMT* expression. Apart from these, several other signaling pathways can regulate DNMT mRNA levels in cancer cells (discussed later in this chapter). We do not yet fully understand the transcriptional and posttranscriptional regulation of DNMTs during the cell cycle, and it is likely that other unknown mechanisms are involved in this process depending on cell type or differentiation state.

B. DNMT Expression During Development

The mammalian nuclear genome undergoes waves of demethylation followed by *de novo* remethylation during early embryogenesis.[17] Once the oocyte has been fertilized, but prior to implantation, the paternal DNA methylation pattern is erased by an unknown active mechanism and the maternal DNA methylation is removed through a passive mechanism mediated by DNA replication.[17] Importantly, methylation patterns of imprinted genes are not changed during this process, and both paternal and maternal DNAs are remethylated by the time of embryonic implantation.[17] There is a second round of demethylation in the primordial germ cells once they enter the gonadal ridge, which occurs at approximately embryonic day E14.[17] At this stage, imprinted genes, in addition to single copy genes, become hypomethylated.[17] This process takes approximately 48 h, and it is unknown whether demethylation occurs through passive or active mechanisms, or to what extent the entire genome becomes hypomethylated.[17] The primordial germ cells then enter mitotic or meiotic arrest, for males and females, respectively.[17] In males, remethylation occurs a few days later at embryonic day E16 or E17 and the primordial germ cells are then able to continue replicating. In females, remethylation in primordial germ cells occurs after birth as oocytes divide.[17]

The understanding of different DNMT expression patterns and their involvement in remethylation and maintenance methylation in preimplantation embryos has evolved as the capabilities of studying this stage of development have improved. Originally, it was believed that only the Dnmt1o isoform (oocyte specific) was expressed in oocytes and preimplantation embryos and was solely responsible for the genomic methylation and imprinting during these stages of development.[18,19] Several reports later indicated that Dnmt1o, Dnmt1, Dnmt3a, and Dnmt3b are all found in unfertilized eggs and/or preimplantation embryos.[20–22] However, it appears that some of the DNMTs are residual maternal proteins suggesting that they are not newly transcribed in the zygote.[22,23]

Dnmt1o is primarily localized in the cytoplasm and, although it is somewhat controversial as later reports could not confirm this,[20,22] is thought to enter the nucleus for a period of time during the 8–16 cell stage of preimplantation.[19] It is during this period of nuclear localization that Dnmt1o is believed to be involved in maintaining maternal imprinting.[19] Interestingly, the Dnmt1o protein is much more stable than full-length Dnmt1, though the basis for this is not understood. The increased stability may allow for retention of the maternally produced ooplasmic Dnmt1o during the first few cell divisions, before it is transported into the nucleus as described above.[23] Dnmt1 (as opposed to Dnmt1o) is now thought to be expressed at low but sufficient levels to maintain DNA methylation in the early embryo.[20,21] Dnmt1 is localized inside the nucleus, especially in the maternal pronucleus during early preimplantation, with punctate localization in the paternal pronucleus of the fertilized oocyte.[20] Further, maternal Dnmt1 is observed only during the one- and two-cell stages, and additional Dnmt1 is produced by the zygote.[21] Similar to Dnmt1, it was thought that the *de novo* methyltransferases were not present in preimplantation embryos.[24,25] However, maternal Dnmt3a and zygotic Dnmt3b were observed during early and later stages of preimplantation, respectively.[22] Thus only Dnmt1 and Dnmt3b appear to be actively transcribed in the preimplantation embryo. Double knockout of Dnmt3a and Dnmt3b caused partial decreases in methylation at specific loci in the preimplantation embryo.[22,26] However, knockout of full-length DNMT1 had a major effect on imprinted genes and genomic DNA methylation patterns during preimplantation.[20,22] Genetic loss of *Dnmt1 or Dnmt3b* is embryonic lethal in mice, with major defects at embryonic day E9.5 and die soon thereafter.[24,27] *Dnmt3a*-deficient mice are not embryonic lethal, but only live to 4 weeks of age.[24] These studies suggest that each active DNMT has a unique role in, and is required for, normal DNA methylation and mammalian development.

Similar to the process of preimplantation, each DNMT displays a unique pattern of expression during the remethylation that occurs in primordial germ cells. Mouse *Dnmt3a and Dnmt3L* proteins are expressed first in the prenatal testis during remethylation of the paternal genome.[28] DNMT1 expression is specifically reduced during remethylation of the male germ cells and then restored during the early postnatal period along with Dnmt3b (where both enzymes presumably function as maintenance methyltransferases).[28] Though all of the Dnmts have similar low-level expression in prenatal oogonia, Dnmt3a, Dnmt3b, and Dnmt1 are expressed in greater quantities during the early postnatal period in the ovary during remethylation in the female germ line.[28] In somatic tissues, Dnmt1 expression is ubiquitous with highest expression in actively mitotic organs, such as the intestinal tract. Mouse Dnmt3a also displays fairly broad expression in adult tissues, while Dnmt3b tends to be expressed only in certain cell types. In comparison to DNMT1, the DNMT3 proteins are expressed at much lower levels and have approximately 20 times less catalytic activity in cell-free assays.[29]

Very little is known about the transcriptional regulation of DNMTs during development. A recent study reported that Dnmt1o repression during embryogenesis may occur through DNA methylation, whereas the promoter region of full-length *Dnmt1* does not become methylated.[30] As gene repression by DNA methylation is considered to be stable, this observation indicates that *Dnmt1o* downregulation is permanent, but repression of full-length Dnmt1 is transient. As described above, the Sp1 protein can transcriptionally activate mouse or human DNMT3B through binding in the proximal promoter region (Fig. 1). While this activation was found to be similar in both differentiated and undifferentiated cells, the *Dnmt3b* distal promoter contains several repressor elements that are nonfunctional in embryonic stem (ES) cells but are utilized in somatic cells.[11] Downregulation of *Dnmt3b* at these repressor elements is thought to function through DNA methylation or chromatin modifications.[11] Based on these studies and the general knowledge of epigenetic changes that occur throughout differentiation and development, it is plausible that *DNMT* transcription, at least in part, is regulated via epigenetic mechanisms during these processes.

II. Regulation of DNMTs by Posttranslational Modifications

Posttranslational histone modifications have long been associated with epigenetic regulation of gene expression. Many nonhistone proteins are also posttranslationally modified, including several epigenetic factors, resulting in alteration or modulation of their biological functions.[31] Recent studies have revealed that the catalytically active DNMTs are posttranslationally modified, and that these modifications can alter their activity and stability.[32–37] Although the posttranslational modification of DNMTs is becoming more commonly studied, it is unlikely that we have identified the full repertoire of these modifications and their downstream effects on DNA methylation, chromatin structure, and gene expression.

A. Interplay Between Lysine Methylation and Demethylation in DNMT1 Stability

SET7 belongs to a large family of protein lysine methyltransferases containing a SET domain.[31] It was originally thought that the function of these proteins was to target and methylate histones, resulting in various chromatin configurations and gene expression patterns. For example, SET7 monomethylates histone H3 lysine 4 (H3K4) and this modification functions as an active chromatin mark. Recently, SET7 has also been identified as a general protein lysine methyltransferase, as it methylates several nonhistone proteins, including p53, ER α, and DNMT1.[31,32] Methylation by SET7 occurs at lysine 142 of the DNMT1 protein, primarily during late S-phase of the cell cycle (Fig. 2).[32] Methylated DNMT1 is less stable and this mark appears to target the protein through ubiquitination for proteasomal degradation.[32]

Lysine methylation was previously considered to be irreversible.[31] However, discovery of protein lysine demethylases suggests that methylation of histones or other cellular proteins may be dynamically regulated depending on physiological variables, including cell cycle or differentiation. It was also previously thought that histones were the only target of the protein lysine-specific demethylase 1 (LSD1), as it was known to demethylate H3K4me1. Recent reports indicate that LSD1 can demethylate nonhistone proteins as well, such as p53.[38] Loss of LSD1 is associated with accelerated DNMT1 degradation resulting in DNA hypomethylation at H19 and IAP elements, suggesting that LSD1 normally functions to remove the methyl marks that target DNMT1 for degradation, possibly antagonizing SET7 activity by removing the methyl group on DNMT1 (Fig. 2).[37] This relationship underlines the complexity of posttranslational modifications and their biological roles as they respond to environmental stimuli.

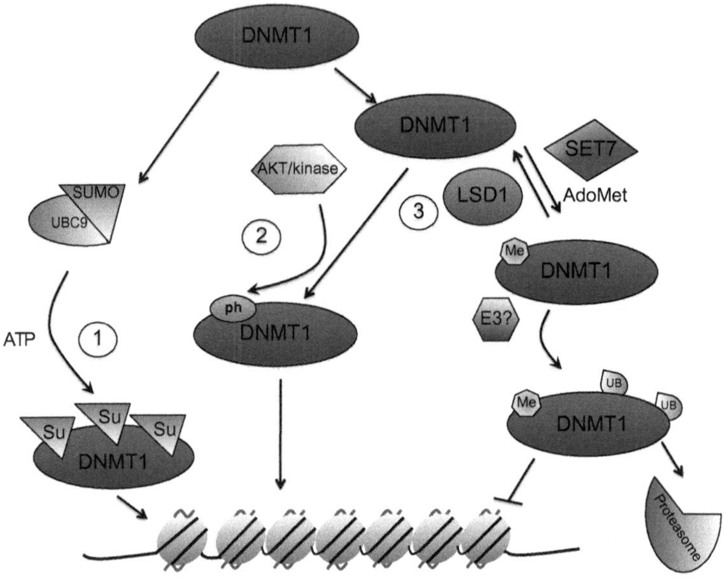

FIG. 2. Posttranslational modifications of DNMT1. DNMT1 is posttranslationally modified by three different mechanisms. (1) DNMT1 sumoylation by UBC9 and SUMO-1 results in an increase of DNMT1 binding to DNA. (2) Phosphorylation of DNMT1 by AKT, and other unknown protein kinases, may change the conformation of DNMT1 and increase its stability and/or activity. (3) Lysine methylation of DNMT1 by SET7, and possible demethylation by LSD1 negating the established SET7 mark on DNMT1, results in a species that is ubiquitinated, destabilized, and targeted for degradation in the proteasome.

B. Methylation–Phosphorylation Switch in DNMT1 Stability and Activity

As stated above, SET7 methylates DNMT1 at lysine 142 and this leads to DNMT1 degradation.[32] Interestingly, the adjacent amino acid residue, serine 143, of DNMT1 was identified in a proteomics analysis to be phosphorylated during mitosis.[39] The role of protein phosphorylation in regulating cell signaling pathways, altering catalytic activity, and affecting DNA or protein-binding capabilities is an area of intense research. In the case of DNMT1, phosphorylation of serine 143 and methylation at lysine 142 appear to be mutually exclusive on the endogenous protein. In addition, a peptide representing serine 143 phosphorylation blocked SET7 methylation in cell-free assays.[40] The phosphorylated form of DNMT1 also displayed an increased half-life as compared to the methylated form (Fig. 2).[40] As the surrounding amino acid residues formed an AKT1 kinase target motif, it was next tested whether this was the main kinase. Indeed, overexpression or pharmacological activation of

AKT1 led to increased phosphorylation of DNMT1.[40] Conversely, treatment of cells with an AKT1 inhibitor caused DNMT1 degradation (presumably through increased methylation by SET7) and global DNA hypomethylation.[40]

Although it is still unknown which other kinase(s) are responsible, DNMT1 also becomes phosphorylated at serine 515, which is located in the replication foci targeting domain.[41] Phosphorylation at this site is suggested to increase catalytic activation of DNMT1 in cell-free assays (Fig. 2).[33] A small peptide containing the phosphorylated serine, and surrounding amino acids, inhibited wild-type Dnmt1 activity to a much greater extent than an unmodified peptide, suggesting that phosphorylation at this site is important for protein folding or domain interactions that were disrupted by the phosphorylated peptide.[33] One controversial aspect of this study was that a serine 515 DNMT1 mutant (S515A) displayed severe loss of catalytic activity though it could still bind to DNA *in vitro*.[33] It was subsequently reported that deletion of the first 580 amino acids of DNMT1 resulted in a catalytically active protein that maintained its preference for hemimethylated DNA in cell-free assays.[42] Based on these studies, it appears that multiple signaling cascades function together to maintain balanced DNMT1 protein levels and catalytic activity. If any of these modifying factors are disrupted, which may occur in disease,[43] DNMT1 expression or activity could be severely affected resulting in aberrant DNA methylation.

C. DNMT Regulation by Sumoylation

Sumoylation is a recently identified posttranslational modification that is chemically similar to ubiquitination but can have very different downstream effects. Sumoylation requires two enzymes, the E1-activating enzyme and the E2-conjugating enzyme, as well as ATP. Ubc9 is one member of the E2 family that is involved in recognizing the sumoylation motif on targeted proteins. Sumoylation is known to function by stabilizing proteins and protecting them from degradation, as well as affecting cellular localization (targeting proteins to the nucleus), protein–protein interactions, and DNA binding. Sumoylation of DNMT1 was first observed as specific shifts in molecular weight as determined by Western blot after incubation in a cell-free sumoylation assay, or following coexpression with Ubc9 and SUMO-1 (a small ubiquitin-like modifier that is conjugated to its substrate upon catalysis; Fig. 2).[35] DNMT1 binds to SUMO-1 and Ubc9 and becomes sumoylated at several lysines throughout the protein. Subsequently, this appears to increase DNA binding by DNMT1, which would hypothetically intensify methyltransferase activity.[35]

In addition to their DNMT activity, DNMT enzymes can act as transcriptional repressors, even when catalytically inactivated by active site mutation. For example, DNMT3A binds to histone deacetylases (HDACs) HDAC1 and HDAC2, through its N-terminal PHD-like domain, at promoter regions, resulting in transcriptional repression.[36] Thus, the many effects that sumoylation can have on protein–protein binding may have important implications for DNMT biological activity. Both DNMT3A and DNMT3B have been shown in separate studies to become sumoylated by SUMO-1.[34,36] Sumoylation of DNMT3A by SUMO-1 disrupted the interaction between DNMT3A and HDACs, allowing transcriptional activation of genes normally repressed by this complex.[36] A very similar observation was made with DNMT3B, which can also be modified by SUMO-1 within its N-terminal region.[34] Although the biological effects of this modification of DNMT3B were not tested, it is possible that, similar to sumoylation of DNMT3A, it would affect binding of DNMT3B to other proteins and DNA. These types of alterations in DNMT activity could have major epigenetic effects without changing the actual protein levels. It is unknown whether sumoylation of DNMTs affects global or locus-specific DNA methylation. Studies that examine this aspect are required to further our understanding of where and when these pathways are utilized in normal growth and development.

III. Altered Regulation of DNMTs During Disease

DNA methylation patterns are disrupted in a number of diseases, including many kinds of cancer, several imprinting disorders, various autoimmune diseases, and multiple neuronal disorders.[44] Of these diseases, cancer and imprinting disorders have been the main focus of research, while the study of autoimmune or neuronal disease has only recently come to the forefront. In general, DNMT expression tends to be upregulated in tumors and associated with altered DNA methylation patterns.[29] The overall observation is that DNA hypermethylation occurs at some promoter regions (tumor suppressor genes) and DNA hypomethylation takes place at other gene promoters (oncogenes) and within repetitive elements in cancer.[45] Although no known DNMT1 mutation has been associated with a given disease, several imprinting disorders are associated with mutations in DNMT3B.[44] Patients with imprinting defects can present with mental retardation as well as other physical abnormalities (see Chapter by Jon F. Wilkins and Francisco Úbeda).[44] Patients with autoimmune disease, such as Lupus, display global hypomethylation as well as promoter-specific hypomethylation similar to what is observed in cancer, but unlike cancer, locus-specific hypermethylation has not been detected in autoimmune diseases.[46] In fact, DNA hypomethylating agents have been found to cause Lupus-like disorders.[46] In contrast, schizophrenia is associated with gene-specific hypermethylation, which reduces expression of proteins involved in normal neuronal function.[47] Therefore, DNMTs and DNA methylation clearly have important and complex roles in disease, and diseased cells have evolved to exploit the many mechanisms that regulate DNMT expression.

A. Regulation of DNMT Expression by miRNAs

MicroRNAs (miRNAs) are small noncoding RNAs, processed through DICER that can bind to mRNA with partial to full complementarity at the 3′UTR. Once bound to target mRNA, miRNAs can either cause degradation of the message within the RISC complex or block protein translation. Whether the mRNA is destroyed or the miRNA merely blocks translation appears to be dependent on the amount of similarity between the miRNA and the target mRNA, but in either case, this leads to decreased protein expression.[48] Several families of miRNA have been shown to affect DNMT expression, either directly or indirectly.

One example of indirect regulation was identified in DICER null cells that display methylation defects, including DNA hypomethylation of telomeric regions.[49] Expression of all three active DNMTs were reduced in these cells, and this downregulation was linked to the RbL2 protein.[49] The authors subsequently found that the Rb proteins, especially RbL2, displayed increased expression in the DICER null cells and that the Rb proteins are normally regulated through the RNAi pathway by the mir290 family of miRNAs.[49] This study holds additional significance as it confirms that not only DNMT1 but also DNMT3A and DNMT3B are regulated to some extent by the Rb-E2F pathway (Fig. 1).

More recently, DNMT1 was shown to be directly affected by IL-6-regulated miRNAs. Two of these miRNAs (miR-148a and miR-152) bound to the 3′UTR of DNMT1 message and decreased DNMT1 enzyme levels in human cholangiocarcinoma cell lines, resulting in demethylation and increased expression of several tumor suppressor genes.[50] Another study demonstrated direct regulation of DNMT3A and DNMT3B by the miR29 family of miRNAs.[51] MiR29 inhibits expression of DNMT3A, and more so DNMT3B, through apparent binding at the 3′UTRs of these genes. The resulting downregulation of DNMT3 proteins led to altered DNA methylation patterns in lung cancer cells, to a similar extent as was seen with DNMT inhibitors, such as 5-aza-2′-deoxycytidine.[51] More recently, miR29b was shown to also decrease expression of Sp1, which is a known transcriptional activator of DNMT1, DNMT3A, and DNMT3B.[52] MiR29b alone could decrease the expression of all three active DNMTs and cause global hypomethylation in several leukemia cell lines.[52]

In a recent attempt to identify unique miRNAs involved in the regulation of DNMT genes, a novel miRNA mechanism was discovered in mammalian cells.[53] As stated above, miRNAs typically bind to 3′UTRs of mRNA and thereby decrease gene expression. Alternatively, miR-148 instead binds to the coding region of three DNMT3B variants, DNMT3B1, DNMT3B2, and DNMT3B4, resulting in their reduced expression.[53] Although this phenomenon is common in plants, this is the first report of a functional miRNA binding site in the coding region of mammalian mRNAs.[53] Therefore, it appears that various noncoding RNAs, which may be disrupted in disease, can modulate *DNMT* expression.

B. Disruption of Tumor Suppressor Genes Alters DNMT Transcription

Many tumor suppressor genes normally function as transcription factors or regulate transcription through cellular signaling pathways. Deletion or mutation of these genes leads to loss of function. This is true for several tumor suppressor proteins that are either normally involved in *DNMT* transcription or upon mutation display increased transcriptional activation of DNMTs. As described previously, two commonly mutated tumor suppressor genes, Rb and p53, can regulate DNMT transcription. DNMT expression has been studied in tumor models that inhibit Rb and p53 either by genetic knockout/mutation or binding of viral antigens, as well as in human tumor samples.[54–56] In the transgenic adenocarcinoma of mouse prostate (TRAMP) model, regulation by Rb and p53 is abrogated through binding of the SV40 Tag, and levels of all three active DNMT enzymes are increased.[57–59] Although part of the increased expression appeared to be a result of increased cell cycling in the tumor, this did not fully account for the level of DNMT expression that was observed in the TRAMP tumors.[59] In another study using human lung cancer samples, mutation of p53 was correlated with increased Sp1 and DNMT1 expression (Fig. 1).[7]

Breast cancer 1 (*BRCA1*) is another well-known tumor suppressor gene that is commonly mutated in breast tumors. It was recently shown that BRCA1 generally activates the transcription of *DNMT1* through binding at an organic cation transporter 1 (*OCT1*) site. BRCA1 deficiency is associated with decreased *DNMT1* expression and global DNA hypomethylation.[60] As *BRCA1* mutations are commonly passed down through the germline, this would indicate that a woman carrying a mutant *BRCA1* allele could have lower levels of DNMT1 and DNA methylation in premalignant breast tissue.[60] Based on several studies using mDnmt1 hypomorphic mouse models, DNA hypomethylation causes genomic instability and promotes tumor initiation.[61–63] Although the scenario observed in BRCA1 mutant cells is unique, as most tumors express DNMTs at higher levels than their normal counterpart, this could still play a role in carcinogenesis. It is also possible that at later stages of progression, these tumors will accumulate other mutations that could result in increased DNMT expression and locus-specific DNA hypermethylation. Further, cancer is a heterogeneous disease and while some patients carry a mutation in a specific gene, such as BRCA1, other patients do not (see Chapter by Minoru Toyota and Eiichiro Yamamoto).

Rat sarcoma (RAS) small GTPase is a protein involved in several signaling cascades of the cell, including the MEK/ERK and the PI3K/JUN/FOS pathway that play various roles in cell growth, differentiation, and survival.[64] As these pathways are important for cancer initiation and development, it is not surprising that *RAS* is commonly mutated during tumorigenesis. The JUN/FOS transcription factors have been shown to activate *Dnmt1* transcription through the AP-1 target motifs.[65] Further, oncogenic RAS-stimulated signal transduction pathways increase *Dnmt1* expression and alter DNA methylation patterns.[66] These observations establish functional disruption of tumor suppressors as a nodal cause of DNMT1-mediated misregulation of DNA methylation.

C. Increased Stability of DNMT Proteins in Cancer

As mentioned above, some breast tumors display BRCA1 mutations and decreased DNMT1 transcription. In contrast, the breast cancer cell line (MCF-7) has wild-type BRCA1, and normal DNMT1 mRNA levels, but increased DNMT1 protein levels.[67] The increased levels were not related to the S-phase stability of DNMT1 protein that is observed in normal cells.[67] Although the authors did not determine the exact mechanism, increased DNMT1 protein stability in MCF-7 cells is responsible for the increased levels and depends on the N-terminal 118 amino acid residues, which when deleted lead to ubiquitination and proteasomal degradation of the protein.[67]

The RAS pathway upregulates Dnmt1 indirectly, through increased cell cycling, and directly via transcriptional activation.[66] The PI3K pathway, which is downstream of RAS, is also involved in DNMT1 stability.[68] PI3K/PKB/Gsk3β signaling increases DNMT1 stability, which is dependent on the first 120 amino acid residues in the DNMT1 sequence (similar to the report described above in MCF-7 cells).[68] Unless the increased stability of DNMT1 is caused by direct binding of these proteins, there is a missing link between the pathways identified here (BRCA1 and RAS) and the presumed modification of DNMT1 protein that is inhibiting degradation by the proteasome. Thus it is likely that other pathways and/or modifying enzymes may be involved in these particular mechanisms.

D. Altered Expression of DNMT3B Variants in Cancer

There are several known transcript variants of DNMT3B that can result in proteins with altered catalytic activity. Initial studies identified two additional DNMT3B variants in tumor cells, DNMT3B4 and DNMT3B5, which lack DNA (cytosine-5) methyltransferase motifs IX and X and thus are thought to be catalytically inactive.[69] Levels of all three active DNMTs were elevated in several tumor types, including bladder, colon, and kidney, as compared to normal adjacent tissue.[69] However, in the panel of tumor samples used in this study, DNMT3B displayed the largest increase in expression of the three genes.[69] Although each expressed variant was not strictly quantified, the tumors appeared to have altered expression of DNMT3B variants as compared to each other and their normal matched tissue.[69]

DNMT3B was later examined for mutations and splice variant expression in precancerous liver diseases and in hepatocellular carcinoma (HCC) tissues.[70] Although no mutations were identified in the DNMT3B gene, there was increased expression of DNMT3B4, especially in conjunction with decreased expression of DNMT3B3, which was correlated with DNA hypomethylation at pericentric satellite regions in precancerous lesions and in HCC.[70] Further, overexpression of DNMT3B4 in HEK293 cells led to hypomethylation of satellite 2 regions of pericentric heterochromatin.[70]

In another study, wherein Dnmt3b variant expression was carefully examined in a variety of cancer cell lines and tissues, a third novel *DNMT3B* transcript was identified, *DNMT3B7*, which contains an insert that is part of intron 11.[71] All cell lines that were examined expressed DNMT3B variants with altered splicing in the 5′ region, except the HCC cells that expressed DNMT3B4 as described above.[70,71] These variants were expressed in both primary tumor tissues and tumor cell lines and all of them lack the C-terminal catalytic domain.[71] Expression of the DNMT3B7 variant in HEK293 cells led to both increased and decreased expression of target genes correlating with promoter hypomethylation or hypermethylation, respectively.[71] This suggests that overexpression of *DNMT3B* variants, through an unknown mechanism, disrupts DNA methylation pathways such that both DNA hypomethylation and hypermethylation can occur. The above scenario closely matches the aberrant DNA methylation observed in cancer, in which normally methylated regions become hypomethylated, in the presence of high levels of DNMT proteins. Hypomethylation, especially in the case of repetitive elements and telomeres, leads to chromosomal instability. In addition, other normally unmethylated loci, such as tumor suppressor gene promoters, become hypermethylated, potentially repressing their expression and thus facilitating tumor development.

E. DNMT1 Expression in Autoimmune and Allergic Disorders

DNA hypomethylation is a hallmark of autoimmune disease, and DNMT inhibitors can cause autoimmune-like disorders, so it might be expected that DNMTs are underexpressed in these types of disease.[72] Atopic dermatitis (AD) is an inherited allergic inflammatory condition that chronically relapses.[73] Patients who suffer from this disorder are very sensitive to environmental allergens and skin irritations. AD lesions are characterized by local infiltration of T-helper 2 (TH2) cells in response to interleukins. These cells are involved in the IgE response and chronic inflammation that are hallmarks of this disease. DNA hypomethylation increases both the activity of TH2 cells and the production of IgE.[74] DNMT1 mRNA expression was slightly decreased in peripheral blood mononuclear cells of AD patients, and this was significant in a group of patients displaying high IgE levels as compared to control samples.[73]

Systemic lupus erythematosus (SLE) is a chronic autoimmune disease that frequently affects several organs, including the heart, joints, skin, kidneys, and nervous system.[75] Multiple immunological pathways and immune cell types are activated in SLE.[76] Global and locus-specific DNA hypomethylation have been found in T cells from SLE patients, and this is correlated with activation of autoimmune-associated genes such as *CD70* and *LFA-1*.[76] Because miRNAs are involved in the regulation of *DNMTs* in normal and cancer cells, it is hypothesized that this phenomenon may be operative in SLE as well.[76] MiR-21 and miR-148a are two upregulated miRNAs in CD4+ T cells from SLE patients and MRL/lpr mice, which are commonly used as a murine model of Lupus.[76] Further, IL-6-dependent increases in miR-148a also affect DNMT1 expression.[50] When miR-21 and miR-148a were transfected into cells, expression of DNMT1 and DNMT3B was decreased. Additionally, *CD70* and *LFA-1* gene promoters became hypomethylated and were expressed.[76] More studies are required to determine the involvement of DNMTs in allergic and autoimmune disease. However, these studies indicate that epigenetic pathways may have a role in autoimmune development, which expands on previous views that these were essentially genetic disorders.

F. DNMT1 Expression in Schizophrenia

Reelin and glutamic acid decarboxylase (GAD_{67}) expression is decreased in the gamma-amino butyric acid (GABA)-ergic interneurons of brain samples from patients with schizophrenia, as compared to normal control samples.[77] There are several pieces of evidence indicating that transcriptional regulation of *Reelin* and GAD_{67} expression may normally occur via promoter DNA hypermethylation. For example, promoter hypermethylation status correlates with decreased expression of these genes and treatment of cells with hypomethylating agents activates *Reelin* expression.[78–80] In addition, Reelin is hypermethylated in GABAergic neurons of schizophrenia patients, as compared to normal neurons.[81] *DNMT1* expression is increased in GABAergic neurons, which inversely correlates with GAD_{67} mRNA levels.[82] Further, gene knockdown experiments in cortical cell cultures link DNMT1 to the hypermethylation and repression of GAD_{67} and *Reelin*.[80] More studies are required to determine the extent of aberrant DNA methylation in the brains of schizophrenic patients, as well as the role of DNMTs in temporal and spatial gene regulation in the brain.

IV. Drug-Induced Reductions in DNMT Levels

The previous section describes altered DNMT expression in various disease. The involvement of DNMTs in the initiation and development of disease has spurred the identification and generation of DNMT inhibitors. There are several DNMT inhibitors that are derived from both natural and synthetic sources.[83,84] The exact mechanism of how these drugs function is still not fully understood. Some of these drugs appear to work by directly binding to DNMTs and thereby inhibiting their catalytic function, as with RG108.[83,84] Procainamide is believed to function by binding to CG-rich sequences in the genome, preventing DNMT binding and methylation.[83,84] The nucleoside analogs 5-azacytidine (5-aza), 5-aza-2-deoxycytidine (5-Aza-dC), and zebularine all have a similar mechanism involving incorporation into DNA during replication, resulting in an irreversible covalent linkage with DNMTs at the 6-carbon position.[83,84] This decreases available enzymes and thus the DNMTs are no longer able to maintain DNA methylation patterns, causing global DNA hypomethylation. Recent reports suggest that these drugs, and other epigenetic targeting molecules, may also lead to degradation of DNMTs, thereby having demethylating effects even in cells that are not cycling.

A. Degradation of DNMTs by Nucleoside Analogs

Three studies report induction of proteasomal degradation by DNMT inhibitors. The first two utilize the nucleoside analogs 5-Aza and 5-Aza-dC, while the third identifies a novel DNMT inhibitor SGI-1027, a quinoline analog, which appears to compete with *S*-adenosyl-L-methionine (AdoMet) for the cofactor-binding site of DNMTs.[85–87] As described above, one known mechanism of 5-Aza and 5-Aza-dC is to deplete the free DNMTs by covalent binding. However, it seems that these drugs also cause DNMT1 degradation through the well-established ubiquitin-dependent proteasomal pathway.[86,87] The ubiquitin protein ligases or E3 enzymes are responsible for identifying target motifs containing lysine residues and then catalyzing the bond between the lysine of the substrate protein and ubiquitin. The E3 ubiquitin ligase CDH1 is involved in regulating protein levels during late M and G_0 transition to anaphase.

Interestingly, the two reports of cytosine analogs causing DNMT1 degradation have some contradictory results. The earlier study reported that DNMT1 physically interacts with CDH1[86] and that targeted degradation through the proteasome does not depend on DNA replication but does require several domains within the DNMT1 protein, including the KEN box motif in the zinc-binding domain, the bromo adjacent homology domain, and the nuclear localization sequence.[86] While both studies are in agreement that the proteasome is involved, the later report could not confirm the original conclusion that CDH1 is the only ubiquitin ligase involved in this particular pathway.[87] In addition, the second study reported that DNA replication-mediated 5-Aza-dC incorporation and *DNMT1* adduct formation are a prerequisite for DNMT1 degradation.[87]

The third study characterizes a novel DNMT inhibitor (SGI-1027) that selectively causes degradation of DNMT1 but not DNMT3A or DNMT3B in cancer cells. This specific degradation also occurs through the proteasomal pathway, similar to degradation caused by cytosine analogs.[85] In fact, the targeting of DNMT1 for proteasomal degradation with SGI-1027 could be due to adduct formation as well. For example, DNMT proteins that are in the process of catalyzing a methyltransfer reaction on DNA and are bound by the SGI-1027 (AdoMet competitor) could form stable stalled reaction intermediates, covalently linked to the DNA. This distinction will be important in identifying drugs that cause DNMT degradation, as it will affect whether they are active in cycling or arrested cells. Cancer cells should be preferentially sensitive to drugs that require cell cycling to be active. Reciprocally, normal cells may be unacceptably sensitive to drugs that cause DNA hypomethylation in nonmitotic cells leading to pronounced side effects with treatment.

B. Destabilization of DNMTs by HDAC Inhibitors

Posttranslational modification of histones plays a central role in epigenetic gene regulation. Histone acetylation is commonly associated with an open chromatin state and gene expression, while histone deacetylation causes chromatin condensation and gene repression.[88] The patterns of histone modifications are disrupted in diseases such as cancer, in addition to the aberrant DNA methylation described above.[88] As HDAC enzymes are responsible for removal of acetyl groups from histones, they have been a focus of inhibitor development. A number of HDAC inhibitors have been identified, and some are in clinical trials for use as therapeutic agents, such as trichostatin A (TSA) and Vorinostat (SAHA).[88]

Because HDACs have also been shown to deacetylate nonhistone proteins, HDAC inhibitors are being studied for alternative effects. For example, the HDAC inhibitors LBH589 and SAHA each leads to ubiquitination and proteasomal degradation of DNMT1.[89] HSP90 is normally bound to DNMT1 and functions as a chaperone, stabilizing DNMT1 and preventing it from becoming ubiquitinated.[89] Acetylation of HSP90 disrupts this interaction, and deacetylation by HDAC1 is required for secure binding to DNMT1.[89] This study underlines the complexity of mechanisms for novel therapeutics and the difficulty in understanding drug effects in normal or diseased cells. Further, this raises the possibility that many commonly used drugs may be functioning through epigenetic pathways in addition to previously established mechanisms. For example, hydralazine, a commonly used antihypertensive drug, and valproic acid, used for epilepsy and psychological disorders, cause DNA hypomethylation either by direct inhibition of DNMT1 or by altering *DNMT1* regulation.[90–93]

V. Concluding Remarks and Future Directions

The DNMTs are essential for mammalian development and in maintaining normal cellular functions. Their expression and activity are regulated on many levels, including transcriptional activation and mRNA stability, as well as protein targeting, localization, activation, and stabilization. Disruption of any of these mechanisms can result in aberrant DNMT expression and activity, as seen in diseases such as cancer and autoimmune disorders. Altered DNMT expression is related to changes in genome-wide DNA methylation patterns that can have potent effects on expression of the large number of genes that are controlled by promoter methylation. Continued study of DNMT enzyme regulation in normal development and deregulation in disease could potentially identify novel biological phenomena and targets for treatment.

In the past several years, researchers have gained some insight into the regulation of DNMT enzymes. However, the cumulative knowledge suggests that the regulation of DNMTs is quite complicated and multifaceted. More studies are required to understand how these enzymes function in multiple cell types, developmental stages, as well as various disease states. With new technologies, we hope to gain better understanding of these mechanisms (see Chapter by Eleanor Wong and Chia-Lin Wei). For example, chromatin immunoprecipitation followed by next-generation sequencing should help us to identify transcription factors, coactivators, and corepressors that are bound to any of the DNMT gene promoters. In fact, it is likely that there are current datasets that include this information already, that only require data mining and follow-up experiments, to expand the list of factors that are involved in activating and repressing transcription of DNMTs.

Purification of protein complexes and use of mass spectrometry have also advanced greatly in recent years and will certainly continue to do so as the demand for better proteomic analysis increases. These types of experiments are not trivial, but there already are several laboratories with expertise in copurifying interacting proteins in sufficient quantities that individual proteins can then be identified by mass spectrometry. Posttranslational modifications can also be identified by mass spectrometry with optimized methods and analytical programs. These types of studies will help us to determine how DNMT activity and protein stability are regulated in the cell. The experiments discussed above and numerous additional intriguing questions promise the continued research of the regulation and activation of DNMTs involved in epigenome maintenance.

Acknowledgments

We thank Pierre Olivier Esteve, Thomas C. Evans, and William Jack for suggestions and advice on the chapter. We thank Drs. Donald G. Comb and Richard J. Roberts, Mr. James V. Ellard, and New England Biolabs, Inc. for supporting the basic research.

References

1. Robertson KD, Keyomarsi K, Gonzales FA, Velicescu M, Jones PA. Differential mRNA expression of the human DNA methyltransferases (DNMTs) 1, 3a and 3b during the G(0)/G (1) to S phase transition in normal and tumor cells. *Nucleic Acids Res* 2000;**28**:2108–13.
2. Szyf M, Bozovic V, Tanigawa G. Growth regulation of mouse DNA methyltransferase gene expression. *J Biol Chem* 1991;**266**:10027–30.
3. Margot JB, Cardoso MC, Leonhardt H. Mammalian DNA methyltransferases show different subnuclear distributions. *J Cell Biochem* 2001;**83**:373–9.
4. Bigey P, Ramchandani S, Theberge J, Araujo FD, Szyf M. Transcriptional regulation of the human DNA methyltransferase (dnmt1) gene. *Gene* 2000;**242**:407–18.
5. McCabe MT, Davis JN, Day ML. Regulation of DNA methyltransferase 1 by the pRb/E2F1 pathway. *Cancer Res* 2005;**65**:3624–32.
6. Calzone L, Gelay A, Zinovyev A, Radvanyi F, Barillot E. A comprehensive modular map of molecular interactions in RB/E2F pathway. *Mol Syst Biol* 2008;**4**:173.
7. Lin RK, Wu CY, Chang JW, Juan LJ, Hsu HS, Chen CY, et al. Dysregulation of p53/Sp1 control leads to DNA methyltransferase-1 overexpression in lung cancer. *Cancer Res* 2010;**70**:5807–17.
8. Peterson EJ, Bogler O, Taylor SM. p53-mediated repression of DNA methyltransferase 1 expression by specific DNA binding. *Cancer Res* 2003;**63**:6579–82.
9. Levine AJ. p53, the cellular gatekeeper for growth and division. *Cell* 1997;**88**:323–31.
10. Torrisani J, Unterberger A, Tendulkar SR, Shikimi K, Szyf M. AUF1 cell cycle variations define genomic DNA methylation by regulation of DNMT1 mRNA stability. *Mol Cell Biol* 2007;**27**:395–410.
11. Ishida C, Ura K, Hirao A, Sasaki H, Toyoda A, Sakaki Y, et al. Genomic organization and promoter analysis of the Dnmt3b gene. *Gene* 2003;**310**:151–9.
12. Jinawath A, Miyake S, Yanagisawa Y, Akiyama Y, Yuasa Y. Transcriptional regulation of the human DNA methyltransferase 3A and 3B genes by Sp3 and Sp1 zinc finger proteins. *Biochem J* 2005;**385**:557–64.
13. Gartel AL, Ye X, Goufman E, Shianov P, Hay N, Najmabadi F, et al. Myc represses the p21 (WAF1/CIP1) promoter and interacts with Sp1/Sp3. *Proc Natl Acad Sci USA* 2001;**98**:4510–5.
14. Gowher H, Stuhlmann H, Felsenfeld G. Vezf1 regulates genomic DNA methylation through its effects on expression of DNA methyltransferase Dnmt3b. *Genes Dev* 2008;**22**:2075–84.
15. Lopez de Silanes I, Gorospe M, Taniguchi H, Abdelmohsen K, Srikantan S, Alaminos M, et al. The RNA-binding protein HuR regulates DNA methylation through stabilization of DNMT3b mRNA. *Nucleic Acids Res* 2009;**37**:2658–71.
16. Wang W, Caldwell MC, Lin S, Furneaux H, Gorospe M. HuR regulates cyclin A and cyclin B1 mRNA stability during cell proliferation. *EMBO J* 2000;**19**:2340–50.
17. Reik W, Dean W, Walter J. Epigenetic reprogramming in mammalian development. *Science* 2001;**293**:1089–93.
18. Howell CY, Bestor TH, Ding F, Latham KE, Mertineit C, Trasler JM, et al. Genomic imprinting disrupted by a maternal effect mutation in the Dnmt1 gene. *Cell* 2001;**104**:829–38.

19. Ratnam S, Mertineit C, Ding F, Howell CY, Clarke HJ, Bestor TH, et al. Dynamics of Dnmt1 methyltransferase expression and intracellular localization during oogenesis and preimplantation development. *Dev Biol* 2002;**245**:304–14.
20. Kurihara Y, Kawamura Y, Uchijima Y, Amamo T, Kobayashi H, Asano T, et al. Maintenance of genomic methylation patterns during preimplantation development requires the somatic form of DNA methyltransferase 1. *Dev Biol* 2008;**313**:335–46.
21. Cirio MC, Ratnam S, Ding F, Reinhart B, Navara C, Chaillet JR. Preimplantation expression of the somatic form of Dnmt1 suggests a role in the inheritance of genomic imprints. *BMC Dev Biol* 2008;**8**:9.
22. Hirasawa R, Chiba H, Kaneda M, Tajima S, Li E, Jaenisch R, et al. Maternal and zygotic Dnmt1 are necessary and sufficient for the maintenance of DNA methylation imprints during preimplantation development. *Genes Dev* 2008;**22**:1607–16.
23. Ding F, Chaillet JR. In vivo stabilization of the Dnmt1 (cytosine-5)-methyltransferase protein. *Proc Natl Acad Sci USA* 2002;**99**:14861–6.
24. Okano M, Bell DW, Haber DA, Li E. DNA methyltransferases Dnmt3a and Dnmt3b are essential for de novo methylation and mammalian development. *Cell* 1999;**99**:247–57.
25. Watanabe D, Suetake I, Tada T, Tajima S. Stage- and cell-specific expression of Dnmt3a and Dnmt3b during embryogenesis. *Mech Dev* 2002;**118**:187–90.
26. Kato Y, Kaneda M, Hata K, Kumaki K, Hisano M, Kohara Y, et al. Role of the Dnmt3 family in de novo methylation of imprinted and repetitive sequences during male germ cell development in the mouse. *Hum Mol Genet* 2007;**16**:2272–80.
27. Li E, Bestor TH, Jaenisch R. Targeted mutation of the DNA methyltransferase gene results in embryonic lethality. *Cell* 1992;**69**:915–26.
28. La Salle S, Mertineit C, Taketo T, Moens PB, Bestor TH, Trasler JM. Windows for sex-specific methylation marked by DNA methyltransferase expression profiles in mouse germ cells. *Dev Biol* 2004;**268**:403–15.
29. Robertson KD. DNA methylation, methyltransferases, and cancer. *Oncogene* 2001;**20**:3139–55.
30. Ko YG, Nishino K, Hattori N, Arai Y, Tanaka S, Shiota K. Stage-by-stage change in DNA methylation status of Dnmt1 locus during mouse early development. *J Biol Chem* 2005;**280**:9627–34.
31. Pradhan S, Chin HG, Esteve PO, Jacobsen SE. SET7/9 mediated methylation of non-histone proteins in mammalian cells. *Epigenetics* 2009;**4**:383–7.
32. Esteve PO, Chin HG, Benner J, Feehery GR, Samaranayake M, Horwitz GA, et al. Regulation of DNMT1 stability through SET7-mediated lysine methylation in mammalian cells. *Proc Natl Acad Sci USA* 2009;**106**:5076–81.
33. Goyal R, Rathert P, Laser H, Gowher H, Jeltsch A. Phosphorylation of serine-515 activates the Mammalian maintenance methyltransferase Dnmt1. *Epigenetics* 2007;**2**:155–60.
34. Kang ES, Park CW, Chung JH. Dnmt3b, de novo DNA methyltransferase, interacts with SUMO-1 and Ubc9 through its N-terminal region and is subject to modification by SUMO-1. *Biochem Biophys Res Commun* 2001;**289**:862–8.
35. Lee B, Muller MT. SUMOylation enhances DNA methyltransferase 1 activity. *Biochem J* 2009;**421**:449–61.
36. Ling Y, Sankpal UT, Robertson AK, McNally JG, Karpova T, Robertson KD. Modification of de novo DNA methyltransferase 3a (Dnmt3a) by SUMO-1 modulates its interaction with histone deacetylases (HDACs) and its capacity to repress transcription. *Nucleic Acids Res* 2004;**32**:598–610.
37. Wang J, Hevi S, Kurash JK, Lei H, Gay F, Bajko J, et al. The lysine demethylase LSD1 (KDM1) is required for maintenance of global DNA methylation. *Nat Genet* 2009;**41**:125–9.
38. Nicholson TB, Chen T. LSD1 demethylates histone and non-histone proteins. *Epigenetics* 2009;**4**:129–32.

39. Dephoure N, Zhou C, Villen J, Beausoleil SA, Bakalarski CE, Elledge SJ, et al. A quantitative atlas of mitotic phosphorylation. *Proc Natl Acad Sci USA* 2008;**105**:10762–7.
40. Esteve PO, Chang Y, Samaranayake M, Upadhyay AK, Horton JR, Feehery GR, et al. Methylation and phosphorylation switch between adjacent lysine and serine determines human DNMT1 stability. *Nat Struct Mol Biol* 2011;**18**:42–8.
41. Glickman JF, Pavlovich JG, Reich NO. Peptide mapping of the murine DNA methyltransferase reveals a major phosphorylation site and the start of translation. *J Biol Chem* 1997;**272**:17851–7.
42. Bacolla A, Pradhan S, Larson JE, Roberts RJ, Wells RD. Recombinant human DNA (cytosine-5) methyltransferase. III. Allosteric control, reaction order, and influence of plasmid topology and triplet repeat length on methylation of the fragile X CGG.CCG sequence. *J Biol Chem* 2001;**276**:18605–13.
43. Vasudevan KM, Garraway LA. AKT signaling in physiology and disease. *Curr Top Microbiol Immunol* 2011;**347**:105–33.
44. Robertson KD. DNA methylation and human disease. *Nat Rev Genet* 2005;**6**:597–610.
45. Feinberg AP, Tycko B. The history of cancer epigenetics. *Nat Rev Cancer* 2004;**4**:143–53.
46. Szyf M. Epigenetic therapeutics in autoimmune disease. *Clin Rev Allergy Immunol* 2010;**39**:62–77.
47. Iwamoto K, Kato T. Epigenetic profiling in schizophrenia and major mental disorders. *Neuropsychobiology* 2009;**60**:5–11.
48. Bartel DP. MicroRNAs: genomics, biogenesis, mechanism, and function. *Cell* 2004;**116**:281–97.
49. Benetti R, Gonzalo S, Jaco I, Munoz P, Gonzalez S, Schoeftner S, et al. A mammalian microRNA cluster controls DNA methylation and telomere recombination via Rbl2-dependent regulation of DNA methyltransferases. *Nat Struct Mol Biol* 2008;**15**:998.
50. Braconi C, Huang N, Patel T. MicroRNA-dependent regulation of DNA methyltransferase-1 and tumor suppressor gene expression by interleukin-6 in human malignant cholangiocytes. *Hepatology* 2010;**51**:881–90.
51. Fabbri M, Garzon R, Cimmino A, Liu Z, Zanesi N, Callegari E, et al. MicroRNA-29 family reverts aberrant methylation in lung cancer by targeting DNA methyltransferases 3A and 3B. *Proc Natl Acad Sci USA* 2007;**104**:15805–10.
52. Garzon R, Liu S, Fabbri M, Liu Z, Heaphy CE, Callegari E, et al. MicroRNA-29b induces global DNA hypomethylation and tumor suppressor gene reexpression in acute myeloid leukemia by targeting directly DNMT3A and 3B and indirectly DNMT1. *Blood* 2009;**113**:6411–8.
53. Duursma AM, Kedde M, Schrier M, le Sage C, Agami R. miR-148 targets human DNMT3b protein coding region. *RNA* 2008;**14**:872–7.
54. Greenberg NM, DeMayo F, Finegold MJ, Medina D, Tilley WD, Aspinall JO, et al. Prostate cancer in a transgenic mouse. *Proc Natl Acad Sci USA* 1995;**92**:3439–43.
55. Harvey M, Vogel H, Lee EY, Bradley A, Donehower LA. Mice deficient in both p53 and Rb develop tumors primarily of endocrine origin. *Cancer Res* 1995;**55**:1146–51.
56. Williams BO, Remington L, Albert DM, Mukai S, Bronson RT, Jacks T. Cooperative tumorigenic effects of germline mutations in Rb and p53. *Nat Genet* 1994;**7**:480–4.
57. McCabe MT, Low JA, Daignault S, Imperiale MJ, Wojno KJ, Day ML. Inhibition of DNA methyltransferase activity prevents tumorigenesis in a mouse model of prostate cancer. *Cancer Res* 2006;**66**:385–92.
58. Morey Kinney SR, Smiraglia DJ, James SR, Moser MT, Foster BA, Karpf AR. Stage-specific alterations of DNA methyltransferase expression, DNA hypermethylation, and DNA hypomethylation during prostate cancer progression in the transgenic adenocarcinoma of mouse prostate model. *Mol Cancer Res* 2008;**6**:1365–74.
59. Morey SR, Smiraglia DJ, James SR, Yu J, Moser MT, Foster BA, et al. DNA methylation pathway alterations in an autochthonous murine model of prostate cancer. *Cancer Res* 2006;**66**:11659–67.

60. Shukla V, Coumoul X, Lahusen T, Wang RH, Xu X, Vassilopoulos A, et al. BRCA1 affects global DNA methylation through regulation of DNMT1. *Cell Res* 2010;**20**:1201–15.
61. Gaudet F, Hodgson JG, Eden A, Jackson-Grusby L, Dausman J, Gray JW, et al. Induction of tumors in mice by genomic hypomethylation. *Science* 2003;**300**:489–92.
62. Kinney SR, Moser MT, Pascual M, Greally JM, Foster BA, Karpf AR. Opposing roles of Dnmt1 in early- and late-stage murine prostate cancer. *Mol Cell Biol* 2010;**30**:4159–74.
63. Yamada Y, Jackson-Grusby L, Linhart H, Meissner A, Eden A, Lin H, et al. Opposing effects of DNA hypomethylation on intestinal and liver carcinogenesis. *Proc Natl Acad Sci USA* 2005;**102**:13580–5.
64. Campbell SL, Khosravi-Far R, Rossman KL, Clark GJ, Der CJ. Increasing complexity of Ras signaling. *Oncogene* 1998;**17**:1395–413.
65. Rouleau J, MacLeod AR, Szyf M. Regulation of the DNA methyltransferase by the Ras-AP-1 signaling pathway. *J Biol Chem* 1995;**270**:1595–601.
66. MacLeod AR, Rouleau J, Szyf M. Regulation of DNA methylation by the Ras signaling pathway. *J Biol Chem* 1995;**270**:11327–37.
67. Agoston AT, Argani P, Yegnasubramanian S, De Marzo AM, Ansari-Lari MA, Hicks JL, et al. Increased protein stability causes DNA methyltransferase 1 dysregulation in breast cancer. *J Biol Chem* 2005;**280**:18302–10.
68. Sun L, Zhao H, Xu Z, Liu Q, Liang Y, Wang L, et al. Phosphatidylinositol 3-kinase/protein kinase B pathway stabilizes DNA methyltransferase I protein and maintains DNA methylation. *Cell Signal* 2007;**19**:2255–63.
69. Robertson KD, Uzvolgyi E, Liang G, Talmadge C, Sumegi J, Gonzales FA, et al. The human DNA methyltransferases (DNMTs) 1, 3a and 3b: coordinate mRNA expression in normal tissues and overexpression in tumors. *Nucleic Acids Res* 1999;**27**:2291–8.
70. Saito Y, Kanai Y, Sakamoto M, Saito H, Ishii H, Hirohashi S. Overexpression of a splice variant of DNA methyltransferase 3b, DNMT3b4, associated with DNA hypomethylation on pericentromeric satellite regions during human hepatocarcinogenesis. *Proc Natl Acad Sci USA* 2002;**99**:10060–5.
71. Ostler KR, Davis EM, Payne SL, Gosalia BB, Exposito-Cespedes J, Le Beau MM, et al. Cancer cells express aberrant DNMT3B transcripts encoding truncated proteins. *Oncogene* 2007;**26**:5553–63.
72. Richardson B. Primer: epigenetics of autoimmunity. *Nat Clin Pract Rheumatol* 2007;**3**:521–7.
73. Nakamura T, Sekigawa I, Ogasawara H, Mitsuishi K, Hira K, Ikeda S, et al. Expression of DNMT-1 in patients with atopic dermatitis. *Arch Dermatol Res* 2006;**298**:253–6.
74. Kuwabara N, Kondo N, Fukutomi O, Fujii H, Orii T. Methylation patterns of I epsilon region in B cells stimulated with interleukin 4 and Epstein-Barr virus in patients with a high level of serum IgE. *Eur J Immunogenet* 1995;**22**:265–75.
75. Manson JJ, Rahman A. Systemic lupus erythematosus. *Orphanet J Rare Dis* 2006;**1**:6.
76. Pan W, Zhu S, Yuan M, Cui H, Wang L, Luo X, et al. MicroRNA-21 and microRNA-148a contribute to DNA hypomethylation in lupus CD4+ T cells by directly and indirectly targeting DNA methyltransferase 1. *J Immunol* 2010;**184**:6773–81.
77. Guidotti A, Auta J, Davis JM, Di-Giorgi-Gerevini V, Dwivedi Y, Grayson DR, et al. Decrease in reelin and glutamic acid decarboxylase67 (GAD67) expression in schizophrenia and bipolar disorder: a postmortem brain study. *Arch Gen Psychiatry* 2000;**57**:1061–9.
78. Chen Y, Sharma RP, Costa RH, Costa E, Grayson DR. On the epigenetic regulation of the human reelin promoter. *Nucleic Acids Res* 2002;**30**:2930–9.
79. Mitchell CP, Chen Y, Kundakovic M, Costa E, Grayson DR. Histone deacetylase inhibitors decrease reelin promoter methylation in vitro. *J Neurochem* 2005;**93**:483–92.

80. Noh JS, Sharma RP, Veldic M, Salvacion AA, Jia X, Chen Y, et al. DNA methyltransferase 1 regulates reelin mRNA expression in mouse primary cortical cultures. *Proc Natl Acad Sci USA* 2005;**102**:1749–54.
81. Abdolmaleky HM, Cheng KH, Russo A, Smith CL, Faraone SV, Wilcox M, et al. Hypermethylation of the reelin (RELN) promoter in the brain of schizophrenic patients: a preliminary report. *Am J Med Genet B Neuropsychiatr Genet* 2005;**134B**:60–6.
82. Veldic M, Caruncho HJ, Liu WS, Davis J, Satta R, Grayson DR, et al. DNA-methyltransferase 1 mRNA is selectively overexpressed in telencephalic GABAergic interneurons of schizophrenia brains. *Proc Natl Acad Sci USA* 2004;**101**:348–53.
83. Brueckner B, Lyko F. DNA methyltransferase inhibitors: old and new drugs for an epigenetic cancer therapy. *Trends Pharmacol Sci* 2004;**25**:551–4.
84. Szyf M. DNA methylation and demethylation probed by small molecules. *Biochim Biophys Acta* 2010;**1799**:750–9.
85. Datta J, Ghoshal K, Denny WA, Gamage SA, Brooke DG, Phiasivongsa P, et al. A new class of quinoline-based DNA hypomethylating agents reactivates tumor suppressor genes by blocking DNA methyltransferase 1 activity and inducing its degradation. *Cancer Res* 2009;**69**:4277–85.
86. Ghoshal K, Datta J, Majumder S, Bai S, Kutay H, Motiwala T, et al. 5-Aza-deoxycytidine induces selective degradation of DNA methyltransferase 1 by a proteasomal pathway that requires the KEN box, bromo-adjacent homology domain, and nuclear localization signal. *Mol Cell Biol* 2005;**25**:4727–41.
87. Patel K, Dickson J, Din S, Macleod K, Jodrell D, Ramsahoye B. Targeting of 5-aza-2′-deoxycytidine residues by chromatin-associated DNMT1 induces proteasomal degradation of the free enzyme. *Nucleic Acids Res* 2010;**38**:4313–24.
88. Bolden JE, Peart MJ, Johnstone RW. Anticancer activities of histone deacetylase inhibitors. *Nat Rev Drug Discov* 2006;**5**:769–84.
89. Zhou Q, Agoston AT, Atadja P, Nelson WG, Davidson NE. Inhibition of histone deacetylases promotes ubiquitin-dependent proteasomal degradation of DNA methyltransferase 1 in human breast cancer cells. *Mol Cancer Res* 2008;**6**:873–83.
90. Deng C, Lu Q, Zhang Z, Rao T, Attwood J, Yung R, et al. Hydralazine may induce autoimmunity by inhibiting extracellular signal-regulated kinase pathway signaling. *Arthritis Rheum* 2003;**48**:746–56.
91. Detich N, Bovenzi V, Szyf M. Valproate induces replication-independent active DNA demethylation. *J Biol Chem* 2003;**278**:27586–92.
92. Milutinovic S, D′Alessio AC, Detich N, Szyf M. Valproate induces widespread epigenetic reprogramming which involves demethylation of specific genes. *Carcinogenesis* 2007;**28**:560–71.
93. Singh N, Duenas-Gonzalez A, Lyko F, Medina-Franco JL. Molecular modeling and molecular dynamics studies of hydralazine with human DNA methyltransferase 1. *ChemMedChem* 2009;**4**:792–9.

Mechanistic and Functional Links Between Histone Methylation and DNA Methylation

Taiping Chen

Developmental and Molecular Pathways,
Novartis Institutes for Biomedical Research,
Cambridge, Massachusetts, USA

DNA methylation is a common mechanism of epigenetic regulation in eukaryotic organisms ranging from fungi to mammals. Genetic studies in model organisms have demonstrated the involvement of DNA methylation in a variety of biological processes. In mammals, DNA methylation patterns are established and maintained by three DNA methyltransferases: Dnmt3a, Dnmt3b, and Dnmt1. The basis of the specificity of the DNA methylation machinery and how DNA methylation patterns are regulated remain poorly understood. However, accumulating evidence suggests complex interplay between DNA methylation and other epigenetic mechanisms. Of particular interest is histone lysine methylation that has been shown to be tightly linked to DNA methylation in various systems. This chapter highlights the findings of several recent studies that provide insights into the mechanistic and functional interactions between histone methylation and DNA methylation.

I. Introduction

DNA methylation, the covalent addition of a methyl (CH_3) group to the nucleotide cytosine, is an epigenetic modification conserved in most major eukaryotic groups, including many fungi, plants, and animals, although it has been lost in some organisms such as the budding yeast *Saccharomyces cerevisiae* and the nematode worm *Caenorhabditis elegans*.[1] The biological significance and functions of DNA methylation vary among different organisms. In mammals, DNA methylation is essential for embryonic development and

Progress in Molecular Biology
and Translational Science, Vol. 101
DOI: 10.1016/B978-0-12-387685-0.00010-X

plays important roles in a variety of biological processes, including gene regulation, suppression of transposable elements, genomic imprinting, and X chromosome inactivation.[2] Aberrant changes in DNA methylation levels and patterns are associated with a number of developmental disorders (such as immunodeficiency, centromeric instability, and facial anomalies (ICF) syndrome, Beckwith–Wiedemann syndrome, Prader–Willi syndrome, and Angelman syndrome) and complex diseases (e.g., cancer, immunological abnormalities, and psychiatric disorders).[3–5]

In animal genomes, methylated cytosines are mostly restricted to CG dinucleotides (although non-CG methylation is prevalent in embryonic stem (ES) cells). This is in contrast to plants and some fungi (e.g., *Neurospora crassa*), in which DNA methylation occurs in both symmetric (CG, CHG; H = A, C, or T) and asymmetric (CHH) contexts.[6,7] Overall, 60–80% of all CpG dinucleotides in a mammalian genome are methylated. However, methylated cytosines are not randomly distributed across the genome. Heterochromatin, including centromeric, pericentric, and subtelomeric regions as well as transposable elements, is heavily methylated, and this contributes to the transcriptionally repressed, highly condensed chromatin structure characteristics of these regions. Most coding regions also show high degrees of DNA methylation. In contrast, many CpG islands at 5′ promoters lack DNA methylation. However, in genomic regions where transcription is stably silenced, such as imprinted genes and the inactive X chromosome, promoter-associated CpG islands are methylated, and this methylation is essential for maintaining the silenced state.[8,9]

Three active DNA methyltransferases have been identified in mammals (see Chapters by Zĕljko M. Svedružic; and Frédéric Chédin). Dnmt1 functions primarily as a maintenance methyltransferase that copies the parental strand CpG methylation pattern onto the daughter strand after each round of DNA replication.[10] Dnmt3a and Dnmt3b function as *de novo* methyltransferases that are responsible for establishing DNA methylation patterns during early embryogenesis and gametogenesis.[11,12] Dnmt3a and Dnmt3b also cooperate with Dnmt1 to maintain the levels and patterns of DNA methylation.[13,14]

DNA methylation is mechanistically and functionally linked to other epigenetic mechanisms, including histone modifications. Of particular importance is histone lysine methylation, which has been shown to regulate DNA methylation or act cooperatively with DNA methylation in a variety of biological processes.[15–21] Methylation has been shown to occur at five lysine residues on histone N-terminal tails (H3K4, H3K9, H3K27, H3K36, and H4K20) and one lysine residue within the globular domain of H3 (K79), and these residues can be mono-, di-, or trimethylated. In general, methylation of H3K4, H3K36, and H3K79 correlates with open and transcriptionally active chromatin, whereas methylation of H3K9, H3K27, and H4K20 is associated with condensed and transcriptionally repressive chromatin.[22,23]

Histone lysine methylation is dynamically regulated by protein lysine methyltransferases (PKMTs) and protein lysine demethylases (PKDMs) (see Chapter by L. Aravind *et al.*). All PKMTs, with the exception of the Dot1 family, contain a characteristic SET domain.[24,25] Dot1 family members, which methylate H3K79, contain conserved sequence motifs characteristic of class I methyltransferases such as Dnmts and protein arginine methyltransferases.[25] Two families of PKDMs have been identified. The amine oxidase family has two members, LSD1 (also known as KDM1A or AOF2) and the newly identified KDM1B (also known as LSD2 or AOF1). Both of them use flavin adenine dinucleotide (FAD) as a cofactor, and demethylate mono- and dimethylated, but not trimethylated, lysines. The Jumonji-C (JmjC) domain family, which has multiple members, uses an Fe^{2+}- and 2-oxoglutarate-dependent dioxygenase mechanism, and is able to demethylate mono-, di-, and trimethylated states.[26,27]

In this chapter, I highlight several recent advances in understanding the interplay between histone methylation and DNA methylation, focusing on findings in mammals.

II. An Evolutionarily Conserved Pathway Between H3K9 Methylation and DNA Methylation

How DNA methylation patterns are generated and regulated is poorly understood. Because the Dnmt1 and Dnmt3 families of DNA methyltransferases do not appear to have intrinsic sequence specificity beyond CpG dinucleotides,[28–30] chromatin structure has been suspected to be involved in the regulation of DNA methylation. Consistent with this notion, mutations of genes encoding components of chromatin-remodeling complexes, such as the SNF2 family members, have been shown to alter DNA methylation in plants and animals.[31–33]

The first evidence that histone methylation controls DNA methylation came from studies with the filamentous fungus *N. crassa*.[15] In *N. crassa*, methylated cytosines are extensively distributed within centromeres, telomeres, transposon relics, and repetitive DNA, all of which are products of the genome defense system repeat-induced point mutation (RIP).[34] Unlike in animals and plants, DNA methylation is not essential in *N. crassa*, thereby facilitating genetic studies of DNA methylation in this organism. In a mutagenesis screen, the Selker group identified multiple loci required for DNA methylation, including *dim-2* (defective in methylation 2), which specifies a DNA methyltransferase that is responsible for all known cytosine methylation in *N. crassa*.[35,36] Interestingly, one of the mutants, in which DNA methylation was entirely eliminated, disrupted a SET domain protein, DIM-5.[15] DIM-5 catalyzes trimethylation of the H3K9 associated with RIP'd DNA.[37] These findings imply that global DNA methylation is dependent on histone H3K9 methylation in *N. crassa*.

Histone H3K9 methylation is also important for some DNA methylation in plants.[16,38] In *Arabidopsis thaliana*, there are three families of DNA methyltransferases. DRM2 (domains rearranged methyltransferase 2), an ortholog of the mammalian Dnmt3 enzymes, appears to carry out all *de novo* methylation.[39] The Dnmt1-like MET1 (methyltransferase 1) primarily maintains CpG methylation,[40,41] while the plant-specific methyltransferase CMT3 (chromomethylase 3), a chromodomain-containing protein, is responsible for the maintenance of CHG and other non-CpG methylation.[42,43] In genetic screens for suppressors of DNA methylation-correlated gene silencing, Jackson et al. and Malagnac et al. isolated mutations in the *KRYPTONITE* (*KYP*) gene (also known as SU(VAR) 3–9 homolog 4 (*SUVH4*)), which produces a histone H3K9 methyltransferase similar to mammalian Suv39h. Loss-of-function *Kyp/suvh4* alleles resembled *CMT3* mutants, showing loss of cytosine methylation primarily at CHG sites and reactivation of endogenous retrotransposon sequences[16,38].

To determine whether a similar connection between histone methylation and DNA methylation is operative in mammals, Lehnertz et al. investigated DNA methylation at various repetitive sequences in *Suv39h1* and *Suv39h2* double null (dn) mouse ES cells.[17] Suv39h1 and Suv39h2 function collaboratively to establish H3K9 trimethylation at pericentric heterochromatin and are required to maintain genome stability.[44] Consistently, the *Suv39h* dn ES cells displayed severe loss of cytosine methylation at pericentric satellite repeats but not at centromeric repeats and other repetitive sequences examined. The authors also showed that H3K9 trimethylation at pericentric heterochromatin is unaltered in *Dnmt1*- or *Dnmt3a/Dnmt3b*-deficient ES cells.[17] These data suggest that DNA methylation at pericentric satellite repeats acts genetically downstream of Suv39h-mediated H3K9 trimethylation. G9a and the closely related GLP/EuHMTase1 form a heteromeric complex and are crucial for H3K9 methylation (mainly mono- and dimethylation) of euchromatin; they also regulate DNA methylation.[45–50] However, G9a appears to recruit Dnmt3a and Dnmt3b independently of its histone methyltransferase activity,[48–50] suggesting a more complex relationship between the histone and DNA methylation systems in mammals.

These studies demonstrated an evolutionarily conserved role of H3K9 methylation in the control of DNA methylation. However, the molecular mechanisms by which histone methylation directs DNA methylation may vary in different organisms. In *N. crassa*, heterochromatin protein 1 (HP1) appears to serve as a bridge between H3K9 methylation and DNA methylation.

HP1 proteins are a highly conserved family involved in heterochromatin formation and gene silencing.[51] These proteins contain a chromodomain, which binds methylated H3K9, and a chromoshadow domain, which interacts with diverse factors.[52] Mutations in the *Neurospora* HP1 gene *hpo* eliminate all detectable DNA methylation, just like null mutations in *dim-2* or *dim-5*.[53] Further, HP1 directly interacts, via its chromoshadow domain, with DIM-2.[54] These data suggest that DNA methylation in *N. crassa* is largely the result of a unidirectional pathway in which DIM-5 methylates histone H3K9 and then the DIM-2/HP1 complex recognizes the resulting trimethyl-H3K9 marks. In *Arabidopsis*, though the only known HP1 ortholog LHP1 interacts with CMT3 *in vitro*, genetic studies revealed that LHP1 is not required for DNA methylation.[16,38,55] However, the chromodomain of CMT3 can interact directly with the histone H3 tail when K9 and K27 are simultaneously methylated.[55] Mammals have three HP1 variants: HP1α, HP1β, and HP1γ. Coimmunoprecipitation experiments show that, in mouse ES cells, Dnmt3b interacts with HP1α and HP1β, but not with HP1γ.[17] In other cell types, Dnmt1, Dnmt3a, and Dnmt3b seem to interact with all three HP1 variants.[56,57] It remains to be determined, however, whether mammalian HP1 proteins are required for DNA methylation.

III. A Role for LSD1 in Coordinating Histone and DNA Methylation?

For a long time, protein lysine methylation was considered to be a permanent modification. This view has changed with the identification of LSD1 and other PKDMs. LSD1 demethylates primarily mono- and dimethyl H3K4, although its substrate specificity appears to be modulated by interacting proteins.[58,59] Recent studies have indicated that the action of LSD1 is not solely directed toward histone proteins. For example, LSD1 has been shown to demethylate p53 at K370 and E2F1 at K185 and regulates the functions of these proteins.[60,61]

Recently, we identified Dnmt1 as a novel substrate for LSD1.[20] We showed that *Lsd1*-deficient mouse ES cells maintained an undifferentiated state when cultured in regular ES medium but underwent cell death upon induction of differentiation and failed to form embryoid bodies and teratomas, a phenotype similar to that of *Dnmt1*- or *Dnmt3a/Dnmt3b*-deficient ES cells.[13,62,63] Indeed, in the absence of LSD1, the cells showed progressive loss of global DNA methylation, suggesting a defect in the maintenance of DNA methylation. Western blot and immunofluorescence analyses showed that the Dnmt1 level was substantially reduced, whereas the levels of Dnmt3a and Dnmt3b were unaltered. The reduction of Dnmt1 was due to enhanced turnover of the

Dnmt1 protein (see Chapter by Shannon R. Morey Kinney and Sriharsa Pradhan). The observation that the deregulation of Dnmt1 in *Lsd1*-deficient cells occurred at the posttranslational level raised the possibility of Dnmt1 being a direct substrate for LSD1. Indeed, metabolic-labeling experiments revealed enhanced methylation of Dnmt1 protein in the absence of LSD1, suggesting that Dnmt1 is subject to LSD1-mediated demethylation *in vivo*. Using a candidate approach, we showed that K1096 (K1094 in human DNMT1), a putative Set7/9 methylation site, can be methylated by Set7/9 and demethylated by LSD1 *in vitro*. However, Dnmt1, which contains more than 120 lysine residues, seems to be methylated at multiple sites, as metabolic-labeling experiments indicated that mutating K1096 in mouse Dnmt1 slightly reduced, but did not abolish, Dnmt1 methylation.[20] A subsequent study indeed showed that Set7/9 can also methylate K142 and reduce the level of Dnmt1 (see Chapter by Shannon R. Morey Kinney and Sriharsa Pradhan).[64] Taken together, these results suggest that Dnmt1 stability is regulated by lysine methylation. LSD1 and Set7/9 (possibly other PKDMs and PKMTs as well), by acting directly on both histones and Dnmt1, may play a role in coordinating histone methylation and DNA methylation (Fig. 1).

Consistent with its role in maintaining DNA methylation, Dnmt1 is ubiquitously expressed in proliferating cells. Although the regulation of Dnmt1 expression is not well understood, multiple factors have been shown to modulate *Dnmt1* transcription. Our finding that the stability of the Dnmt1 protein is regulated by lysine methylation adds additional complexity in the control of Dnmt1 level. Regulation of Dnmt1 at the posttranslational level might be relevant in various biological processes, including embryogenesis and tumorigenesis. In preimplantation embryos, maternally derived Dnmt1o (oocyte

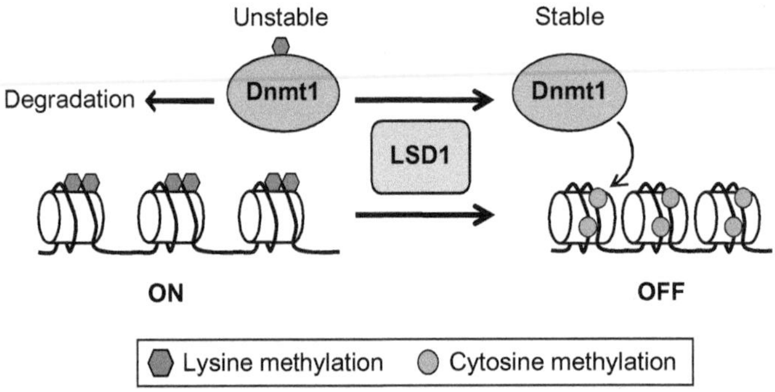

FIG. 1. *A possible role for LSD1 in coordinating histone methylation and DNA methylation.* Methylated Dnmt1 is metabolically unstable. LSD1, by acting directly on both histone H3 and Dnmt1, causes H3K4 demethylation and increased Dnmt1 and DNA methylation, resulting in chromatin condensation and gene silencing.

specific) is the major Dnmt1 variant, which has been shown to be highly stable.[65] Cancer cells often show increased levels of Dnmt1, and enhanced Dnmt1 stability appears to be partly responsible.[66,67]

IV. H3K4 Demethylation and Genomic Imprinting

In contrast to H3K9 methylation, which usually correlates with and promotes DNA methylation, H3K4 methylation seems to protect genomic regions from DNA methylation. Our finding that demethylation of H3K4 is required for the establishment of maternal genomic imprints provides a good example.[21]

Genomic imprinting is an epigenetic process in the germ line that leads to differential modification of the genome in the male and female gametes, resulting in parent-of-origin-specific expression of a small subset of genes ($\sim$100 imprinted genes have been identified in mammals; see Chapter by Jon F. Wilkins and Francisco Úbeda). DNA methylation is believed to be the primary imprinting signal, as all imprinted genes show allele-specific DNA methylation at one or more regulatory regions and deletion of such differentially methylated regions (DMRs) causes loss of imprinting.[68–72] Most methylation imprints are inherited from the mother and are put onto the genome in oocytes by Dnmt3a.[12] Genetic studies demonstrated that *Dnmt3L*, which encodes a protein similar to Dnmt3a and Dnmt3b but without methyltransferase activity, is also essential for germline imprinting.[73,74] Dnmt3L has been shown to interact with Dnmt3a and stimulates its activity.[74–78] A recent study showed that Dnmt3L binds the N-terminal tail of histone H3 and the interaction is inhibited by methylation at H3K4, suggesting that H3K4 methylation may play a role in the establishment of DNA methylation imprints.[79]

In an attempt to test the hypothesis that H3K4 methylation is involved in germline imprinting, we carried out bioinformatic analyses to identify PKMTs and PKDMs that are expressed in germ cells. These analyses, as well as experimental data generated subsequently, indicated that, in adult mice, the LSD1 family member KDM1B (also known as LSD2 or AOF1) is almost exclusively expressed in growing oocytes.[21] Biochemical experiments demonstrated that KDM1B is an active demethylase specific for mono- and dimethyl H3K4.[21,80–82] Targeted disruption of the gene encoding KDM1B (i.e., *Aof1*) showed no effect on mouse development. However, KDM1B-deficient females showed a maternal effect lethal phenotype, similar to mice deficient for *Dnmt3a* or *Dnmt3L*. Indeed, in the absence of KDM1B, growing oocytes showed a substantial increase in H3K4 methylation and failed to set up DNA methylation marks at DMRs of four of seven imprinted genes examined. As a result, embryos derived from KDM1B-deficient oocytes exhibited biallelic expression or biallelic silencing of the affected genes and could not survive beyond midgestation.[21] Based on these findings, we propose that, during oogenesis, H3K4 methylation needs to be removed to allow *de novo* DNA methylation at imprinted loci (Fig. 2).

Interestingly, phylogenetic analysis indicates that KDM1B is present in mammals and plants but not in insects[83] (see Part II of this volume). This raises the possibility that KDM1B may have evolved specifically for genomic imprinting. However, KDM1B deficiency affects some, but not all, maternal imprints, suggesting that *de novo* DNA methylation at some imprinted loci is either controlled by other H3K4-specific PKDMs (e.g., LSD1) or independent of H3K4 methylation. Although KDM1B does not control the establishment of paternal imprints during spermatogenesis, H3K4 methylation is absent at paternally methylated DMRs at stages preceding the global histone-to-protamine exchange, raising the possibility that removal of H3K4 methylation is also required for setting up imprints in the male germ line.[84] KDM1B overexpression or deficiency had dramatic effects on global H3K4 methylation,

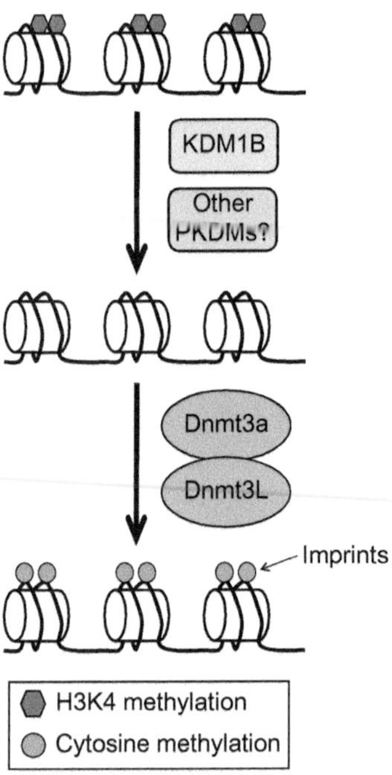

FIG. 2. *A model for the establishment of genomic imprints in female germ cells*. In this model, demethylation of H3K4 by KDM1B (and perhaps other PKDMs) creates docking sites for Dnmt3L, which recruits and/or activates Dnmt3a, and Dnmt3a puts methyl groups on DNA at imprinted loci.

suggesting that KDM1B acts not only on imprinted loci but on many other chromatin regions as well. Yet, KDM1B controls *de novo* DNA methylation at imprinted genes, without affecting global DNA methylation.[21] Therefore, the specificity is probably conferred by the *de novo* DNA methylation machinery. Indeed, the Dnmt3a–Dnmt3L complex has been shown to preferentially methylate DNA sequences with CpG sites 8–10 bp apart.[85] Interestingly, an 8- to 10-bp CpG periodicity is present within DMRs of maternally imprinted genes, although such a structural feature may not be specific to DMRs.[85,86] It is also possible that other histone modifications are required to guide *de novo* DNA methylation at imprinted genes. Further, transcription across imprinting control regions has been implicated in acquisition of maternal imprints.[87]

KDM1B is not highly expressed in most somatic tissues and is not required for mouse development.[21] However, human KDM1B is upregulated in multiple types of cancer, due to amplification of chromosome 6p22, where human *AOF1* is located.[88,89] Indeed, KDM1B has been implicated in gene transcription in human cancer cells.[82] It could be informative to determine whether overexpression of KDM1B plays a role in tumor formation and progression. Genes like *AOF1*, which are not essential for development and physiology but may be critical for cancer cells, are appealing candidate targets for cancer therapy.

V. Concluding Remarks

Over the past two decades, great progress has been made in elucidating the functions of DNA methylation in mammals. Genetic manipulations of DNA methyltransferases have demonstrated the involvement of DNA methylation in a variety of biological processes. However, much less is known about how DNA methylation patterns are generated and regulated. While there is evidence that DNA methylation and other epigenetic mechanisms, such as histone modifications, function collaboratively, the mechanistic links between these systems, in most cases, remain to be determined. In addition, it is largely unknown how DNA methylation and histone modifications work coordinately in the context of development. The finding that KDM1B is required for the establishment of maternal genomic imprints represents a rare example of the intricate interplay between histone methylation and DNA methylation in a developmental process. Moreover, although a large number of histone-modifying enzymes have been identified, investigation of their biological functions has lagged behind. Many of these factors have been implicated in cancer and other diseases. Understanding their roles in normal development and physiology and their links to various diseases will likely open up new avenues for the diagnosis, treatment, and prevention of these diseases (see Part VI of this volume).

In the future, we expect to see more studies that address these issues. Gene targeting in mice, in combination with the ever-advancing "omics" technologies, will continue to be a powerful approach.

References

1. Colot V, Rossignol JL. Eukaryotic DNA methylation as an evolutionary device. *Bioessays* 1999;**21**:402–11.
2. Chen T, Li E. DNA methylation regulates genomic imprinting, X inactivation, and gene expression during mammalian development. In: Ma J, editor. *Gene expression and regulation*. Beijing: High Education Press & Springer, Beijing; 2005. p. 377–91.
3. Feinberg AP. Phenotypic plasticity and the epigenetics of human disease. *Nature* 2007;**447**:433–40.
4. Jones PA, Baylin SB. The epigenomics of cancer. *Cell* 2007;**128**:683–92.
5. Petronis A. Epigenetics as a unifying principle in the aetiology of complex traits and diseases. *Nature* 2010;**465**:721–7.
6. Suzuki MM, Bird A. DNA methylation landscapes: provocative insights from epigenomics. *Nat Rev Genet* 2008;**9**:465–76.
7. Lister R, Pelizzola M, Dowen RH, Hawkins RD, Hon G, Tonti-Filippini J, et al. Human DNA methylomes at base resolution show widespread epigenomic differences. *Nature* 2009;**462**:315–22.
8. Bird A. DNA methylation patterns and epigenetic memory. *Genes Dev* 2002;**16**:6–21.
9. Weber M, Schubeler D. Genomic patterns of DNA methylation: targets and function of an epigenetic mark. *Curr Opin Cell Biol* 2007;**19**:273–80.
10. Chen T, Li E. Structure and function of eukaryotic DNA methyltransferases. *Curr Top Dev Biol* 2004;**60**:55–89.
11. Okano M, Bell DW, Haber DA, Li E. DNA methyltransferases Dnmt3a and Dnmt3b are essential for de novo methylation and mammalian development. *Cell* 1999;**99**:247–57.
12. Kaneda M, Okano M, Hata K, Sado T, Tsujimoto N, Li E, et al. Essential role for de novo DNA methyltransferases Dnmt3a in paternal and maternal imprinting. *Nature* 2004;**429**:900–3.
13. Chen T, Ueda Y, Dodge JE, Wang Z, Li E. Establishment and maintenance of genomic methylation patterns in mouse embryonic stem cells by Dnmt3a and Dnmt3b. *Mol Cell Biol* 2003;**23**:5594–605.
14. Liang G, Chan MF, Tomigahara Y, Tsai YC, Gonzales FA, Li E, et al. Cooperativity between DNA methyltransferases in the maintenance methylation of repetitive elements. *Mol Cell Biol* 2002;**22**:480–91.
15. Tamaru H, Selker EU. A histone H3 methyltransferase controls DNA methylation in Neurospora crassa. *Nature* 2001;**414**:277–83.
16. Jackson JP, Lindroth AM, Cao X, Jacobsen SE. Control of CpNpG DNA methylation by the KRYPTONITE histone H3 methyltransferase. *Nature* 2002;**416**:556–60.
17. Lehnertz B, Ueda Y, Derijck AA, Braunschweig U, Perez-Burgos L, Kubicek S, et al. Suv39h-mediated histone H3 lysine 9 methylation directs DNA methylation to major satellite repeats at pericentric heterochromatin. *Curr Biol* 2003;**13**:1192–200.
18. Vire E, Brenner C, Deplus R, Blanchon L, Fraga M, Didelot C, et al. The Polycomb group protein EZH2 directly controls DNA methylation. *Nature* 2006;**439**:871–4.
19. Esteve PO, Chin HG, Smallwood A, Feehery GR, Gangisetty O, Karpf AR, et al. Direct interaction between DNMT1 and G9a coordinates DNA and histone methylation during replication. *Genes Dev* 2006;**20**:3089–103.

20. Wang J, Hevi S, Kurash JK, Lei H, Gay F, Bajko J, et al. The lysine demethylase LSD1 (KDM1) is required for maintenance of global DNA methylation. *Nat Genet* 2009;**41**:125–9.
21. Ciccone DN, Su H, Hevi S, Gay F, Lei H, Bajko J, et al. KDM1B is a histone H3K4 demethylase required to establish maternal genomic imprints. *Nature* 2009;**461**:415–8.
22. Jenuwein T, Allis CD. Translating the histone code. *Science* 2001;**293**:1074–80.
23. Kouzarides T. Chromatin modifications and their function. *Cell* 2007;**128**:693–705.
24. Marmorstein R. Structure of SET domain proteins: a new twist on histone methylation. *Trends Biochem Sci* 2003;**28**:59–62.
25. Cheng X, Collins RE, Zhang X. Structural and sequence motifs of protein (histone) methylation enzymes. *Annu Rev Biophys Biomol Struct* 2005;**34**:267–94.
26. Shi Y, Whetstine JR. Dynamic regulation of histone lysine methylation by demethylases. *Mol Cell* 2007;**25**:1–14.
27. Klose RJ, Zhang Y. Regulation of histone methylation by demethylimination and demethylation. *Nat Rev Mol Cell Biol* 2007;**8**:307–18.
28. Yoder JA, Soman NS, Verdine GL, Bestor TH. DNA (cytosine-5)-methyltransferases in mouse cells and tissues. Studies with a mechanism-based probe. *J Mol Biol* 1997;**270**:385–95.
29. Okano M, Xie S, Li E. Cloning and characterization of a family of novel mammalian DNA (cytosine-5) methyltransferases. *Nat Genet* 1998;**19**:219–20.
30. Dodge J, Ramsahoye BH, Wo ZG, Okano M, Li E. De novo methylation of MMLV provirus in embryonic stem cells: CpG versus non-CpG methylation. *Gene* 2002;**289**:41–8.
31. Jeddeloh JA, Stokes TL, Richards EJ. Maintenance of genomic methylation requires a SWI2/SNF2-like protein. *Nat Genet* 1999;**22**:94–7.
32. Gibbons RJ, McDowell TL, Raman S, O'Rourke DM, Garrick D, Ayyub H, et al. Mutations in ATRX, encoding a SWI/SNF-like protein, cause diverse changes in the pattern of DNA methylation. *Nat Genet* 2000;**24**:368–71.
33. Dennis K, Fan T, Geiman T, Yan Q, Muegge K. Lsh, a member of the SNF2 family, is required for genome-wide methylation. *Genes Dev* 2001;**15**:2940–4.
34. Selker EU, Tountas NA, Cross SH, Margolin BS, Murphy JG, Bird AP, et al. The methylated component of the Neurospora crassa genome. *Nature* 2003;**422**:893–7.
35. Foss HM, Roberts CJ, Claeys KM, Selker EU. Abnormal chromosome behavior in Neurospora mutants defective in DNA methylation. *Science* 1993;**262**:1737–41.
36. Kouzminova E, Selker EU. Dim-2 encodes a DNA methyltransferase responsible for all known cytosine methylation in Neurospora. *EMBO J* 2001;**20**:4309–23.
37. Tamaru H, Zhang X, McMillen D, Singh PB, Nakayama J, Grewal SI, et al. Trimethylated lysine 9 of histone H3 is a mark for DNA methylation in Neurospora crassa. *Nat Genet* 2003;**34**:75–9.
38. Malagnac F, Bartee L, Bender J. An Arabidopsis SET domain protein required for maintenance but not establishment of DNA methylation. *EMBO J* 2002;**21**:6842–52.
39. Cao X, Jacobsen SE. Role of the Arabidopsis DRM methyltransferases in de novo DNA methylation and gene silencing. *Curr Biol* 2002;**12**:1138–44.
40. Finnegan EJ, Peacock WJ, Dennis ES. Reduced DNA methylation in Arabidopsis thaliana results in abnormal plant development. *Proc Natl Acad Sci USA* 1996;**93**:8449–54.
41. Ronemus MJ, Galbiati M, Ticknor C, Chen J, Dellaporta SL. Demethylation-induced developmental pleiotropy in Arabidopsis. *Science* 1996;**273**:654–7.
42. Bartee L, Malagnac F, Bender J. Arabidopsis cmt3 chromomethylase mutations block non-CG methylation and silencing of an endogenous gene. *Genes Dev* 2001;**15**:1753–8.
43. Lindroth AM, Cao X, Jackson JP, Zilberman D, McCallum CM, Henikoff S, et al. Requirement of CHROMOMETHYLASE3 for maintenance of CpXpG methylation. *Science* 2001;**292**:2077–80.

44. Peters AH, O'Carroll D, Scherthan H, Mechtler K, Sauer S, Schofer C, et al. Loss of the Suv39h histone methyltransferases impairs mammalian heterochromatin and genome stability. *Cell* 2001;**107**:323–37.
45. Tachibana M, Sugimoto K, Nozaki M, Ueda J, Ohta T, Ohki M, et al. G9a histone methyltransferase plays a dominant role in euchromatic histone H3 lysine 9 methylation and is essential for early embryogenesis. *Genes Dev* 2002;**16**:1779–91.
46. Tachibana M, Ueda J, Fukuda M, Takeda N, Ohta T, Iwanari H, et al. Histone methyltransferases G9a and GLP form heteromeric complexes and are both crucial for methylation of euchromatin at H3-K9. *Genes Dev* 2005;**19**:815–26.
47. Ikegami K, Iwatani M, Suzuki M, Tachibana M, Shinkai Y, Tanaka S, et al. Genome-wide and locus-specific DNA hypomethylation in G9a deficient mouse embryonic stem cells. *Genes Cells* 2007;**12**:1–11.
48. Dong KB, Maksakova IA, Mohn F, Leung D, Appanah R, Lee S, et al. DNA methylation in ES cells requires the lysine methyltransferase G9a but not its catalytic activity. *EMBO J* 2008;**27**:2691–701.
49. Tachibana M, Matsumura Y, Fukuda M, Kimura H, Shinkai Y. G9a/GLP complexes independently mediate H3K9 and DNA methylation to silence transcription. *EMBO J* 2008;**27**:2681–90.
50. Epsztejn-Litman S, Feldman N, Abu-Remaileh M, Shufaro Y, Gerson A, Ueda J, et al. De novo DNA methylation promoted by G9a prevents reprogramming of embryonically silenced genes. *Nat Struct Mol Biol* 2008;**15**:1176–83.
51. Fanti L, Pimpinelli S. HP1: a functionally multifaceted protein. *Curr Opin Genet Dev* 2008;**18**:169–74.
52. Grewal SI, Jia S. Heterochromatin revisited. *Nat Rev Genet* 2007;**8**:35–46.
53. Freitag M, Hickey PC, Khlafallah TK, Read ND, Selker EU. HP1 is essential for DNA methylation in neurospora. *Mol Cell* 2004;**13**:427–34.
54. Honda S, Selker EU. Direct interaction between DNA methyltransferase DIM-2 and HP1 is required for DNA methylation in Neurospora crassa. *Mol Cell Biol* 2008;**28**:6044–55.
55. Lindroth AM, Shultis D, Jasencakova Z, Fuchs J, Johnson L, Schubert D, et al. Dual histone H3 methylation marks at lysines 9 and 27 required for interaction with CHROMOMETHYLASE3. *EMBO J* 2004;**23**:4286–96.
56. Fuks F, Hurd PJ, Deplus R, Kouzarides T. The DNA methyltransferases associate with HP1 and the SUV39H1 histone methyltransferase. *Nucleic Acids Res* 2003;**31**:2305–12.
57. Smallwood A, Esteve PO, Pradhan S, Carey M. Functional cooperation between HP1 and DNMT1 mediates gene silencing. *Genes Dev* 2007;**21**:1169–78.
58. Shi Y, Lan F, Matson C, Mulligan P, Whetstine JR, Cole PA, et al. Histone demethylation mediated by the nuclear amine oxidase homolog LSD1. *Cell* 2004;**119**:941–53.
59. Metzger E, Wissmann M, Yin N, Muller JM, Schneider R, Peters AH, et al. LSD1 demethylates repressive histone marks to promote androgen-receptor-dependent transcription. *Nature* 2005;**437**:436–9.
60. Huang J, Sengupta R, Espejo AB, Lee MG, Dorsey JA, Richter M, et al. p53 is regulated by the lysine demethylase LSD1. *Nature* 2007;**449**:105–8.
61. Kontaki H, Talianidis I. Lysine methylation regulates E2F1-induced cell death. *Mol Cell* 2010;**39**:152–60.
62. Lei H, Oh SP, Okano M, Juttermann R, Goss KA, Jaenisch R, et al. De novo DNA cytosine methyltransferase activities in mouse embryonic stem cells. *Development* 1996;**122**:3195–205.
63. Tucker KL, Talbot D, Lee MA, Leonhardt H, Jaenisch R. Complementation of methylation deficiency in embryonic stem cells by DNA methyltransferase minigene. *Proc Natl Acad Sci USA* 1996;**93**:12920–5.

64. Esteve PO, Chin HG, Benner J, Feehery GR, Samaranayake M, Horwitz GA, et al. Regulation of DNMT1 stability through SET7-mediated lysine methylation in mammalian cells. *Proc Natl Acad Sci USA* 2009;**106**:5076–81.
65. Ding F, Chaillet JR. In vivo stabilization of the Dnmt1 (cytosine-5)-methyltransferase protein. *Proc Natl Acad Sci USA* 2002;**99**:14861–6.
66. Agoston AT, Argani P, Yegnasubramanian S, De Marzo AM, Ansari-Lari MA, Hicks JL, et al. Increased protein stability causes DNA methyltransferase 1 dysregulation in breast cancer. *J Biol Chem* 2005;**280**:18302–10.
67. Sun L, Zhao H, Xu Z, Liu Q, Liang Y, Wang L, et al. Phosphatidylinositol 3-kinase/protein kinase B pathway stabilizes DNA methyltransferase I protein and maintains DNA methylation. *Cell Signal* 2007;**19**:2255–63.
68. Tremblay KD, Saam JR, Ingram RS, Tilghman SM, Bartolomei MS. A paternal-specific methylation imprint marks the alleles of the mouse H19 gene. *Nat Genet* 1995;**9**:407–13.
69. Thorvaldsen JL, Duran KL, Bartolomei MS. Deletion of the H19 differentially methylated domain results in loss of imprinted expression of H19 and Igf2. *Genes Dev* 1998;**12**:3693–702.
70. Yoon BJ, Herman H, Sikora A, Smith LT, Plass C, Soloway PD. Regulation of DNA methylation of Rasgrf1. *Nat Genet* 2002;**30**:92–6.
71. Wutz A, Smrzka OW, Schweifer N, Schellander K, Wagner EF, Barlow DP. Imprinted expression of the Igf2r gene depends on an intronic CpG island. *Nature* 1997;**389**:745–9.
72. Shemer R, Birger Y, Riggs AD, Razin A. Structure of the imprinted mouse Snrpn gene and establishment of its parental-specific methylation pattern. *Proc Natl Acad Sci USA* 1997;**94**:10267–72.
73. Bourc'his D, Xu GL, Lin CS, Bollman B, Bestor TH. Dnmt3L and the establishment of maternal genomic imprints. *Science* 2001;**294**:2536–9.
74. Hata K, Okano M, Lei H, Li E. Dnmt3L cooperates with the Dnmt3 family of de novo DNA methyltransferases to establish maternal imprints in mice. *Development* 2002;**129**:1983–93.
75. Chedin F, Lieber MR, Hsieh CL. The DNA methyltransferase-like protein DNMT3L stimulates de novo methylation by Dnmt3a. *Proc Natl Acad Sci USA* 2002;**99**:16916–21.
76. Margot JB, Ehrenhofer-Murray AE, Leonhardt H. Interactions within the mammalian DNA methyltransferase family. *BMC Mol Biol* 2003;**4**:7–15.
77. Suetake I, Shinozaki F, Miyagawa J, Takeshima H, Tajima S. DNMT3L stimulates the DNA methylation activity of Dnmt3a and Dnmt3b through a direct interaction. *J Biol Chem* 2004;**279**:27816–23.
78. Gowher H, Liebert K, Hermann A, Xu G, Jeltsch A. Mechanism of stimulation of catalytic activity of Dnmt3A and Dnmt3B DNA-(cytosine-C5)-methyltransferases by Dnmt3L. *J Biol Chem* 2005;**280**:13341–8.
79. Ooi SK, Qiu C, Bernstein E, Li K, Jia D, Yang Z, et al. DNMT3L connects unmethylated lysine 4 of histone H3 to de novo methylation of DNA. *Nature* 2007;**448**:714–7.
80. Karytinos A, Forneris F, Profumo A, Ciossani G, Battaglioli E, Binda C, et al. A novel mammalian flavin-dependent histone demethylase. *J Biol Chem* 2009;**284**:17775–82.
81. Yang Z, Jiang J, Stewart DM, Qi S, Yamane K, Li J, et al. AOF1 is a histone H3K4 demethylase possessing demethylase activity-independent repression function. *Cell Res* 2010;**20**:276–87.
82. Fang R, Barbera AJ, Xu Y, Rutenberg M, Leonor T, Bi Q, et al. Human LSD2/KDM1b/AOF1 regulates gene transcription by modulating intragenic H3K4me2 methylation. *Mol Cell* 2010;**39**:222–33.
83. Zhou X, Ma H. Evolutionary history of histone demethylase families: distinct evolutionary patterns suggest functional divergence. *BMC Evol Biol* 2008;**8**:294–309.
84. Delaval K, Govin J, Cerqueira F, Rousseaux S, Khochbin S, Feil R. Differential histone modifications mark mouse imprinting control regions during spermatogenesis. *EMBO J* 2007;**26**:720–9.

85. Jia D, Jurkowska RZ, Zhang X, Jeltsch A, Cheng X. Structure of Dnmt3a bound to Dnmt3L suggests a model for de novo DNA methylation. *Nature* 2007;**449**:248–51.
86. Ferguson-Smith AC, Greally JM. Epigenetics: perceptive enzymes. *Nature* 2007;**449**:148–9.
87. Chotalia M, Smallwood SA, Ruf N, Dawson C, Lucifero D, Frontera M, et al. Transcription is required for establishment of germline methylation marks at imprinted genes. *Genes Dev* 2009;**23**:105–17.
88. Orlic M, Spencer CE, Wang L, Gallie BL. Expression analysis of 6p22 genomic gain in retinoblastoma. *Genes Chromosom Cancer* 2006;**45**:72–82.
89. Heidenblad M, Lindgren D, Jonson T, Liedberg F, Veerla S, Chebil G, et al. Tiling resolution array CGH and high density expression profiling of urothelial carcinomas delineate genomic amplicons and candidate target genes specific for advanced tumors. *BMC Med Genomics* 2008;**1**:3–14.

Section IV

DNA Methylation and Silencing

RNA-Mediated Silencing Mechanisms in Mammalian Cells

Anton Wutz

Wellcome Trust Centre for Stem Cell Research, University of Cambridge, Cambridge, United Kingdom

Noncoding RNAs are a structural component of the nuclear scaffold and have been implicated in controlling gene expression. In mammals, long noncoding RNAs contribute to the regulation of imprinted gene expression, dosage compensation, development, and tumorigenesis. RNA is also a component of pericentric heterochromatin, and transcripts have been identified at the chromosomal telomeres. The functions of noncoding RNAs are likely diverse, and their underlying mechanisms are just beginning to be understood. Several noncoding RNAs interact with chromatin-modifying complexes and might have a role in targeting chromatin modifications to specific regions of the genome. This suggests a prominent function of RNA in establishing histone modification and DNA methylation patterns in development. Studies on model systems such as X inactivation, the regulation of the *Hox* clusters, and genomic imprinting have begun to shed light on the role of noncoding RNAs in chromosomal organization and regulation of gene expression. Well-studied examples of noncoding RNAs include *Xist*, *Air*, *Kcnq1ot1*, *HOTAIR*, and *Tsix*. Here, a concise review of noncoding RNA function in mammals is given, and the present understanding and future directions of the field are summarized.

Progress in Molecular Biology
and Translational Science, Vol. 101
DOI: 10.1016/B978-0-12-387685-0.00011-1

1877-1173/11 $35.00

I. Introduction

Structurally diverse classes of RNA perform a wide range of functions in many biological processes. Messenger RNAs have a prominent role as the carriers of genomic information to the ribosomes, where the process of translation replenishes the proteins of the cell. In addition, a large number of transcripts act as structural and/or catalytic RNAs, such as the ribosomal RNAs (rRNAs) and tRNAs. Examples of RNA-mediated complex assembly can be found in splicing, where RNAs play an important role in splice site recognition and intron excision. Beyond that, RNAs have been found in many nuclear and cytoplasmic ribonucleoprotein (RNP) particles. Some of these have been assigned functions in the export of mRNAs or the regulation of translation and degradation of RNAs. RNAs direct the processing of rRNAs and tRNAs. Noncoding RNAs have also been shown to regulate gene expression, thus emphasizing that RNAs are not merely an end product of transcription but, curiously, reciprocally regulate the expression of genetic information. RNA functions are implicated in the normal organization of the nucleus and the structure of the chromosomes. Small RNAs provide sequence specificity for targeting chromatin modifications and gene regulation in a wide range of organisms. In mammals, the role of long noncoding transcripts in gene regulation is appreciated. Long noncoding RNAs are not exclusive to mammals and are also found in other organisms such as the *roX* RNAs in the fly *Drosophila melanogaster*. In this chapter, an overview of the current understanding of the function of noncoding RNA-mediated repression of transcription in mammals is presented. Structurally different RNAs engage diverse mechanisms that lead to different regulatory outcomes. RNA-guided genome modifications include DNA methylation and other epigenetic modifications such as posttranslational modifications of histone proteins. Parallels and clear differences between these RNAs are discussed and an outlook on future avenues of research is suggested.

II. RNA Constitutes a Structural Component of the Mammalian Cell Nucleus

RNAs are a major component of the cell nucleus. Diverse RNP particles are formed by a heterogeneous population of RNAs, with potential implications in controlling gene expression or as a structural component of the nuclear scaffold (reviewed in Ref. 1). It has long been recognized that chromatin preparations contain RNA.[2] Also, treatment of mammalian nuclei with RNA degrading enzymes leads to changes in the fine structure of the remaining chromatin-free nuclear scaffold.[3] The role of proteins in gene regulation is widely studied. Examples include the histones for packaging DNA and transcription factors as instructive molecules to key steps in development. In contrast, only a small number of functional noncoding RNAs have been well characterized, and only recently has a more comprehensive understanding of the role of RNA in gene regulation begun to emerge. One avenue of research has attempted to understand the nuclear matrix as an RNA-based structure. Until recently, approaches have been mainly confined to structural and biochemical studies mainly due to technical limitations. It has been difficult to obtain crucial molecular and genetic evidence that would allow a more functional understanding of its role in gene regulation.

However, several studies have uncovered a large number of non(-protein)-coding transcripts and shown their function in gene regulation (for reviews, see Refs. 4–6). A variety of nuclear RNA-containing complexes have been well characterized.[7] These molecular data provide some links between genetic evidence of gene regulation and the biochemical composition of the nuclear RNPs. Consequently, aspects of a nuclear matrix concept have become useful for understanding the function of regulatory noncoding transcripts.

III. A Role for RNAs in Regulating Chromatin Modifications and Organization

Noncoding RNAs have been implicated in the regulation of imprinted genes, dosage compensation, and a number of genes in mammalian development (reviewed in Refs. 4–6). Further, a role for RNA processing and RNA-mediated regulation is beginning to be understood in the context of human disease.[7] Long noncoding RNAs were discovered by searches for genes that regulate epigenetic phenomena. The noncoding *Xist* RNA was identified as a gene that was exclusively expressed from the inactive X chromosome in female human and mouse cells.[8–10] *Xist* was shown to function as an RNA by observations that it localizes specifically to the inactive X chromosome.[11] The large noncoding *Air* RNA was identified as a repressor of the paternally inherited allele of the imprinted *Igf2r* gene in mice.[12] *Xist* and *Air* provide two demonstrated examples of noncoding transcripts that affect gene repression. Other examples include the noncoding *Tsix* RNA that regulates *Xist* expression in mice[13,14] and the *Kcnq1ot1* RNA that has been implicated in regulating the imprinted *Kcnq1* domain (associated with Beckwith–Wiedemann syndrome)[15,16] (see Chapter by Jon F. Wilkins and Francisco Úbeda). These findings have led to the view that long noncoding transcripts perform functions in gene repression in *cis* acting on a single gene locus (*Tsix*), a chromosomal domain (*Air* and *Kcnq1ot1*), or an entire chromosome (*Xist*). However, more recent evidence suggests that noncoding RNAs also act in *trans*. The noncoding *HOTAIR* RNA is expressed from the *HOXC* locus and mediates gene repression in the *HOXD* cluster.[17] Collectively, these findings suggest that RNAs can function in different modes to exert an effect on gene expression.

A. Evidence That Small RNAs Guide Repression in Mammals

In addition to long noncoding RNAs, small RNAs confer the ability to silence expression by leading to chromatin modifications and DNA methylation in mammals. An evolutionarily conserved mechanism for the repression of transposable elements relies on small RNAs that associate with proteins of the PIWI family[18]. piRNAs are 25- to 28-nt RNAs that are derived from transposable elements and function in a wide range of species to protect the germ line from transposon mobilization and insertional mutagenesis. These piRNA complexes direct the cleavage of transcripts originating from transposable elements, and in mammals also direct transcriptional silencing. In mice, the PIWI family proteins Miwi, Mili, and Miwi2 bind piRNAs.[19] These proteins and piRNA production are required for transcriptional silencing and DNA methylation of transposable elements in the male germ line (for review, see Ref. 20). Mutations in *Miwi*,[21] *Mili*,[22] and *Miwi2*[23] cause male infertility in mice, associated with transposon mobilization in male germ cells and consequently failure of spermatogenesis.[19,23] The DNA cytosine methyltransferase genes *Dnmt3a* and *Dnmt3b* are essential for DNA methylation of transposon sequences and for transcriptional silencing.[24,25] In the male germ line, *Dnmt3a* and *Dnmt3b* function together with *Dnmt3L*, a protein that has sequence similarity to methyltransferases but no catalytic activity[26] (see Chapter by Frédéric Chédin). These findings provide a link between piRNAs and DNA methylation, whereby piRNA complexes provide sequence specificity that targets DNA methylation to transposable elements.

In mice, dispersed transposon insertion sites start the production of primary piRNAs, which associate with MILI in the cytoplasm and thereby initiate the host defense response (Fig. 1). MIWI2 associates with secondary piRNAs that are antisense to the transposable element transcripts.[19] Secondary piRNAs result from MILI-mediated cleavage of transcripts derived from bidirectional transcription over transposable elements. Importantly, piRNA association with MIWI2 and localization to the nucleus requires MILI, suggesting a hierarchical assembly process. Studies of the mechanism of piRNA production in flies have suggested a ping-pong mechanism, in which secondary piRNAs are produced through cleavage of antisense transposon transcripts by MILI complexes containing primary piRNAs.[27]

At present, the molecular mechanism linking piRNA-containing complexes to DNA methylation is not clear. As piRNAs also exert transcriptional silencing in organisms with minimal DNA methylation, such as *D. melanogaster*,[27] it is

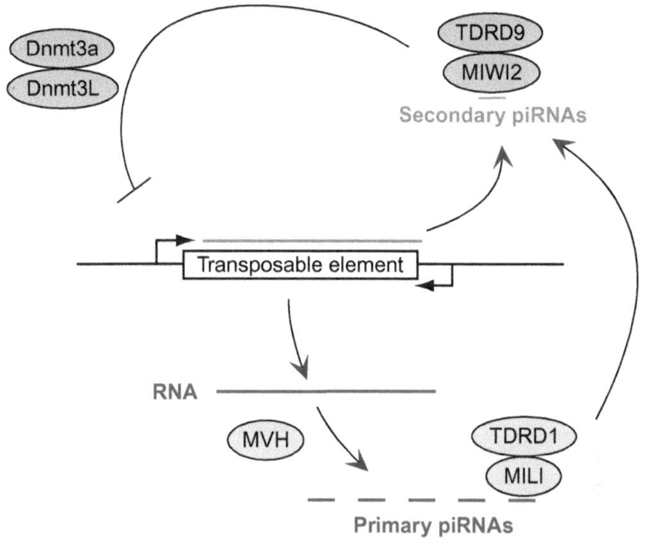

FIG. 1. *piRNAs mediate silencing of transposons.* In the male germ line, 25- to 28-nt long small RNAs are cleaved from transposon transcripts. In mice, these primary piRNAs associate with the PIWI domain protein MILI. This process requires the MVH helicase. piRNA-loaded MILI complexes localize to the cytoplasm and mediate sequence-specific cleavage of antisense transcripts derived from bidirectional transposon transcription. The resulting secondary piRNAs associate with the MIWI2 protein, which enters the nucleus and targets DNA methylation and causes transcriptional repression of the transposons. TDRD1 and TDR9 bind to methylated arginine residues on MILI and MIWI2 proteins and are essential for repression of a subset of transposons. DNA methylation requires the DNA methyltransferase Dnmt3A and the structural protein Dnmt3L.

likely that histone modifications could be targeted by piRNAs in a sequence-specific manner (see Chapter by Veiko Krauss and Gunter Reuter). A number of proteins have been implicated in transposon silencing in the male germ line, which seem to act in a pathway with piRNAs. In mice, the RNA helicase MVH is required for the early steps in the production of piRNAs.[28] Mutations in *TDRD1*[29] and *TDRD9*[30] affect suppression of a subclass of transposable elements including mouse LINE 1 non-LTR retrotransposons. This observation suggests some divergence in the pathways repressing different classes of transposable elements in mammals. TDRD1 and TDRD9 bind to symmetrically methylated arginine residues in MILI and MIWI2 through their Tudor domains and are essential for the proper localization and the loading of the piRNAs into these complexes (reviewed in Ref. 31).

In the female germ line, transposon silencing is less efficient and transposable elements are transcribed in mammalian oocytes.[20] It has been speculated that the cell cycle arrest of oocytes might provide a natural barrier to transposon insertions and, thus, the female germ cells are less vulnerable to mobile element transposition. Notably, mutations in the piRNA pathway do not affect oogenesis in mice. In the female germ line, siRNAs have been detected that are derived from transposon sequences. However, it is unclear if these siRNAs have a similar role to piRNAs in the male germ line, and if siRNAs affect DNA methylation in mammals.

B. Long Noncoding Transcripts in Gene Regulation

A large number of long noncoding RNAs have been identified by genome-wide approaches.[32,33] Transcription of many noncoding RNAs is developmentally regulated, for example, by the Oct4 and Nanog transcription factors in pluripotent cells of the early embryo.[34] Developmentally regulated expression and evolutionary conservation suggest that many of these noncoding RNAs could be functionally relevant.[32] Noncoding RNAs have been implicated in nuclear organization and compartmentalization based on their localization pattern.[35] A large fraction of noncoding RNAs associate with chromatin-modifying complexes.[36] For some of these RNAs, an effect on gene regulation has been demonstrated by elegant experiments interfering with their function. This suggests that noncoding transcription could play a prominent role in establishing gene expression patterns in development. Several mechanisms have been considered through which noncoding RNAs could function.[37] One idea is that RNAs have a structural role in complex assembly and target chromatin-modifying activities to specific genes. A well-studied example of this class of RNAs is *Xist*. Another potential mechanism is by the act of transcription across a gene promoter, such as has been proposed for the *Tsix* noncoding RNA.[38] Transcription over a gene promoter is also associated with the establishment of methylation marks in the female germ line.[39] Establishment of differentially methylated regions at the *Gnas* locus in oocytes requires transcripts from an upstream *Nesp* promoter. This has led to the idea that transcription over a promoter can lead to *Dnmt3L–Dnmt3a*-dependent DNA methylation in the female germ line.[39] A recent genome-wide study found that the association of sense–antisense transcript pairs with DNA methylation on CpG islands is more widespread and extends beyond imprinted genes.[40] This suggests a general mechanism for gene regulation by antisense transcription over a gene promoter leading to DNA methylation.

1. Chromosome-Wide Gene Silencing and Polycomb Recruitment by *Xist* RNA

The mammalian X inactivation center (*Xic*) is a well-studied genetic locus that specifies a number of noncoding RNAs.[8,9,41] *Xic* controls the silencing of one of the two X chromosomes in female mammals, for dosage compensation between the sexes. X inactivation is initiated in the early embryo and involves a number of regulatory RNAs such as *Xist* and *Tsix* (Fig. 2). *Xist* expression is the trigger for chromosome-wide gene repression and is controlled in mice by the antisense *Tsix* RNA.

The 17- to 19-kb long noncoding *Xist* RNA[8–10] accumulates within the territory of the inactive X chromosome and spreads chromatin modifications and gene repression from the *Xic* locus in *cis* over the entire X chromosome.[11] It is one of the most clear examples of RNA acting in organizing and regulating chromatin and has been the subject of a number of studies. Gene deletion has demonstrated that *Xist* is required for chromosome-wide gene repression.[42,43] Further, transgenic expression of *Xist* from autosomes can recapitulate gene silencing and heterochromatin formation on chromosomes carrying *Xic* or *Xist* transgenes.[44–48] This has led to abundant evidence that *Xist* RNA is a functional and structural component of the X inactivation process. Its mechanism is only beginning to be understood, and a number of factors have been identified recently.

Chromosome-wide localization of *Xist* RNA is intimately linked with heterogeneous nuclear RNA-binding protein U (hnRNP U) and scaffold attachment factor A (SAF-A). SAF-A is enriched on the inactive X chromosome (Xi), which

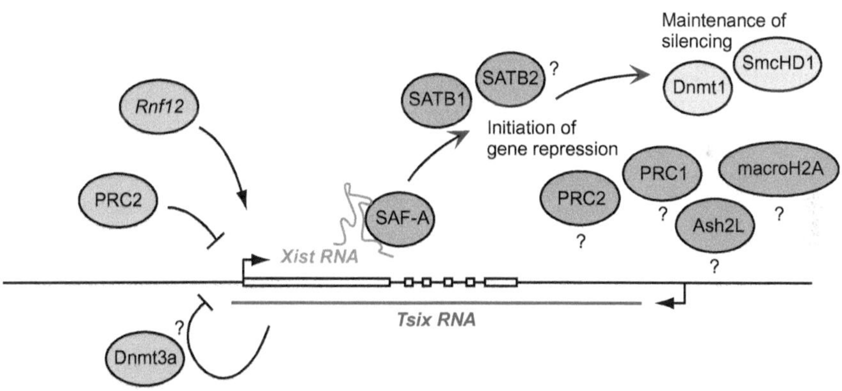

FIG. 2. *The noncoding Xist and Tsix RNAs regulate X inactivation in mammals*. The X inactivation center in mammals contains the elements for regulating dosage compensation and contains noncoding RNAs. *Xist* is a noncoding RNA that associates with the inactive X chromosome and triggers chromosome-wide gene repression. SAF-A has been implicated in *Xist* spreading over the chromosome. Factors such as SATB1 and SATB2 have been implicated in the initiation of gene repression, and SmcHD1 in the maintenance of DNA methylation and gene repression. Chromatin-modifying activities such as Polycomb complexes (PRC1 and PRC2) and chromosomal proteins are recruited in an *Xist*-dependent manner. In mice, *Xist* is regulated by the antisense *Tsix* RNA. *Tsix* acts as a repressor of *Xist* and causes DNA methylation of the *Xist* promoter. Interaction of *Tsix* with the DNA methyltransferase Dnmt3A has been reported (see text). In addition, *Xist* expression is induced by the ubiquitin E3 ligase *Rnf12* and repressed by PRC2. However, the molecular links between these factors and *Xist* promoter activity remain unknown.

requires the RNA-binding RGG domain of SAF-A[49] and depends on the expression of *Xist*.[50] Dynamics measurements of fluorescently marked SAF-A in living cells have shown that it forms a stable structure on the Xi, suggesting a role in stabilizing the inactive state of the chromosome.[51] Recent results, using RNAi-mediated depletion, show that hnRNP U is required for localization of *Xist* RNA to the inactive X territory in certain types of cells.[52] SAF-A/hnRNP U was previously identified as a component of abundant nuclear and cytoplasmic RNP complexes.[53] This suggests that interactions similar to the one between SAF-A and *Xist* might be more generally involved in gene regulation by noncoding RNAs.

Xist-mediated gene repression on the inactive X chromosome requires a small repeated RNA sequence motif on the 5′-end of *Xist*. Deletion of this "repeat A" causes the remaining *Xist* RNA to localize to the X chromosome without triggering gene repression.[54] Though *Xist* repeat A RNA associates with Polycomb proteins[55] and splicing factors,[56] the mechanism(s) by which it causes gene repression remains unclear. Initiation of chromosome-wide gene repression is also dependent on the differentiation state of the cell. *Xist* can initiate gene silencing in cells of the early embryo but not in differentiated cells.[48] A context for gene silencing by *Xist* has also been documented in specific differentiation stages in the blood system,[57] and in a certain type of lymphoma.[58] Using a mouse tumor model, it has been shown that the special AT-rich sequence-binding protein 1 (SATB1) is required for the initiation of gene repression by *Xist* in T cell lymphoma cells.[58] This result provides a first link to the molecular basis for *Xist*-mediated silencing. In T cells, SATB1 forms a nuclear cage-like structure, in which it appears to localize to the bases of chromatin loops.[59] Gene disruption has demonstrated that SATB1 is required for the regulation of several T cell-expressed genes.[60] A role has been proposed for SATB1 in organizing chromatin for coordinate gene regulation. Notably, SATB1 can be associated with the activation and repression of genes. In X inactivation, SATB1 might function in combination with *Xist* RNA to repress genes.[61] SATB1 is not expressed in all cell types in which *Xist* can initiate silencing, suggesting that other factors with similar function might be present in different lineages. In development, SATB1-related factors might be involved in the initiation of X inactivation, and it has been suggested that SATB2 might provide a function similar to SATB1.[58]

Though both SATB1 and SAF-A have been implicated in chromatin organization and bind to scaffold attachment regions, they do not overlap on the inactive X chromosome. SAF-A binding is observed in the center of the chromosome territory, while SATB1 localizes to the periphery and is thought to associate with chromatin containing genes. These observations highlight differences in the underlying mechanisms of chromatin organization and hint at a potential distinction between mechanisms for the regulation of chromatin containing genes and chromatin not containing genes.

Spreading of *Xist* is accompanied by the recruitment of Polycomb complexes and their associated chromatin modifications to the Xi. The involvement of Polycomb complexes in X inactivation was first noted when a female-specific phenotype was uncovered in mice carrying a mutation in the Polycomb group gene *Eed*.[62] Eed is a component of the catalytically active Polycomb repressive complex 2 (PRC2), which mediates histone H3 lysine 27 (H3K27) di- and trimethylation. Recruitment of Polycomb proteins was first demonstrated on the Xi in trophoblast stem cells.[63] Subsequently, it was discovered that a large number of Polycomb proteins are recruited to the Xi in a developmental manner.[64] Conceivably, the entire set of mammalian Polycomb group proteins might be recruited to the Xi at different differentiation stages. This also includes a further active complex (PRC1) that mediates histone H2A monoubiquitination.[65] Recruitment of Polycomb group proteins and their histone-modifying activities to the Xi require the presence of *Xist* RNA.[66–68] However, Polycomb recruitment does not require the gene repression activity of *Xist* and can be observed when a mutated *Xist* lacking repeat A is expressed.[66–68] While a function for *Eed* in maintaining X inactivation and *Xist* expression has been observed in the extraembryonic lineages,[69] the role of Polycomb complexes on the Xi in the embryo is still unclear.[68,70] X inactivation represents a prominent example of the recruitment of chromatin-modifying activities by long noncoding RNAs and provides a model for studying the underlying mechanism.

The composition of Xi chromatin differs from that of other chromosomes, and a number of Xi-specific modifications and factors have been described. The histone variant macroH2A[71] and the Trithorax group protein Ash2L[50] are recruited to the Xi in differentiated cells. All of the known chromatin modifications of the Xi are dependent on *Xist* expression and are lost if *Xist* expression is blocked.[50] The notable exceptions are hypoacetylation of histone H4 and gene silencing, which are stable without *Xist* expression in differentiated cells.[72] Recently, high-resolution mapping data for chromatin modifications on the Xi have become available. Interestingly, two studies have found a near-uniform distribution of macroH2A[73] and histone H3K27 trimethylation[74] over the length of the chromosome. This could indicate that chromatin on the Xi has a homogeneous composition. However, DNA methylation patterns do not support such a notion, as genes are preferentially methylated on the Xi relative to nongenic chromatin.[75] Separation of chromatin modifications from gene silencing by a repeat A mutant *Xist* also suggests that a distinction between genes and nongenic sequences is likely, at least at the initiation of X inactivation. At present, it seems likely that genes and the nongenic core of the Xi chromosome territory are initially regulated by different chromatin modifications. The nongenic chromatin gives rise to a repressive compartment, into which genes are recruited in a manner that requires repeat A of *Xist*.[76] Once genes have become silenced and are associated with this repressive compartment,

Xi chromatin may become more uniformly modified. Notably, the primary interaction points of *Xist* with the chromosome have not been identified in previous studies and are an important issue to resolve in the future.

In differentiated cells, gene silencing on the Xi is stable and does not require the continuous presence of *Xist* RNA.[77] A transition from a reversibly silenced chromosome to a stable maintenance state has been documented during the differentiation of mouse embryonic stem cells.[48] This transition requires the DNA methylation system. In mice carrying a mutation of the DNA cytosine methyltransferase gene *Dnmt1*, reactivation of genes on the Xi has been observed.[78] Recently, the SmcHD1 protein has been identified as specifically required for the maintenance of DNA methylation and repression of promoters on the Xi.[79] Interestingly, SMC-related genes that show some structural similarity to SmcHD1 have also been linked to RNA-guided DNA methylation in *Arabidopsis*.[80,81] These findings suggest that a conserved mechanism involving SmcHD1 and DNA methyltransferases mediates the stabilization of gene silencing in X inactivation.[82]

X inactivation might provide a unique possibility to unravel the link between the nuclear matrix and gene regulation. *Xist* highlights the complexity of this mode of regulation and exemplifies the role of chromatin-modifying activities such as the Polycomb group of proteins or DNA methylation pathways.

2. *HOTAIR* RNA Regulates Polycomb Recruitment and Repression of *Hox* Genes

Polycomb group complexes were originally identified in the regulation of homeotic gene expression in the fly. In mammals, homeobox transcription factors (HOX) are organized into four clusters: *HOXA*, *HOXB*, *HOXC*, and *HOXD*. Recently, several noncoding RNAs were described within the *HOX* clusters[17] that are expressed in a developmentally organized fashion. One noncoding RNA, *HOTAIR*, is expressed from the *HOXC* locus and mediates silencing of genes in the *HOXD* cluster.[17] *HOTAIR* is approximately 2 kb long and binds PRC2. It is required for regulating histone H3 methylation at the *HOXD* locus. *HOTAIR* also interacts with a second complex that contains LSD1, CoREST, and REST.[83] This finding shows that multiple chromatin-modifying complexes are targeted by *HOTAIR*, suggesting that *HOTAIR* acts as a scaffold and bridges between PRC2 and the LSD1/CoREST/REST complex (Fig. 3). Expression of *HOTAIR* has also been associated with cancer metastasis.[84] Elevated expression of *HOTAIR* is observed in primary and metastatic breast cancer. Further, depletion of *HOTAIR* from cancer cells leads to a reduced invasiveness of cells that express a high level of Polycomb proteins.[84] These findings suggest that noncoding RNA-mediated targeting of Polycomb complexes is a crucial event in breast tumorigenesis (see also Chapter by Minoru Toyota and Eiichiro Yamamoto).

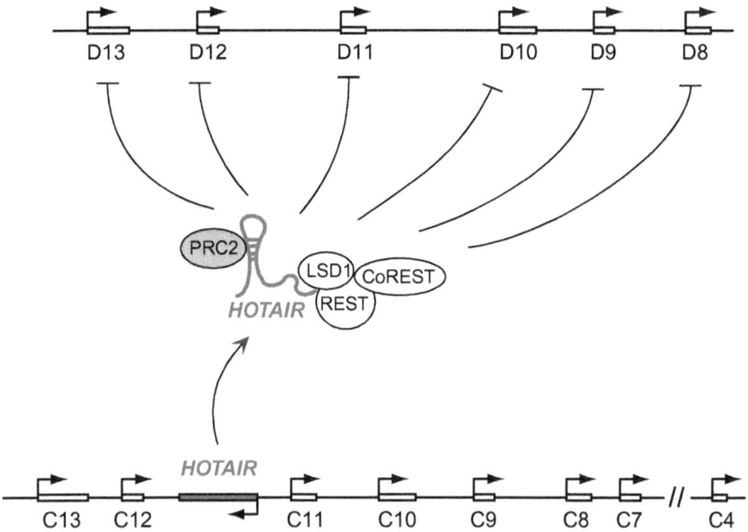

FIG. 3. *Noncoding RNA in HOX gene regulation.* HOTAIR is a noncoding RNA that is transcribed in antisense orientation to *HOX* genes in the *HOXC* cluster. *HOTAIR* binds PRC2 and a complex that contains LSD1, CoREST, and REST. Expression of *HOTAIR* targets PRC2 to the *HOXD* cluster and represses *HOXD* genes. Of potential clinical importance, HOTAIR expression in breast cancer correlates with the likelihood of metastasis and tumor cell invasiveness.

Several noncoding RNAs have been detected in the *HOX* clusters, which might have similar effects on gene regulation as *HOTAIR*.[17,32,36] In addition, a genome-wide study identified a large set of regulatory noncoding RNAs that associate with chromatin-modifying activities, such as PRC2 or the histone methyltransferase G9a.[36] Together, these findings suggest that RNA-mediated targeting of Polycomb complexes is not limited to *HOX* gene regulation but might contribute to the regulation of a wide range of genes. It remains to be seen if noncoding RNAs are a general feature of Polycomb complex recruitment in mammals.

3. *Kcnq1ot1* RNA Recruits Polycomb Complexes and Leads to DNA Methylation in the Imprinted *Kcnq1* Domain

Kcnq1ot1 is a noncoding RNA that directs silencing of genes in the imprinted *Kcnq1* domain.[15] At the mouse *Kcnq1* imprinted cluster, paternal-specific silencing of genes depends on *Kcnq1ot1* transcription.[85,86] *Kcnq1ot1* is a 90-kb long noncoding RNA that interacts with the histone methyltransferases G9a and PRC2 (Fig. 4). This interaction is observed in placental tissues but is absent in the fetal liver.[15] Consequently, *Kcnq1ot1* expression leads to

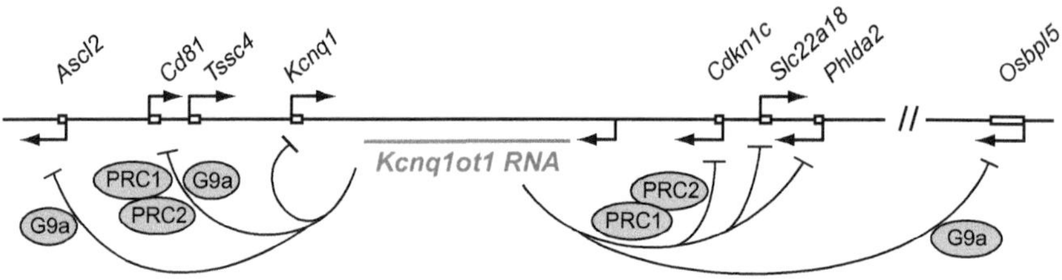

FIG. 4. *The imprinted Kcnq1 domain is regulated by a noncoding RNA.* The noncoding *Kcnq1ot1* RNA represses imprinted genes on the paternally inherited chromosome. *Kcnq1ot1* binds PRC2 and G9a and acts to establish chromatin modifications in a domain in the placenta. Mutations in PRC2 components and G9a affect repression of a subset of the genes in the imprinted domain. In the embryonic lineages, *Kcnq1ot1* associates with Dnmt1 and mediates DNA methylation.

chromatin modifications in a lineage-specific manner.[15] It has been proposed that *Kcnq1ot1* induces the formation of a nuclear domain characterized by histone modifications.[15,87] *Kcnq1ot1* forms a compartment in the early embryo that lacks RNA polymerase and is associated with the Polycomb group proteins Ezh2 and Rnf2.[16] Polycomb complexes are required for imprinted silencing and direct genomic contraction of the *Kcnq1ot1* domain.[16] These data suggested that *Kcnq1ot1* shares certain mechanistic aspects with the noncoding *Xist* RNA. Both RNAs form domains in *cis* to their transcription site and recruit Polycomb complexes. Recently, an interaction of *Kcnq1ot1* with the DNA methyltransferase Dnmt1 has also been observed.[88] *Kcnq1ot1* is required for differential DNA methylation of imprinted genes in the *Kcnq1* domain in embryonic tissues. In summary, *Kcnq1ot1* appears to be a noncoding RNA that associates with chromatin-modifying complexes and acts in a lineage-specific context. In contrast to *HOTAIR*, *Kcnq1ot1* acts strictly in *cis* to its site of expression and does not affect sequences on other chromosomes.

4. *Air* RNA Regulates Imprinted Expression of the Mouse *Igf2r* Gene

In mice, the *Igf2r* gene is imprinted and maternally expressed (Fig. 5) (see also Chapter by Jon F. Wilkins and Francisco Úbeda). Repression of the paternally inherited *Igf2r* locus depends on an intronic CpG island, which carries a differential methylation mark on the maternally inherited chromosome.[89] This intronic CpG island is required for imprinted expression.[90] The unmethylated, paternally inherited CpG island acts as a promoter for the *Air* antisense transcript.[91] *Air* is a > 100-kb noncoding RNA that is not spliced and is retained predominantly in the nucleus.[92] *Air* transcription is required for repression of several imprinted genes on the paternal chromosome.[12]

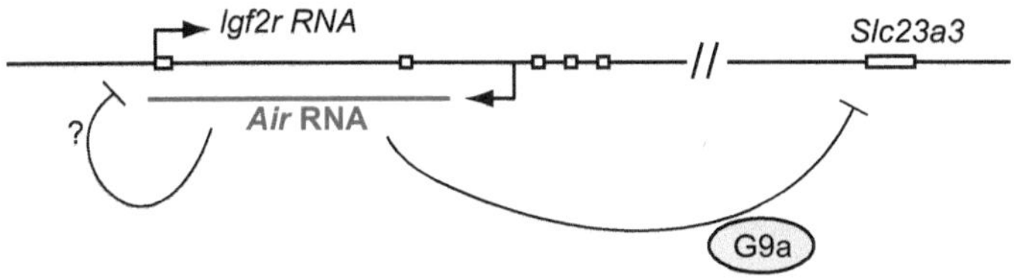

FIG. 5. *Genomic imprinting at the mouse Igf2r locus.* In mice, the paternally inherited *Igf2r* is repressed, and only the maternally inherited allele is expressed. Repression of *Igf2r* is mediated by the noncoding *Air* RNA that is expressed in opposite orientation and overlaps the *Igf2r* promoter. *Air* expression is controlled by a CpG island promoter that is methylated and silenced on the maternally inherited allele. In the placenta, *Air* regulates additional genes such as *Slc22a3*. *Air* binds to the histone methyltransferase G9a, that is, thought to establish chromatin modifications in a domain in extraembryonic lineages. In neurons, *Air* is not expressed and biallelic *Igf2r* expression can be observed.

In embryonic tissues, *Air* transcription overruns the *Igf2r* promoter and leads to DNA methylation.[89] However, in the placenta, *Air* represses additional genes that do not acquire DNA methylation and do not require overlapping transcription.[93] This suggests that *Air* acts through different mechanisms in a lineage-specific manner. In extraembryonic tissues of the placenta, *Air* RNA recruits the histone methyltransferase G9a to the *Slc22a3* promoter and directs histone H3 lysine 9 (H3K9) methylation and gene silencing.[94] This finding shows that chromatin-modifying activities are recruited in *cis* by the noncoding *Air* RNA, similar to the action of *Kcnq1ot1* RNA. However, there is no evidence for an involvement of Polycomb group complexes at the imprinted *Igf2r* domain.

5. The Noncoding *Tsix* RNA Is an Antisense Regulator of *Xist* in Mice

The mammalian *Xic* contains elements that regulate inactivation of one of the two X chromosomes in female cells. The relevant mechanism thus distinguishes between two X chromosomes in the same nucleus and treats them in very different ways. This involves regulators that can act in *cis*. One of these is *Xist* RNA, which spreads inactivation over an X chromosome from which it is transcribed but does not affect the homologous second chromosome. In mice, *Xist* is regulated by another noncoding RNA (Fig. 2). This antisense RNA regulator was named *Tsix* and acts as a repressor of *Xist* on the same chromosome.[13,95] Disruption of *Tsix* in female cells causes activation of *Xist* from the X chromosome carrying the *Tsix* deletion and, as a consequence, leads to preferential inactivation of this chromosome.[95] In contrast, forced *Tsix* expression blocks *Xist* accumulation and blocks inactivation of the X chromosome that expresses *Tsix*.[14,96] Similar to *Xist*, *Tsix* functions in *cis* on the chromosome from which it is produced. However, *Tsix* has a single target, the *Xist* gene. *Tsix* expression ensures that one X chromosome remains active, while all other X chromosomes are inactivated in female cells. This suggests that *Tsix* enables greater precision over controlling X inactivation in mice.[97]

Regulation of expression within the *Xic* is complex. In mice, an X-linked activator has been proposed to explain the fact that X inactivation initiates in female but not in male cells. There is evidence that *trans*-chromosomal interactions could play a role in this process. Pairing of *Xic* regions from two chromosomes has been observed before X inactivation is initiated.[98,99] A locus upstream of *Xist*, *Xpr*,[100] and the *Tsix* promoter region[101] have been identified as elements that can mediate pairing. *Xpr* pairing has been implicated in sensing the number of X chromosomes before inactivation. It is thought that *Xpr* pairing may make the cell competent to undergo X inactivation, before the choice is made of which X chromosome is to be inactivated.[100] Recently, *Rnf12* was identified as an X-linked *trans*-acting factor.[102] In the presence of two copies of *Rnf12*, *Xist* expression occurs with a greater probability. Once one X chromosome becomes silenced, one copy of the *Rnf12* is also repressed and, thus, the threshold for triggering *Xist* activation is lowered.

Other noncoding transcripts have been identified within the *Xic* that have been implicated in regulating *Tsix* and *Xist*. A role of small RNAs in X inactivation and the involvement of the *Dicer* nuclease have been proposed.[103] This is a controversial point at present, as these results were not recapitulated in other studies.[104] The *Xist* promoter was reported to be upregulated indirectly, as a result of reduced DNA methylation in *Dicer* deficient cells.[105] Therefore, it has been suggested that *Dicer* and small RNAs might play only a minor role for X inactivation, and that the initial study might have observed indirect effects of *Dicer* deficiency.[106]

On the molecular level, *Tsix* binds the DNA methyltransferase Dnmt3a and also associates with Ezh2 which is the catalytic component of PRC2. The fact that regulation of X inactivation is not perturbed in cells that lack Dnmt3a or PRC2 function suggests that more work is needed to assess the specificity and function of these proposed interactions.[5] The observation that regulation within the *Xic* predominantly depends on long noncoding RNAs might suggest a potential role of transcripts in *cis* regulation. Unlike proteins and small RNAs, large ncRNAs remain tethered to the site of transcription and can therefore uniquely direct allelic regulation.[5]

IV. Repression of Ribosomal DNA Repeats by the NoRC Repressor Complex Is Regulated by IGS Noncoding RNAs

rRNA genes are arranged in clusters, with rDNA sequences interspersed among long intergenic sequence stretches. Transcripts originating from these intergenic spacer sequences (IGS) are required for establishing and maintaining a specific heterochromatic configuration at a subset of rDNA arrays.[107] rDNA clusters are located on different chromosomes and form nucleolar organizer regions. In interphase, several nucleolar organizers aggregate to form nucleoli as the primary sites of rRNA transcription and processing.[108] *rRNA* genes are transcribed by RNA polymerase I and therefore might involve a substantially different regulatory process from the transcription of genes by RNA polymerase II. However, there appear to be common principles to both instances of transcriptional regulation.[109] Some of the *rDNA* genes are silenced and modified by DNA methylation. Repression of a subset (in mouse and human cells about 40% of the rDNA) is thought to be relevant for regulating ribosomal biosynthesis and for stabilizing the rDNA repeats. Transcripts produced from the IGS are processed to smaller 150- to 300-nt long pRNAs, that are sequence homologous to the rRNA promoter.[107] pRNAs have been shown to interact with a nucleolar chromatin remodeling complex (NoRC) that consists of TIP5 (TTF-I interacting protein 5) and the ATPase subunit SNF2h (Fig. 6). pRNAs are stabilized by binding to TIP5 and are required for NoRC targeting to the rDNA repeats and for repression of rRNA transcription.[107,110] A recent study also suggests that the folding and the structure of the IGS transcripts are important for interaction with NoRC and may also regulate the stability and conformation of the NoRC complex.[110] This suggests that mechanisms for repression of RNA polymerase I transcription might overlap with mechanisms for noncoding RNA-guided gene repression. rDNA silencing involves DNA methylation and histone modifications that have also been associated with repression of RNA polymerase II transcription.

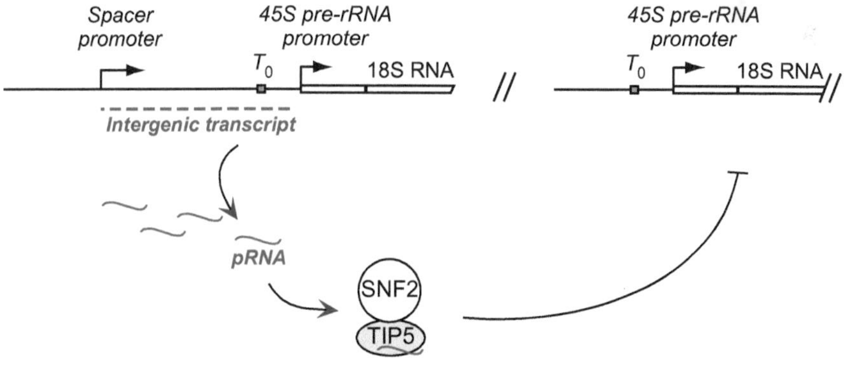

FIG. 6. *Regulation of the rDNA clusters.* The *rRNA* genes are arranged in clusters that contain spacer sequences. A fraction of these rRNA repeats is silenced by a process involving a noncoding RNA that is transcribed from the intergenic spacers. Cleavage of the spacer transcript produces a pRNA that binds to the nucleolar chromatin remodeling complex NoRC. NoRC consists of two subunits TIP5 and SNF2. pRNA binding to TIP5 induces a conformational shift and is required for NoRC targeting to the rRNA repeat. T_0, upstream terminator.

V. Pervasive Transcription: RNA at Centromeric and Telomeric Chromatin

Constitutive heterochromatin on highly repeated sequences contributes to genomic stability by suppressing recombination.[111] The centromeric and telomeric repeats are two prominent structural parts of the chromosomes that have specific pathways of chromatin modification. In mice, sequences around the centromeres are characterized by DNA methylation and chromatin modifications.[112] Pericentric chromatin is enriched for histone H3K9 and histone H4 lysine 20 (H4K20) trimethylation.[112–114] The histone methyltransferases that catalyze these modifications have been identified.[112,114] It is thought that HP1 binding to trimethylated H3K9 is essential for repression of transcription of the pericentric repeats.[111] Further, the DNA methyltransferase Dnmt3b is specifically required for methylation of centromeric minor satellite repeats in mice.[25] Genomic instability in human Immunodeficiency, Centromere instability and Facial anomalies (ICF) syndrome patients has also been associated with mutations in *DNMT3B*.[115] These observations demonstrate that maintenance of the heterochromatin is important for proper function of the centromeres. An RNA component has been implicated in the chromatin configuration at the centromeric repeats.[116] Recent work in the tammar wallaby has shown that retroviral insertions drive centromeric transcription and lead to the accumulation of 34- to 42-nt long RNAs.[117] Interestingly, the NoRC has also been implicated in heterochromatin formation at centromeric repeats.[118] These findings suggest that an RNA-based targeting of chromatin-modifying activities might be operating at the centromeric heterochromatin, similar to the silent nucleolar rDNA clusters. Undoubtedly, future work will provide insights into the extent of this overlap between nucleolar and centromeric heterochromatin.

Similar to the centromeres, heterochromatin at the chromosome ends is thought to be essential for normal telomere function and suppression of illegitimate recombination between different chromosomes. Transcripts of the mammalian telomeres (TERRA RNA) have been identified.[119] Developmentally regulated transcripts of the telomeres are thought to regulate telomerase activity.[120] TERRA RNA is also upregulated in ICF syndrome patients, who have a mutation in the DNA methyltransferase *DNMT3B*, and might contribute to the chromosomal instability phenotype.[121] This suggests that DNA methylation acts as repressor of telomeric transcription. The present evidence suggests that telomeric transcription acts as a feedback mechanism for inhibiting telomerase on long telomeres.[120] A contribution of telomeric RNA to heterochromatin formation is possible and awaits further investigation.

VI. Aspects of the Evolution of Regulatory RNAs

Several noncoding RNAs appear to use overlapping mechanisms (Table I). Polycomb complexes are engaged by *Xist*, *Kcnq1ot1*, and *HOTAIR*. NoRC seems to play a role at the rDNA clusters and centromeric repeats, in which both have been associated with noncoding RNAs and DNA methylation. A striking feature among noncoding RNAs is their poor sequence conservation, which has hindered the identification of motifs. However, lack of sequence conservation does not indicate a lack of function.[122] Could lack of overall conservation be due to only a small content of functional motifs, or to discontinuous motifs? Most of the sequence of noncoding RNAs may simply reflect the genomic region of origin. This could make it very difficult to identify relevant similarity between noncoding RNAs.

A surprising observation regarding noncoding RNAs is their rapid appearance and disappearance in evolution. *Tsix* is a prominent antisense regulator of *Xist* in mice. However, in humans, *TSIX* is apparently not functional and is coexpressed with *XIST*.[123] The noncoding *Air* RNA is required for imprinting and paternal repression of the *Igf2r* gene in mice. But *Air* is absent in dogs, where *Igf2r* is not imprinted.[124] Also the human *IGF2R* gene is predominantly biallelically expressed. Human *AIR* has been detected but appears less functional and does not induce *IGF2R* promoter methylation.[125] Interestingly, also in mice, *Air* is not expressed in neurons where biallelic *Igf2r* expression can be observed.[126] Further, *Xist* has been identified in all placental mammals but has not been found in marsupial mammals or more distant vertebrates.[127] This is also reflected in different strategies for dosage compensation, and consequently the inactive X chromosome in marsupials carries different histone modifications such as HP1 and histone H3K9 trimethylation.[128]

The rapid emergence of some noncoding RNAs suggests that they adopt mechanisms from older evolutionary pathways. A recent study exemplifies this idea by showing that two chromatin-modifying complexes are brought together by *HOTAIR*. Conceivably, noncoding RNAs could link specific complexes to generate new functions. Together with the ability to target in a sequence-specific manner, a distinct regulatory mechanism could then be established within a relatively small evolutionary period. This idea would make it interesting to decipher modular mechanisms that serve as the building blocks of noncoding RNA function. A natural source for RNA-based mechanisms is the nucleolus. The pathways regulating rRNA gene transcription and structure could be a resource for mechanisms that might be adopted by evolving noncoding RNAs.

TABLE I

Interactions of Noncoding RNAs with Chromatin-Modifying Complexes

	DNA methylation					Histone methylation				Chromatin remodeling complex		RNA-binding proteins			
Noncoding RNA	Dnmt3a	Dnmt3L	Dnmt3b	Dnmt1	SmcHD1	PRC1	PRC2	Suvar39h	G9a	NoRC	LSD1/ REST/ CoREST	SAF-A	Mili	Miwi2	Tip5
piRNA	x	x											x	x	
Xist				x	x	x	x					x			
Air RNA									x						
Kcnq1ot1	x					x	x		x						
HOTAIR							x				x				
Tsix	x														
IGS RNA										x					x
Centromere RNA			x					x		x					

Different noncoding RNAs share certain interactions with specific proteins and chromatin modifications. Other interactions appear specific to a single noncoding RNA.

VII. Concluding Remarks and Future Directions in RNA Research

The function of noncoding RNA in development is far less well understood than the action of protein factors. Here, it is suggested that the inability to efficiently classify the variety of regulatory noncoding RNAs might impede a more comprehensive understanding. Most noncoding RNAs are studied in their immediate biological context, with very little cross-referencing between different RNAs. This is in contrast to the classification and "functionalization" of proteins based on domain homologies. The definition of protein families has greatly accelerated experimental exploration of function by enabling to pull information together from many different phenomena.

Despite the small number of noncoding RNAs that have thus far been studied in detail, the data suggest that noncoding transcription could be key for the regulation of several biological systems. Even though not all noncoding RNAs can be expected to be master regulators such as *Xist* is in X inactivation, it appears that RNAs might be equally important as protein transcription factors in the developmental control of gene expression.

It seems that if noncoding RNA research in gene regulation were to be transformed, a novel means of classification is needed. Sequence seems to be an unsuitable criterion due to the rapid evolution and general lack of conservation (though see Ref. 129). This makes database searches, that are standard for proteins or coding regions, relatively powerless. RNA structure predictions further appear useful only within given complexes such as the ribosome. It is unclear if noncoding transcripts fold to thermodynamic minimal energies. Induced fit binding of RNA is well documented and indicates that RNA folding must consider kinetics and structural influences of interacting proteins. The observation that noncoding RNAs carry motifs interspersed with less-functional sequences derived from the same genetic locus suggests that a classification based on the complexes or proteins they bind might be a promising approach. Even if RNA itself is hard to classify, the proteins to which it binds might provide a hint of its function. Approaches for systematic analysis of RNA binding to regulatory complexes have been pioneered.[36] In the future, innovative biochemistry combined with genetics will undoubtedly contribute to the functional classification of noncoding RNAs, and this promises to provide a deeper and more integrated understanding of the role of noncoding RNAs in the regulation of gene expression.

Acknowledgment

A. W. is the recipient of a Wellcome Trust Senior Research Fellowship (Grant reference 087530/Z/08/A).

References

1. Nickerson J. Experimental observations of a nuclear matrix. *J Cell Sci* 2001;**114**:463–74.
2. Paul J, Duerksen JD. Chromatin-associated RNA content of heterochromatin and euchromatin. *Mol Cell Biochem* 1975;**9**:9–16.
3. Nickerson JA, Krochmalnic G, Wan KM, Penman S. Chromatin architecture and nuclear RNA. *Proc Natl Acad Sci USA* 1989;**86**:177–81.
4. Koziol MJ, Rinn JL. RNA traffic control of chromatin complexes. *Curr Opin Genet Dev* 2010;**20**:142–8.
5. Lee JT. Lessons from X-chromosome inactivation: long ncRNA as guides and tethers to the epigenome. *Genes Dev* 2009;**23**:1831–42.
6. Mohammad F, Mondal T, Kanduri C. Epigenetics of imprinted long noncoding RNAs. *Epigenetics* 2009;**4**:277–86.
7. Cooper TA, Wan L, Dreyfuss G. RNA and disease. *Cell* 2009;**136**:777–93.
8. Borsani G, Tonlorenzi R, Simmler MC, Dandolo L, Arnaud D, Capra V, et al. Characterization of a murine gene expressed from the inactive X chromosome. *Nature* 1991;**351**:325–9.
9. Brockdorff N, Ashworth A, Kay GF, Cooper P, Smith S, McCabe VM, et al. Conservation of position and exclusive expression of mouse Xist from the inactive X chromosome. *Nature* 1991;**351**:329–31.
10. Brown CJ, Ballabio A, Rupert JL, Lafreniere RG, Grompe M, Tonlorenzi R, et al. A gene from the region of the human X inactivation centre is expressed exclusively from the inactive X chromosome. *Nature* 1991;**349**:38–44.
11. Clemson CM, McNeil JA, Willard HF, Lawrence JB. XIST RNA paints the inactive X chromosome at interphase: evidence for a novel RNA involved in nuclear/chromosome structure. *J Cell Biol* 1996;**132**:259–75.
12. Sleutels F, Zwart R, Barlow DP. The non-coding Air RNA is required for silencing autosomal imprinted genes. *Nature* 2002;**415**:810–3.
13. Lee JT, Davidow LS, Warshawsky D. Tsix, a gene antisense to Xist at the X-inactivation centre. *Nat Genet* 1999;**21**:400–4.
14. Luikenhuis S, Wutz A, Jaenisch R. Antisense transcription through the Xist locus mediates Tsix function in embryonic stem cells. *Mol Cell Biol* 2001;**21**:8512–20.
15. Pandey RR, Mondal T, Mohammad F, Enroth S, Redrup L, Komorowski J, et al. Kcnq1ot1 antisense noncoding RNA mediates lineage-specific transcriptional silencing through chromatin-level regulation. *Mol Cell* 2008;**32**:232–46.
16. Terranova R, Yokobayashi S, Stadler MB, Otte AP, van Lohuizen M, Orkin SH, et al. Polycomb group proteins Ezh2 and Rnf2 direct genomic contraction and imprinted repression in early mouse embryos. *Dev Cell* 2008;**15**:668–79.
17. Rinn JL, Kertesz M, Wang JK, Squazzo SL, Xu X, Brugmann SA, et al. Functional demarcation of active and silent chromatin domains in human HOX loci by noncoding RNAs. *Cell* 2007;**129**:1311–23.
18. Malone CD, Hannon GJ. Small RNAs as guardians of the genome. *Cell* 2009;**136**:656–68.
19. Aravin AA, Sachidanandam R, Bourc'his D, Schaefer C, Pezic D, Toth KF, et al. A piRNA pathway primed by individual transposons is linked to de novo DNA methylation in mice. *Mol Cell* 2008;**31**:785–99.
20. Zamudio N, Bourc'his D. Transposable elements in the mammalian germline: a comfortable niche or a deadly trap? *Heredity* 2010;**105**:92–104.
21. Deng W, Lin H. miwi, a murine homolog of piwi, encodes a cytoplasmic protein essential for spermatogenesis. *Dev Cell* 2002;**2**:819–30.

22. Kuramochi-Miyagawa S, Kimura T, Ijiri TW, Isobe T, Asada N, Fujita Y, et al. Mili, a mammalian member of piwi family gene, is essential for spermatogenesis. *Development* 2004;**131**:839–49.
23. Carmell MA, Girard A, van de Kant HJ, Bourc'his D, Bestor TH, de Rooij DG, et al. MIWI2 is essential for spermatogenesis and repression of transposons in the mouse male germline. *Dev Cell* 2007;**12**:503–14.
24. Kato Y, Kaneda M, Hata K, Kumaki K, Hisano M, Kohara Y, et al. Role of the Dnmt3 family in de novo methylation of imprinted and repetitive sequences during male germ cell development in the mouse. *Hum Mol Genet* 2007;**16**:2272–80.
25. Okano M, Bell DW, Haber DA, Li E. DNA methyltransferases Dnmt3a and Dnmt3b are essential for de novo methylation and mammalian development. *Cell* 1999;**99**:247–57.
26. Bourc'his D, Bestor TH. Meiotic catastrophe and retrotransposon reactivation in male germ cells lacking Dnmt3L. *Nature* 2004;**431**:96–9.
27. Brennecke J, Aravin AA, Stark A, Dus M, Kellis M, Sachidanandam R, et al. Discrete small RNA-generating loci as master regulators of transposon activity in *Drosophila*. *Cell* 2007;**128**:1089–103.
28. Kuramochi-Miyagawa S, Watanabe T, Gotoh K, Takamatsu K, Chuma S, Kojima-Kita K, et al. MVH in piRNA processing and gene silencing of retrotransposons. *Genes Dev* 2010;**24**:887–92.
29. Reuter M, Chuma S, Tanaka T, Franz T, Stark A, Pillai RS. Loss of the Mili-interacting Tudor domain-containing protein-1 activates transposons and alters the Mili-associated small RNA profile. *Nat Struct Mol Biol* 2009;**16**:639–46.
30. Shoji M, Tanaka T, Hosokawa M, Reuter M, Stark A, Kato Y, et al. The TDRD9-MIWI2 complex is essential for piRNA-mediated retrotransposon silencing in the mouse male germline. *Dev Cell* 2009;**17**:775–87.
31. Siomi MC, Mannen T, Siomi H. How does the royal family of Tudor rule the PIWI-interacting RNA pathway? *Genes Dev* 2010;**24**:636–46.
32. Guttman M, Amit I, Garber M, French C, Lin MF, Feldser D, et al. Chromatin signature reveals over a thousand highly conserved large non-coding RNAs in mammals. *Nature* 2009;**458**:223–7.
33. Marques AC, Ponting CP. Catalogues of mammalian long noncoding RNAs: modest conservation and incompleteness. *Genome Biol* 2009;**10**:R124.
34. Sheik Mohamed J, Gaughwin PM, Lim B, Robson P, Lipovich L. Conserved long noncoding RNAs transcriptionally regulated by Oct4 and Nanog modulate pluripotency in mouse embryonic stem cells. *RNA* 2010;**16**:324–37.
35. Sone M, Hayashi T, Tarui H, Agata K, Takeichi M, Nakagawa S. The mRNA-like noncoding RNA Gomafu constitutes a novel nuclear domain in a subset of neurons. *J Cell Sci* 2007;**120**:2498–506.
36. Khalil AM, Guttman M, Huarte M, Garber M, Raj A, Rivea Morales D, et al. Many human large intergenic noncoding RNAs associate with chromatin-modifying complexes and affect gene expression. *Proc Natl Acad Sci USA* 2009;**106**:11667–72.
37. Pauler FM, Koerner MV, Barlow DP. Silencing by imprinted noncoding RNAs: is transcription the answer? *Trends Genet* 2007;**23**:284–92.
38. Ohhata T, Hoki Y, Sasaki H, Sado T. Crucial role of antisense transcription across the Xist promoter in Tsix-mediated Xist chromatin modification. *Development* 2008;**135**:227–35.
39. Chotalia M, Smallwood SA, Ruf N, Dawson C, Lucifero D, Frontera M, et al. Transcription is required for establishment of germline methylation marks at imprinted genes. *Genes Dev* 2009;**23**:105–17.

40. Watanabe Y, Numata K, Murata S, Osada Y, Saito R, Nakaoka H, et al. Genome-wide analysis of expression modes and DNA methylation status at sense-antisense transcript loci in mouse. *Genomics* 2010;**96**:333–41.
41. Brown CJ, Lafreniere RG, Powers VE, Sebastio G, Ballabio A, Pettigrew AL, et al. Localization of the X inactivation centre on the human X chromosome in Xq13. *Nature* 1991;**349**:82–4.
42. Penny GD, Kay GF, Sheardown SA, Rastan S, Brockdorff N. Requirement for *Xist* in X chromosome inactivation. *Nature* 1996;**379**:131–7.
43. Marahrens Y, Panning B, Dausman J, Strauss W, Jaenisch R. *Xist*-deficient mice are defective in dosage compensation but not spermatogenesis. *Genes Dev* 1997;**11**:156–66.
44. Heard E, Mongelard F, Arnaud D, Avner P. Xist yeast artificial chromosome transgenes function as X-inactivation centers only in multicopy arrays and not as single copies. *Mol Cell Biol* 1999;**19**:3156–66.
45. Herzing LB, Romer JT, Horn JM, Ashworth A. Xist has properties of the X-chromosome inactivation centre. *Nature* 1997;**386**:272–5.
46. Lee JT, Jaenisch R. Long-range cis effects of ectopic X-inactivation centres on a mouse autosome. *Nature* 1997;**386**:275–9.
47. Lee JT, Strauss WM, Dausman JA, Jaenisch R. A 450 kb transgene displays properties of the mammalian X-inactivation center. *Cell* 1996;**86**:83–94.
48. Wutz A, Jaenisch R. A shift from reversible to irreversible X inactivation is triggered during ES cell differentiation. *Mol Cell* 2000;**5**:695–705.
49. Helbig R, Fackelmayer FO. Scaffold attachment factor A (SAF-A) is concentrated in inactive X chromosome territories through its RGG domain. *Chromosoma* 2003;**112**:173–82.
50. Pullirsch D, Hartel R, Kishimoto H, Leeb M, Steiner G, Wutz A. The Trithorax group protein Ash2l and Saf-A are recruited to the inactive X chromosome at the onset of stable X inactivation. *Development* 2010;**137**:935–43.
51. Fackelmayer FO. A stable proteinaceous structure in the territory of inactive X chromosomes. *J Biol Chem* 2005;**280**:1720–3.
52. Hasegawa Y, Brockdorff N, Kawano S, Tsutui K, Tsutui K, Nakagawa S. The matrix protein hnRNP U is required for chromosomal localization of Xist RNA. *Dev Cell* 2010;**19**:469–76.
53. Fackelmayer FO, Dahm K, Renz A, Ramsperger U, Richter A. Nucleic-acid-binding properties of hnRNP-U/SAF-A, a nuclear-matrix protein which binds DNA and RNA in vivo and in vitro. *Eur J Biochem* 1994;**221**:749–57.
54. Wutz A, Rasmussen TP, Jaenisch R. Chromosomal silencing and localization are mediated by different domains of Xist RNA. *Nat Genet* 2002;**30**:167–74.
55. Zhao J, Sun BK, Erwin JA, Song JJ, Lee JT. Polycomb proteins targeted by a short repeat RNA to the mouse X chromosome. *Science* 2008;**322**:750–6.
56. Royce-Tolland ME, Andersen AA, Koyfman HR, Talbot DJ, Wutz A, Tonks ID, et al. The A-repeat links ASF/SF2-dependent Xist RNA processing with random choice during X inactivation. *Nat Struct Mol Biol* 2010;**17**:948–54.
57. Savarese F, Flahndorfer K, Jaenisch R, Busslinger M, Wutz A. Hematopoietic precursor cells transiently reestablish permissiveness for X inactivation. *Mol Cell Biol* 2006;**26**:7167–77.
58. Agrelo R, Souabni A, Novatchkova M, Haslinger C, Leeb M, Komnenovic V, et al. SATB1 defines the developmental context for gene silencing by Xist in lymphoma and embryonic cells. *Dev Cell* 2009;**16**:507–16.
59. Cai S, Lee CC, Kohwi-Shigematsu T. SATB1 packages densely looped, transcriptionally active chromatin for coordinated expression of cytokine genes. *Nat Genet* 2006;**38**:1278–88.
60. Alvarez JD, Yasui DH, Niida H, Joh T, Loh DY, Kohwi-Shigematsu T. The MAR-binding protein SATB1 orchestrates temporal and spatial expression of multiple genes during T-cell development. *Genes Dev* 2000;**14**:521–35.
61. Brockdorff N. SAT in silence. *Dev Cell* 2009;**16**:483–4.

62. Wang J, Mager J, Chen Y, Schneider E, Cross JC, Nagy A, et al. Imprinted X inactivation maintained by a mouse Polycomb group gene. *Nat Genet* 2001;**28**:371–5.
63. Mak W, Baxter J, Silva J, Newall AE, Otte AP, Brockdorff N. Mitotically stable association of polycomb group proteins eed and enx1 with the inactive x chromosome in trophoblast stem cells. *Curr Biol* 2002;**12**:1016–20.
64. Plath K, Talbot D, Hamer KM, Otte AP, Yang TP, Jaenisch R, et al. Developmentally regulated alterations in Polycomb repressive complex 1 proteins on the inactive X chromosome. *J Cell Biol* 2004;**167**:1025–35.
65. de Napoles M, Mermoud JE, Wakao R, Tang YA, Endoh M, Appanah R, et al. Polycomb group proteins Ring1A/B link ubiquitylation of histone H2A to heritable gene silencing and X inactivation. *Dev Cell* 2004;**7**:663–76.
66. Kohlmaier A, Savarese F, Lachner M, Martens J, Jenuwein T, Wutz A. A chromosomal memory triggered by Xist regulates histone methylation in X inactivation. *PLoS Biol* 2004;**2**: E171.
67. Plath K, Fang J, Mlynarczyk-Evans SK, Cao R, Worringer KA, Wang H, et al. Role of histone H3 lysine 27 methylation in X inactivation. *Science* 2003;**300**:131–5.
68. Schoeftner S, Sengupta AK, Kubicek S, Mechtler K, Spahn L, Koseki H, et al. Recruitment of PRC1 function at the initiation of X inactivation independent of PRC2 and silencing. *EMBO J* 2006;**25**:3110–22.
69. Kalantry S, Mills KC, Yee D, Otte AP, Panning B, Magnuson T. The Polycomb group protein Eed protects the inactive X-chromosome from differentiation-induced reactivation. *Nat Cell Biol* 2006;**8**:195–202.
70. Leeb M, Wutz A. Ring1B is crucial for the regulation of developmental control genes and PRC1 proteins but not X inactivation in embryonic cells. *J Cell Biol* 2007;**178**:219–29.
71. Costanzi C, Pehrson JR. Histone macroH2A1 is concentrated in the inactive X chromosome of female mammals. *Nature* 1998;**393**:599–601.
72. Csankovszki G, Panning B, Bates B, Pehrson JR, Jaenisch R. Conditional deletion of Xist disrupts histone macroH2A localization but not maintenance of X inactivation. *Nat Genet* 1999;**22**:323–4.
73. Mietton F, Sengupta AK, Molla A, Picchi G, Barral S, Heliot L, et al. Weak but uniform enrichment of the histone variant macroH2A1 along the inactive X chromosome. *Mol Cell Biol* 2009;**29**:150–6.
74. Marks H, Chow JC, Denissov S, Francoijs KJ, Brockdorff N, Heard E, et al. High-resolution analysis of epigenetic changes associated with X inactivation. *Genome Res* 2009;**19**:1361–73.
75. Hellman A, Chess A. Gene body-specific methylation on the active X chromosome. *Science* 2007;**315**:1141–3.
76. Chaumeil J, Le Baccon P, Wutz A, Heard E. A novel role for Xist RNA in the formation of a repressive nuclear compartment into which genes are recruited when silenced. *Genes Dev* 2006;**20**:2223–37.
77. Brown CJ, Willard HF. The human X-inactivation centre is not required for maintenance of X-chromosome inactivation. *Nature* 1994;**368**:154–6.
78. Sado T, Fenner MH, Tan SS, Tam P, Shioda T, Li E. X inactivation in the mouse embryo deficient for Dnmt1: distinct effect of hypomethylation on imprinted and random X inactivation. *Dev Biol* 2000;**225**:294–303.
79. Blewitt ME, Gendrel AV, Pang Z, Sparrow DB, Whitelaw N, Craig JM, et al. SmcHD1, containing a structural-maintenance-of-chromosomes hinge domain, has a critical role in X inactivation. *Nat Genet* 2008;**40**:663–9.
80. Ausin I, Mockler TC, Chory J, Jacobsen SE. IDN1 and IDN2 are required for de novo DNA methylation in Arabidopsis thaliana. *Nat Struct Mol Biol* 2009;**16**:1325–7.

81. Kanno T, Bucher E, Daxinger L, Huettel B, Bohmdorfer G, Gregor W, et al. A structural-maintenance-of-chromosomes hinge domain-containing protein is required for RNA-directed DNA methylation. *Nat Genet* 2008;**40**:670–5.
82. Heard E, Colot V. Chromosome structural proteins and RNA-mediated epigenetic silencing. *Dev Cell* 2008;**14**:813–4.
83. Tsai MC, Manor O, Wan Y, Mosammaparast N, Wang JK, Lan F, et al. Long noncoding RNA as modular scaffold of histone modification complexes. *Science* 2010;**329**:689–93.
84. Gupta RA, Shah N, Wang KC, Kim J, Horlings HM, Wong DJ, et al. Long non-coding RNA HOTAIR reprograms chromatin state to promote cancer metastasis. *Nature* 2010;**464**:1071–6.
85. Fitzpatrick GV, Soloway PD, Higgins MJ. Regional loss of imprinting and growth deficiency in mice with a targeted deletion of KvDMR1. *Nat Genet* 2002;**32**:426–31.
86. Mancini-Dinardo D, Steele SJ, Levorse JM, Ingram RS, Tilghman SM. Elongation of the Kcnq1ot1 transcript is required for genomic imprinting of neighboring genes. *Genes Dev* 2006;**20**:1268–82.
87. Redrup L, Branco MR, Perdeaux ER, Krueger C, Lewis A, Santos F, et al. The long noncoding RNA Kcnq1ot1 organises a lineage-specific nuclear domain for epigenetic gene silencing. *Development* 2009;**136**:525–30.
88. Mohammad F, Mondal T, Guseva N, Pandey GK, Kanduri C. Kcnq1ot1 noncoding RNA mediates transcriptional gene silencing by interacting with Dnmt1. *Development* 2010;**137**:2493–9.
89. Stoger R, Kubicka P, Liu CG, Kafri T, Razin A, Cedar H, et al. Maternal-specific methylation of the imprinted mouse Igf2r locus identifies the expressed locus as carrying the imprinting signal. *Cell* 1993;**73**:61–71.
90. Wutz A, Theussl HC, Dausman J, Jaenisch R, Barlow DP, Wagner EF. Non-imprinted Igf2r expression decreases growth and rescues the Tme mutation in mice. *Development* 2001;**128**:1881–7.
91. Lyle R, Watanabe D, te Vruchte D, Lerchner W, Smrzka OW, Wutz A, et al. The imprinted antisense RNA at the Igf2r locus overlaps but does not imprint Mas1. *Nat Genet* 2000;**25**:19–21.
92. Seidl CI, Stricker SH, Barlow DP. The imprinted Air ncRNA is an atypical RNAPII transcript that evades splicing and escapes nuclear export. *EMBO J* 2006;**25**:3565–75.
93. Sleutels F, Tjon G, Ludwig T, Barlow DP. Imprinted silencing of Slc22a2 and Slc22a3 does not need transcriptional overlap between Igf2r and Air. *EMBO J* 2003;**22**:3696–704.
94. Nagano T, Mitchell JA, Sanz LA, Pauler FM, Ferguson-Smith AC, Feil R, et al. The Air noncoding RNA epigenetically silences transcription by targeting G9a to chromatin. *Science* 2008;**322**:1717–20.
95. Lee JT, Lu N. Targeted mutagenesis of Tsix leads to nonrandom X inactivation. *Cell* 1999;**99**:47–57.
96. Stavropoulos N, Lu N, Lee JT. A functional role for Tsix transcription in blocking Xist RNA accumulation but not in X-chromosome choice. *Proc Natl Acad Sci USA* 2001;**98**:10232–7.
97. Wutz A, Gribnau J. X inactivation Xplained. *Curr Opin Genet Dev* 2007;**17**:387–93.
98. Bacher CP, Guggiari M, Brors B, Augui S, Clerc P, Avner P, et al. Transient colocalization of X-inactivation centres accompanies the initiation of X inactivation. *Nat Cell Biol* 2006;**8**:293–9.
99. Xu N, Tsai CL, Lee JT. Transient homologous chromosome pairing marks the onset of X inactivation. *Science* 2006;**311**:1149–52.
100. Augui S, Filion GJ, Huart S, Nora E, Guggiari M, Maresca M, et al. Sensing X chromosome pairs before X inactivation via a novel X-pairing region of the Xic. *Science* 2007;**318**:1632–6.
101. Lee JT. Regulation of X-chromosome counting by Tsix and Xite sequences. *Science* 2005;**309**:768–71.

102. Jonkers I, Barakat TS, Achame EM, Monkhorst K, Kenter A, Rentmeester E, et al. RNF12 is an X-Encoded dose-dependent activator of X chromosome inactivation. *Cell* 2009;**139**:999–1011.
103. Ogawa Y, Sun BK, Lee JT. Intersection of the RNA interference and X-inactivation pathways. *Science* 2008;**320**:1336–41.
104. Kanellopoulou C, Muljo SA, Dimitrov SD, Chen X, Colin C, Plath K, et al. X chromosome inactivation in the absence of Dicer. *Proc Natl Acad Sci USA* 2009;**106**:1122–7.
105. Nesterova TB, Popova BC, Cobb BS, Norton S, Senner CE, Tang YA, et al. Dicer regulates Xist promoter methylation in ES cells indirectly through transcriptional control of Dnmt3a. *Epigenetics Chromatin* 2008;**1**:2.
106. Kota SK. RNAi in X inactivation: contrasting findings on the role of interference. *Bioessays* 2009;**31**:1280–3.
107. Mayer C, Schmitz KM, Li J, Grummt I, Santoro R. Intergenic transcripts regulate the epigenetic state of rRNA genes. *Mol Cell* 2006;**22**:351–61.
108. McStay B, Grummt I. The epigenetics of rRNA genes: from molecular to chromosome biology. *Annu Rev Cell Dev Biol* 2008;**24**:131–57.
109. Schlesinger S, Selig S, Bergman Y, Cedar H. Allelic inactivation of rDNA loci. *Genes Dev* 2009;**23**:2437–47.
110. Mayer C, Neubert M, Grummt I. The structure of NoRC-associated RNA is crucial for targeting the chromatin remodelling complex NoRC to the nucleolus. *EMBO Rep* 2008;**9**:774–80.
111. Peters AH, O'Carroll D, Scherthan H, Mechtler K, Sauer S, Schofer C, et al. Loss of the Suv39h histone methyltransferases impairs mammalian heterochromatin and genome stability. *Cell* 2001;**107**:323–37.
112. Schotta G, Lachner M, Sarma K, Ebert A, Sengupta R, Reuter G, et al. A silencing pathway to induce H3-K9 and H4-K20 trimethylation at constitutive heterochromatin. *Genes Dev* 2004;**18**:1251–62.
113. Martens JH, O'Sullivan RJ, Braunschweig U, Opravil S, Radolf M, Steinlein P, et al. The profile of repeat-associated histone lysine methylation states in the mouse epigenome. *EMBO J* 2005;**24**:800–12.
114. Rea S, Eisenhaber F, O'Carroll D, Strahl BD, Sun ZW, Schmid M, et al. Regulation of chromatin structure by site-specific histone H3 methyltransferases. *Nature* 2000;**406**:593–9.
115. Xu GL, Bestor TH, Bourc'his D, Hsieh CL, Tommerup N, Bugge M, et al. Chromosome instability and immunodeficiency syndrome caused by mutations in a DNA methyltransferase gene. *Nature* 1999;**402**:187–91.
116. Maison C, Bailly D, Peters AH, Quivy JP, Roche D, Taddei A, et al. Higher-order structure in pericentric heterochromatin involves a distinct pattern of histone modification and an RNA component. *Nat Genet* 2002;**30**:329–34.
117. Carone DM, Longo MS, Ferreri GC, Hall L, Harris M, Shook N, et al. A new class of retroviral and satellite encoded small RNAs emanates from mammalian centromeres. *Chromosoma* 2009;**118**:113–25.
118. Guetg C, Lienemann P, Sirri V, Grummt I, Hernandez-Verdun D, Hottiger MO, et al. The NoRC complex mediates the heterochromatin formation and stability of silent rRNA genes and centromeric repeats. *EMBO J* 2010;**29**:2135–46.
119. Azzalin CM, Reichenbach P, Khoriauli L, Giulotto E, Lingner J. Telomeric repeat containing RNA and RNA surveillance factors at mammalian chromosome ends. *Science* 2007;**318**:798–801.
120. Schoeftner S, Blasco MA. Developmentally regulated transcription of mammalian telomeres by DNA-dependent RNA polymerase II. *Nat Cell Biol* 2008;**10**:228–36.

121. Deng Z, Campbell AE, Lieberman PM. TERRA, CpG methylation and telomere heterochromatin: lessons from ICF syndrome cells. *Cell Cycle* 2010;**9**:69–74.
122. Pang KC, Frith MC, Mattick JS. Rapid evolution of noncoding RNAs: lack of conservation does not mean lack of function. *Trends Genet* 2006;**22**:1–5.
123. Migeon BR, Chowdhury AK, Dunston JA, McIntosh I. Identification of TSIX, encoding an RNA antisense to human XIST, reveals differences from its murine counterpart: implications for X inactivation. *Am J Hum Genet* 2001;**69**:951–60.
124. O'Sullivan FM, Murphy SK, Simel LR, McCann A, Callanan JJ, Nolan CM. Imprinted expression of the canine IGF2R, in the absence of an anti-sense transcript or promoter methylation. *Evol Dev* 2007;**9**:579–89.
125. Yotova IY, Vlatkovic IM, Pauler FM, Warczok KE, Ambros PF, Oshimura M, et al. Identification of the human homolog of the imprinted mouse Air non-coding RNA. *Genomics* 2008;**92**:464–73.
126. Yamasaki Y, Kayashima T, Soejima H, Kinoshita A, Yoshiura K, Matsumoto N, et al. Neuron-specific relaxation of Igf2r imprinting is associated with neuron-specific histone modifications and lack of its antisense transcript Air. *Hum Mol Genet* 2005;**14**:2511–20.
127. Duret L, Chureau C, Samain S, Weissenbach J, Avner P. The Xist RNA gene evolved in eutherians by pseudogenization of a protein-coding gene. *Science* 2006;**312**:1653–5.
128. Rens W, Wallduck MS, Lovell FL, Ferguson-Smith MA, Ferguson-Smith AC. Epigenetic modifications on X chromosomes in marsupial and monotreme mammals and implications for evolution of dosage compensation. *Proc Natl Acad Sci USA* 2010;**107**:17657–62.
129. Brown JW, Birmingham A, Griffiths PE, Jossinet F, Kachouri-Lafond R, Knight R, et al. The RNA structure alignment ontology. *RNA* 2009;**15**:1623–31.

Biological Functions of Methyl-CpG-Binding Proteins

Pierre-Antoine Defossez* and Irina Stancheva†

*CNRS UMR7216, Université Paris 7, Paris Cedex 13, France

†Wellcome Trust Centre for Cell Biology, University of Edinburgh, Edinburgh, United Kingdom

DNA methylation is a stable epigenetic mark in plant and vertebrate genomes; it is implicated in regulation of higher order chromatin structure, maintenance of genome integrity, and stable patterns of gene expression. Biological effects of DNA methylation are, at least in part, mediated by proteins that preferentially bind to methylated DNA. It is now recognized that several structurally unrelated protein folds have the ability to recognize methylated CpGs *in vitro* and *in vivo*. In this chapter, we focus on the three major families of methyl-CpG-binding proteins: the MBD protein family, Kaiso and Kaiso-like proteins, and SRA domain proteins. We discuss the structural bases of methyl-CpG recognition, the function and specific properties of individual proteins, and their role in human disease such as Rett syndrome and cancer.

Progress in Molecular Biology
and Translational Science, Vol. 101
DOI: 10.1016/B978-0-12-387685-0.00012-3

I. Introduction

DNA methylation at carbon five of cytosine is an abundant modification in plant and vertebrate genomes. In most vertebrate somatic cells, methylation occurs on DNA symmetrically in the context of CpG dinucleotides, while in plants DNA methylation can be found at CpG as well as non-CpG sites, for example, CpNpG. Mammalian genomes tend to be globally methylated with the exception of stretches of DNA 500–2000 bp long known as CpG islands.[1–3] Unlike the rest of the genome, these sequences are CpG-rich and approximately 50% of them map to promoters for protein-coding genes.[3] Nonpromoter CpG islands can be found in inter- and intragenic sequences and are likely to represent either alternative transcription start sites of protein-coding genes or noncoding RNAs.[4–6] Over half of mouse (51%) and human (62%) protein-coding genes are associated with CpG islands, and most of these are constitutively unmethylated in all somatic cells. Non-CpG island promoters show variable, and often tissue-specific, DNA methylation patterns.[4,7–9] Taken together, approximately 70–80% of all CpGs are methylated in mammalian cells, which represents about 4×10^7 individual methylated sites in the case of the human genome.

The biological roles of DNA methylation include regulation of parent-of-origin expression of imprinted genes, X chromosome inactivation in female mammals, maintenance of genome stability, and repression of genes and retrotransposons.[10,11] These functions of DNA methylation are discussed in more detail in Chapters by Anton Wutz; Mary G. Goll *et al.*; Jon F. Wilkins and Francisco Úbeda; and Minoru Toyota and Eiichiro Yamamoto of this volume. In general, it is clear that most methylated CpGs are not located within genes, given the large size of most vertebrate genomes in relation to their protein-coding or transcribed counterpart. Therefore, an attractive hypothesis put forward two decades ago, but never rigorously tested experimentally, is that DNA methylation may serve to suppress "transcriptional noise" that may arise from promiscuous binding of abundant transcription factors to DNA in large eukaryotic genomes.[12]

There is substantial evidence, accumulated from *in vitro* and *in vivo* studies, that DNA methylation is inhibitory to transcription. Experiments performed with methylated or unmethylated plasmids microinjected into *Xenopus leavis* oocytes or transfected into mammalian cells have shown that DNA methylation becomes inhibitory only after the methylated plasmid assembles into chromatin.[13,14] This suggested that DNA methylation in conjunction with histone proteins and possibly other factors generates a repressive chromatin structure.

It is now well established that DNA methylation can have a direct as well as indirect negative impact on transcription. Direct transcriptional repression is achieved when recognition sites for transcription activators become methylated. For example, N-Myc and STAT1 do not bind their recognition sites on DNA when those are methylated.[15–17] An indirect transcriptional repression by DNA methylation involves proteins that bind preferentially to methylated DNA, and in turn inhibit transcription (Fig. 1). In the following parts of this chapter, we will describe the known families of methyl-CpG-binding proteins, their properties, and basic biological functions. We will also discuss the significance of past and ongoing research into the roles of these proteins in understanding complex human disease conditions such as Rett syndrome (RTT) and cancer.

II. Families of Methyl-CpG-Binding Proteins

In retrospect, it was not immediately obvious that physiological effects of DNA methylation should be mediated by proteins that bind methylated CpGs. One could imagine that nucleosomes would position differently on methylated and unmethylated DNA and thus result in different chromatin structure and higher order folding, in addition to the direct mode of transcriptional inhibition mentioned above. Therefore, the discovery that there are proteins which bind methylated CpGs with higher affinity than the unmethylated ones was somewhat accidental but led to a rapid expansion of the field and many important subsequent findings.

Currently, there are three known distinct families of proteins that bind methylated DNA. These include the methyl-binding domain (MBD) family, Kaiso and Kaiso-like proteins, and SRA domain proteins.

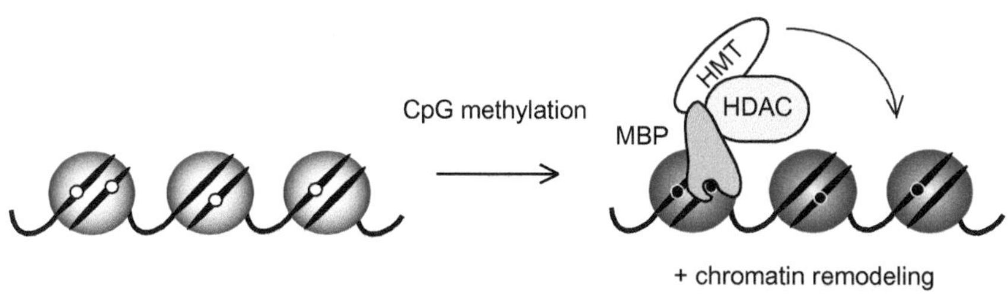

FIG. 1. *Methyl-CpG-binding proteins translate DNA methylation into chromatin modifications*. DNA is wrapped around nucleosomes (large spheres) and contains CpG dinucleotides (white circles). When CpGs are methylated (black circles), they can attract methyl-CpG-binding proteins (MBP), which themselves interact with enzymes that deacetylate histones (HDAC), methylate histones (HMT), or remodel chromatin.

A. MBD Family

It has been known since the late 1980s that certain proteins bind to symmetrically methylated double-stranded DNA.[18,19] Electrophoretic mobility shift assays, with radioactively labeled methylated probes incubated with mouse nuclear extracts, detected two major shifted bands. These were named MeCP1 and MeCP2, for methylated CpG-binding protein 1 and 2, respectively.[20] It took Adrian Bird and colleagues another 5 years to purify and clone the gene for one of these, MeCP2.[21] Subsequent deletion analyses defined the region of MeCP2 required for binding to methylated DNA, named MBD[22] (Fig. 2A) and a transcriptional repression domain (TRD).[23] EST database searches identified four additional mammalian proteins that contain an MBD domain: MBD1, MBD2, MBD3, and MBD4 (Fig. 2A).[24] Homologs of these proteins exist in all vertebrate species sequenced to date.[25] Together with MeCP2, they are currently known as the MBD family of

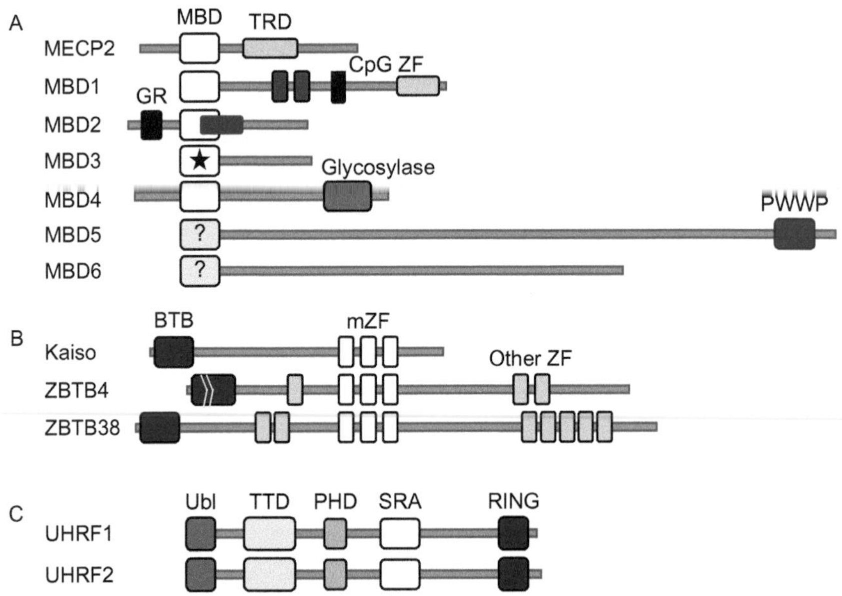

FIG. 2. *The three families of methyl-CpG-binding proteins in mammals and their organization.* (A) MBD domain family of proteins: MBD, methyl-CpG-binding domain; TRD, transcriptional repression domain; CpG ZF, unmethylated-CpG-binding zinc finger; glycosylase, DNA glycosylase activity. MBD3 has an MBD domain that does not bind methylated DNA (indicated by star). (B) Kaiso and Kaiso-like proteins: BTB, BTB/POZ domain; mZF, methyl-CpG-binding zinc fingers. (C) SRA domain proteins: Ubl, ubiquitin-like domain; TTD, tandem tudor domain; PHD, PHD finger domain; SRA, SRA domain; RING, RING finger domain.

proteins. All of them, apart from MBD3,[25] can bind to DNA probes carrying a single symmetrically methylated CpG *in vitro*, albeit with different affinities. MBD2, a relatively ubiquitous protein in mammalian cells, is likely to account for the MeCP1.

Unlike the animal MBD family, 13 MBD proteins have been identified in *Arabidopsis thaliana* and 17 in rice.[26] Of these, only five, atMBD4, atMBD5, atMBD6, atMBD7, and atMBD11, have been reported to bind methylated DNA.[26,27] It should be noted that, in mammals, additional proteins carry a MBD-like domain, but none of them interact stably with methylated DNA *in vitro* or *in vivo*: examples include MBD5, MBD6, histone methylases SETDB1 and SETDB2 as well as BAZ1A and BAZ1B, which are components of chromatin-remodeling complexes ACF and SNF2h, respectively.[25,28] As orthologs of these proteins exist in species that do not have DNA methylation, and some of them bind RNA,[29,30] the MBD-like domain appears to be an ancient protein fold that has evolved to bind methylated DNA.

The structures of the MBD domains of MBD1 and MeCP2, bound to methylated DNA, were solved by nuclear magnetic resonance (NMR) and X-ray crystallography, respectively. They share a similar sandwich fold, composed of stacked β-strands and an α-helix[31,32] (Fig. 3A and B). The MBD domain interacts with methylated CpGs in the major groove, and this

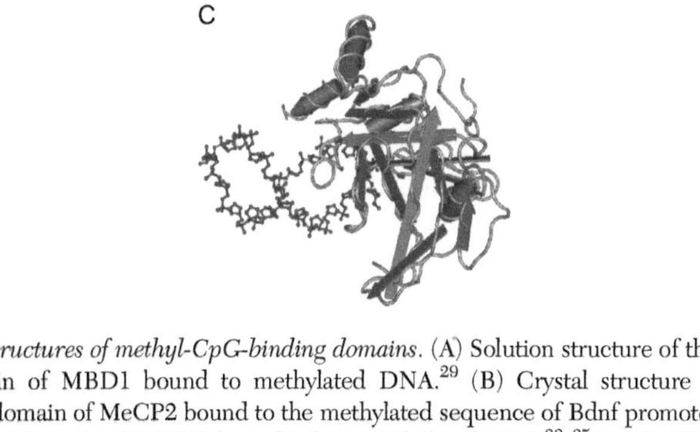

FIG. 3. *Structures of methyl-CpG-binding domains*. (A) Solution structure of the methyl-CpG-binding domain of MBD1 bound to methylated DNA.[29] (B) Crystal structure of the methyl-CpG-binding domain of MeCP2 bound to the methylated sequence of Bdnf promoter.[31] (C) Crystal structure of SRA domain of UHRF1 bound to hemimethylated DNA.[33–35] In (A)–(C), only the sugar-phosphate backbone of DNA is shown.

interaction is held together by hydrogen bonds, salt bridges, and structured water molecules.[31] In addition to the MBD, individual members of the family contain other specific motifs such as AT-hooks, GR-repeats, CXXC zinc fingers, and, in the case of MBD4, a functional glycosylase domain (Fig. 2A).

Mammalian MBD2 and MBD3 proteins are very similar to one another, sharing 80% similarity outside the MBD domain. They likely have descended via a gene duplication event from an ancestral MBD2/MBD3-like protein, such as that in *Drosophila melanogaster* (whose genome has negligible amount of methylated cytosine detectable early in development[36,37]). In the mouse and human genomes, MBD1 and MBD2 genes reside on the same chromosome about 3.5 Mb apart (Fig. 4). The closest neighbor of MBD1 is CXXC1, also known as CGBP and CFP1, a zinc finger protein which binds preferentially to unmethylated CpGs.[38,39] In addition to the MBD domain, MBD1 contains two to three CXXC motifs depending on the isoform, and the third CXXC can bind nonmethylated DNA,[40] suggesting that MBD1 may have descended from a partial duplication of the MBD2 gene while borrowing the CXXC zinc fingers from CXXC1. The evolutionary origins of MeCP2 and MBD4 are more difficult to trace; however, it is clear that the rapid expansion of the MBD family has corresponded with the increase in the methylated fraction of the genome—from the mosaic DNA methylation found in chordates such as *Ciona intestinalis* to the genome-wide methylation in mammals.[2,41]

This brings about the question of whether the MBD family proteins are functionally redundant.[42] Genetic studies in mice have shown that individual MBD genes are not essential for viability or embryo development.[43–47] However, the phenotypes of these knockouts animals are clearly different (Fig. 5). MBD1-null mice are mildly autistic and have impaired learning skills.[43,48]

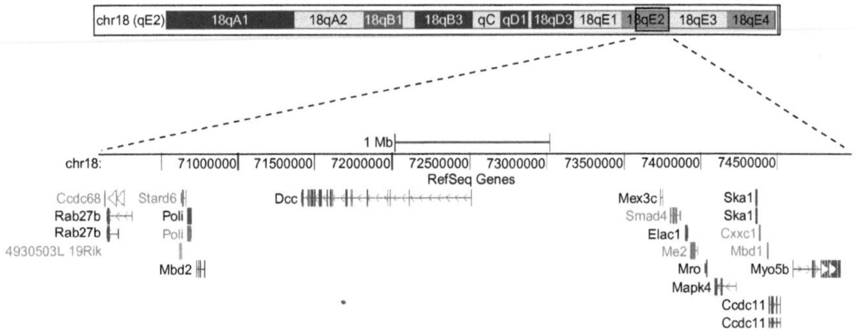

FIG. 4. *Mbd1, Cfp1, and Mbd2 genes reside in close proximity on mouse chromosome 18.* A map of 18qE2 region of mouse chromosome 18. Note that *Mbd1* and *Cfp1(CxxC1)* are neighboring genes, while *Mbd2* is located 3 Mb away.

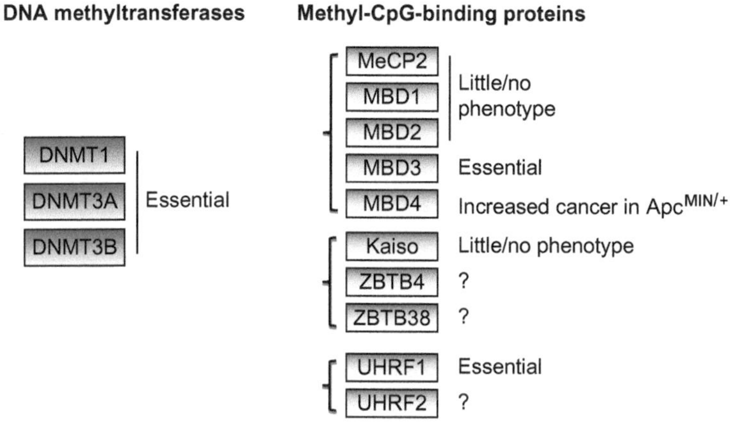

FIG. 5. *Results of genetic studies in mouse.* The phenotypic consequences of deleting the indicated genes are shown. See main text for references.

MBD2-null mice have defects in the differentiation of immune cells, and the females have a defect in offspring nurturing.[44,49] MeCP2-deficient mice develop a neurological disorder, which mimics the RTT in humans.[45,47] These phenotypes may be in part determined by the expression patterns of MBD proteins in different tissues. Most MBD proteins are ubiquitously expressed; however, MBD1 is more abundant in lung and liver, while MBD2 is highly expressed in the dorsal striatum in the brain and MBD4 in the immune system. In the case of MeCP2, it is clear that its singular abundance in postmitotic neurons (6.4×10^7 or one MeCP2 molecule per every 200 bp of DNA) is essential for normal brain function.[50]

Biochemical studies have indicated that MBD1 and MeCP2 can bind methylated DNA sequence specifically *in vitro*.[51,52] On DNA probes containing a single methylated CpG, MBD1 prefers to bind TmeCGCA or TGmeCGCA sites, while MeCP2 prefers methylated CpG followed by a run of four or more A/T bases. However, the specificity can be overridden either by clustering of multiple methylated CpGs or high protein concentration.[50,52] For example, in neurons, MeCP2 bindings seem to track the distribution of methylated CpGs.[50] Nevertheless, the estimated number of meCpG sites per diploid nucleus is approximately 4×10^7, while the number of MBD protein molecules is typically much lower in most tissues. Therefore, the affinity of MBD1, MeCP2 as well as other methyl-CpG-binding proteins for specific methylated sites may be important in determining how MBD proteins are distributed on methylated DNA *in vivo*.

The capacity of MBD proteins to bind methylated DNA with high affinity and specificity has led to their use as experimental tools. In essence, the recombinant MBD domain can be used as an affinity column to purify methylated DNA from a mixture of unmethylated and methylated fragments.[4,53] The methylated complement can then be analyzed by a variety of downstream approaches (such as hybridization to microarrays, PCR, or high-throughput sequencing). This elegant idea has given rise to many technological variations,[54] which allowed valuable observations to be made regarding the distribution of DNA methylation.[55] In theory, this principle can also be adapted to monitor the DNA methylation status of known sites *in vitro* or *in vivo*[56] (see also Chapter by Eleanor Wong and Chia-Lin Wei).

The replacement of methylated cytosine by hydroxymethylated cytosine causes a 50-fold drop in the affinity of MeCP2 for its targets.[57] This could be important, as hydroxymethylation is a likely mechanism for the removal of DNA methylation in early development.

B. Kaiso and Kaiso-Like Proteins

Certain protein domains other than the MBD can also permit the preferential binding of methylated CpG over the identical unmethylated sequence. This realization came from biochemical experiments: a protein activity, present in murine leukemia cells, and binding to a methylated intron of the S100A4 gene, was purified and found not to correspond to any of the MBD proteins. Peptide sequencing revealed that the protein was in fact a zinc finger transcription factor, Kaiso[58] (Fig. 2B). This protein had been independently identified earlier, in a two-hybrid screen for proteins binding p120 catenin,[59] as will be discussed later. Experiments with methylated reporter plasmids confirmed that Kaiso does repress transcription from methylated templates.[58]

Kaiso resembles the MBD proteins in that it prefers to bind methylated CpGs in certain sequence contexts,[58,60] but it also differs from them in an important way: it binds not only methylated CpGs but also certain unmethylated sequences which, in fact, do not even contain a CpG and are not methylatable.[60,61] The nonmethylated Kaiso targets are called Kaiso-binding sites, or KBSs. They contain a TpG core, and the structural similarity of T with methylated C probably explains why these two types of sequences can be bound. The structural characterization of Kaiso has not been reported as of this writing, but it should yield important insight into its mode of binding methylated DNA.

Kaiso is a member of the ZBTB family; these proteins contain, always at the N-terminus, a BTB/POZ domain, and a variable number of Kruppel-like C2H2 zinc fingers. The BTB/POZ domain permits protein–protein interactions.[62] The BTB domain of Kaiso recruits the corepressor NCoR.[63] Kaiso has been studied mainly in two biological systems: *Xenopus laevis* and mammals. The knockdown of Kaiso in *Xenopus* embryos, through the use of morpholino oligonucleotides, leads to the premature transcription of many genes and to developmental arrest and apoptosis.[64] This phenotype is quite similar to the consequences that follow removal of the DNA methylating enzyme DNMT1 itself,[65] suggesting that Kaiso is indeed a methylation-dependent repressor.

Kaiso was also studied in mouse models. Unlike the *Xenopus* results, deleting Kaiso from the mouse genome does not cause major phenotypic disturbances: the Kaiso mutant mice are viable, fertile, and live as long as normal mice.[66]

A homology search led to the identification of two proteins closely related to Kaiso, ZBTB4 and ZBTB38, that share many of its properties[67] (Fig. 2B). They bind not only methylated DNA but also certain unmethylated sequences[60] and repress transcription, although by a different set of corepressors than Kaiso (ZBTB38 interacts with CtBP1[68]). The three proteins are expressed in many of the same tissues and cell types and could conceivably be redundant. Interestingly, these proteins have undergone rapid recent evolution as ZBTB4 has not been identified outside of mammals.[69] The genetic study of ZBTB4 and ZBTB38 in mouse has not yet been reported but should yield valuable insight into the role of these proteins, and their possible redundancy with Kaiso.

The human genome encodes about 700 proteins containing zinc fingers. Kaiso and its paralogs apparently recognize methylated DNA by a canonical mechanism,[60] raising the possibility that additional zinc finger proteins will be found to bind methylated DNA in the future. ZFP57, a protein that controls genomic imprinting,[70] is an interesting candidate for this function. Recently, a biochemical approach has been used to identify proteins that preferentially bind methylated or unmethylated DNA,[71] some of these proteins contain zinc fingers, and may be novel methyl-binding proteins.

C. SRA Domain Proteins

A third group of methyl-CpG-binding proteins was discovered in much the same way that Kaiso was. A protein activity binding an enhancer only in its methylated state was purified and found to differ from the known methyl-CpG-binding proteins.[72] This protein, UHRF1, was previously found in a one-hybrid screen for factors binding a nonmethylated regulatory sequence in the promoter of the topoisomerase II alpha gene.[73] The methyl-binding function of UHRF1 is now far more documented than its possible role in binding nonmethylated DNA. UHRF1 is also called ICBP90 in humans and Np95 in mouse, but we will only use its systematic name from here on.

UHRF1 contains five recognizable protein domains[74] (Fig. 2C). A ubiquitin-like fold with unknown function is located at the N-terminus. By analogy with other proteins, it might be involved in interacting with the proteasome.[75] The following region, a tandem tudor domain, binds histone H3 tails trimethylated on residue K9 and is likely involved in tethering the protein to heterochromatic regions.[76–78] The third domain is the SRA, which mediates the recognition of methylated DNA. The protein also carries a PHD finger, which in other proteins, was shown to bind modified histones.[79] The role of the UHRF1 PHD finger, however, is unknown. Finally, UHRF1 contains a RING finger at its C-terminus and is endowed with catalytic activity as an E3 ubiquitin ligase. This activity is directed toward histone H3,[80] and possibly other substrates. The functional relevance of this ubiquitination activity is unclear at present. At any rate, UHRF1, like MBD4, is a methyl-binding protein that also bears an enzymatic activity.

The structure of UHRF1 bound to methylated DNA was solved by X-ray crystallography[33–35] (Fig. 3C). The protein flips the methylated cytosine out of the DNA double helix, in a mechanism reminiscent of that used by DNA methyltransferases.[81] The flipped methylcytosine is accepted into a hydrophobic cavity, where the methyl group on the cytosine contributes several important interactions, and this explains why methylated cytosine is bound better than unmethylated cytosine. On the opposite strand, the presence of a methyl on the cytosine is disadvantageous for binding, explaining why UHRF1 has better affinity for hemimethylated DNA than for fully methylated DNA. The nucleotides flanking the CpG do not make specific contacts with the protein, and this predicts that UHRF1 should bind hemimethylated CpG irrespective of their sequence context.

UHRF1 is unique among all methylcytosine-binding proteins in being essential in mouse (Fig. 5); its deletion leads to rapid developmental arrest.[82] This effect is likely due to a key role of UHRF1 in DNA methylation maintenance, which will be discussed in a following section. Also, UHRF1 can act as a transcriptional repressor, by recruiting the histone methyltransferase G9a.[83]

In mammalian genomes, UHRF1 has only one close paralog: UHRF2, which contains the same domains in the same order (Fig. 2C). It can also bind methylated DNA,[72] but its function has not been investigated in detail. The genome of *A. thaliana* encodes many proteins with SRA domains and, interestingly, some of those are involved in binding non-CpG methylation.[84,85] For instance, VIM1 binds both meCpG and meCpHpG[86] (H = A, C, or T). Cytosine methylation in the CpA and CpT context is abundant in ES cells,[87] and it is tempting to speculate that UHRF2 might recognize this mark.

III. Biological Functions of Methyl-CpG-Binding Proteins

A. Maintenance of Heterochromatin

All methyl-binding proteins identified to date associate with chromatin-modifying activities that establish or maintain heterochromatin. For instance, MBD1 interacts with SETDB1,[88] MBD2 with the NuRD corepressor complex,[89,90] and MeCP2 with SIN3A/HDAC1 or HDAC2.[23] These associations predict that the methyl-binding proteins may be involved in heterochromatin maintenance, either as a general maintenance mechanism or to ensure transcriptional regulation.

A particularly clear example is provided by UHRF1: the phenotype observed upon removal of UHRF1 from mammalian cells is a widespread loss of DNA methylation in the genome.[82,91] Mechanistically, this has been explained by the fact that the protein can bind hemimethylated DNA with high affinity and that it also interacts with DNMT1, therefore targeting it to the hemimethylated sites that need to be remethylated after DNA replication (Fig. 6). UHRF1 is involved not just in the maintenance of heterochromatin but also in its establishment, as ES cells lacking UHRF1 cannot efficiently silence exogenous reporter genes.[92] The role in heterochromatin maintenance is not limited to UHRF1, as MBD1 has also been evidenced to maintain H3K9 methylation.[88]

B. Regulation of Gene Expression

Relatively few genes have been positively shown to be bound and repressed by methyl-binding proteins, and most of the data so far imply proteins of the MBD family. A first example concerns the *Xist* gene, which is methylated and inactive on the active X chromosome of cells from female mammals. It has been shown that this gene is repressed by MBD2, but not by MBD1, MeCP2, or Kaiso, showing that at least certain functions of the MBPs in transcriptional

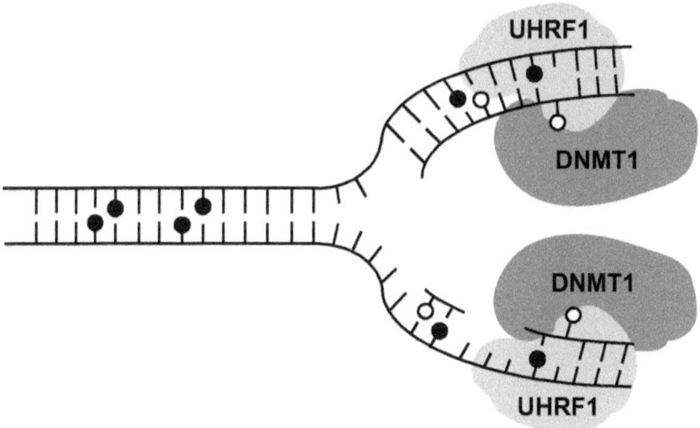

FIG. 6. *The role of UHRF1 in maintaining DNA methylation*. CpG dinucleotides are often symmetrically methylated (indicated by the black circle on each DNA strand). After DNA replication, each CpG becomes hemimethylated (indicated by white circle on one strand, black on the other strand). The protein UHRF1 recognizes hemimethylated CpGs with high affinity, flips out the unmethylated base, and recruits the enzyme DNMT1, which methylates the unmethylated cytosine.

regulation are not redundant.[93] The genes that confer pluripotency to cells, such as *Oct4*, are also inactivated by DNA methylation in differentiated cells. By analogy with *Xist*, they might be silenced by the action of one or several methyl-binding proteins. Germ line genes have CpG islands that are methylated in somatic tissues, and they are repressed by methylation.[2,3,8,9] However, the nature of the MBPs that represses their transcription is unknown.

A handful of MBD1 targets have been identified by genomic approaches. Some of these contribute to the role of MBD1 in neuronal differentiation,[94,95] and others, which are not regulated by MBD2, further exemplify the idea that the methylcytosine-binding proteins act nonredundantly.[52]

MBD2 regulates the expression of IL4 in T lymphocytes,[49] and the expression of rho globin in red blood cells.[96] In addition, the removal of MBD2 causes the expression of several genes in the inappropriate segment of the mouse gastrointestinal tract,[97] suggesting that it may have a general role in ensuring the accuracy of tissue-specific transcription. However, this function is not strictly essential, as $Mbd2^{-/-}$ mice are viable.

MeCP2 has a global effect on transcription in neurons (see below), and targets of Kaiso have been identified in the context of the Wnt pathway (see Section IV).

C. Control of Repetitive Sequences

Repetitive sequences make up about half of mammalian genomes. Some of these sequences are ancient and have mutated away from being functional, but others are more recent and have the potential to be transcribed and produce proteins. Their expression can be deleterious not only by causing misexpression of neighboring genes but also by promoting mutagenic transposition events. There is a clear evolutionary advantage for repetitive sequences to be repressed, and DNA methylation is a major mechanism used to that end[98] (see Chapters by Anton Wutz; and Mary G. Goll and Marnie E. Halpern).

Several methyl-binding proteins have been shown to bind repetitive sequences and inhibit their expression. MeCP2 has an important role in repressing the transposition of L1 retrotransposons in the brain.[99,100] The mouse IAP repeats are reactivated in the absence of MBD1 in neural stem cells,[43] and in the absence of MeCP2 in mature neurons.[50] Kaiso binds IAP sequences as well but is dispensable for their repression,[66] and so is MBD2.[44] As for UHRF1, it has a crucial role in silencing the pericentric major satellite repeats.[101]

D. Methyl-Binding Proteins as Effectors of Signaling Pathways

Transcription factors often respond to signaling pathways to adapt the cellular transcriptional output to environmental cues. Two of the methyl-binding proteins have been clearly placed in signaling cascades, MeCP2 and Kaiso.

Investigations in *Xenopus* have shown that MeCP2 represses the transcription of the *xHairy2a* gene, a neuronal determinant. Upon activation of the Notch signaling pathway, MeCP2 is released, permitting the transcription of *xHairy2a* and neuronal differentiation. A mutant form of MeCP2 found in RTT is refractory to this Notch-mediated inhibition, which may account for some of its properties.[102] In mammalian neurons, MeCP2 has been shown to be phosphorylated by CaMKII, which is itself activated by the calcium influx that follows neuronal stimulation.[103] Phosphorylated MeCP2 is released from the promoter of BDNF, a crucial neurotrophin, which can then be expressed.[104]

The Wnt pathway regulates cell proliferation and differentiation. Notably, it controls the number and activity of stem cells in the hematopoietic system and colon.[105] Some arguments suggest that Kaiso may repress target genes of the Wnt pathway and be itself inactivated when the Wnt pathway is turned on. The first hint in this direction came from the discovery of Kaiso as an interactor of p120 catenin,[59] a submembrane protein that plays roles analogous to beta-catenin. This finding, and subsequent experiments in cultured cells, led to a model in which p120 catenin can translocate to the nucleus and modulate the activity of Kaiso there.[106–108] Other investigators found that in *Xenopus*, Kaiso is a player in the noncanonical Wnt pathway; it represses target genes by binding to KBSs in their promoters and is itself displaced when the noncanonical Wnt pathway is activated.[109,110] This role has been called into question, as other investigators pointed out limitations of the techniques used, and put forward differing interpretations of the data.[111–113]

E. Control of DNA Demethylation

One important area of investigation concerns the removal of DNA methylation. This can occur by passive dilution if the mark is not replaced after each round of replication, but it can also be an active process involving enzymatic activities (see Chapter by Taiping Chen). DNA demethylation events take place in several situations, including global demethylation of the paternal genome after fertilization, global demethylation in primordial germ cells, and promoter-specific DNA demethylation during gene activation. MBD4 has been involved in promoter-specific demethylation events,[114] as well as in genome-wide demethylation.[115,116]

IV. Methyl-CpG-Binding Proteins and Human Disease

It is currently well established that defects in DNA methylation associate with variety of human pathological conditions including imprinting disorders, neonatal diabetes, mental retardation, and cancer. The roles played by methyl-CpG-binding proteins in some forms of human disease are better understood than others.

A. Rett Syndrome

RTT is a severe human neurological disorder affecting 1:12,500 females by the age of 12 and is commonly lethal in males (see also Chapter by Jon F. Wilkins and Francisco Úbeda). RTT is manifested by loss of verbal skills, repetitive hand movement, mental retardation, breading abnormalities, and often seizures. Most RTT-associated mutations are sporadic, and 95% of them map to the X-linked *MECP2* gene.[117,118] Mice carrying targeted deletion of *Mecp2* or mutations occurring in RTT recapitulate to great extent the symptoms of the human disease and are widely used as models to study RTT.[45,47,119] Unlike humans, $Mecp2^{-/y}$ males are viable but have smaller brain size, develop disease symptoms at about 6 weeks after birth, and die within 12 weeks. The $Mecp2^{-/-}$ female mice are viable and fertile with significantly later and milder onset of the disease.

The precise molecular mechanism(s) for the cause of the RTT are yet unclear. Expression studies carried out with brains of MeCP2-deficient and MeCP2-overexpressing mice indicate that many transcripts are downregulated in the absence of MeCP2 and, conversely, upregulated when MeCP2 is in excess, leading to the hypothesis that MeCP2 may act as an activator of transcription.[120] Given that MeCP2 is extremely abundant in postmitotic neurons and seems to track the distribution of methylated CpGs in the genome,[50] an alternative hypothesis could be that MeCP2 increases the fidelity of transcription in neurons by globally suppressing transcriptional noise.

Interestingly, reactivation of MeCP2 expression in brain or in symptomatic male mice carrying conditional $Mecp2^{lox\text{-}stop/y}$ allele can largely revert the RTT symptoms arguing that the lack of MeCP2 does not cause fatal and irreversible structural damage to neuronal cells.[121] Potentially, these studies provide an avenue for future gene therapy applications. Recently, neuronal cells have been produced by reprogramming of fibroblasts from Rett patients and have been used to grow human Rett neurons in culture.[122] This will be a useful tool for the study of this disease.

B. Cancer

Cancer is characterized by a variety of genetic mutations that lead to unrestricted growth, proliferation in the absence of growth stimuli, evasion of apoptosis, and often metastasis from the actual site of appearance to other tissues. This is often accompanied by epigenetic changes, in particular, aberrant DNA methylation of gene promoters leading to inactivation of tumor suppressor and other genes in cancer.[123,124] A global depletion of methylated CpGs from the genome often leading to expression of oncogenes and genomic instability is also common in tumors[125,126] (see Chapter by Minoru Toyota and Eiichiro Yamamoto).

Experiments aiming to map the binding of MBD proteins to gene promoters in cancer cell lines have indicated that several MBD proteins can bind to heavily methylated CpG islands.[127,128] This is in contrast with normal cells where occupancy of methylated sites by multiple MBD proteins is rarely detected.[51] As discussed previously, occurrence of multiple methylated CpGs on a short stretch of DNA can override sequence-specific interaction of MBD proteins with methylated DNA, suggesting that silencing of tumor suppressor genes would be achieved either by cooperative binding of several MBD proteins at methylated promoters or by a single protein that happen to be most abundant in the particular tumor cell type.

Genetic experiments in mice indicate that lack of MBD2 and Kaiso proteins, similar to partial loss of function DNMT1 alleles, can suppress the formation of intestinal tumors in $Apc^{+/Min}$ mouse model of colorectal cancer.[66,129,130] Whether MBD1-deficient mice are also resistant to tumorigenesis is yet to be determined. However, lack of MBD4 glycosylase involved in methylation-dependent mismatch repair promotes tumor formation in $Apc^{+/Min}$ mice.[46] One might suspect that the levels of MBD2 and MBD4, their binding to DNA, or interaction with corepressors might be regulated in cells in response to growth or proliferation signals. However, this has never been investigated in detail. In the case of MBD1, SUMOylation by PIAS1 and PIAS3 SUMO ligases prevents the assembly of MBD1-associated silencing complex and leads to derepression of MBD1 target genes.[131] As PIAS proteins are often downregulated in tumors, this may leads to an increase of repressive MBD1 complexes that may be required to silence multiple methylated promoters.

Despite the prevalent view that the role of methyl-CpG-binding proteins in cancer must be mediated by their interaction with methylated DNA, a methylation-independent role of ZBTB4 protein in promoting cellular proliferation and survival has been uncovered. Thus, ZBTB4 cooperates with MIZ1 protein and Sin3/HDAC complex to repress transcription of p53 target gene p21CIP1 in a DNA methylation-independent fashion.[132] Downregulation of ZBTB4 mediated by HIP2 kinase phosphorylation is concomitant with p53 activation and leads to p21 upregulation, cell cycle arrest, and apoptosis.[133] This suggests that the levels of ZBTB4 may be critical in determining the cellular responses to p53 activation. Taken together, it is apparent that methyl-CpG-binding proteins play vital roles in cancer formation.

V. Concluding Remarks

There are currently three known families of proteins that bind methylated DNA, comprising a total of nine proteins in mammals. Intensive research in many different laboratories has contributed greatly to elucidating their important biological functions. However, there are also a number of outstanding questions to be addressed in the near future. It is very likely that there are additional proteins that bind methylated DNA and those are yet to be identified. Detailed maps of methyl-CpG-binding protein profiles across the genome are yet to be generated, and last, but not least, systems biology approaches and mathematical modeling are yet to be implemented to understand how the multitude of methyl-CpG-binding proteins is used in different cell types to fulfill specific functions.

Acknowledgments

Work in the laboratory of P.-A. D. is supported by Institut National du Cancer, Association pour la Recherche sur le Cancer, and Ligue Nationale contre le Cancer. P.-A. D. thanks Sophie Laget for her help with the figures presented in this chapter. Work in I. S. laboratory is supported by Cancer Research UK.

References

1. Antequera F, Macleod D, Bird AP. Specific protection of methylated CpGs in mammalian nuclei. *Cell* 1989;**58**:509–17.
2. Suzuki MM, Bird A. DNA methylation landscapes: provocative insights from epigenomics. *Nat Rev Genet* 2008;**9**:465–76.
3. Illingworth RS, Bird AP. CpG islands—'a rough guide'. *FEBS Lett* 2009;**583**:1713–20.
4. Illingworth R, Kerr A, Desousa D, Jorgensen H, Ellis P, Stalker J, et al. A novel CpG island set identifies tissue-specific methylation at developmental gene loci. *PLoS Biol* 2008;**6**:e22.
5. Maunakea AK, Nagarajan RP, Bilenky M, Ballinger TJ, D'Souza C, Fouse SD, et al. Conserved role of intragenic DNA methylation in regulating alternative promoters. *Nature* 2010;**466**:253–7.
6. Illingworth RS, Gruenewald-Schneider U, Webb S, Kerr AR, James KD, Turner DJ, et al. Orphan CpG islands identify numerous conserved promoters in the mammalian genome. *PLoS Genet* 2010;**6**(9):e1001134.
7. Weber M, Hellmann I, Stadler MB, Ramos L, Paabo S, Rebhan M, et al. Distribution, silencing potential and evolutionary impact of promoter DNA methylation in the human genome. *Nat Genet* 2007;**39**:457–66.
8. Mohn F, Weber M, Rebhan M, Roloff TC, Richter J, Stadler MB, et al. Lineage-specific polycomb targets and de novo DNA methylation define restriction and potential of neuronal progenitors. *Mol Cell* 2008;**30**:755–66.
9. Borgel J, Guibert S, Li Y, Chiba H, Schubeler D, Sasaki H, et al. Targets and dynamics of promoter DNA methylation during early mouse development. *Nat Genet* 2010;**42**:1093–100.
10. Bird A. DNA methylation patterns and epigenetic memory. *Genes Dev* 2002;**16**:6–21.
11. Ooi SK, O'Donnell AH, Bestor TH. Mammalian cytosine methylation at a glance. *J Cell Sci* 2009;**122**:2787–91.
12. Bird AP. Gene number, noise reduction and biological complexity. *Trends Genet* 1995;**11**:94–100.

13. Buschhausen G, Wittig B, Graessmann M, Graessmann A. Chromatin structure is required to block transcription of the methylated herpes simplex virus thymidine kinase gene. *Proc Natl Acad Sci USA* 1987;**84**:1177–81.
14. Kass SU, Landsberger N, Wolffe AP. DNA methylation directs a time-dependent repression of transcription initiation. *Curr Biol* 1997;**7**:157–65.
15. Chen B, He L, Savell VH, Jenkins JJ, Parham DM. Inhibition of the interferon-gamma/signal transducers and activators of transcription (STAT) pathway by hypermethylation at a STAT-binding site in the p21WAF1 promoter region. *Cancer Res* 2000;**60**:3290–8.
16. Perini G, Diolaiti D, Porro A, Della Valle G. In vivo transcriptional regulation of N-Myc target genes is controlled by E-box methylation. *Proc Natl Acad Sci USA* 2005;**102**:12117–22.
17. Iguchi-Ariga SM, Schaffner W. CpG methylation of the cAMP-responsive enhancer/promoter sequence TGACGTCA abolishes specific factor binding as well as transcriptional activation. *Genes Dev* 1989;**3**:612–9.
18. Huang LH, Wang R, Gama-Sosa MA, Shenoy S, Ehrlich M. A protein from human placental nuclei binds preferentially to 5-methylcytosine-rich DNA. *Nature* 1984;**308**:293–5.
19. Wang RY, Zhang XY, Khan R, Zhou YW, Huang LH, Ehrlich M. Methylated DNA-binding protein from human placenta recognizes specific methylated sites on several prokaryotic DNAs. *Nucleic Acids Res* 1986;**14**:9843–60.
20. Meehan RR, Lewis JD, McKay S, Kleiner EL, Bird AP. Identification of a mammalian protein that binds specifically to DNA containing methylated CpGs. *Cell* 1989;**58**:499–507.
21. Lewis JD, Meehan RR, Henzel WJ, Maurer-Fogy I, Jeppesen P, Klein F, et al. Purification, sequence, and cellular localization of a novel chromosomal protein that binds to methylated DNA. *Cell* 1992;**69**:905–14.
22. Nan X, Meehan RR, Bird A. Dissection of the methyl-CpG binding domain from the chromosomal protein MeCP2. *Nucleic Acids Res* 1993;**21**:4886–92.
23. Nan X, Ng HH, Johnson CA, Laherty CD, Turner BM, Eisenman RN, et al. Transcriptional repression by the methyl-CpG-binding protein MeCP2 involves a histone deacetylase complex. *Nature* 1998;**393**:386–9.
24. Hendrich B, Bird A. Identification and characterization of a family of mammalian methyl-CpG binding proteins. *Mol Cell Biol* 1998;**18**:6538–47.
25. Hendrich B, Tweedie S. The methyl-CpG binding domain and the evolving role of DNA methylation in animals. *Trends Genet* 2003;**19**:269–77.
26. Grafi G, Zemach A, Pitto L. Methyl-CpG-binding domain (MBD) proteins in plants. *Biochim Biophys Acta* 2007;**1769**:287–94.
27. Scebba F, Bernacchia G, De Bastiani M, Evangelista M, Cantoni RM, Cella R, et al. Arabidopsis MBD proteins show different binding specificities and nuclear localization. *Plant Mol Biol* 2003;**53**:715–31.
28. Laget S, Joulie M, Le Masson F, Sasai N, Christians E, Pradhan S, et al. The human proteins MBD5 and MBD6 associate with heterochromatin but they do not bind methylated DNA. *PLoS ONE* 2010;**5**(8):e11982.
29. Mayer C, Schmitz KM, Li J, Grummt I, Santoro R. Intergenic transcripts regulate the epigenetic state of rRNA genes. *Mol Cell* 2006;**22**:351–61.
30. Jeffery L, Nakielny S. Components of the DNA methylation system of chromatin control are RNA-binding proteins. *J Biol Chem* 2004;**279**:49479–87.
31. Ho KL, McNae IW, Schmiedeberg L, Klose RJ, Bird AP, Walkinshaw MD. MeCP2 binding to DNA depends upon hydration at methyl-CpG. *Mol Cell* 2008;**29**:525–31.
32. Ohki I, Shimotake N, Fujita N, Jee J, Ikegami T, Nakao M, et al. Solution structure of the methyl-CpG binding domain of human MBD1 in complex with methylated DNA. *Cell* 2001;**105**:487–97.

33. Arita K, Ariyoshi M, Tochio H, Nakamura Y, Shirakawa M. Recognition of hemi-methylated DNA by the SRA protein UHRF1 by a base-flipping mechanism. *Nature* 2008;**455**:818–21.
34. Avvakumov GV, Walker JR, Xue S, Li Y, Duan S, Bronner C, et al. Structural basis for recognition of hemi-methylated DNA by the SRA domain of human UHRF1. *Nature* 2008;**455**:822–5.
35. Hashimoto H, Horton JR, Zhang X, Bostick M, Jacobsen SE, Cheng X. The SRA domain of UHRF1 flips 5-methylcytosine out of the DNA helix. *Nature* 2008;**455**:826–9.
36. Lyko F, Ramsahoye BH, Jaenisch R. DNA methylation in Drosophila melanogaster. *Nature* 2000;**408**:538–40.
37. Marhold J, Kramer K, Kremmer E, Lyko F. The Drosophila MBD2/3 protein mediates interactions between the MI-2 chromatin complex and CpT/A-methylated DNA. *Development* 2004;**131**:6033–9.
38. Carlone DL, Hart SR, Ladd PD, Skalnik DG. Cloning and characterization of the gene encoding the mouse homologue of CpG binding protein. *Gene* 2002;**295**:71–7.
39. Thomson JP, Skene PJ, Selfridge J, Clouaire T, Guy J, Webb S, et al. CpG islands influence chromatin structure via the CpG-binding protein Cfp1. *Nature* 2010;**464**:1082–6.
40. Jorgensen HF, Ben-Porath I, Bird AP. Mbd1 is recruited to both methylated and nonmethylated CpGs via distinct DNA binding domains. *Mol Cell Biol* 2004;**24**:3387–95.
41. Simmen MW, Leitgeb S, Charlton J, Jones SJ, Harris BR, Clark VH, et al. Nonmethylated transposable elements and methylated genes in a chordate genome. *Science* 1999;**283**:1164–7.
42. Clouaire T, Stancheva I. Methyl-CpG binding proteins: specialized transcriptional repressors or structural components of chromatin? *Cell Mol Life Sci* 2008;**65**:1509–22.
43. Zhao X, Ueba T, Christie BR, Barkho B, McConnell MJ, Nakashima K, et al. Mice lacking methyl-CpG binding protein 1 have deficits in adult neurogenesis and hippocampal function. *Proc Natl Acad Sci USA* 2003;**100**:6777–82.
44. Hendrich B, Guy J, Ramsahoye B, Wilson VA, Bird A. Closely related proteins MBD2 and MBD3 play distinctive but interacting roles in mouse development. *Genes Dev* 2001;**15**:710–23
45. Guy J, Hendrich B, Holmes M, Martin JE, Bird A. A mouse Mecp2-null mutation causes neurological symptoms that mimic Rett syndrome. *Nat Genet* 2001;**27**:322–6.
46. Millar CB, Guy J, Sansom OJ, Selfridge J, MacDougall E, Hendrich B, et al. Enhanced CpG mutability and tumorigenesis in MBD4-deficient mice. *Science* 2002;**297**:403–5.
47. Chen RZ, Akbarian S, Tudor M, Jaenisch R. Deficiency of methyl-CpG binding protein-2 in CNS neurons results in a Rett-like phenotype in mice. *Nat Genet* 2001;**27**:327–31.
48. Allan AM, Liang X, Luo Y, Pak C, Li X, Szulwach KE, et al. The loss of methyl-CpG binding protein 1 leads to autism-like behavioral deficits. *Hum Mol Genet* 2008;**17**:2047–57.
49. Hutchins AS, Mullen AC, Lee HW, Sykes KJ, High FA, Hendrich BD, et al. Gene silencing quantitatively controls the function of a developmental trans-activator. *Mol Cell* 2002;**10**:81–91.
50. Skene PJ, Illingworth RS, Webb S, Kerr AR, James KD, Turner DJ, et al. Neuronal MeCP2 is expressed at near histone-octamer levels and globally alters the chromatin state. *Mol Cell* 2010;**37**:457–68.
51. Klose RJ, Sarraf SA, Schmiedeberg L, McDermott SM, Stancheva I, Bird AP. DNA binding selectivity of MeCP2 due to a requirement for A/T sequences adjacent to methyl-CpG. *Mol Cell* 2005;**19**:667–78.
52. Clouaire T, de Las Heras JI, Merusi C, Stancheva I. Recruitment of MBD1 to target genes requires sequence-specific interaction of the MBD domain with methylated DNA. *Nucleic Acids Res* 2010;**38**:4620–34.
53. Cross SH, Charlton JA, Nan X, Bird AP. Purification of CpG islands using a methylated DNA binding column. *Nat Genet* 1994;**6**:236–44.
54. Rauch TA, Pfeifer GP. The MIRA method for DNA methylation analysis. *Methods Mol Biol* 2009;**507**:65–75.

55. Selker EU, Tountas NA, Cross SH, Margolin BS, Murphy JG, Bird AP, et al. The methylated component of the Neurospora crassa genome. *Nature* 2003;**422**:893–7.
56. Stains CI, Furman JL, Segal DJ, Ghosh I. Site-specific detection of DNA methylation utilizing mCpG-SEER. *J Am Chem Soc* 2006;**128**:9761–5.
57. Valinluck V, Tsai HH, Rogstad DK, Burdzy A, Bird A, Sowers LC. Oxidative damage to methyl-CpG sequences inhibits the binding of the methyl-CpG binding domain (MBD) of methyl-CpG binding protein 2 (MeCP2). *Nucleic Acids Res* 2004;**32**:4100–8.
58. Prokhortchouk A, Hendrich B, Jorgensen H, Ruzov A, Wilm M, Georgiev G, et al. The p120 catenin partner Kaiso is a DNA methylation-dependent transcriptional repressor. *Genes Dev* 2001;**15**:1613–8.
59. Daniel JM, Reynolds AB. The catenin p120(ctn) interacts with Kaiso, a novel BTB/POZ domain zinc finger transcription factor. *Mol Cell Biol* 1999;**19**:3614–23.
60. Sasai N, Nakao M, Defossez PA. Sequence-specific recognition of methylated DNA by human zinc-finger proteins. *Nucleic Acids Res* 2010;**38**:5015–22.
61. Daniel JM, Spring CM, Crawford HC, Reynolds AB, Baig A. The p120(ctn)-binding partner Kaiso is a bi-modal DNA-binding protein that recognizes both a sequence-specific consensus and methylated CpG dinucleotides. *Nucleic Acids Res* 2002;**30**:2911–9.
62. Perez-Torrado R, Yamada D, Defossez PA. Born to bind: the BTB protein-protein interaction domain. *Bioessays* 2006;**28**:1194–202.
63. Yoon HG, Chan DW, Reynolds AB, Qin J, Wong J. N-CoR mediates DNA methylation-dependent repression through a methyl CpG binding protein Kaiso. *Mol Cell* 2003;**12**:723–34.
64. Ruzov A, Dunican DS, Prokhortchouk A, Pennings S, Stancheva I, Prokhortchouk E, et al. Kaiso is a genome-wide repressor of transcription that is essential for amphibian development. *Development* 2004;**131**:6185–94.
65. Stancheva I, Meehan RR. Transient depletion of xDnmt1 leads to premature gene activation in Xenopus embryos. *Genes Dev* 2000;**14**:313–27.
66. Prokhortchouk A, Sansom O, Selfridge J, Caballero IM, Salozhin S, Aithozhina D, et al. Kaiso-deficient mice show resistance to intestinal cancer. *Mol Cell Biol* 2006;**26**:199–208.
67. Filion GJ, Zhenilo S, Salozhin S, Yamada D, Prokhortchouk E, Defossez PA. A family of human zinc finger proteins that bind methylated DNA and repress transcription. *Mol Cell Biol* 2006;**26**:169–81.
68. Sasai N, Matsuda E, Sarashina E, Ishida Y, Kawaichi M. Identification of a novel BTB-zinc finger transcriptional repressor, CIBZ, that interacts with CtBP corepressor. *Genes Cells* 2005;**10**:871–85.
69. Sasai N, Defossez PA. Many paths to one goal? The proteins that recognize methylated DNA in eukaryotes. *Int J Dev Biol* 2009;**53**:323–34.
70. Li X, Ito M, Zhou F, Youngson N, Zuo X, Leder P, et al. A maternal-zygotic effect gene, Zfp57, maintains both maternal and paternal imprints. *Dev Cell* 2008;**15**:547–57.
71. Bartke T, Vermeulen M, Xhemalce B, Robson SC, Mann M, Kouzarides T. Nucleosome-interacting proteins regulated by DNA and histone methylation. *Cell* 2010;**143**:470–84.
72. Unoki M, Nishidate T, Nakamura Y. ICBP90, an E2F-1 target, recruits HDAC1 and binds to methyl-CpG through its SRA domain. *Oncogene* 2004;**23**:7601–10.
73. Hopfner R, Mousli M, Jeltsch JM, Voulgaris A, Lutz Y, Marin C, et al. ICBP90, a novel human CCAAT binding protein, involved in the regulation of topoisomerase II alpha expression. *Cancer Res* 2000;**60**:121–8.
74. Hashimoto H, Horton JR, Zhang X, Cheng X. UHRF1, a modular multi-domain protein, regulates replication-coupled crosstalk between DNA methylation and histone modifications. *Epigenetics* 2009;**4**:8–14.

75. Su V, Lau AF. Ubiquitin-like and ubiquitin-associated domain proteins: significance in proteasomal degradation. *Cell Mol Life Sci* 2009;**66**:2819–33.
76. Karagianni P, Amazit L, Qin J, Wong J. ICBP90, a novel methyl K9 H3 binding protein linking protein ubiquitination with heterochromatin formation. *Mol Cell Biol* 2008;**28**:705–17.
77. Papait R, Pistore C, Grazini U, Babbio F, Cogliati S, Pecoraro D, et al. The PHD domain of Np95 (mUHRF1) is involved in large-scale reorganization of pericentromeric heterochromatin. *Mol Biol Cell* 2008;**19**:3554–63.
78. Rottach A, Frauer C, Pichler G, Bonapace IM, Spada F, Leonhardt H. The multi-domain protein Np95 connects DNA methylation and histone modification. *Nucleic Acids Res* 2010;**38**:1796–804.
79. Taverna SD, Li H, Ruthenburg AJ, Allis CD, Patel DJ. How chromatin-binding modules interpret histone modifications: lessons from professional pocket pickers. *Nat Struct Mol Biol* 2007;**14**:1025–40.
80. Citterio E, Papait R, Nicassio F, Vecchi M, Gomiero P, Mantovani R, et al. Np95 is a histone-binding protein endowed with ubiquitin ligase activity. *Mol Cell Biol* 2004;**24**:2526–35.
81. Klimasauskas S, Kumar S, Roberts RJ, Cheng X. HhaI methyltransferase flips its target base out of the DNA helix. *Cell* 1994;**76**:357–69.
82. Sharif J, Muto M, Takebayashi S, Suetake I, Iwamatsu A, Endo TA, et al. The SRA protein Np95 mediates epigenetic inheritance by recruiting Dnmt1 to methylated DNA. *Nature* 2007;**450**:908–12.
83. Kim JK, Esteve PO, Jacobsen SE, Pradhan S. UHRF1 binds G9a and participates in p21 transcriptional regulation in mammalian cells. *Nucleic Acids Res* 2009;**37**:493–505.
84. Johnson LM, Bostick M, Zhang X, Kraft E, Henderson I, Callis J, et al. The SRA methyl-cytosine-binding domain links DNA and histone methylation. *Curr Biol* 2007;**17**:379–84.
85. Johnson LM, Law JA, Khattar A, Henderson IR, Jacobsen SE. SRA-domain proteins required for DRM2-mediated de novo DNA methylation. *PLoS Genet* 2008;**4**:e1000280.
86. Woo HR, Dittmer TA, Richards EJ. Three SRA-domain methylcytosine-binding proteins cooperate to maintain global CpG methylation and epigenetic silencing in Arabidopsis. *PLoS Genet* 2008;**4**:e1000156.
87. Lister R, Pelizzola M, Dowen RH, Hawkins RD, Hon G, Tonti-Filippini J, et al. Human DNA methylomes at base resolution show widespread epigenomic differences. *Nature* 2009;**462**:315–22.
88. Sarraf SA, Stancheva I. Methyl-CpG binding protein MBD1 couples histone H3 methylation at lysine 9 by SETDB1 to DNA replication and chromatin assembly. *Mol Cell* 2004;**15**:595–605.
89. Ng HH, Zhang Y, Hendrich B, Johnson CA, Turner BM, Erdjument-Bromage H, et al. MBD2 is a transcriptional repressor belonging to the MeCP1 histone deacetylase complex. *Nat Genet* 1999;**23**:58–61.
90. Feng Q, Zhang Y. The MeCP1 complex represses transcription through preferential binding, remodeling, and deacetylating methylated nucleosomes. *Genes Dev* 2001;**15**:827–32.
91. Bostick M, Kim JK, Esteve PO, Clark A, Pradhan S, Jacobsen SE. UHRF1 plays a role in maintaining DNA methylation in mammalian cells. *Science* 2007;**317**:1760–4.
92. Meilinger D, Fellinger K, Bultmann S, Rothbauer U, Bonapace IM, Klinkert WE, et al. Np95 interacts with de novo DNA methyltransferases, Dnmt3a and Dnmt3b, and mediates epigenetic silencing of the viral CMV promoter in embryonic stem cells. *EMBO Rep* 2009;**10**:1259–64.
93. Barr H, Hermann A, Berger J, Tsai HH, Adie K, Prokhortchouk A, et al. Mbd2 contributes to DNA methylation-directed repression of the Xist gene. *Mol Cell Biol* 2007;**27**:3750–7.
94. Li X, Barkho BZ, Luo Y, Smrt RD, Santistevan NJ, Liu C, et al. Epigenetic regulation of the stem cell mitogen Fgf-2 by Mbd1 in adult neural stem/progenitor cells. *J Biol Chem* 2008;**283**:27644–52.

95. Liu C, Teng ZQ, Santistevan NJ, Szulwach KE, Guo W, Jin P, et al. Epigenetic regulation of miR-184 by MBD1 governs neural stem cell proliferation and differentiation. *Cell Stem Cell* 2010;**6**:433–44.
96. Kransdorf EP, Wang SZ, Zhu SZ, Langston TB, Rupon JW, Ginder GD. MBD2 is a critical component of a methyl cytosine-binding protein complex isolated from primary erythroid cells. *Blood* 2006;**108**:2836–45.
97. Berger J, Sansom O, Clarke A, Bird A. MBD2 is required for correct spatial gene expression in the gut. *Mol Cell Biol* 2007;**27**:4049–57.
98. Yoder JA, Walsh CP, Bestor TH. Cytosine methylation and the ecology of intragenomic parasites. *Trends Genet* 1997;**13**:335–40.
99. Yu F, Zingler N, Schumann G, Stratling WH. Methyl-CpG-binding protein 2 represses LINE-1 expression and retrotransposition but not Alu transcription. *Nucleic Acids Res* 2001;**29**:4493–501.
100. Muotri AR, Marchetto MC, Coufal NG, Oefner R, Yeo G, Nakashima K, et al. L1 retrotransposition in neurons is modulated by MeCP2. *Nature* 2010;**468**:443–6.
101. Papait R, Pistore C, Negri D, Pecoraro D, Cantarini L, Bonapace IM. Np95 is implicated in pericentromeric heterochromatin replication and in major satellite silencing. *Mol Biol Cell* 2007;**18**:1098–106.
102. Stancheva I, Collins AL, Van den Veyver IB, Zoghbi H, Meehan RR. A mutant form of MeCP2 protein associated with human Rett syndrome cannot be displaced from methylated DNA by notch in Xenopus embryos. *Mol Cell* 2003;**12**:425–35.
103. Zhou Z, Hong EJ, Cohen S, Zhao WN, Ho HY, Schmidt L, et al. Brain-specific phosphorylation of MeCP2 regulates activity-dependent Bdnf transcription, dendritic growth, and spine maturation. *Neuron* 2006;**52**:255–69.
104. Chen WG, Chang Q, Lin Y, Meissner A, West AE, Griffith EC, et al. Derepression of BDNF transcription involves calcium-dependent phosphorylation of MeCP2. *Science* 2003;**302**:885–9.
105. Radtke F, Clevers H, Riccio O. From gut homeostasis to cancer. *Curr Mol Med* 2006;**6**:275–89.
106. van Roy FM, McCrea PD. A role for Kaiso-p120ctn complexes in cancer? *Nat Rev Cancer* 2005;**5**:956–64.
107. Soubry A, van Hengel J, Parthoens E, Colpaert C, Van Marck E, Waltregny D, et al. Expression and nuclear location of the transcriptional repressor Kaiso is regulated by the tumor microenvironment. *Cancer Res* 2005;**65**:2224–33.
108. Daniel JM. Dancing in and out of the nucleus: p120(ctn) and the transcription factor Kaiso. *Biochim Biophys Acta* 2007;**1773**:59–68.
109. Park JI, Kim SW, Lyons JP, Ji H, Nguyen TT, Cho K, et al. Kaiso/p120-catenin and TCF/beta-catenin complexes coordinately regulate canonical Wnt gene targets. *Dev Cell* 2005;**8**:843–54.
110. Park JI, Ji H, Jun S, Gu D, Hikasa H, Li L, et al. Frodo links dishevelled to the p120-catenin/Kaiso pathway: distinct catenin subfamilies promote Wnt signals. *Dev Cell* 2006;**11**:683–95.
111. Ruzov A, Savitskaya E, Hackett JA, Reddington JP, Prokhortchouk A, Madej MJ, et al. The non-methylated DNA-binding function of Kaiso is not required in early Xenopus laevis development. *Development* 2009;**136**:729–38.
112. Ruzov A, Hackett JA, Prokhortchouk A, Reddington JP, Madej MJ, Dunican DS, et al. The interaction of xKaiso with xTcf3: a revised model for integration of epigenetic and Wnt signalling pathways. *Development* 2009;**136**:723–7.
113. Iioka H, Doerner SK, Tamai K. Kaiso is a bimodal modulator for Wnt/beta-catenin signaling. *FEBS Lett* 2009;**583**:627–32.

114. Kim MS, Kondo T, Takada I, Youn MY, Yamamoto Y, Takahashi S, et al. DNA demethylation in hormone-induced transcriptional derepression. *Nature* 2009;**461**:1007–12.
115. Rai K, Huggins IJ, James SR, Karpf AR, Jones DA, Cairns BR. DNA demethylation in zebrafish involves the coupling of a deaminase, a glycosylase, and gadd45. *Cell* 2008;**135**:1201–12.
116. Rai K, Sarkar S, Broadbent TJ, Voas M, Grossmann KF, Nadauld LD, et al. DNA demethylase activity maintains intestinal cells in an undifferentiated state following loss of APC. *Cell* 2010;**142**:930–42.
117. Wan M, Lee SS, Zhang X, Houwink-Manville I, Song HR, Amir RE, et al. Rett syndrome and beyond: recurrent spontaneous and familial MECP2 mutations at CpG hotspots. *Am J Hum Genet* 1999;**65**:1520–9.
118. Amir RE, Van den Veyver IB, Wan M, Tran CQ, Francke U, Zoghbi HY. Rett syndrome is caused by mutations in X-linked MECP2, encoding methyl-CpG-binding protein 2. *Nat Genet* 1999;**23**:185–8.
119. Shahbazian M, Young J, Yuva-Paylor L, Spencer C, Antalffy B, Noebels J, et al. Mice with truncated MeCP2 recapitulate many Rett syndrome features and display hyperacetylation of histone H3. *Neuron* 2002;**35**:243–54.
120. Chahrour M, Jung SY, Shaw C, Zhou X, Wong ST, Qin J, et al. MeCP2, a key contributor to neurological disease, activates and represses transcription. *Science* 2008;**320**:1224–9.
121. Guy J, Gan J, Selfridge J, Cobb S, Bird A. Reversal of neurological defects in a mouse model of Rett syndrome. *Science* 2007;**315**:1143–7.
122. Marchetto MC, Carromeu C, Acab A, Yu D, Yeo GW, Mu Y, et al. A model for neural development and treatment of rett syndrome using human induced pluripotent stem cells. *Cell* 2010;**143**:527–39.
123. Egger G, Liang G, Aparicio A, Jones PA. Epigenetics in human disease and prospects for epigenetic therapy. *Nature* 2004;**429**:457–63.
124. Chen WY, Baylin SB. Inactivation of tumor suppressor genes: choice between genetic and epigenetic routes. *Cell Cycle* 2005;**4**:10–2.
125. Ehrlich M, Jiang G, Fiala E, Dome JS, Yu MC, Long TI, et al. Hypomethylation and hypermethylation of DNA in Wilms tumors. *Oncogene* 2002;**21**:6694–702.
126. Feinberg AP. The epigenetics of cancer etiology. *Semin Cancer Biol* 2004;**14**:427–32.
127. Ballestar E, Paz MF, Valle L, Wei S, Fraga MF, Espada J, et al. Methyl-CpG binding proteins identify novel sites of epigenetic inactivation in human cancer. *EMBO J* 2003;**22**:6335–45.
128. Lopez-Serra L, Ballestar E, Fraga MF, Alaminos M, Setien F, Esteller M. A profile of methyl-CpG binding domain protein occupancy of hypermethylated promoter CpG islands of tumor suppressor genes in human cancer. *Cancer Res* 2006;**66**:8342–6.
129. Sansom OJ, Berger J, Bishop SM, Hendrich B, Bird A, Clarke AR. Deficiency of Mbd2 suppresses intestinal tumorigenesis. *Nat Genet* 2003;**34**:145–7.
130. Eads CA, Nickel AE, Laird PW. Complete genetic suppression of polyp formation and reduction of CpG-island hypermethylation in Apc(Min/+) Dnmt1-hypomorphic mice. *Cancer Res* 2002;**62**:1296–9.
131. Lyst MJ, Nan X, Stancheva I. Regulation of MBD1-mediated transcriptional repression by SUMO and PIAS proteins. *EMBO J* 2006;**25**:5317–28.
132. Weber A, Marquardt J, Elzi D, Forster N, Starke S, Glaum A, et al. Zbtb4 represses transcription of P21CIP1 and controls the cellular response to p53 activation. *EMBO J* 2008;**27**:1563–74.
133. Yamada D, Perez-Torrado R, Filion G, Caly M, Jammart B, Devignot V, et al. The human protein kinase HIPK2 phosphorylates and downregulates the methyl-binding transcription factor ZBTB4. *Oncogene* 2009;**28**:2535–44.

Section V

DNA Methylation in Medicine

Diseases Associated with Genomic Imprinting

Jon F. Wilkins* and Francisco Úbeda†

*Santa Fe Institute, Santa Fe, New Mexico, USA

†Department of Ecology and Evolutionary Biology, University of Tennessee, Knoxville, TN, USA

Genomic imprinting is the phenomenon where the expression of a locus differs between the maternally and paternally inherited alleles. Typically, this manifests as transcriptional silencing of one of the alleles, although many genes are imprinted in a tissue- or isoform-specific manner. Diseases associated with imprinted genes include various cancers, disorders of growth and metabolism, and disorders in neurodevelopment, cognition, and behavior, including certain major psychiatric disorders. In many cases, the disease phenotypes associated

Progress in Molecular Biology
and Translational Science, Vol. 101
DOI: 10.1016/B978-0-12-387685-0.00013-5

with dysfunction at particular imprinted loci can be understood in terms of the evolutionary processes responsible for the origin of imprinting. Imprinted gene expression represents the outcome of an intragenomic evolutionary conflict, where natural selection favors different expression strategies for maternally and paternally inherited alleles. This conflict is reasonably well understood in the context of the early growth effects of imprinted genes, where paternally inherited alleles are selected to place a greater demand on maternal resources than are maternally inherited alleles. Less well understood are the origins of imprinted gene expression in the brain, and their effects on cognition and behavior.

This chapter reviews the genetic diseases that are associated with imprinted genes, framed in terms of the evolutionary pressures acting on gene expression at those loci. We begin by reviewing the phenomenon and evolutionary origins of genomic imprinting. We then discuss diseases that are associated with genetic or epigenetic defects at particular imprinted loci, many of which are associated with abnormalities in growth and/or feeding behaviors that can be understood in terms of the asymmetric pressures of natural selection on maternally and paternally inherited alleles. We next described the evidence for imprinted gene effects on adult cognition and behavior, and the possible role of imprinted genes in the etiology of certain major psychiatric disorders. Finally, we conclude with a discussion of how imprinting, and the evolutionary–genetic conflicts that underlie it, may enhance both the frequency and morbidity of certain types of diseases.

I. Overview of Genomic Imprinting

A. What Is an Imprinted Gene?

The term *genomic imprinting* is typically used to refer to the phenomenon where the pattern of expression of an allele depends on its parental origin.[1] In the simplest cases, one of the two alleles is transcriptionally silenced, while the other is expressed. Often, however, imprinted genes exhibit complex patterns of tissue- and isoform-specific imprinting.[2–6] Some researchers will refer to a gene being "maternally imprinted" or "paternally imprinted." However, these phrases are used inconsistently in the literature, leading to a degree of confusion. In some contexts, the phrase "maternally imprinted" is used to mean "maternally silenced," while in other contexts it means "maternally modified," where that modification could be either silencing or activating.

It is preferable to refer to a locus as being imprinted if maternally and paternally inherited alleles at the locus exhibit systematic expression differences, and to explicitly describe the pattern of silencing, expression, and modification at a given locus. For example, in the mouse, the imprinted gene *Grb10* is paternally expressed in brain, but maternally expressed in the placenta and most embryonic tissues.[7] In contexts such as this, use of phrases such as "maternally imprinted" and "paternally imprinted" leads to unnecessary confusion.

B. How Common Is Imprinting?

It is common to think of genomic imprinting as a specifically mammalian phenomenon. Consistent with this view, many of the key components and features of the imprinting system that we find in humans appear to have arisen before the split between marsupial and eutherian (placental) mammals.[8–15] However, imprinted genes have also been identified in angiosperms (flowering plants), where imprinted gene expression has many similarities to what we observe in mammals, suggesting that an analogous phenomenon has evolved independently.[16]

There are also imprinting-like phenomena that have been described in various insects,[17] where the term "imprinting" was originally coined.[18,19] As for other taxa, such as birds and fish, it is at this point unclear whether any genes are imprinted. For several species, studies have looked at the expression of orthologs of certain genes known to be imprinted in mammals, typically the canonical pair of imprinted genes *Igf2* (insulin-like growth factor type 2) and *Igf2r* (insulin-like growth factor type 2 receptor). These studies have shown specific genes to be unimprinted in monotremes,[9,10,20] amphibians,[21] birds,[12,13] and fish,[22,23] leading some to conclude that imprinting does not exist in those species. However, there have been no systematic efforts to identify imprinted genes in most species, and it remains possible that other genes are imprinted in some or all of those species.

In humans, it is thought that somewhere between one and a few percent of the genome is subject to imprinting, although the exact number is unknown. The standard catalogs of imprinted genes,[24,25] including only those loci for which there is strong, direct empirical evidence, typically include fewer than a hundred entries. However, computational studies have identified much larger numbers of "predicted" imprinted genes: 600 in mice[26] and more than 150 in humans.[27] Further, a pair of studies measuring the allele-specific expression levels in the mouse brain identified approximately 1300 genes with monoallelic or strongly biased gene expression, suggesting widespread imprinting in that tissue.[28,29] Therefore, it seems likely that the total number of imprinted genes in humans will be greater than what is suggested by the current lists, but exactly how much greater remains to be determined.

C. How Does Imprinting Work?

Genomic imprinting relies on the existence of differential epigenetic modifications on the maternally and paternally derived alleles at a locus. This typically involves differential DNA methylation at CpG dinucleotides, as well as differential modification of histones (acetylation, methylation, etc.). These epigenetic modifications are established during gametogenesis, with different marks being established in the male and female germ lines. After fertilization, these differential marks are propagated across cell divisions in an allele-specific manner, allowing different expression to be maintained throughout development. Epigenetic propagation involves the action of the maintenance methyltransferase Dnmt1, which specifically recognizes the hemimethylated form of CpG dinucleotides that results from DNA replication. (In the hemimethylated state, the cytosine on one strand is methylated, while the cytosine on the newly synthesized strand is unmethylated.)

Throughout development, particularly in the earliest stages, these epigenetic marks are also subject to various modifications and reprogramming. Most striking is the large-scale demethylation of the paternally derived genome that occurs after fertilization, but before fusion of the two pronuclei.[30,31] Imprinted loci are also often subject to epigenetic spreading in *cis*, resulting in coordinated imprinted expression among clusters of loci. Thus, many of these clusters are defined by a suite of parent-of-origin-specific epigenetic modifications along an entire chromosomal region, but most of these modifications will derive from a single imprinting control region (ICR) that is differentially methylated during gametogenesis. Often, secondary epigenetic differences are not established until after fertilization.

D. Why Are There Imprinted Genes?

The discovery of genomic imprinting in mammals has triggered a proliferation of evolutionary theories.[32,33] The theory that has received the greatest amount of attention, and which provides the best explanation for the phenotypic consequences, direction of silencing, and taxonomic distribution of imprinted genes is the kinship theory of imprinting.[34–38] According to this theory, imprinting is the result of an intragenomic conflict, where natural selection acts differently on maternally and paternally derived alleles at the same locus. The asymmetric action of selection is often thought of in terms of inclusive-fitness effects: what matters in terms of natural selection is the total number of copies of an allele that are passed on to future generations, independent of whether those copies are passed on directly by the focal individual, or by a relative of the focal individual who is carrying identical copies of the allele.

This framework was developed initially in the context of imprinted gene effects on fetal growth, where natural selection acts differently on maternally and paternally inherited alleles at a locus that affects the magnitude of the fetal demand on maternal resources. From the perspective of an allele, the optimal level of resource demand is determined by a trade-off between the benefit derived from acquiring additional resources from the mother, and fitness cost that results from reducing the quantity of resources available to the mother's other offspring. The magnitude of the fitness cost is determined by the relatedness of the focal allele to those other offspring, or the probability that those offspring inherit an identical copy of the allele. For a maternally inherited allele, that probability is 0.5, while for a paternally inherited allele, it is somewhat less, depending on the degree of polyandry in the population (the probability that the mother's other offspring have a different father).

Thus, at the margins, a paternally inherited allele will favor greater demand on maternal resources, while a maternally inherited allele will favor reduced demand, preserving more resources for the other offspring. At an unimprinted locus, alleles are constrained to exhibit a single pattern of expression, irrespective of whether they are inherited from a male or a female. In that circumstance, we expect natural selection to settle on a demand level somewhere between the maternal and paternal optima. However, at an imprinted locus, where alleles acquire the ability to take on two different conditional expression strategies, the evolutionary dynamics resulting from the intragenomic conflict lead to the transcriptional silencing of one of the two alleles. At a locus where increasing gene expression results in a greater demand on maternal resources (e.g., growth factor like Igf2), it is the maternally inherited allele that becomes silenced, while the paternally inherited allele is expressed at the level that maximizes its (inclusive) fitness. At a locus where higher gene expression reduces demand (e.g., a growth suppressor like Igf2r), the opposite pattern results, with paternal silencing and maternal expression.

In recent years, this theory has been extended to include other types of interactions among related individuals. In particular, the interaction between father and offspring within the nuclear family[39] and social interactions in a population with limited dispersal.[40,41] The kinship theory of genomic imprinting was originally formulated within the context of mother–offspring interactions leaving the father outside of the picture. In mammals, fathers start contributing resources after weaning and even if the amount of resources contributed by the father might be less than the amount contributed by the mother, it can reverse the direction of the imprint.[39]

Recent work takes the kinship theory beyond the nuclear family into a social context.[40–43] These models no longer consider interactions between "mum, dad, and baby" for the allocation of parental resources, but interactions between brothers and cousins in a viscous population competing for resources at different developmental stages.[40–43] The later models provide the theoretical foundation for the evolution of genomic imprinting the postinfant brain. Models for the evolution of imprinting through social interactions require that demographic patterns (migration, reproductive success, life expectancy) differ between males and females.[40,41] When females tend to migrate more than males, a juvenile in the population is more related to her siblings, cousins, aunts, and uncles via her paternally inherited copy than her maternally inherited copy. Thus the maternally inherited allele is selected to be more egoistic, while the paternally inherited copy is selected to be more altruistic. Similar conclusions can be derived when females show greater reproductive success and when the expected life of females is shorter than the expected life of males.[40,41]

II. Disorders Associated with Particular Imprinted Genes and Regions

Much of our understanding of the phenotypic effects of imprinted genes in humans comes from the clinical manifestations of uniparental disomies (UPDs), where both copies of a chromosome are inherited from the same parent. These individuals are karyotypically normal, and, in the absence of genomic imprinting, we might expect UPDs to be without phenotypic effect. There are, however, two ways in which UPD can be associated with disease. First, uniparental isodisomy (where two copies of the same chromosome are inherited) can result in the unmasking of deleterious recessive mutations. Second, if a chromosome harbors one or more imprinted genes, a UPD will be associated with overexpression from imprinted loci, underexpression, or a combination of the two. Most imprinted genes occur in clusters, such that a UPD will typically encompass multiple imprinted genes. Thus, evidence linking disorders to a particular UPD may be suggestive of a role for imprinted genes, but this evidence becomes compelling only when systematic patterns emerge regarding the parental origin of the UPD, or when other evidence provides a direct link to one or more specific imprinted loci.

Imprinted genes are also subject to epigenetic dysregulation, such as hypomethylation or hypermethylation of regulatory elements. Clustered imprinted genes are often intricately coregulated, such that a single epimutation may alter expression of multiple imprinted genes. For imprinted loci where a complete loss of expression is lethal, certain epimutations may produce less severe phenotypes.

In this section, we describe the diseases that are associated with particular chromosomal regions.

A. Chromosome 20: Pseudohypoparathyroidism and Disorders of the *GNAS* Locus

1. Forms of Pseudohypoparathyroidism

Pseudohypoparathyroidism (PHP) is associated with end-organ resistance to parathyroid hormone (PTH).[44,45] That is, PTH levels are not reduced (as in hypoparathyroidism), but the response to PTH is diminished in a subset of its target cells. In fact, PHP is associated with elevated serum levels of PTH, as well as elevated serum phosphate and reduced serum calcium. PTH normally regulates serum calcium through its action on bone and kidney, via the G_s-coupled receptor PTHR1, and secretion of PTH from the parathyroid gland is stimulated by low serum calcium.[46,47] PTH acts on the renal proximal tubule to increase the level of 25-hydroxyvitamin D1-α-hydroxylase, which leads to elevated 1,25-dihydroxyvitamin D3, and thus to enhanced intestinal absorption of calcium and phosphate, and also mobilizes calcium and phosphate through its action on bone.

In patients with PHP, resistance to PTH appears to be limited to the renal proximal tubule, while the action of the hormone on bone and other tissues is unaffected.[48–50] Clinically, PHP is divided into two types, based on urinary excretion following diagnostic administration of PTH. In PHP type I, excretion of both cAMP and phosphate are blunted, while in PHP type II, only phosphate excretion is blunted.[51]

PHP-II is relatively rare, and the molecular and genetic basis for this variant remains poorly understood. PHP-I is much more common and is associated with maternally inherited heterozygous defects at the *GNAS* locus, which encodes the α subunit of the stimulatory G-protein ($G_s\alpha$).[52–55] The clinical manifestation of PHP-I and related disorders depends on both the nature of the genetic (or epigenetic) defect, and on the parental inheritance of the affected allele. PHP-I is further divided into two subclasses, PHP-Ia and PHP-Ib, based on the presence or absence of physical features that define Albright's hereditary osteodystrophy (AHO). PTH resistance coupled with AHO is categorized as PHP-Ia, whereas PTH resistance alone defines PHP-Ib. The physical features associated with AHO include short stature, mild mental retardation, obesity, and characteristic bone deformations, including shortening of the fourth and fifth metacarpals (brachydactyly).

2. Transcripts at the *GNAS* Locus

The complexity of both the clinical manifestations and heritability of these disorders derives from the extreme transcriptional complexity of the *GNAS* locus. *GNAS* is located on chromosome 20q[56,57] and is responsible for the production of numerous transcripts, the expression of which depends on both cell type and allelic parent of origin[2–6] (see Fig. 1).

Several of the GNAS transcripts share a common set of downstream exons (2–13), but originate from different promoters, and incorporate alternate versions of exon 1.[58–60] The furthest downstream promoter is responsible for production of the $G_s\alpha$ transcript and will be referred to here as the $G_s\alpha$ promoter. Through alternate splicing, this transcript produces long and short versions ($G_s\alpha$-L and $G_s\alpha$-S), which differ in the inclusion or exclusion of 45 nucleotides from exon 3.[61] This transcript also produces the truncated $G_s\alpha$-N1,

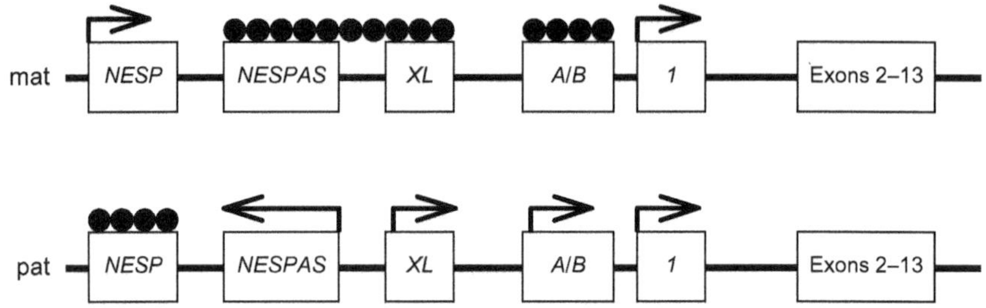

FIG. 1. *Structure of the complex GNAS locus.* This figure illustrates the methylation and expression patterns on the maternally (top) and paternally (bottom) inherited alleles. Arrows indicate expression, and filled circles indicate the three differentially methylated regions. Distances along the chromosome are not drawn to scale.

includes exons 1–3 and exon N1, which contains an in-frame stop codon. $G_s\alpha$-L and $G_s\alpha$-S perform similar functions but exhibit slight differences that remain incompletely understood.[62] $G_s\alpha$-N1 lacks many of $G_s\alpha$ functional domains, and its function is unknown, but its mouse homolog is highly expressed in the brain.[63]

In most tissues, expression of $G_s\alpha$ is biallelic, although the paternal copy is partially or completely silenced in renal cortex,[64] thyroid, pituitary, and ovaries.[65–67] This biallelic expression likely accounts for the fact that maternally inherited loss-of-function mutations are not lethal, as heterozygous expression of $G_s\alpha$ is sufficient to maintain normal function in those tissues. For instance, the fact that PHP does not affect the action of PTH on bone results from the biallelic expression of $G_s\alpha$ in that tissue.[68]

Approximately 2.5 kb upstream from the $G_s\alpha$ promoter is a second promoter that is responsible for production of the A/B transcript (homologous to the 1A transcript in mice).[69,70] The A/B transcript produces an alternate first exon, which is spliced to exon 2, but this first exon does not contain an in-frame translation-initiation codon, though transcription may start from within the shared exon 2, leading to a truncated $G_s\alpha$ variant.[69] Alternatively, the A/B transcript may be noncoding and may function primarily in a regulatory role in *cis*.

The A/B promoter lies within a differentially methylated region (DMR). The promoter is methylated and repressed on the maternally derived copy, and is unmethylated and active on the paternally derived copy.[69–72] Loss of methylation from the maternally inherited allele acts not only to activate transcription of A/B from the allele but also to repress $G_s\alpha$ transcription in *cis*. Thus, the expression of these two transcripts is reciprocally regulated, but the mechanism of regulation is not understood.

The next promoter upstream from $G_s\alpha$ produces the extra-large $G_s\alpha$ variant ($G_s\alpha$-XL), which shares a long C-terminal sequence with $G_s\alpha$, but differs in the large N-terminal region encoded by the XL alternate first exon.[2,3,71,72] Like

$G_s\alpha$, $G_s\alpha$-XL produces long and short variants through inclusion or exclusion of exon 3, as well as a truncated version that incorporates the N1 exon.[2,3] Like the A/B promoter, the XL promoter lies within a DMR and is maternally silenced. Unlike A/B, which exhibits a complex pattern of tissue-specific and partial silencing, $G_s\alpha$-XL is exclusively expressed from the paternal copy[2,3,73,74] (but see Ref. 75).

The XL promoter is also responsible for a small protein produced from a second open-reading frame located entirely within the XL exon 1.[76,77] The protein, ALEX, has been shown to interact with $G_s\alpha$-XL *in vitro*, but the function of this gene product *in vivo* remains to be understood.

Furthest upstream is the *NESP* promoter. As with the other GNAS promoters, the *NESP* exon 1 is spliced to the shared exons 2–13. However, the entire protein-coding region for this transcript lies within the first exon so that this protein shares no sequence with the $G_s\alpha$ variants.[2,3,74] The gene product (NESP55) is a neuroendocrine secretory protein expressed in neuroendocrine tissues and the peripheral and central nervous systems.[78] *Nesp* knockout mice appear phenotypically normal, but suffer from certain behavioral abnormalities.[79] The NESP promoter lies within a paternally methylated DMR, and expression of this transcript is exclusively from the maternally derived copy.[2,3,74]

The GNAS locus is also host to a noncoding antisense RNA transcript known as NESPAS (or GNASAS). The NESPAS promoter lies within the XL DMR, and its transcript is produced only from the paternally derived allele.[73,80,81] Elimination of the promoter results in derepression of NESP and demethylation of the NESP DMR, suggesting that transcription of NESPAS is the primary mechanism by which maternal expression of NESP is enforced.[82]

Still further upstream is the *STX16* locus, encoding syntaxin 16. This locus is not imprinted, nor is it considered part of the GNAS complex locus. However, it appears that *STX16* may harbor a long-range *cis*-acting element that participates in regulation of the GNAS transcripts. Microdeletions within *STX16* have been associated with dysregulation of the A/B and $G_s\alpha$ transcripts, as these microdeletions cause PHP-Ib, but only when maternally inherited.[83] As *STX16* itself is not imprinted, this suggests that the cause is a *cis*-acting regulatory interaction with nearby imprinted genes.

3. Establishment of Epigenetic Marks at GNAS

The GNAS locus contains three distinct DMRs, and in each case, methylation covering the promoter region is associated with transcriptional repression. The furthest downstream DMR covers the *A/B* promoter and is methylated on the maternally derived allele.[71,72,84] Methylation of this DMR is responsible not only for the maternal silencing of the A/B transcript but also for the preferential paternal expression of the $G_s\alpha$ transcript in certain tissues.[85–87]

Further upstream is a second maternally methylated DMR that covers the *$G_s\alpha$-XL* and *NESPAS* promoters, driving paternal-specific expression of both transcripts.[88–90] Furthest upstream is a DMR covering the *NESP* promoter that is paternally methylated, causing maternal expression of *NESP*.[74,89,91]

The methylation patterns at these three DMRs are not independent, however. Methylation at the *NESP* DMR does not occur until after fertilization[71,72] and depends on transcription from the paternally inherited NESPAS,[82] as targeted deletion of NESPAS results in loss of methylation and biallelic expression of NESP when paternally inherited. Interestingly, this NESPAS deletion also leads to partial methylation of the paternal A/B promoter, which results in decreased A/B expression and increased $G_s\alpha$ expression.[82] Thus, it appears that the NESPAS DMR is the element primarily responsible for control of imprinted gene expression in this cluster.

4. Diseases at the Locus

PHP-Ia results from maternal inheritance of loss-of-function mutations at the $G_s\alpha$ locus, and the tissue-specific resistance patterns associated with PHP-Ia are explained by the tissue-specific patterns of imprinting at the locus.[44] Maternally inherited loss of function results in complete or nearly complete loss of transcription in cell types where the paternal allele is completely or partially silenced. In cell types with biallelic expression, the result is simply a 50% reduction in transcription, which does not appear to substantially affect the PTH response in those cells.[68]

PHP-Ib is also inherited maternally, but is not due to inactivating mutations in $G_s\alpha$. Rather, this disease subtype is associated with broad epigenetic defects at the *GNAS* locus. A diverse set of genetic lesions have been associated with PHP-Ib, but in each case, the mutation causes loss of imprinting (derepression) of the A/B transcript.[71,72,85,86] In cell types, where $G_s\alpha$ and A/B are reciprocally coregulated, the derepression of A/B reduces the expression of $G_s\alpha$, resulting in the PTH-resistant phenotype. However, the expression of these two transcripts does not appear to be coupled in all cell types, as A/B expression exists in some tissues in the absence of $G_s\alpha$ imprinting.[85–87] Presumably, derepression in these tissues does not diminish $G_s\alpha$ expression, and it is PTH resistance in those tissues that are responsible for AHO, which is present in PHP-Ia, but absent in PHP-Ib.

In the related disorder of pseudopseudohypoparathyroidism (PPHP), the physical characteristics associated with AHO are present, but without the resistance to PTH and other hormones.[92] Like PHP-Ia, PPHP results from $G_s\alpha$-inactivating mutations, and, in fact, these two diseases can arise from the same genetic defect, and both are often found in the same families.[55,93] Whereas PHP-Ia results from maternal inheritance of these defects, PPHP is paternally inherited.[94,95] This pattern suggests that the hormone resistance associated with PHP is attributable to the loss of maternal expression of $G_s\alpha$, whereas the AHO component of the disease is the result of haploinsufficiency of $G_s\alpha$ in tissues where it is normally biallelically expressed, but is independent of parental origin.

Mutations causing constitutive $G_s\alpha$ activity are also associated with various diseases, but are lethal if inherited, and are therefore typically of somatic origin. Activating mutations have been described in various tumors, including particularly endocrine adenomas.[96] Activating $G_s\alpha$ mutations occurring early in development (giving rise to mosaic constitutive activity) lead to McCune–Albright syndrome, which involves abnormalities of the skin, bone, and endocrine organs.[97,98] To the best of our knowledge, the possibility of systematic phenotypic differences depending on the parental origin of the constitutively active allele has not been examined.

B. Chromosomes 7 and 11: Silver–Russell and Beckwith–Wiedemann Syndromes

1. Silver–Russell Syndrome

Silver–Russell syndrome (SRS) is a growth disorder defined by intrauterine growth restriction (IUGR) in combination with a subset of other abnormalities that can include hypoglycemia, feeding problems, lack of subcutaneous fat, and early onset of puberty, among others. Individuals with SRS are typically small for gestational age, often weighing less than 3 kg at birth, and the average height for adults with SRS is less than 5 ft.[99–101]

SRS does not have a single genetic basis, and genetic associations have been reported for chromosomes 1, 7, 8, 11, 15, 17, 18, and X.[102] For most of these chromosomal associations, SRS has been observed in a small number of patients exhibiting either trisomy or a large-scale deletion or translocation, and the mechanism through which these defects lead to SRS remains poorly understood. The genetic (and epigenetic) defects on chromosomes 7 and 11 are most commonly associated with SRS, and have been most studied. Both of these chromosomes are host to clusters of imprinted genes that appear to play a role in the etiology of the disease, and it is these defects that are the focus of this section.

2. Chromosome 7

Approximately 5–10% of SRS cases are associated with maternal UPD at chromosome 7 (MatUPD7), where the individual is karyotypically normal, but both copies of chromosome 7 have been inherited from the mother, and therefore exhibit the maternal-specific epigenetic modifications.[102,103] Three regions of chromosome 7 contain clusters of imprinted genes, and any combination of these might contribute SRS. The three regions, 7p11.2–13, 7q21, and

7q32, all contain imprinted genes that are expected to contribute to growth restriction when maternally duplicated. In addition, there is some evidence from smaller genetic lesions that provides some insight as to how these different regions might contribute to other aspects of the SRS phenotype.

The 7p11.2–13 region includes the *GRB10* (growth factor receptor bound protein 10) locus, which may produce as many as 13 transcripts, most of which are thought to be noncoding, and which include maternal, paternal, and biallelic expression in different tissues.[7,104–106] In particular, the maternally expressed γ1 transcript has been identified in placental tissues, while other splice variants are paternally expressed in the brain.[7] The genes neighboring *GRB10* are thought to be unimprinted in humans,[107] and *GRB10* has been shown to reduce the size and efficiency of the placenta.[108]

These patterns suggest that the contribution of this chromosomal region to the growth-restriction aspects of the SRS phenotype in the MatUPD7 cases is likely mediated through increased expression of the γ1 form in placental tissues. Loss of paternal expression of other forms in the brain may additionally contribute to the cognitive aspects of SRS. However, identification of a family in which maternal inheritance of a segmental duplication covering this region is associated with mental retardation[109] suggests that this loss of paternal expression may not be the only way in which this locus affects cognition in SRS patients, as these individuals possess a normal paternally inherited chromosome 7.

Four imprinted genes have been identified in the 7q21 region: the maternally expressed tissue factor pathway inhibitor 2 (*TFPI2*) locus and the paternally expressed epsilon-sarcoglycan (*SGCE*) and *PEG10* loci.[110–112] The *CALCR* locus appears to be monoallelically expressed in the brain, but which allele is silenced has not yet been definitively established,[110] though the mouse homolog *Calcr* is maternally expressed in brain.[113] Other transcripts in the region are imprinted in the mouse, but are either unimprinted or have uncertain imprinting status in humans.

PEG10 is a retrotransposon-derived gene that plays an important role in placental development,[114] and loss of *PEG10* expression is a likely contributor to growth restriction in SRS. *TFPI2* is a putative tumor suppressor,[115] suggesting that it may interfere with cell proliferation. It is maternally expressed in extraembryonic tissues, and thus increased expression in MatUPD7 may also contribute to growth restriction. Mutations in SGCE are a major cause of myoclonus-dystonia syndrome (MDS).[116] MDS is a movement disorder characterized by rapid muscle contractions and with twisting and repetitive movements producing abnormal postures. SRS patients often present with low muscle tone, but the connection between these phenotypes is not transparent.

The 7q32 chromosomal region contains the paternally expressed *MEST*, *MEST1T1* (antisense to *MEST*), and *COPG21T1* (an intronic transcript found within the biallelically expressed *COPG2* gene) loci, as well as two maternally expressed loci, *CPA4* and *KLF14*.[117] Knocking out the mouse ortholog of *MEST* (*Peg1/Mest*) results in IUGR, as well as a suite of behavioral abnormalities relating to maternal care for offspring, such as pup retrieval, nest building, and placentophagia.[37,38,40,41,118] The absence of a functional *MEST* therefore seems a likely contributor to the undergrowth phenotype (but see Ref. 119). *KLF14* specifies a transcription factor, and has been undergoing accelerated evolution in the human lineage.[120] These features make it an interesting candidate, but do not suggest any specific mechanism through which overexpression in MatUPD7 might contribute to SRS.

Located nearby in the 7q31.2 region is the *FOXP2* locus, mutations of which are associated with developmental verbal dyspraxia (DVD).[121–123] One study has suggested that this disorder may result specifically from the absence of a functional paternally inherited copy of the gene.[124] If *FOXP2* is, in fact, subject to parent-of-origin effects, the loss of a paternally inherited copy in MatUPD7 may contribute to the speech effects associated with SRS patients, many of whom exhibit DVD.

3. Chromosome 11

Chromosome 11 contains two clusters of imprinted genes (see Fig. 2), both located in the 11p15.5 region, but regulated by separate imprinting control regions (ICRs). The more telomeric of the two ICRs, ICR1, controls expression of the reciprocally imprinted *IGF2* (insulin-like growth factor type 2) and *H19* loci. Normally, *IGF2* is paternally expressed,[125,126] while *H19* is maternally

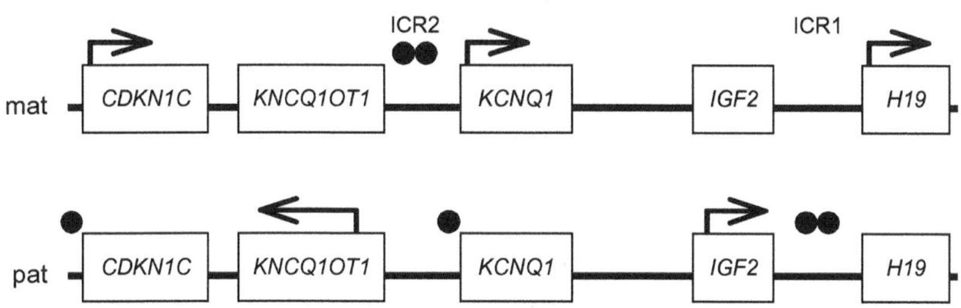

FIG. 2. *Structure of the 11p15.5 imprinted gene clusters.* This figure illustrates the methylation and expression patterns on the maternally (top) and paternally (bottom) inherited alleles. Arrows indicate expression, and filled circles indicate the ICR1 and ICR2 differentially methylated regions. Distances not to scale.

expressed.[127] This pattern is controlled by epigenetic differences between the two alleles at the *H19* DMR.[128–130] When unmethylated (as on the maternally inherited copy), the *H19* DMR binds to CTCF, which serves as an insulator, isolating *IGF2* from a downstream enhancer element, which interacts instead with the *H19* promoter region. Methylation of the paternally inherited copy blocks CTCF binding, thereby eliminating the insulator activity and allowing the enhancer to interact instead with the *IGF2* promoter.

IGF2 is a major contributor to growth in early development, and approximately 50–60% of SRS patients exhibit epigenetic defects in the 11p15.5 region.[102,131] Particularly common is hypomethylation at ICR1, which results in the epigenetic silencing of *IGF2* from both alleles.[103,132] Further, the degree of hypomethylation correlates with the clinical severity of the SRS phenotype.[133] Thus, it appears that loss of *IGF2* expression is sufficient to generate all key aspects of the SRS phenotype, particularly those directly related to growth.

ICR2 controls a cluster of imprinted transcripts, most of which are maternally expressed and are associated with negative growth effects. Normally, maternal methylation of the KvDMR silences maternal expression of the *KCNQ1OT1* noncoding RNA transcript. Expression of *KCNQ1OT1* from the paternally inherited copy acts in *cis* to suppress expression of a number of nearby genes, including *SLC22A18*, *PHLDA2*, *CDKN1C*, and *KCNQ1*.[134–136] Note that the imprinted region in mouse extends further, including the *Osbpl5*, *Tssc4*, and *Nap1l4* loci, which are biallelically expressed in humans.[137] At the moment, the potential contributions of genes in this region to the SRS phenotype remain unclear.

A prospective study identified a number of clinical features for which SRS patients with MatUPD7 and hypomethylation at ICR1 differ statistically, either in the likelihood of displaying that aspect of the disease phenotype, or in the clinical severity.[138] MatUPD7 patients were more likely to display developmental delays, to require speech therapy, and to exhibit certain craniofacial features, such as a triangular face and low-set ears. Patients with hypomethylation at ICR1 were more likely to exhibit developmental asymmetries and cognitive defects.

The remaining 30–40% of SRS cases have not been definitively associated with specific genetic or epigenetic defects, and it is possible that many of those cases are related to loci (imprinted or not) on chromosomes other than 7 and 11. This apparent causal heterogeneity, along with the subtle phenotypic differences among patients with different underlying causes, suggests that the bulk of the clinical features associated with SRS may be relatively generic consequences of undergrowth, particularly during prenatal development. It is also possible that in the future, SRS may be differentiated into subtypes based on genetic and epigenetic etiology.

4. Beckwith–Wiedemann Syndrome

Beckwith–Wiedemann syndrome (BWS) is associated with overgrowth and is in many ways genetically and phenotypically reciprocal to SRS [139]. BWS is associated with macroglossia (enlargement of the tongue), large prenatal and childhood body mass (> 90th percentile), and defects in the abdominal wall. BWS also results in extreme placental overgrowth, with placentas that are approximately 2× normal weight.[140–142]

Like SRS, BWS is associated with a heterogeneous genetic etiology but is most often associated with epigenetic defects covering the 11p15.5 imprinted region, which account for 60–70% of cases.[143–147] Over half of BWS patients exhibit hypomethylation at ICR2, resulting in loss of expression of *SLC22A18*, *PHLDA2*, *CDKN1C*, and *KCNQ1* from the maternally inherited copy. In approximately 5% of cases, patients show hypermethylation at ICR1, which results in aberrant expression of *IGF2* from the normally silenced maternally inherited copy. Another ~15% of cases are accounted for by paternal UPD covering 11p15.5, eliminating maternal expression (and increasing paternal expression) from all imprinted genes in the region. A small fraction (5–10%) of cases are associated with mutations in *CDKN1C* (previously *p57*(*KIP2*)), which specifies a cyclin-dependent kinase, a tumor suppressor that exerts its negative effects on cell proliferation by inhibiting progression through the cell cycle.[148]

The overall pattern observed in BWS is qualitatively analogous to what is seen in SRS. The syndrome can result from a heterogeneous collection of underlying genetic and epigenetic defects, but most cases are associated with dysregulation of one or both of two loci with broad effects on cell proliferation and growth: *IGF2* and *CDKN1C*. This pattern suggests that many of the features associated with BWS are generic consequences of an overgrowth phenotype. At the same time, certain patterns have emerged that point toward subtle clinical distinctions associated with different molecular etiologies. For example, certain features of BWS may be overrepresented in patients with *CDKN1C* mutations, including polydactyly, extra nipple, and cleft palate.[148] Eventually, patterns like this may make it possible to disentangle the contributions of various loci in the 11p15.5 region to this syndrome.

Given the reciprocal phenotypes associated with SRS and BWS, and the reciprocal epigenetic defects in the 11p15.5 imprinted region that are associated with the two syndromes, it seems reasonable to expect that BWS might also be associated with PatUPD7. In mice, PatUPD of chromosome 11, which is syntenic with human chromosome 7, results in offspring that are 30% larger than their littermates.[149] However, in humans, the consequences of paternal isodisomy in this region are unclear. In four reported cases of PatUPD7, three show normal growth,[150–152] and one shows overgrowth.[153] Two of these patients (one of which showed overgrowth) were screened due to the fact that they had cystic fibrosis.

C. Chromosome 14: UPD14

UPDs of chromosome 14, first described in 1991,[154,155] are thought to represent a relatively rare disorder. However, the frequency is not well estimated, particularly for the maternal UPD (MatUPD14), due to the facts that it has a relatively nonspecific phenotype and molecular testing is not routine. MatUPD14 syndrome is associated with growth retardation, hypotonia (muscle weakness), joint laxity, early onset of puberty, and mild dysmorphism of the hands, feet, and face.[156]

Paternal UPD14 (PatUPD14) syndrome is substantially less common, and is associated with a much more extreme phenotype, including polyhydromnios, premature labor, skeletal abnormalities, respiratory and neurodevelopmental problems, and often early death.[156]

Both UPD14 syndromes are thought to be associated with altered gene expression in the 14q32 region, which contains a cluster of imprinted genes, including the paternally expressed *DLK1*, *RTL1* (*PEG11*), and *DIO3* along with the maternally expressed *GTL2* (*MEG3*), *RTL1as*, *MEG8*, and *BEGAIN* (see Fig. 3). Imprinting in these regions is controlled by two different DMRs: the DLK1-GTL2 intergenic DMR (IG-DMR) and the *GTL2*-DMR.[157,158] The two DMRs appear to function hierarchically and in a tissue-specific fashion.[159,160]

The centrality of this region is supported by patients displaying the MatUPD14 clinical phenotype in the absence of a chromosomal UPD. Loss of methylation at the paternal IG-DMR produces the MatUPD14 phenotype.[161,162] Similarly, the PatUPD14 phenotype has been observed in a patient with a segmental paternal UPD spanning the 14q32–14q32.33 region.[163] In each case, however, the observed defects are associated with aberrant expression of the entire cluster of imprinted genes, and the relative contributions of individual genes to the disease phenotypes are not understood.

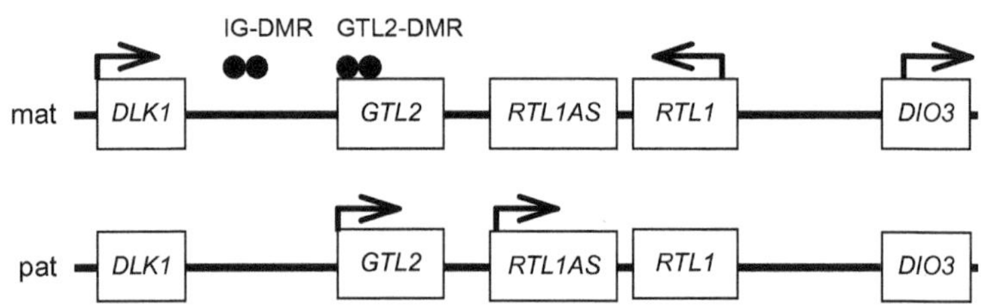

FIG. 3. Structure of the 14q32 imprinted gene cluster. This figure illustrates the methylation and expression patterns on the maternally (top) and paternally (bottom) inherited alleles. Arrows indicate expression, and filled circles indicate the intergenic and *GTL2* differentially methylated regions. Distances not to scale.

D. Chromosome 15: Prader–Willi and Angelman Syndromes

Prader–Willi syndrome (PWS) and Angelman syndrome (AS) were the first known examples of human diseases involving imprinted genes. They occur with a frequency of 1:15,000 and 1:25,000 live births, respectively, and are caused by alterations in region 15q11–13 of chromosome 15. This chromosomal region contains a cluster of imprinted genes that are expressed from the paternally inherited or the maternally inherited chromosome only (see Fig. 4). The parent-of-origin expression of genes in this cluster is regulated by an ICR.

The paternally expressed genes in region 15q11–13 are *MKRN3*, *MAGEL2*, *NDN*, *C15orf2*, SNURF-SNRPN, and a group of *snoRNA* genes. Expression of paternally inherited genes *MKRN3*, *NDN*, and *SNURF-SNRPN* is regulated by differential methylation of the promoter regions of each gene. *C15orf2* is paternally expressed in the fetal brain but biallelically expressed in other organs. The relative contribution of each of these genes to the PWS clinical phenotype is yet to be determined.

The maternally expressed genes in region 15q11–13 are *UBE3A* and *ATP10C*. Expression of maternally inherited genes *UBE3A* and *ATP10C* is not achieved through differential methylation of the promoter regions of each gene. Silencing of the paternally inherited copy of *UBE3A* is achieved through differential expression of the 3′ end of the *SNURF-SNRPN* transcript acting as an antisense transcript.[164] The imprinted expression of gene *UBE3A*

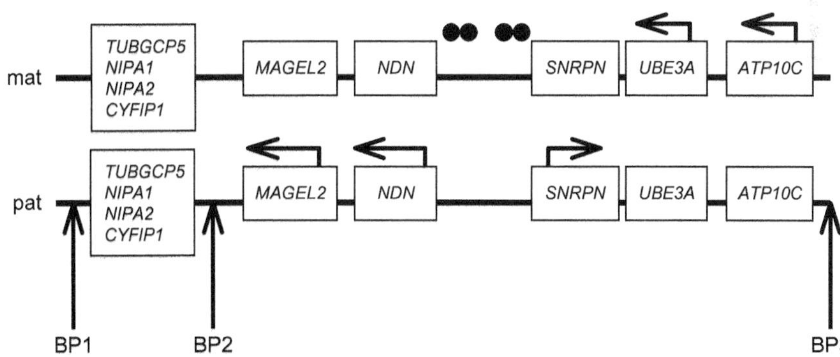

FIG. 4. *Structure of the AS-PWS imprinted region.* This figure illustrates the methylation and expression patterns on the maternally (top) and paternally (bottom) inherited alleles. Horizontal arrows indicate expression, and filled circles indicate the differentially methylated region. Distances not to scale. Vertical arrows indicate the relative locations of the three breakpoints described in the text. The genes lying between breakpoints one and two (at the far left side of the figure) are all unimprinted.

is tissue specific and restricted to some types of cells in the brain. *UBE3A* is the critical gene leading to the AS clinical phenotype. *ATP10C* is maternally expressed in the brain but biallelically expressed in other organs.[165] The orthologous gene in mouse (*Atp10a*) is not imprinted.[166]

The ICR regulates in *cis* imprint resetting and maintenance in the whole cluster of imprinted genes.[167] It consists of two critical elements the PWS-SRO and the AS-SRO.[168] PWS-SRO controls the maintenance of the paternal imprint during early embryonic development. AS-SRO controls the establishment of the maternal imprint in the female germ line.

PWS and AS result from complete or partial deletion of chromosomal region 15q11–13, UPD (inheritance of the two copies of a chromosomes from the same father) of chromosome 15. These are imprinting defects that may or may not be caused by deletions in the imprinting center of chromosomal region 15q11–13.

Seventy percent of all PWS cases are due to the paternal inheritance of a *de novo* interstitial deletion of a region of chromosome 15. This region includes the cluster of imprinted genes and several nonimprinted genes. Deletions are caused by nonhomologous recombination events and can be of two kinds: class I deletions affect the region comprised between break point 1 (BP1) and break point 3 (BP3), and class II deletions affect the region comprised between break point 2 (BP2) and BP3. Paternally inherited deletions result in the lack of expression of imprinted genes that are active when paternally inherited

Between 25% and 30% of all PWS cases are due to maternal UPD. These UPDs are caused by maternal meiotic nondisjunction followed by mitotic loss of paternal chromosome 15 after fertilization. Maternal UPDs result in the lack of expression of imprinted genes that are active when paternally inherited and up to a twofold increment in expression of genes that are active when maternally inherited.

At most 3% of all PWS cases are due to imprinting defects that result in the paternal chromosome carrying a maternal imprint. Imprinting defects caused by deletions affecting the ICR are very rare, while imprinting defects caused by epimutations affecting the IR are more common. Epimutations can occur during imprint erasure in primordial germ cells, or during imprint establishment or maintenance after fertilization. If the epimutation occurs after fertilization, it may result in mosaicism. In PWS patients, the paternal chromosome that carries an incorrect maternal imprint is always derived from the paternal grandmother,[169] which suggests that the incorrect imprint in the PWS patients results from failure of the paternal germ line to erase the grandmaternal imprint. Supporting this observation, mosaicism in PWS patients due to an imprinting defect are very rare. Imprinting defects result in gene silencing of paternally expressed genes.

Seventy percent of all AS cases are due to the maternal inheritance of the same deletions described for PWS affecting the cluster of imprinted genes in region 15q11–13.

Ten percent of all AS cases are due to mutations in gene *UBE3A*.[170] Another 2–5% of AS cases are due to paternal UPD covering the 15q11–13 region. These UPDs are caused by maternal nondisjunction with postzygotic duplication of chromosome 15 inherited via sperm.

Between 2% and 4% of all AS cases are due to imprinting defects that result in the maternal chromosome carrying a paternal imprint. Imprinting defects caused by deletions affecting the ICR are very rare while imprinting defects caused by epimutations affecting the ICR are more common. In AS patients, the maternal chromosome carrying an incorrect paternal imprint is inherited either from the maternal grandfather or from the maternal grandmother.[169,171] This finding suggests that the imprinting defect occurs after erasure of the parental imprints and results from an error in imprint establishment or imprint maintenance. Corroborating this observation, more than 40% of AS patients with an imprinting defect are found to have somatic mosaicism. The remaining approximately 15% of AS cases are caused by genetic defects of unknown nature.

Patients suffering from PWS present a clinical phenotype that affects feeding, weight, and growth among others. The clinical phenotype corresponding to these features is markedly biphasic with either weaning or menarche (which is still debated) separating both phases.[172] Early infants present low birth weight, severe hypotonia, and feeding difficulties. Late infants show hyperphagia (insatiable and/or nondiscriminatory appetite) and obesity. Accompanying features are short stature, small hands and feet, almond-shaped eyes, triangular mouth, and hypogonadism in both sexes.

AS patients show distinctive behavior with temper tantrums, obsessive-compulsive behavior, and sometimes psychiatric disturbance. Mild to moderate mental retardation is also observed. Patients with class I deletion have generally more behavioral and psychological problems than individuals with class II deletion.[173]

Patients suffering from AS also present clinical phenotype that affects feeding, and growth among others. They present prolonged sucking although poorly coordinated and microcephaly. In contrast with PWS, the clinical phenotype of AS patients is not biphasic. The behavior of AS patients is also affected showing sleep disorders, happy demeanor, that includes inappropriate laughter and excitability, and limited speech. Severe mental retardation is also observed.

E. X Chromosome: Turner and Klinefelter Syndromes

Turner syndrome (TS) results from the absence of all or part of one of the X chromosomes in females (45, XO females, with "45" referring to the total number of nuclear chromosomes, as opposed to the normal, 46-chromosome karyotype). Individuals with TS typically display short stature with broad chests, low-set ears, and webbed necks and are often subject to cardiovascular and renal defects.[174,175] Klinefelter syndrome (KS) is a condition in males in which they inherit two X chromosomes in addition to a Y chromosome (47, XXY males). Individuals with KS often have small testicles and reduced fertility, but the phenotypic manifestations are highly variable, with many individuals having few detectable symptoms.[176]

Neither TS nor KS is an imprinting disorder per se, but both are potentially subject to influence by imprinted genes. A cluster of imprinted genes has been identified on the mouse X chromosome, and at least one of those genes is associated with effects on cognition and behavior.[177,178] This raises the possibility of phenotypically relevant imprinted X-linked genes in humans.

In TS, individuals inherit a single X chromosome. Normally in mammals, males have one X chromosome, while females have two. The Y chromosome contains many fewer genes than the X, and dosage compensation is achieved through epigenetic silencing of one of the two X chromosomes in females.[179] In some contexts, X inactivation itself is imprinted, with the paternally inherited X undergoing inactivation in marsupials and in the extraembryonic tissues of some eutherians, including mice.[180]

However, not all of the genes on the X chromosome are silenced, as approximately 15–20% escape inactivation,[181,182] and many of the features associated with TS are likely due to haploinsufficiency at those loci. For example, the *SHOX* locus, located in the pseudoautosomal region, is thought to be the most significant contributor to the stature effects in TS.[183]

The single X chromosome inherited by someone with TS will be either maternally or paternally inherited (X_mO or X_pO, respectively), and a number of studies have looked for phenotypic differences between these two subsets of TS patients. Some studies have failed to find any significant imprinting effects on the physical manifestations of TS, including stature, body mass index, cardiac, renal, skeletal, lymphatic, aural, or ocular systems,[184,185] though one study has found that X_mO patients were more likely to have kidney malformations, had lower LDL cholesterol, and were less likely to have ocular abnormalities,[186] and there is some evidence for imprinting effects on the response to treatment with growth hormone.[187]

While the effect of X-linked imprinted genes on the physical features in TS are at present unclear, there is strong evidence pointing toward cognitive differences between X_mO and X_pO females. The first study to focus on these differences found evidence that X_pO females had better verbal skills, less social-cognitive impairment, and better behavioral inhibition and planning skills.[188]

More recent brain-imaging studies have identified systematic differences in brain structure that suggest a role for X-linked imprinted genes in neurodevelopment. X_pO females were found to have a larger volume of gray matter in the caudate nuclei, and a larger volume of white matter bilaterally in the temporal lobes.[189] Another study has found that X_mO females have increased gray matter in the left superior temporal gyrus.[190] Other studies have failed to find significant imprinting effects on brain structure in TS patients.[191–193]

Findings suggestive of functional differences have also been identified in subsequent studies, although the magnitude of the effects is often quite small. X_mO females appear to exhibit enhanced forgetting in verbal contexts, while forgetting is more pronounced for X_pO females in spatial contexts.[194] Other studies have suggested that X_mO females suffer greater impairment in verbal cognition[195] and arithmetic function.[196] The consequences of TS for brain structure and function, and the evidence for and against a significant effect of imprinted genes, have been the subject of two recent reviews.[197,198]

Similarly, in KS, the supernumerary X chromosome can be either maternally or paternally inherited, such that there are two distinct groups of KS individuals: X_mX_mY males and X_mX_pY males. Studies on imprinting effects in KS have been more limited, but one study has found that X_mX_pY males had increased body size parameters for some measurements, consistent with a growth-enhancing effect of one or more imprinted genes on the X chromosome.[199] This study also found that X_mX_pY males were significantly more likely to have impaired speech and motor developmental problems. A second study reported an association between inheritance of a paternally derived X chromosome and later onset of puberty.[200]

F. Chromosome 6: Transient Neonatal Diabetes

The 6q24 region is associated with transient neonatal diabetes mellitus type 1 (TNDM1),[201] and contains two imprinted genes where a subset of transcripts is maternally silenced in at least some tissues,[202] *PLAGL1* (a.k.a. *ZAC1* or *LOT1*), a zinc-finger containing transcription factor involved in apoptosis and cell-cycle control,[203] and *HYMAI*, which produces a noncoding RNA. Overexpression from these loci due to genetic or epigenetic abnormalities in the 6q24 region account for approximately 70% of cases of TNDM1,[204] often accompanied by macroglossia. Sources of overexpression include PatUPD6, duplication of the paternal 6q24 region, and loss of methylation at the maternally inherited TND DMR.[204,205]

The phenotype associated with paternalization of the 6q24 region is puzzling in two respects, both relating to the fact that the known imprinted genes in the region are maternally silenced. First, there are no reported phenotypic effects associated with maternalization of the locus (through, e.g., MatUPD6), despite the fact that this would result in a complete loss of function in cell types where these genes are maternally silenced. Second, based on theoretical analysis and the patterns observed with other imprinted loci, we expect maternal silencing to arise at loci with growth-enhancing effects.

Contrary to this expectation, *PLAGL1* appears to be a tumor suppressor,[206] and paternalization of the locus results in IUGR in >95% of TNDM1 cases,[207] rather than overgrowth.

III. Psychiatric Disorders and Other Behavioral Effects

Most of the disorders described in the previous section are characterized primarily by their effects on growth and metabolism, and in some cases effects on behaviors that relate directly to resource acquisition. The phenotypic effects associated with disruption or duplication of particular imprinted genes in these contexts are consistent with predictions from simple evolutionary models (with a few exceptions). However, many imprinted genes are also expressed in the adult brain and affect cognitive and behavioral traits in ways that are not as easily understood.

Nevertheless, it appears that there are certain systematic patterns in the phenotypic effects of imprinted gene expression in the brain, suggesting that the function of these genes has been shaped, in part, by intragenomic conflict. Further, many imprinted genes appear to contribute substantially to a number of common psychiatric disorders. Evidence for this contribution comes primarily from two sources: (1) psychiatric problems that are associated with known imprinting-related disorders and (2) genetic studies that have identified statistical associations of particular disorders with known imprinted genes, or have found parent-of-origin effects associated with particular genetic markers.

We begin this section with a brief summary of what is known regarding the roles of maternally and paternally expressed imprinted genes in the brain, and what these patterns suggest regarding the evolutionary pressures acting on these genes. We next discuss, the evidence for the contribution of imprinted genes to the etiologies of schizophrenia and autism, and describe the oppositional model of these disorders that is suggested by this evidence. Finally, we briefly survey the evidence for imprinted gene effects in other psychiatric disorders.

A. Imprinting Effects on Brain Structure and Function

In previous sections, we have already encountered evidence, in the context of specific disorders, that imprinted genes play an important role in brain development and may have systematic effects on behavior and cognition. The behavioral phenotypes associated with AS and PWS have been interpreted in terms of intragenomic conflict over the distribution of parental resources,[39,208] where paternally inherited alleles favor greater resource acquisition prior to weaning, when the resource demand falls primarily on the mother, but maternally inherited alleles increasingly favor greater demand as the paternal resource contribution grows.[39] An alternative explanation has been made in terms of intragenomic conflict over egoistic and altruistic behaviors,[40–42] where paternally inherited alleles favor greater egoistic behavior in interactions with nuclear family members, but maternally inherited alleles favor greater egoistic behavior in social interactions.[40–42]

The patterns observed in some of the studies on TS are suggestive of an intragenomic conflict over the allocation of neural resources to different cognitive tasks, although the effects are small and have not been observed consistently. The patterns that have been observed are consistent with imprinted genes on the paternally inherited X favoring greater investment in verbal and social cognition, while those on the maternally inherited X favor greater investment in spatial cognition.[188,194,195]

Evidence for intragenomic conflicts over brain structure has also been derived from parthenogenetic (PG) and androgenetic (AG) chimeras in mice. These chimeras consist of a mixture of normal, biparental cells, and cells that contain either two maternally derived (PG) or two paternally derived (AG) sets of chromosomes.[209] The PG chimeras had an increased brain volume relative to body size, while the AG chimeras had a reduced brain to body size ratio. Further, cortical areas were particularly enlarged (relative to other brain structures) in PG chimeras, and the PG cells were particularly enriched in those areas. Conversely, AG chimeras showed relative enlargement of limbic structures, and enrichment of AG cells in those areas, including hypothalamic, septal, and preoptic structures.

The patterns of brain structure and cell deposition in the chimera experiment is suggestive of a conflict in which maternally derived alleles favor greater investment in cortical functions, while paternally inherited alleles favor relatively more investment in limbic functions, although it is worth noting that this interpretation is not necessarily consistent with the apparent patterns suggested by the TS comparisons. The recent genome-wide study of imprinted gene expression in the mouse brain found approximately 1300 imprinted transcripts, and found dynamic changes in the patterns of imprinted gene expression through development.[28,29] For example, the majority of imprinted genes identified are maternally expressed early in development, while in the adult brain, the majority are paternally expressed.

One study on the inheritance of human cognitive abilities found a potential imprinting effect in normal cognition. The cognitive abilities of children were found to be highly correlated with their mothers' abilities for tasks associated with the frontal, parietal, and temporal lobes, while the effects of both parents were equally important for tasks associated with the occipital lobe.[210] This pattern is consistent with the distributions of PG and AG cells in the mouse chimeras.

Other behavioral effects associated with imprinted genes have been described in mice. Deletion of the paternally expressed *Peg1/Mest* and *Peg3* loci in adult females are each associated with deficits in specific maternal behaviors.[118,211] Deletion of the maternally expressed *Ube3a* produces defects in context-dependent memory.[212] Deletion of the maternally expressed *Rasgrf1* causes defects in memory consolidation,[213] and may contribute to depression.[214] Deletion of the paternally expressed *Ndn* actually results in enhanced spatial learning,[215] while deletion of the maternally expressed *Nesp* produces abnormal reactivity to novel environments.[79]

What is clear at this point is that imprinted genes play a significant role in brain development and function, but that the influence of those genes is complex. Several of the empirical observations are suggestive of systematic patterns in the phenotypic effects of maternally and paternally expressed imprinted genes, but those patterns are often based on small effects observed for small numbers of loci. First steps have been taken to construct an overarching theoretical framework for understanding imprinted gene effects in the brain that would be analogous to the framework existing for growth effects.[40,41] Significantly more research—both empirical and theoretical—is needed in this area.

B. Imprinted Gene Contributions to Schizophrenia and Autism

There has been a recent concerted effort to understand the role of imprinted genes in behavior and cognition, specifically in the context of schizophrenia and autism. A number of comprehensive reviews have collected the evidence for a role of imprinted genes in the etiology of both disorders.[216–219] In fact, imprinted genes account for some of the most significant associations of these diseases with particular loci or chromosomal regions. For example, a recent meta-analysis of GWAS analyses of schizophrenia found only one locus that showed statistically significant association at the genome-wide level.[220] This locus includes the imprinted gene *LRRTM1*, which is maternally silenced, and shows high expression during development throughout the cortical plate, as well as the septum caudate, putamen, dorsolateral thalamus, and lateral geniculate body.[221] Interestingly, *LRRTM1* is also associated with handedness,[222] suggesting that its effect on susceptibility to schizophrenia may be mediated through its effects on brain lateralization.

Many of these imprinted gene effects follow systematic patterns in which schizophrenia and autism correlate with imbalances in maternal and paternal genetic contributions to the individual. Schizophrenia is associated with excess maternal contribution (e.g., loss of function of a paternally expressed gene, or duplication of a maternally expressed gene), while autism is associated with an excess of paternal expression.[217]

This pattern suggests a model of cognitive/behavioral phenotypes in which schizophrenia and autism can productively be thought of as oppositional disorders. That is, it appears that they represent opposite extreme values along some phenotypic axis, and that there may be an intragenomic conflict between maternally and paternally derived genes with respect to the optimal cognitive/behavioral phenotype along that axis. While both optima are presumably well within the normal range (far from the extreme phenotype values associated with either of these two disorders), they differ such that the patrilineal optimum is slightly closer to the autism end of the spectrum, while the matrilineal optimum is slightly closer to the schizophrenia end.

Evolutionary theory predicts that psychotic-spectrum disorders will be linked to a clinical phenotype called the "hyper-egoistic brain," while autistic-spectrum disorders will be linked to the "hyper-altruistic brain" clinical phenotype.[40,41] The behavioral phenotypes associated with hyper-altruistic or hyper-egoistic brains need not (and generally will not) be functionally altruistic or egoistic, respectively. These disorders represent major disruptions at the level of the promiate mechanisms underlying social behavior and are not well-honed adaptations operating for the good of either the maternal or paternal gene copy.

TS is associated with elevated rates of autism, but, curiously, autism appears to be more common in X_mO patients than in X_pO patients,[223,224] contrary to what might be expected based on extrapolation from the patterns observed with imprinted autosomal loci. One possibility is that imprinted genes on the X chromosome are under strong selection based on sex differences (since only females normally inherit a paternally derived X), and that this is confounding the other selective pressures on these loci. Recall that the apparent imprinting effects on certain aspects of cognition in TS also appear to be at odds with the general patterns of influence of imprinted genes. Unraveling the effects of imprinting and sex differences for X-linked and autosomal loci will require additional research.

C. Imprinted Gene Effects in Other Psychiatric Disorders

More limited evidence points toward a contribution of imprinted genes to the etiology of other specific psychopathologies,[225] although in each case, the potential molecular and genetic mechanisms have yet to be fully elucidated, and attempts to understand the evolutionary origins are purely speculative at this point.

1. Obsessive-Compulsive Disorder

Obsessive-compulsive disorder (OCD) is associated not only with obsessive and compulsive behaviors, but also with temper issues, externalizing behavior, and emotional problems.[226] OCD is extremely common, estimated to affect as many as 5 million people in the United States,[227] and shows a strong genetic component.[228] OCD is comorbid with Prader–Willi syndrome and occurs in PWS-like patients.[229] One hypothesis is that the absence of imprinted small nucleolar RNAs (SnoRNAs) that normally interact with Serotonin 2C receptor subtypes may contribute to the etiology of OCD.[230]

2. Attention-Deficit Hyperactivity Disorder

Attention deficit hyperactivity disorder (ADHD)[231] is also extremely common and highly heritable and occurs at high frequency in conjunction with PWS.[232] Imprinting effects on hyperactivity have been reported in mice,[149] and parent-of-origin effects have been reported in disorders that are comorbid with ADHD, such as Tourette's syndrome,[233] and a specific polymorphism in the gene encoding brain-derived neurotrophic factor (BDNF) has been specifically associated with susceptibility to ADHD.[234]

3. Bipolar Affective Disorder

Bipolar affective disorder (BPAD) and other mood disorders are highly comorbid with ADHD,[235–237] and cyclical depression has been reported in conjunction with PWS.[238,239] The severity of symptoms in BPAD in conjunction with ADHD shows dependence on parent of origin,[240] and several genes that affect the dopaminergic and serotinergic systems that are common targets of therapeutic intervention show evidence for imprinting effects, including dopa decarboxylase (DDC),[241] tryptophan hydroxylase 2 (TPH2),[242] and BDNF.[243,244]

IV. The Cost of Imprinting

In some ways, diseases associated with imprinted genes are no different from other diseases with a genetic basis. Mutations or epimutations occur in the germ line or the soma and produce the disease phenotype. One obvious difference is that, since imprinted genes are typically expressed from only one of the two alleles, only one loss-of-function mutation is required to effectively knock out the gene. Thus, at least for genes where loss-of-function mutations would normally be recessive, the monoallelic expression associated with imprinted genes adds a degree of penetrance to mutations. Further, imprinted genes are subject to certain mutations or epimutations that result in transcriptional reactivation of the normally silenced allele, often referred to as "loss-of-imprinting" mutations. This reactivation results in an increase in the overall expression level, and is associated with a number of diseases.[245–247]

Thus, there is a certain cost, in elevated penetrance of mutations and epimutations associated with imprinted genes, that is really associated with their monoallelic expression. However, there is a more subtle, but potentially much more significant, cost associated with genomic imprinting, that derives from how the imprinted genes alter the evolutionary dynamics of the systems in which they appear.

In general, the effect of natural selection is not identical for maternally and paternally inherited alleles. The magnitude of the selection asymmetry may be greatest in the context of fetal growth effects, decreasing significantly for postnatal behavioral effects, and may be quite subtle for many cognitive and behavioral phenotypes in adults. However, at loci where imprinted gene expression has been established, even subtle selective effects can have significant consequences over sufficiently long time scales.

In particular, consider a pair of oppositely imprinted loci (one maternally expressed and one paternally expressed), where the phenotypic effect of increasing gene expression at one locus is opposed to the effect of increasing expression from the other locus. If the matrilineal and patrilineal optima differ even slightly for this phenotype, these two loci will become engaged in an evolutionary arms race, with each under selection to increase its level of expression from the active allele.

In the simplest possible model, this escalation will go on forever, so that each locus is producing an infinite amount of gene product. Clearly, this is not realistic, and at some point, some other effect will limit the escalation. Among the possibilities for this limiting effect are metabolic cost associated with increased gene expression, mechanistic limitations on expression from one of the loci, and deleterious side effects associated with increased expression. The extent to which having imprinted genes is deleterious depends, in part, on which of these limiting factors dominates in practice. However, in each case, we expect to find pairs or groups of genes that have opposing phenotypic effects, and that are expressed at a level higher than what would be expected in the absence of imprinting. These elevated, oppositional patterns of expression have a number of potential consequences.

A. Mutational Effects

We have already noted that imprinted genes are more susceptible to loss-of-function mutations than their unimprinted counterparts, owing to their monoallelic expression. In addition, if the wild-type expression level is elevated due to intragenomic conflict, the phenotypic consequences of a loss-of-function mutation will be more dramatic at an imprinted locus than it would have been in the absence of the conflict-driven escalation in gene expression.

B. Epimutations

In many cases, transcriptional inactivation of the silenced allele at an imprinted locus is achieved through the application of DNA methylation and/or histone modifications. At such a locus, the level of expression from the active allele will be determined largely by *cis*-acting regulatory elements encoded in the DNA itself. This arrangement produces a vulnerability to epimutations, where the silencing epigenetic marks are lost, resulting in a dramatic increase in the overall level of gene expression from the locus (as the normally silenced allele will have approximately the same *cis*-acting regulatory motifs, which become active in the absence of the epigenetic silencing).

The possibility of such reactivating epimutations imposes a twofold cost on systems of imprinted genes as compared to their unimprinted counterparts. Any locus may, in principle, be subjected to mutations that increase the gene dosage (e.g., a mutation that increases the copy number). Imprinted genes are susceptible to those mutation processes as well as to epimutations (which occur at substantially higher frequencies than other classes of mutation[248]). Further, due to the evolutionary escalation in expression level expected among imprinted genes, the effect of doubling the number of active gene copies in the cell may be more pronounced than would be the case for an unimprinted locus.

C. Imprinting and Cancer

We have noted that imprinted genes have an increased susceptibility to mutations and epimutations that increase or eliminate gene expression from the locus. In addition, the resulting change in the absolute gene expression level will tend to be greater at an imprinted locus than at an unimprinted one. Another feature of imprinted genes is that they are typically associated either with growth-enhancing or growth-suppressing functions in early development.

It is not surprising, then, that dysregulation of imprinted genes is found in many cancers. A locus with a growth-enhancing effect in early development will often maintain a mitogenic effect in adult somatic cells, and reactivation of the silenced allele can contribute to uncontrolled cell proliferation. At the same time, many (maternally expressed) imprinted genes have evolved a growth-suppressing function. Many of these genes may then act as *de facto* tumor suppressors in adult tissues. However, these genes will differ from many other tumor suppressors in the fact that there is only a single active copy, which reduces the number of somatic mutations required to eliminate the tumor-suppressing activity of the locus.

The contribution of epigenetic dysregulation to cancer is treated in detail elsewhere in this volume (Chapter 14), and will not be covered further here.

D. Pleiotropic Effects

In reality, pairs of antagonistically coevolving genes will not be perfectly aligned in terms of their phenotypic consequences. The space of possible phenotypes occupies a large number of dimensions, and the marginal effect of a small change in gene expression from a locus can be pictured as a vector in this high-dimensional space. In the previous sections, we have discussed the escalation among imprinted genes in terms of an evolutionary conflict over a single aspect of the phenotype (e.g., fetal growth rate). In general, changes in gene expression will affect not only the aspect of the phenotype that is the object of the evolutionary conflict, but other aspects of the phenotype as well, even if the maternally and paternally inherited alleles share a common phenotypic optimum for those other aspects. As a result of these pleiotropic effects, the escalation that is driven by a conflict over one aspect of the phenotype can force those other aspects of the phenotype away from their shared optima.

In a simple, linear model of the antagonistic coevolution of imprinted genes with pleiotropic effects, it is possible to quantify the magnitude of the phenotypic deviation at the evolutionarily stable state.[249] In general, conflict will result in the fixation of suboptimal phenotypes. Except for a vanishingly small set of special cases, the equilibrium phenotype in the presence of imprinting will deviate from that which maximizes the overall fitness of the organism (or the average fitness of the alleles it is carrying), even for those aspects of the phenotype for which all of the alleles in the organism share a common optimum.

With respect to the particular aspect of the phenotype that is the basis of the conflict, we might naively expect that the evolutionarily stable phenotype value would lie somewhere between the matrilineal and patrilineal optima. However, in the presence of pleiotropic effects of the imprinted genes, this expectation does not necessarily hold. In a simple model, it is predicted that roughly half of the time the equilibrium phenotype value along the phenotypic axis of conflict will lie outside of the range defined by the matrilineal and patrilineal optima. Thus, the combination of intragenomic conflict and pleiotropic effects of imprinted genes create a situation where natural selection will often produce a phenotype that is more extreme than what is favored by either of the conflicting loci.

E. Decanalization

Another consequence of increasing the level of expression from a locus is that it will tend to generate an increase in the expression variance. Under widely differing circumstances, there seems to be a relatively simple relationship, where the variance in gene expression scales roughly as the square of the mean. This relationship can be seen in yeast,[250] where this variance represents stochastic variation among genetically identical cells, as well as in human lymphoblasts,[251] where it represents stochastic variation, as well as the consequences of interindividual genetic variation at other loci. In both cases, the scaling relationship is robust over multiple orders of magnitude of the absolute expression level.

Many biological processes are characterized by the phenomenon of canalization, which refers to mechanisms that reduce the phenotypic variation in the face of underlying genetic or environmental variation.[252–254] The escalation that results from intragenomic, interlocus conflict among imprinted genes can, under some circumstances, lead to the undermining of these canalization mechanisms (conflict-induced decanalization), resulting in an increased frequency of extreme phenotypes, even if those phenotypes are associated with disease states.[255]

F. The Imprinting Load

The set of phenomena described here combines to produce the "imprinting load," which can be thought of as the average fitness cost associated with imprinted gene expression. More formally, we consider the average fitness associated with a system that includes imprinted genes at its evolutionary equilibrium. This is compared to the average fitness of the same system, but in the absence of genomic imprinting. The imprinting load is simply the difference between the two average fitnesses.

The imprinting load is a quantity that is difficult to calculate for real systems. However, one can calculate the imprinting load for particular models, and this can provide insight into which factors are most important in determining the magnitude of the fitness reduction. For example, in the simple models of pleiotropy and decanalization described above, the imprinting load scales roughly as the square of the magnitude of the conflict between the matrilineal and patrilineal phenotypic optima.[249,255] For example, assume imprinting load is ξ in a system where the matrilineal and patrilineal optimal phenotypes differ by a quantity α. In a system that was identical, but where the optima differed by 2α, the imprinting load would be approximately 4ξ.

Interestingly, in both models, the magnitude of the imprinting load is much more sensitive to other parameters of the model. In the pleiotropy model, the most important factor is the relationship between the pleiotropic effects of the two loci. In the decanalization model, the most important factor is the way in which the gene products interact to generate the phenotype.

In both cases, these other, dominant factors can be interpreted broadly as aspects of "mechanism." Thus, the simple models suggest that the addition of genomic imprinting to a system generically results in a reduction in fitness.

However, there does not appear to be a general answer to the question of how large this effect is in the absence of consideration of certain details of how the system is constructed.

V. Conclusions

Genomic imprinting affects on the order of 1% of the genome and contributes to many parent-of-origin effects in heritable disease. The evolutionary forces responsible for the origin of imprinted gene expression help to explain many of the phenotypic consequences of imprinting-related disorders, including the growth effects and certain consequences for feeding behaviors and adult metabolism. In many cases, evolutionary reasoning also explains the direction of the parent-of-origin effects. Imprinting also plays an important role in neural development, and defects in imprinted genes are associated with numerous cognitive and behavioral consequences, including possible associations with major psychiatric disorders including autism and schizophrenia. Our evolutionary understanding of the cognitive and behavioral effects of imprinted genes is much less well developed than our understanding of the effects on growth and metabolism.

The existence of imprinted genes potentially increases the penetrance of genetic defects, as a single mutation is sufficient to induce complete loss of function at an imprinted locus. Similarly, the existence of the molecular machinery responsible for imprinting creates the opportunity for epimutations that result in dysregulation of expression, and may occur as orders of magnitude more frequently than mutations to the DNA sequence. Perhaps, more important than either of these effects, however, are the consequences of antagonistic coevolution among imprinted genes that can lead to the accumulation of maladaptive phenotypes, and may contribute to elevated frequencies of certain disease states.

References

1. Genomic imprinting. Wilkins JF, editor. *Advances in experimental medicine and biology*. New York: Springer; 2008.
2. Hayward BE, Kamiya M, Strain L, Moran V, Campbell R, Hayashizaki Y, et al. The human GNAS1 gene is imprinted and encodes distinct paternally and biallelically expressed G proteins. *Proc Natl Acad Sci USA* 1998;**95**(17):10038–43.
3. Hayward BE, Moran V, Strain L, Bonthron DT, et al. Bidirectional imprinting of a single gene: GNAS1 encodes maternally, paternally, and biallelically derived proteins. *Proc Natl Acad Sci USA* 1998;**95**(26):15475–80.

4. Holmes R, Williamson C, Peters J, Paul D, Wells C, Group RIKENGE, et al. A comprehensive transcript map of the mouse Gnas imprinted complex. *Genome Res* 2003;**13**(6B):1410–5.
5. Kelsey G. Epigenetics and imprinted genes: insights from the imprinted Gnas locus. *Horm Res* 2009;**71**(Suppl. 2):22–9.
6. Peters J, Williamson CM. Control of imprinting at the Gnas cluster. *Adv Exp Med Biol* 2008;**626**:16–26.
7. Monk D, Arnaud P, Frost J, Hills FA, Stanier P, Feil R, et al. Reciprocal imprinting of human GRB10 in placental trophoblast and brain: evolutionary conservation of reversed allelic expression. *Hum Mol Genet* 2009;**18**(16):3066–74.
8. Killian JK, Byrd JC, Jirtle JV, Munday BL, Stoskopf MK, MacDonald RG, et al. M6P/IGF2R imprinting evolution in mammals. *Mol Cell* 2000;**5**(4):707–16.
9. Killian JK, Nolan CM, Stewart N, Munday BL, Andersen NA, Nicol S, et al. Monotreme IGF2 expression and ancestral origin of genomic imprinting. *J Exp Zool* 2001;**291**(2):205–12.
10. Killian JK, Nolan CM, Wylie AA, Li T, Vu TH, Hoffman AR, et al. Divergent evolution in M6P/IGF2R imprinting from the Jurassic to the Quaternary. *Hum Mol Genet* 2001;**10**(17):1721–8.
11. Lawton BR, Carone BR, Obergfell CJ, Ferreri GC, Gondolphi CM, VandeBerg JL, et al. Genomic imprinting of IGF2 in marsupials is methylation dependent. *BMC Genomics* 2008;**9**:205.
12. Nolan CM, Killian JK, Petitte JN, Jirtle RL, et al. Imprint status of M6P/IGF2R and IGF2 in chickens. *Dev Genes Evol* 2001;**211**(4):179–83.
13. O'Neill MJ, Ingram RS, Vrana PB, Tilghman SM, et al. Allelic expression of IGF2 in marsupials and birds. *Dev Genes Evol* 2000;**210**(1):18–20.
14. Renfree MB, Hore TA, Shaw G, Graves JAM, Pask AJ, et al. Evolution of genomic imprinting: insights from marsupials and monotremes. *Annu Rev Genomics Hum Genet* 2009;**10**:241–62.
15. Suzuki S, Renfree MB, Pask AJ, Shaw G, Shin K, Kohda T, et al. Genomic imprinting of IGF2, p57(KIP2) and PEG1/MEST in a marsupial, the tammar wallaby. *Mech Dev* 2005;**122**(2):213–22.
16. Garnier O, Laouiellé-Duprat S, Spillane C, et al. Genomic imprinting in plants. *Adv Exp Med Biol* 2008;**626**:89–100.
17. Goday C, Esteban MR. Chromosome elimination in sciarid flies. *Bioessays* 2001;**23**(3):242–50.
18. Crouse HV. The controlling element in sex chromosome behavior in *Sciara*. *Genetics* 1960;**45**:1429–43.
19. Stern C. The nucleus and somatic cell variation. *J Cell Physiol* 1958;**52**(Suppl. 1):1–27 discussion 27–34.
20. Weidman JR, Murphy SK, Nolan CM, Dietrich FS, Jirtle RL, et al. Phylogenetic footprint analysis of IGF2 in extant mammals. *Genome Res* 2004;**14**(9):1726–32.
21. Yamada Y, Hagiwara Y, Shiokawa K, Sakaki Y, Ito T, et al. Spatiotemporal, allelic, and enforced expression of Ximpact, the Xenopus homolog of mouse imprinted gene impact. *Biochem Biophys Res Commun* 1999;**256**(1):162–9.
22. Lawton BR, Sevigny L, Obergfell C, Reznick D, O'Neill RJ, O'Neill MJ, et al. Allelic expression of IGF2 in live-bearing, matrotrophic fishes. *Dev Genes Evol* 2005;**215**(4):207–12.
23. Tsalavouta M, Astudillo O, Byrnes L, Nolan CM, et al. Regulation of expression of zebrafish (Danio rerio) insulin-like growth factor 2 receptor: implications for evolution at the IGF2R locus. *Evol Dev* 2009;**11**(5):546–58.
24. Morison IM, Ramsay JP, Spencer HG, et al. A census of mammalian imprinting. *Trends Genet* 2005;**21**(8):457–65.
25. Morison IM, Reeve AE. A catalogue of imprinted genes and parent-of-origin effects in humans and animals. *Hum Mol Genet* 1998;**7**(10):1599–609.

26. Luedi PP, Hartmink AJ, Jirtle RL, et al. Genome-wide prediction of imprinted murine genes. *Genome Res* 2005;**15**(6):875–84.
27. Luedi PP, Dietrich FS, Weidman JR, Bosko JM, Jirtle RL, Hartemink AJ, et al. Computational and experimental identification of novel human imprinted genes. *Genome Res* 2007;**17**(12):1723–30.
28. Gregg C, Zhang J, Butler JE, Haig D, Dulac C, et al. Sex-specific parent-of-origin allelic expression in the mouse brain. *Science* 2010;**329**(5992):682–5.
29. Gregg C, Zhang J, Weissbourd B, Luo S, Schroth GP, Haig D, et al. High-resolution analysis of parent-of-origin allelic expression in the mouse brain. *Science* 2010;**329**(5992):643–8.
30. Mayer W, Niveleau A, Walter J, Fundele R, Haaf T, et al. Demethylation of the zygotic paternal genome. *Nature* 2000;**403**(6769):501–2.
31. Oswald J, Engemann S, Lane N, Mayer W, Olek A, Fundele R, et al. Active demethylation of the paternal genome in the mouse zygote. *Curr Biol* 2000;**10**(8):475–8.
32. Moore T, Mills W. Evolutionary theories of imprinting—enough already. *Adv Exp Med Biol* 2008;**626**:116–22.
33. Wood AJ, Oakey RJ. Genomic imprinting in mammals: emerging themes and established theories. *PLoS Genet* 2006;**2**(11):e147.
34. Haig D. The kinship theory of genomic imprinting. *Annu Rev Ecol Syst* 2000;**31**:9–32.
35. Haig D, Graham C. Genomic imprinting and the strange case of the insulin-like growth factor II receptor. *Cell* 1991;**64**(6):1045–6.
36. Moore T, Haig D. Genomic imprinting in mammalian development: a parental tug-of-war. *Trends Genet* 1991;**7**(2):45–9.
37. Wilkins J, Haig D. Inbreeding, maternal care and genomic imprinting. *J Theor Biol* 2003;**221**(4):559–64.
38. Wilkins J, Haig D. What good is genomic imprinting: the function of parent-specific gene expression. *Nat Rev Genet* 2003;**4**(5):359–68.
39. Úbeda F. Evolution of genomic imprinting with biparental care: implications for Prader-Willi and Angelman syndromes. *PLoS Biol* 2008;**6**(8):e208.
40. Úbeda F, Gardner A. A model for genomic imprinting in the social brain: adults. *Evolution* 2010;**65**(2):462–75.
41. Úbeda F, Gardner A. A model for genomic imprinting in the social brain: juveniles. *Evolution* 2010;**64**(9):2587–600.
42. Brandvain Y. Matrisibs, patrisibs, and the evolution of imprinting on autosomes and sex chromosomes. *Am Nat* 2010;**176**(4):511–21.
43. van Cleve J, Feldman MW, Lehmann L, et al. How demography, life history, and kinship shape the evolution of genomic imprinting. *Am Nat* 2010;**176**(4):440–55.
44. Bastepe M. The GNAS locus and pseudohypoparathyroidism. *Adv Exp Med Biol* 2008;**626**:27–40.
45. Persani L, Calebiro D, Cordella D, Weber G, Gelmini G, Libri D, et al. Genetics and phenomics of hypothyroidism due to TSH resistance. *Mol Cell Endocrinol* 2010;**322**(1–2):72–82.
46. Gensure RC, Gardella TJ, Jüppner H, et al. Parathyroid hormone and parathyroid hormone-related peptide, and their receptors. *Biochem Biophys Res Commun* 2005;**328**(3):666–78.
47. Potts JT. Parathyroid hormone: past and present. *J Endocrinol* 2005;**187**(3):311–25.
48. Ish-Shalom S, Rao LG, Levine MA, Fraser D, Kooh SW, Josse RG, et al. Normal parathyroid hormone responsiveness of bone-derived cells from a patient with pseudohypoparathyroidism. *J Bone Miner Res* 1996;**11**(1):8–14.
49. Murray TM, Rao LG, Wong MM, Waddell JP, McBroom R, Tam CS, et al. Pseudohypoparathyroidism with osteitis fibrosa cystica: direct demonstration of skeletal responsiveness to parathyroid hormone in cells cultured from bone. *J Bone Miner Res* 1993;**8**(1):83–91.

50. Stone MD, Hosking DJ, Garcia-Himmelstine C, White DA, Rosenblum D, Worth HG, et al. The renal response to exogenous parathyroid hormone in treated pseudohypoparathyroidism. *Bone* 1993;**14**(5):727–35.
51. Chase LR, Melson GL, Aurbach GD, et al. Pseudohypoparathyroidism: defective excretion of 3', 5'-AMP in response to parathyroid hormone. *J Clin Invest* 1969;**48**(10):1832–44.
52. Aldred MA, Aftimos S, Hall C, Waters KS, Thakker RV, Trembath RC, et al. Constitutional deletion of chromosome 20q in two patients affected with albright hereditary osteodystrophy. *Am J Med Genet* 2002;**113**(2):167–72.
53. Lania AG, Mantovani G, Spada A, et al. Mechanisms of disease: mutations of G proteins and G-protein-coupled receptors in endocrine diseases. *Nat Clin Pract Endocrinol Metab* 2006;**2**(12):681–93.
54. Mantovani G, Spada A. Mutations in the Gs alpha gene causing hormone resistance. *Best Pract Res Clin Endocrinol Metab* 2006;**20**(4):501–13.
55. Weinstein LS, Gejman PV, Friedman E, Kadowaki T, Collins RM, Gershon ES, et al. Mutations of the Gs alpha-subunit gene in Albright hereditary osteodystrophy detected by denaturing gradient gel electrophoresis. *Proc Natl Acad Sci USA* 1990;**87**(21):8287–90.
56. Levine MA, Modi WS, O'Brien SJ, et al. Mapping of the gene encoding the alpha subunit of the stimulatory G protein of adenylyl cyclase (GNAS1) to 20q13.2–q13.3 in human by in situ hybridization. *Genomics* 1991;**11**(2):478–9.
57. Rao VV, Schnittger S, Hansmann I, et al. G protein Gs alpha (GNAS 1), the probable candidate gene for Albright hereditary osteodystrophy, is assigned to human chromosome 20q12-q13.2. *Genomics* 1991;**10**(1):257–61.
58. Bray P, Carter A, Simons C, Guo V, Puckett C, Kamholz J, et al. Human cDNA clones for four species of G alpha s signal transduction protein. *Proc Natl Acad Sci USA* 1986;**83**(23):8893–7.
59. Kozasa T, Itoh H, Tsukamoto T, Kaziro Y, et al. Isolation and characterization of the human Gs alpha gene. *Proc Natl Acad Sci USA* 1988;**85**(7):2081–5.
60. Robishaw JD, Smigel MD, Gilman AG, et al. Molecular basis for two forms of the G protein that stimulates adenylate cyclase. *J Biol Chem* 1986;**261**(21):9587–90
61. Graziano MP, Freissmuth M, Gilman AG, et al. Expression of Gs alpha in *Escherichia coli*. Purification and properties of two forms of the protein. *J Biol Chem* 1989;**264**(1):409–18.
62. Thiele S, Werner R, Ahrens W, Hübner A, Hinkel KG, Höppner W, et al. A disruptive mutation in exon 3 of the GNAS gene with albright hereditary osteodystrophy, normocalcemic pseudohypoparathyroidism, and selective long transcript variant Gsalpha-L deficiency. *J Clin Endocrinol Metab* 2007;**92**(5):1764–8.
63. Crawford JA, Mutchler KJ, Sullivan BE, Lanigan TM, Clark MS, Russo AF, et al. Neural expression of a novel alternatively spliced and polyadenylated Gs alpha transcript. *J Biol Chem* 1993;**268**(13):9879–85.
64. Yu S, Yu D, Lee E, Eckhaus M, Lee R, Corria Z, et al. Variable and tissue-specific hormone resistance in heterotrimeric Gs protein alpha-subunit (Gsalpha) knockout mice is due to tissue-specific imprinting of the gsalpha gene. *Proc Natl Acad Sci USA* 1998;**95**(15):8715–20.
65. Germain-Lee EL, Ding CL, Deng Z, Crane JL, Saji M, Ringel MD, et al. Paternal imprinting of Galpha(s) in the human thyroid as the basis of TSH resistance in pseudohypoparathyroidism type 1a. *Biochem Biophys Res Commun* 2002;**296**(1):67–72.
66. Hayward BE, Barlier A, Korbonits M, Grossman AB, Jacquet P, Enjalbert A, et al. Imprinting of the G(s)alpha gene GNAS1 in the pathogenesis of acromegaly. *J Clin Invest* 2001;**107**(6):R31–6.
67. Mantovani G, Ballare E, Giammona E, Beck-Peccoz P, Spada A, et al. The gsalpha gene: predominant maternal origin of transcription in human thyroid gland and gonads. *J Clin Endocrinol Metab* 2002;**87**(10):4736–40.

68. Montovani G, Bondioni S, Locatelli M, Pedroni C, Lania AG, Ferrante E, et al. Biallelic expression of the Gsalpha gene in human bone and adipose tissue. *J Clin Endocrinol Metab* 2004;**89**(12):6316–9.
69. Ishikawa Y, Bianchi C, Nadal-Ginard B, Homcy CJ, et al. Alternative promoter and 5′ exon generate a novel Gs alpha mRNA. *J Biol Chem* 1990;**265**(15):8458–62.
70. Swaroop A, Agarwal N, Gruen JR, Bick D, Weissman SM, et al. Differential expression of novel Gs alpha signal transduction protein cDNA species. *Nucleic Acids Res* 1991;**19**(17):4725–9.
71. Liu J, Litman D, Rosenberg MJ, Yu S, Biesecker LG, Weinstein LS, et al. A GNAS1 imprinting defect in pseudohypoparathyroidism type IB. *J Clin Invest* 2000;**106**(9):1167–74.
72. Liu J, Yu S, Litman D, Chen W, Weinstein LS, et al. Identification of a methylation imprint mark within the mouse Gnas locus. *Mol Cell Biol* 2000;**20**(16):5808–17.
73. Li T, Vu TH, Zeng ZL, Nguyen BT, Hayward BE, Bonthron DT, et al. Tissue-specific expression of antisense and sense transcripts at the imprinted Gnas locus. *Genomics* 2000;**69**(3):295–304.
74. Peters J, Wroe SF, Wells CA, Miller HJ, Bodie D, Beechey CV, et al. A cluster of oppositely imprinted transcripts at the Gnas locus in the distal imprinting region of mouse chromosome 2. *Proc Natl Acad Sci USA* 1999;**96**(7):3830–5.
75. Michienzi S, Cherman N, Holmbeck K, Funari A, Collins MT, Bianco P, et al. GNAS transcripts in skeletal progenitors: evidence for random asymmetric allelic expression of Gs alpha. *Hum Mol Genet* 2007;**16**(16):1921–30.
76. Freson K, Jaeken J, Van Helvoirt M, de Zegher F, Wittevrongel C, Thys C, et al. Functional polymorphisms in the paternally expressed XLalphas and its cofactor ALEX decrease their mutual interaction and enhance receptor-mediated cAMP formation. *Hum Mol Genet* 2003;**12**(10):1121–30.
77. Klemke M, Kehlenbach RH, Huttner WB, et al. Two overlapping reading frames in a single exon encode interacting proteins–a novel way of gene usage. *EMBO J* 2001;**20**(14):3849–60.
78. Ischia R, Lovisetti-Scamihorn P, Hogue-Angeletti R, Wolkersdorfer M, Winkler H, Fischer-Colbrie R, et al. Molecular cloning and characterization of NESP55, a novel chromogranin-like precursor of a peptide with 5-HT1B receptor antagonist activity. *J Biol Chem* 1997;**272**(17):11657–62.
79. Plagge A, Isles AR, Gordon E, Humby T, Dean W, Gritsch S, et al. Imprinted Nesp55 influences behavioral reactivity to novel environments. *Mol Cell Biol* 2005;**25**(8):3019–26.
80. Hayward BE, Bonthron DT. An imprinted antisense transcript at the human GNAS1 locus. *Hum Mol Genet* 2000;**9**(5):835–41.
81. Wroe S, Kelsey G, Skinner JA, Bodie D, Ball ST, Beechey CV, et al. An imprinted transcript, antisense to Nesp, adds complexity to the cluster of imprinted genes at the mouse Gnas locus. *Proc Natl Acad Sci USA* 2000;**97**(7):3342.
82. Williamson CM, Turner MD, Ball ST, Nottinghom WT, Glenister P, Fray M, et al. Identification of an imprinting control region affecting the expression of all transcripts in the Gnas cluster. *Nat Genet* 2006;**38**(3):350–5.
83. Bastepe M, Fröhlich LF, Hendy GN, Indridason OS, Josse RG, Koshiyama H, et al. Autosomal dominant pseudohypoparathyroidism type Ib is associated with a heterozygous microdeletion that likely disrupts a putative imprinting control element of GNAS. *J Clin Invest* 2003;**112**(8):1255–63.
84. Sakamoto A, Liu J, Greene A, Chen M, Weinstein LS, et al. Tissue-specific imprinting of the G protein Gsalpha is associated with tissue-specific differences in histone methylation. *Hum Mol Genet* 2004;**13**(8):819–28.

85. Liu J, Chen M, Deng C, Bourc'his D, Nealon JG, Erlichman B, et al. Identification of the control region for tissue-specific imprinting of the stimulatory G protein alpha-subunit. *Proc Natl Acad Sci USA* 2005;**102**(15):5513–8.
86. Liu J, Nealon JG, Weinstein LS, et al. Distinct patterns of abnormal GNAS imprinting in familial and sporadic pseudohypoparathyroidism type IB. *Hum Mol Genet* 2005;**14**(1):95–102.
87. Williamson CM, Ball ST, Nottingham WT, Skinner JA, Plagge A, Turner MD, et al. A cis-acting control region is required exclusively for the tissue-specific imprinting of Gnas. *Nat Genet* 2004;**36**(8):894–9.
88. Choo JH, Kim JD, Chung JH, Stubbs L, Kim J, et al. Allele-specific deposition of macroH2A1 in imprinting control regions. *Hum Mol Genet* 2006;**15**(5):717–24.
89. Coombes C, Arnaud P, Gordon E, Dean W, Coar EA, Williamson CM, et al. Epigenetic properties and identification of an imprint mark in the Nesp-Gnasxl domain of the mouse Gnas imprinted locus. *Mol Cell Biol* 2003;**23**(16):5475–88.
90. Kim JD, Hinz AK, Bergmann A, Huang JM, Ovcharenko I, Stubbs L, et al. Identification of clustered YY1 binding sites in imprinting control regions. *Genome Res* 2006;**16**(7):901–11.
91. Kelsey G, Bodle D, Miller HJ, Beechey CV, Coombes C, Peters J, et al. Identification of imprinted loci by methylation-sensitive representational difference analysis: application to mouse distal chromosome 2. *Genomics* 1999;**62**(2):129–38.
92. Albright F, Forbes AP, Henneman PH, et al. Pseudo-pseudohypoparathyroidism. *Trans Assoc Am Physicians* 1952;**65**:337–50.
93. Levine MA, Jap TS, Mauseth RS, Downs RW, Spiegel AM, et al. Activity of the stimulatory guanine nucleotide-binding protein is reduced in erythrocytes from patients with pseudohypoparathyroidism and pseudopseudohypoparathyroidism: biochemical, endocrine, and genetic analysis of Albright's hereditary osteodystrophy in six kindreds. *J Clin Endocrinol Metab* 1986;**62**(3):497–502.
94. Davies SJ, Hughes HE. Imprinting in Albright's hereditary osteodystrophy. *J Med Genet* 1993;**30**(2):101–3.
95. Wilson LC, Oude Luttikhuis ME, Clayton PT, Fraser WD, Trembath RC, et al. Parental origin of Gs alpha gene mutations in Albright's hereditary osteodystrophy. *J Med Genet* 1994;**31**(11):835–9.
96. Weinstein LS, Yu S, Warner DR, Liu J, et al. Endocrine manifestations of stimulatory G protein alpha-subunit mutations and the role of genomic imprinting. *Endocr Rev* 2001;**22**(5):675–705.
97. Schwindinger WF, Francomano CA, Levine MA, et al. Identification of a mutation in the gene encoding the alpha subunit of the stimulatory G protein of adenylyl cyclase in McCune-Albright syndrome. *Proc Natl Acad Sci USA* 1992;**89**(11):5152–6.
98. Weinstein LS, Shenker A, Gejman PV, Merino MJ, Friedman E, Spiegel AM, et al. Activating mutations of the stimulatory G protein in the McCune-Albright syndrome. *N Engl J Med* 1991;**325**(24):1688–95.
99. Russell A. A syndrome of intra-uterine dwarfism recognizable at birth with cranio-facial dysostosis, disproportionately short arms, and other anomalies (5 examples). *Proc R Soc Med* 1954;**47**(12):1040–4.
100. Silver HK, Kiyasu W, George J, Deamer WC, et al. Syndrome of congenital hemihypertrophy, shortness of stature, and elevated urinary gonadotropins. *Pediatrics* 1953;**12**(4):368–76.
101. Wollmann HA, Kirchner T, Enders H, Preece MA, Ranke MB, et al. Growth and symptoms in Silver-Russell syndrome: review on the basis of 386 patients. *Eur J Pediatr* 1995;**154**(12):958–68.
102. Abu-Amero S, Monk D, Frost J, Preece M, Stanier P, Moore GE, et al. The genetic aetiology of Silver-Russell syndrome. *J Med Genet* 2008;**45**(4):193–9.

103. Abu-Amero S, Wakeling EL, Preece M, Whittaker J, Stanier P, Moore GE, et al. Epigenetic signatures of Silver-Russell syndrome. *J Med Genet* 2010;**47**(3):150–4.
104. Arnaud P, Monk D, Hitchins M, Gordon E, Dean W, Beechey CV, et al. Conserved methylation imprints in the human and mouse GRB10 genes with divergent allelic expression suggests differential reading of the same mark. *Hum Mol Genet* 2003;**12**(9):1005–19.
105. Blagitko N, Mergenthaler S, Schulz U, Wollmann HA, Craigen W, Eggermann T, et al. Human GRB10 is imprinted and expressed from the paternal and maternal allele in a highly tissue- and isoform-specific fashion. *Hum Mol Genet* 2000;**9**(11):1587–95.
106. Hitchins MP, Monk D, Bell GM, Ali Z, Preece MA, Stanier P, et al. Maternal repression of the human GRB10 gene in the developing central nervous system; evaluation of the role for GRB10 in Silver-Russell syndrome. *Eur J Hum Genet* 2001;**9**(2):82–90.
107. Hitchins MP, Bentley L, Monk D, Beechey C, Peters J, Kelsey G, et al. DDC and COBL, flanking the imprinted GRB10 gene on 7p12, are biallelically expressed. *Mamm Genome* 2002;**13**(12):686–91.
108. Charalambous M, Cowley M, Geoghegan F, Smith FM, Radford EJ, Marlow BP, et al. Maternally-inherited Grb10 reduces placental size and efficiency. *Dev Biol* 2010;**337**(1):1–8.
109. Leach NT, Chudoba I, Stewart TV, Holmes LB, Weremowicz S, et al. Maternally inherited duplication of chromosome 7, dup(7)(p11.2p12), associated with mild cognitive deficit without features of Silver-Russell syndrome. *Am J Med Genet A* 2007;**143A**(13):1489–93.
110. Monk D, Wagschal A, Arnaud P, Müller PS, Parker-Katiraee L, Bourc'his D, et al. Comparative analysis of human chromosome 7q21 and mouse proximal chromosome 6 reveals a placental-specific imprinted gene, TFPI2/Tfpi2, which requires EHMT2 and EED for allelic-silencing. *Genome Res* 2008;**18**(8):1270–81.
111. Ono R, Kobayashi S, Wagatsuma H, Aisaka K, Kohda T, Kaneko-Ishino T, et al. A retrotransposon-derived gene, PEG10, is a novel imprinted gene located on human chromosome 7q21. *Genomics* 2001;**73**(2):232–7.
112. Piras G, El K, harroubi A, Kozlov S, Escalante-Alcalde D, Hernandez L, Copeland NG, et al. Zac1 (Lot1), a potential tumor suppressor gene, and the gene for epsilon-sarcoglycan are maternally imprinted genes: identification by a subtractive screen of novel uniparental fibroblast lines. *Mol Cell Biol* 2000;**20**(9):3308–15.
113. Hoshiya H, Meguro M, Kashiwagi A, Okita C, Oshimura M, et al. Calcr, a brain-specific imprinted mouse calcitonin receptor gene in the imprinted cluster of the proximal region of chromosome 6. *J Hum Genet* 2003;**48**(4):208–11.
114. Ono R, Nakamura K, Inoue K, Naruse M, Usami T, Wakisaka-Saito N, et al. Deletion of Peg10, an imprinted gene acquired from a retrotransposon, causes early embryonic lethality. *Nat Genet* 2006;**38**(1):101–6.
115. Ribarska T, Ingenwerth M, Goering W, Engers R, Schulz WA, et al. Epigenetic inactivation of the placentally imprinted tumor suppressor gene TFPI2 in prostate carcinoma. *Cancer Genomics Proteomics* 2010;**7**(2):51–60.
116. Zimprich A, Grabowski M, Asmus F, Naumann M, Berg D, Bertram M, et al. Mutations in the gene encoding epsilon-sarcoglycan cause myoclonus-dystonia syndrome. *Nat Genet* 2001;**29**(1):66–9.
117. Riesewijk AM, Hu L, Schulz U, Tariverdian G, Höglund P, Kere J, et al. Monoallelic expression of human PEG1/MEST is paralleled by parent-specific methylation in fetuses. *Genomics* 1997;**42**(2):236–44.
118. Lefebvre L, Viville S, Barton SC, Ishino F, Keverne EB, Surani MA, et al. Abnormal maternal behaviour and growth retardation associated with loss of the imprinted gene Mest. *Nat Genet* 1998;**20**(2):163–9.
119. Meyer E, Wollmann HA, Eggermann T, et al. Searching for genomic variants in the MESTIT1 transcript in Silver-Russell syndrome patients. *J Med Genet* 2003;**40**(5):e65.

120. Parker-Katiraee L, Carson AR, Yamada T, Arnaud P, Feil R, Abu-Amero SN, et al. Identification of the imprinted KLF14 transcription factor undergoing human-specific accelerated evolution. *PLoS Genet* 2007;**3**(5):e65.
121. Hurst JA, Baraitser M, Auger E, Graham F, Norell S, et al. An extended family with a dominantly inherited speech disorder. *Dev Med Child Neurol* 1990;**32**(4):352–5.
122. Lai CS, Fisher SE, Hurst JA, Levy ER, Hodgson S, Fox M, et al. The SPCH1 region on human 7q31: genomic characterization of the critical interval and localization of translocations associated with speech and language disorder. *Am J Hum Genet* 2000;**67**(2):357–68.
123. Lai CS, Fisher SE, Hurst JA, Vargha-Khadem F, Monaco AP, et al. A forkhead-domain gene is mutated in a severe speech and language disorder. *Nature* 2001;**413**(6855):519–23.
124. Feuk L, Kalervo A, Lipsanen-Nyman M, Skaug J, Nakabayashi K, Finucane B, et al. Absence of a paternally inherited FOXP2 gene in developmental verbal dyspraxia. *Am J Hum Genet* 2006;**79**(5):965–72.
125. DeChiara TM, Efstratiadis A, Robertson EJ, et al. A growth-deficiency phenotype in heterozygous mice carrying an insulin-like growth factor II gene disrupted by targeting. *Nature* 1990;**345**(6270):78–80.
126. Hu JF, Vu TH, Hoffman AR, et al. Differential biallelic activation of three insulin-like growth factor II promoters in the mouse central nervous system. *Mol Endocrinol* 1995;**9**(5):628–36.
127. Bartolomei MS, Zemel S, Tilghman SM, et al. Parental imprinting of the mouse H19 gene. *Nature* 1991;**351**(6322):153–5.
128. Pant V, Mariano P, Kanduri C, Mattsson A, Lobanenkov V, Heuchel R, et al. The nucleotides responsible for the direct physical contact between the chromatin insulator protein CTCF and the H19 imprinting control region manifest parent of origin-specific long-distance insulation and methylation-free domains. *Genes Dev* 2003;**17**(5):586–90.
129. Schoenherr CJ, Levorse JM, Tilghman SM, et al. CTCF maintains differential methylation at the Igf2/H19 locus. *Nat Genet* 2003;**33**(1):66–9.
130. Thorvaldsen JL, Duran KL, Bartolomei MS, et al. Deletion of the H19 differentially methylated domain results in loss of imprinted expression of H19 and Igf2. *Genes Dev* 1998;**12**(23):3693–702.
131. Peñaherrera MS, Weindler S, Van llen MI, Yong SL, Metzger DL, McGillivray B, et al. Methylation profiling in individuals with Russell-Silver syndrome. *Am J Med Genet A* 2010;**152A**(2):347–55.
132. Netchine I, Rossignol S, Dufourg MN, Azzi S, Rousseau A, Perin L, et al. 11p15 imprinting center region 1 loss of methylation is a common and specific cause of typical Russell-Silver syndrome: clinical scoring system and epigenetic-phenotypic correlations. *J Clin Endocrinol Metab* 2007;**92**(8):3148–54.
133. Bruce S, Hannula-Jouppi K, Katarlina P, Mari F, Ingegerd S, Kalle OJ, et al. Clinically distinct epigenetic subgroups in Silver-Russell syndrome: the degree of H19 hypomethylation associates with phenotype severity and genital and skeletal anomalies. *J Clin Endocrinol Metab* 2009;**94**(2):579–87.
134. Lewis A, Mitsuya K, Umlauf D, Smith P, Dean W, Walter J, et al. Imprinting on distal chromosome 7 in the placenta involves repressive histone methylation independent of DNA methylation. *Nat Genet* 2004;**36**(12):1291–5.
135. Redrup L, Branco MR, Perdeaux MR, Krueger C, Lewis A, Santos F, et al. The long noncoding RNA Kcnq1ot1 organises a lineage-specific nuclear domain for epigenetic gene silencing. *Development* 2009;**136**(4):525–30.
136. Umlauf D, Goto Y, Cao R, Cerqueira F, Wagschal A, Zhang Y, et al. Imprinting along the Kcnq1 domain on mouse chromosome 7 involves repressive histone methylation and recruitment of Polycomb group complexes. *Nat Genet* 2004;**36**(12):1296–300.

137. Frost JM, Udayashankar R, Frost JM, Udayashankar R, Moore HD, Moore GE, et al. Telomeric NAP1L4 and OSBPL5 of the KCNQ1 cluster, and the DECORIN gene are not imprinted in human trophoblast stem cells. *PLoS ONE* 2010;**5**(7):e11595.
138. Wakeling EL, Abu-Amero S, Alders M, Bliek J, Forsyth E, Kumar S, et al. Epigenotype-phenotype correlations in Silver-Russell syndrome. *J Med Genet* 2010;**47**:766–8.
139. Eggermann T, Eggerman K, Schönherr N, et al. Growth retardation versus overgrowth: Silver-Russell syndrome is genetically opposite to Beckwith-Wiedemann syndrome. *Trends Genet* 2008;**24**(4):195–204.
140. Elliott M, Bayly R, Elliott M, Bayly R, Cole T, Temple IK, Maher ER, et al. Clinical features and natural history of Beckwith-Wiedemann syndrome: presentation of 74 new cases. *Clin Genet* 1994;**46**(2):168–74.
141. Elliott M, Maher ER. Beckwith-Wiedemann syndrome. *J Med Genet* 1994;**31**(7):560–4.
142. Pettenati MJ, Haines JL, Higgins RR, Wappner RS, Palmer CG, Weaver DD, et al. Wiedemann-Beckwith syndrome: presentation of clinical and cytogenetic data on 22 new cases and review of the literature. *Hum Genet* 1986;**74**(2):143–54.
143. Choufani S, Schuman C, Weksberg R, et al. Beckwith-Wiedemann syndrome. *Am J Med Genet C Semin Med Genet* 2010;**154C**(3):343–54.
144. Cooper WN, Luharia A, Evans GA, Raza H, Haire AC, Grundy R, et al. Molecular subtypes and phenotypic expression of Beckwith-Wiedemann syndrome. *Eur J Hum Genet* 2005;**13**(9):1025–32.
145. Enklaar T, Zabel BU, Prawitt D, et al. Beckwith-Wiedemann syndrome: multiple molecular mechanisms. *Expert Rev Mol Med* 2006;**8**(17):1–19.
146. Maher ER, Reik W. Beckwith-Wiedemann syndrome: imprinting in clusters revisited. *J Clin Invest* 2000;**105**(3):247–52.
147. Weksberg R, Shuman C, Smith AC, et al. Beckwith-Wiedemann syndrome. *Am J Med Genet C Semin Med Genet* 2005;**137C**(1):12–23.
148. Romanelli V, Belinchón A, Benito-Sanz S, Martinez-Glez V, Gracia-Bouthelier R, Heath KE, et al. CDKN1C (p57(Kip2)) analysis in Beckwith-Wiedemann syndrome (BWS) patients: Genotype-phenotype correlations, novel mutations, and polymorphisms. *Am J Med Genet A* 2010;**152A**(6):1390–7.
149. Cattanach BM, Kirk M. Differential activity of maternally and paternally derived chromosome regions in mice. *Nature* 1985;**315**(6019):496–8.
150. Höglund P, Holmberg C, de la C, hapelle A, Kere J, et al. Paternal isodisomy for chromosome 7 is compatible with normal growth and development in a patient with congenital chloride diarrhea. *Am J Hum Genet* 1994;**55**(4):747–52.
151. Le Caignec C, Isidor B, de Pontbriand U, David V, Audrezet MP, Ferec C, et al. Third case of paternal isodisomy for chromosome 7 with cystic fibrosis: a new patient presenting with normal growth. *Am J Med Genet A* 2007;**143A**(22):2696–9.
152. Pan Y, McCaskill CD, Thompson KH, Hicks J, Casey B, Shaffer LG, et al. Paternal isodisomy of chromosome 7 associated with complete situs inversus and immotile cilia. *Am J Hum Genet* 1998;**62**(6):1551–5.
153. Fares F, David M, Lerner A, Diukman R, Lerer I, Abeliovich D, et al. Paternal isodisomy of chromosome 7 with cystic fibrosis and overgrowth. *Am J Med Genet A* 2006;**140**(16):1785–8.
154. Temple IK, Cockwell A, Hassold T, Pettay D, Jacobs P, et al. Maternal uniparental disomy for chromosome 14. *J Med Genet* 1991;**28**(8):511–4.
155. Wang JC, Passage MB, Yen PH, Shapiro LJ, Mohandas TK, et al. Uniparental heterodisomy for chromosome 14 in a phenotypically abnormal familial balanced 13/14 Robertsonian translocation carrier. *Am J Hum Genet* 1991;**48**(6):1069–74.

156. Sutton VR, Shaffer LG. Search for imprinted regions on chromosome 14: comparison of maternal and paternal UPD cases with cases of chromosome 14 deletion. *Am J Med Genet* 2000;**93**(5):381–7.
157. da Rocha ST, Edwards CA, Ito M, Ogata T, Ferguson-Smith AC, et al. Genomic imprinting at the mammalian Dlk1-Dio3 domain. *Trends Genet* 2008;**24**(6):306–16.
158. Kagami M, Sekita Y, Nishimura G, Irie M, Kato F, Okada M, et al. Deletions and epimutations affecting the human 14q32.2 imprinted region in individuals with paternal and maternal upd (14)-like phenotypes. *Nat Genet* 2008;**40**(2):237–42.
159. Kagami M, O'Sullivan MJ, Green AJ, Watabe Y, Arisaka O, Masawa N, et al. The IG-DMR and the MEG3-DMR at human chromosome 14q32.2: hierarchical interaction and distinct functional properties as imprinting control centers. *PLoS Genet* 2010;**6**(6):e1000992.
160. Lin S-P, Coan P, da Rocha ST, Seitz H, Cavaille J, Teng P-W, et al. Differential regulation of imprinting in the murine embryo and placenta by the Dlk1-Dio3 imprinting control region. *Development* 2007;**134**(2):417–26.
161. Hosoki K, Ogata T, Kagami M, Tanaka T, Saitoh S, et al. Epimutation (hypomethylation) affecting the chromosome 14q32.2 imprinted region in a girl with upd(14)mat-like phenotype. *Eur J Hum Genet* 2008;**16**(8):1019–23.
162. Temple IK, Shrubb V, Lever M, Bullman H, Mackay DJG, et al. Isolated imprinting mutation of the DLK1/GTL2 locus associated with a clinical presentation of maternal uniparental disomy of chromosome 14. *J Med Genet* 2007;**44**(10):637–40.
163. Irving MD, Buiting K, Kanber D, Donaghue C, Schulz R, Offiah A, et al. Segmental paternal uniparental disomy (patUPD) of 14q32 with abnormal methylation elicits the characteristic features of complete patUPD14. *Am J Med Genet A* 2010;**152A**(8):1942–50.
164. Runte Färber M, Lich C, Zeschnigk M, Buchholz T, Smith A, et al. Comprehensive methylation analysis in typical and atypical PWS and AS patients with normal biparental chromosomes 15. *Eur J Hum Genet* 2001;**9**(7):519–26.
165. Meguro M, Mitsuya K, Nomura N, Kohda M, Kashiwagi A, Nishigaki R, et al. Large-scale evaluation of imprinting status in the Prader Willi syndrome region: an imprinted direct repeat cluster resembling small nucleolar RNA genes. *Hum Mol Genet* 2001;**10**(4):383–94.
166. DuBose AJ, Johnstone KA, Smith EY, Hallett RAE, Resnick JL, et al. Atp10a, a gene adjacent to the PWS/AS gene cluster, is not imprinted in mouse and is insensitive to the PWS-IC. *Neurogenetics* 2010;**11**(2):145–51.
167. Sutcliffe JS, Nakao M, Christian S, Orstavik KH, Tommerup N, Ledbetter DH, et al. Deletions of a differentially methylated CpG island at the SNRPN gene define a putative imprinting control region. *Nat Genet* 1994;**8**(1):52–8.
168. Buiting K, Saitoh S, Gross S, Dittrich B, Schwartz S, Nicholls RD, et al. Inherited microdeletions in the Angelman and Prader-Willi syndromes define an imprinting centre on human chromosome 15. *Nat Genet* 1995;**9**(4):395–400.
169. Buiting K, Gross S, Lich C, Gillessen-Kaesbach G, el-Maarri O, Horsthemke B, et al. Epimutations in Prader-Willi and Angelman syndromes: a molecular study of 136 patients with an imprinting defect. *Am J Hum Genet* 2003;**72**(3):571–7.
170. Kishino T, Lalande M, Wagstaff J, et al. UBE3A/E6-AP mutations cause Angelman syndrome. *Nat Genet* 1997;**15**(1):70–3.
171. Buiting K, Dittrich B, Gross S, Lich S, Färber C, Buchholz T, et al. Sporadic imprinting defects in Prader-Willi syndrome and Angelman syndrome: implications for imprint-switch models, genetic counseling, and prenatal diagnosis. *Am J Hum Genet* 1998;**63**(1):170–80.
172. Stefan M, Ji H, Simmons RA, Cummings DE, Ahima RS, Friedman MI, et al. Hormonal and metabolic defects in a prader-willi syndrome mouse model with neonatal failure to thrive. *Endocrinology* 2005;**146**(10):4377–85.

173. Doornbos M, Sikkema-Raddatz B, Ruijvenkamp CAL, Dijkhuizen T, Bijlsma EK, Gijsbers ACJ, et al. Nine patients with a microdeletion 15q11.2 between breakpoints 1 and 2 of the Prader-Willi critical region, possibly associated with behavioural disturbances. *Eur J Med Genet* 2009;**52**(2–3):108–15.
174. Kesler SR. Turner syndrome. *Child Adolesc Psychiatr Clin N Am* 2007;**16**(3):709–22.
175. Lippe B. Turner syndrome. *Endocrinol Metab Clin North Am* 1991;**20**(1):121–52.
176. Giltay JC, Maiburg MC. Klinefelter syndrome: clinical and molecular aspects. *Expert Rev Mol Diagn* 2010;**10**(6):765–76.
177. Davies W, Isles A, Smith R, Karunadasa D, Burrmann D, Humby T, et al. Xlr3b is a new imprinted candidate for X-linked parent-of-origin effects on cognitive function in mice. *Nat Genet* 2005;**37**(6):625–9.
178. Raefski AS, O'Neill MJ. Identification of a cluster of X-linked imprinted genes in mice. *Nat Genet* 2005;**37**(6):620–4.
179. Lyon MF. Gene action in the X-chromosome of the mouse (*Mus musculus* L.). *Nature* 1961;**190**:372–3.
180. Huynh KD, Lee JT. Imprinted X inactivation in eutherians: a model of gametic execution and zygotic relaxation. *Curr Opin Cell Biol* 2001;**13**(6):690–7.
181. Carrel L, Cottle AA, Goglin KC, Willard HF, et al. A first-generation X-inactivation profile of the human X chromosome. *Proc Natl Acad Sci USA* 1999;**96**(25):14440–4.
182. Disteche CM. Escapees on the X chromosome. *Proc Natl Acad Sci USA* 1999;**96**(25):14180–2.
183. Blaschke RJ, Rappold GA. SHOX: growth, Léri-Weill and Turner syndromes. *Trends Endocrinol Metab* 2000;**11**(6):227–30.
184. Bondy CA, Matura LA, Wooten N, Troendle J, Zinn AR, Bakalov VK, et al. The physical phenotype of girls and women with Turner syndrome is not X-imprinted. *Hum Genet* 2007;**121**(3–4):469–74.
185. Ko JM, Kim J-M, Kim GH, Lee BH, Yoo HW, et al. Influence of parental origin of the X chromosome on physical phenotypes and GH responsiveness of patients with Turner syndrome. *Clin Endocrinol (Oxf)* 2010;**73**(1):66–71.
186. Sagi L, Zuckerman-Levin N, Gawlik A, Ghizzoni L, Buyukgebiz A, Rakover Y, et al. Clinical significance of the parental origin of the X chromosome in turner syndrome. *J Clin Endocrinol Metab* 2007;**92**(3):846–52.
187. Hamelin CE, Anglin G, Quigley CA, Deal CL, et al. Genomic imprinting in Turner syndrome: effects on response to growth hormone and on risk of sensorineural hearing loss. *J Clin Endocrinol Metab* 2006;**91**(8):3002–10.
188. Skuse DH, James RS, Bishop DV, Coppin B, Dalton P, Aamodt-Leeper G, et al. Evidence from Turner's syndrome of an imprinted X-linked locus affecting cognitive function. *Nature* 1997;**387**(6634):705–8.
189. Cutter WJ, Daly EM, Robertson DMW, Chitnis XA, van melsvoort TAMJA, Simmons A, et al. Influence of X chromosome and hormones on human brain development: a magnetic resonance imaging and proton magnetic resonance spectroscopy study of Turner syndrome. *Biol Psychiatry* 2006;**59**(3):273–83.
190. Kesler SR, Blasey CM, Brown WE, Yankowitz J, Zeng SM, Bender BG, et al. Effects of X-monosomy and X-linked imprinting on superior temporal gyrus morphology in Turner syndrome. *Biol Psychiatry* 2003;**54**(6):636–46.
191. Brown WE, Kesler SR, Eliez S, Warsofsky IS, Haberecht M, Patwardhan A, et al. Brain development in Turner syndrome: a magnetic resonance imaging study. *Psychiatry Res* 2002;**116**(3):187–96.
192. Good CD, Lawrence K, Thomas NS, Price CJ, Ashburner J, Friston KJ, et al. Dosage-sensitive X-linked locus influences the development of amygdala and orbitofrontal cortex, and fear recognition in humans. *Brain* 2003;**126**(Pt 11):2431–46.

193. Kesler SR, Garrett A, Bender B, Yankowitz J, Zeng SM, Reiss AL, et al. Amygdala and hippocampal volumes in Turner syndrome: a high-resolution MRI study of X-monosomy. *Neuropsychologia* 2004;**42**(14):1971–8.
194. Bishop DV, Canning E, Elgar K, Morris E, Jacobs PA, Skuse DH, et al. Distinctive patterns of memory function in subgroups of females with Turner syndrome: evidence for imprinted loci on the X-chromosome affecting neurodevelopment. *Neuropsychologia* 2000;**38**(5):712–21.
195. Loesch DZ, Bui QM, Kelso W, Huggins RM, Slater H, Warne G, et al. Effect of Turner's syndrome and X-linked imprinting on cognitive status: analysis based on pedigree data. *Brain Dev* 2005;**27**(7):494–503.
196. Ergür AT, Ocal G, Berberoglu M, Tekin M, Kiliç BG, Aycan Z, et al. Paternal X could relate to arithmetic function; study of cognitive function and parental origin of X chromosome in Turner syndrome. *Pediatr Int* 2008;**50**(2):172–4.
197. Burnett AC, Reutens DC, Wood AG, et al. Social cognition in Turner's Syndrome. *J Clin Neurosci* 2010;**17**(3):283–6.
198. Mullaney R, Murphy D. Turner syndrome: neuroimaging findings: structural and functional. *Dev Disabil Res Rev* 2009;**15**(4):279–83.
199. Stemkens D, Roza T, Verrij L, Swaab H, van Werkhoven MK, Alizadeh BZ, et al. Is there an influence of X-chromosomal imprinting on the phenotype in Klinefelter syndrome? A clinical and molecular genetic study of 61 cases. *Clin Genet* 2006;**70**(1):43–8.
200. Wikström AM, Painter JM, Raivio Aittomäki K, Dunkel L, et al. Genetic features of the X chromosome affect pubertal development and testicular degeneration in adolescent boys with Klinefelter syndrome. *Clin Endocrinol (Oxf)* 2006;**65**(1):92–7.
201. Mackay DJG, Temple IK. Transient neonatal diabetes mellitus type 1. *Am J Med Genet C Semin Med Genet* 2010;**154C**(3):335–42.
202. Valleley EM, Cordery SF, Bonthron DT, et al. Tissue-specific imprinting of the ZAC/PLAGL1 tumour suppressor gene results from variable utilization of monoallelic and biallelic promoters. *Hum Mol Genet* 2007;**16**(8):972–81.
203. Abdollahi A. LOT1 (ZAC1/PLAGL1) and its family members: mechanisms and functions. *J Cell Physiol* 2007;**210**(1):16–25.
204. Flanagan SE, Patch A-M, Mackay DJG, Edghill EL, Gloyn AL, Robinson D, et al. Mutations in ATP-sensitive K+ channel genes cause transient neonatal diabetes and permanent diabetes in childhood or adulthood. *Diabetes* 2007;**56**(7):1930–7.
205. Temple IK, Shield JPH. 6q24 transient neonatal diabetes. *Rev Endocr Metab Disord* 2010;**11**:199–204.
206. Abdollahi A, Pisarcik D, Roberts D, Weinstein J, Cairns P, Hamilton TC, et al. LOT1 (PLAGL1/ZAC1), the candidate tumor suppressor gene at chromosome 6q24-25, is epigenetically regulated in cancer. *J Biol Chem* 2003;**278**(8):6041–9.
207. Diatloff-Zito C, Nicole A, Marcelin G, Labit H, Marquis E, Bellanné-Chantelot C, et al. Genetic and epigenetic defects at the 6q24 imprinted locus in a cohort of 13 patients with transient neonatal diabetes: new hypothesis raised by the finding of a unique case with hemizygotic deletion in the critical region. *J Med Genet* 2007;**44**(1):31–7.
208. Haig D, Wharton R. Prader-Willi syndrome and the evolution of human childhood. *Am J Hum Biol* 2003;**15**(3):320–9.
209. Keverne EB, Fundele R, Narasimha M, Barton SC, Surani MA, et al. Genomic imprinting and the differential roles of parental genomes in brain development. *Brain Res Dev Brain Res* 1996;**92**(1):91–100.
210. Goos LM, Silverman I. The inheritance of cognitive skills: does genomic imprinting play a role? *J Neurogenet* 2006;**20**(1–2):19–40.
211. Li L, Keverne EB, Aparicio SA, Ishino F, Barton SC, Surani MA, et al. Regulation of maternal behavior and offspring growth by paternally expressed Peg3. *Science* 1999;**284**(5412):330–3.

212. Jiang YH, Armstrong D, Albrecht U, Atkins CM, Noebels JL, Eichele G, et al. Mutation of the Angelman ubiquitin ligase in mice causes increased cytoplasmic p53 and deficits of contextual learning and long-term potentiation. *Neuron* 1998;**21**(4):799–811.
213. Brambilla R, Gnesutta N, Minichiello L, White G, Roylance AJ, Herron CE, et al. A role for the Ras signalling pathway in synaptic transmission and long-term memory. *Nature* 1997;**390** (6657):281–6.
214. Li S, Tian X, Hartley DM, Feig LA, et al. Distinct roles for Ras-guanine nucleotide-releasing factor 1 (Ras-GRF1) and Ras-GRF2 in the induction of long-term potentiation and long-term depression. *J Neurosci* 2006;**26**(6):1721–9.
215. Muscatelli F, Abrous DN, Massacrier A, Boccaccio I, Le M, oal M, Cau P, et al. Disruption of the mouse Necdin gene results in hypothalamic and behavioral alterations reminiscent of the human Prader-Willi syndrome. *Hum Mol Genet* 2000;**9**(20):3101–10.
216. Badcock C, Crespi B. Imbalanced genomic imprinting in brain development: an evolutionary basis for the aetiology of autism. *J Evol Biol* 2006;**19**(4):1007–32.
217. Crespi B. Genomic imprinting in the development and evolution of psychotic spectrum conditions. *Biol Rev Camb Philos Soc* 2008;**83**(4):441–93.
218. Crespi B, Badcock C. Psychosis and autism as diametrical disorders of the social brain. *Behav Brain Sci* 2008;**31**(3):241–61 discussion 261–320.
219. Schanen NC. Epigenetics of autism spectrum disorders. *Hum Mol Genet* 2006;**15**(2): R138–50.
220. Lewis CM, Levinson DF, Wise LH, DeLisi LE, Straub RE, Hovatta I, et al. Genome scan meta-analysis of schizophrenia and bipolar disorder, part II: Schizophrenia. *Am J Hum Genet* 2003;**73**(1):34–48.
221. Francks C, Maegawa S, Laurén J, Abrahams BS, Velayos-Baeza A, Medland SE, et al. LRRTM1 on chromosome 2p12 is a maternally suppressed gene that is associated paternally with handedness and schizophrenia. *Mol Psychiatry* 2007;**12**(12):1129–39 1057.
222. Francks C, DeLisi LE, Shaw SH, Fisher SE, Richardson AJ, Stein JF, et al. Parent-of-origin effects on handedness and schizophrenia susceptibility on chromosome 2p12-q11. *Hum Mol Genet* 2003;**12**(24):3225–30.
223. Donnelly SL, Wolpert CM, Menold MM, Bass MP, Gilbert JR, Cuccaro ML, et al. Female with autistic disorder and monosomy X (Turner syndrome): parent-of-origin effect of the X chromosome. *Am J Med Genet* 2000;**96**(3):312–6.
224. Marco EJ, Skuse DH. Autism-lessons from the X chromosome. *Soc Cogn Affect Neurosci* 2006;**1**(3):183–93.
225. Goos LM, Ragsdale G. Genomic imprinting and human psychology: cognition, behavior and pathology. *Adv Exp Med Biol* 2008;**626**:71–88.
226. Cassidy SB, Morris CA. Behavioral phenotypes in genetic syndromes: genetic clues to human behavior. *Adv Pediatr* 2002;**49**:59–86.
227. Karno M, Golding JM, Sorenson SB, Burnam MA, et al. The epidemiology of obsessive-compulsive disorder in five US communities. *Arch Gen Psychiatry* 1988;**45**(12):1094–9.
228. Alsobrook JP, Zohar AH, Leboyer M, Chabane N, Ebstein RP, Pauls DL, et al. Association between the COMT locus and obsessive-compulsive disorder in females but not males. *Am J Med Genet* 2002;**114**(1):116–20.
229. State MW, Dykens EM, Rosner B, Martin A, King BH, et al. Obsessive-compulsive symptoms in Prader-Willi and "Prader-Willi-Like" patients. *J Am Acad Child Adolesc Psychiatry* 1999;**38** (3):329–34.
230. Cavaillé J, Buiting K, Kiefmann M, Lalande M, Brannan CI, Horsthemke B, et al. Identification of brain-specific and imprinted small nucleolar RNA genes exhibiting an unusual genomic organization. *Proc Natl Acad Sci USA* 2000;**97**(26):14311–6.

231. Swanson JM, Sergeant JA, Taylor E, Sonuga-Barke EJ, Jensen PS, Cantwell DP, et al. Attention-deficit hyperactivity disorder and hyperkinetic disorder. *Lancet* 1998;**351**(9100):429–33.
232. Wigren M, Hansen S. ADHD symptoms and insistence on sameness in Prader-Willi syndrome. *J Intellect Disabil Res* 2005;**49**(Pt 6):449–56.
233. Lichter DG, Jackson LA, Schachter M, et al. Clinical evidence of genomic imprinting in Tourette's syndrome. *Neurology* 1995;**45**(5):924–8.
234. Kent L, Green E, Hawi Z, Kirley A, Dudbridge F, Lowe N, et al. Association of the paternally transmitted copy of common Valine allele of the Val66Met polymorphism of the brain-derived neurotrophic factor (BDNF) gene with susceptibility to ADHD. *Mol Psychiatry* 2005;**10**(10):939–43.
235. Blackman GL, Ostrander R, Herman KC, et al. Children with ADHD and depression: a multisource, multimethod assessment of clinical, social, and academic functioning. *J Atten Disord* 2005;**8**(4):195–207.
236. Doyle AE, Faraone SV. Familial links between attention deficit hyperactivity disorder, conduct disorder, and bipolar disorder. *Curr Psychiatry Rep* 2002;**4**(2):146–52.
237. Faraone SV, Biederman J. Neurobiology of attention-deficit hyperactivity disorder. *Biol Psychiatry* 1998;**44**(10):951–8.
238. Dykens EM, Cassidy SB. Correlates of maladaptive behavior in children and adults with Prader-Willi syndrome. *Am J Med Genet* 1995;**60**(6):546–9.
239. Watanabe H, Ohmori O, Abe K, et al. Recurrent brief depression in Prader-Willi syndrome: a case report. *Psychiatr Genet* 1997;**7**(1):41–4.
240. McMahon FJ, Stine OC, Meyers DA, Simpson SG, DePaulo JR, et al. Patterns of maternal transmission in bipolar affective disorder. *Am J Hum Genet* 1995;**56**(6):1277–86.
241. Børglum AD, Kirov G, Craddock N, Mors O, Muir W, Murray V, et al. Possible parent-of-origin effect of Dopa decarboxylase in susceptibility to bipolar affective disorder. *Am J Med Genet B Neuropsychiatr Genet* 2003;**117B**(1):18–22.
242. Zill P, Baghai TC, Zwanzger P, Schüle C, Eser D, Rupprecht R, et al. SNP and haplotype analysis of a novel tryptophan hydroxylase isoform (TPH2) gene provide evidence for association with major depression. *Mol Psychiatry* 2004;**9**(11):1030–6.
243. Koponen E, Rantamäki T, Voikar V, Saarelainen T, MacDonald E, Castrén E, et al. Enhanced BDNF signaling is associated with an antidepressant-like behavioral response and changes in brain monoamines. *Cell Mol Neurobiol* 2005;**25**(6):973–80.
244. Mattson MP, Maudsley S, Martin B, et al. BDNF and 5-HT: a dynamic duo in age-related neuronal plasticity and neurodegenerative disorders. *Trends Neurosci* 2004;**27**(10):589–94.
245. Jirtle RL, Sander M, Barrett JC, et al. Genomic imprinting and environmental disease susceptibility. *Environ Health Perspect* 2000;**108**(3):271–8.
246. Moss TJ, Wallrath LL. Connections between epigenetic gene silencing and human disease. *Mutat Res* 2007;**618**(1–2):163–74.
247. Úbeda F, Wilkins JF. Imprinted genes and human disease: an evolutionary perspective. *Adv Exp Med Biol* 2008;**626**:101–15.
248. Jablonka E, Lamb MJ. Epigenetic inheritance in evolution. *J Evol Biol* 1998;**11**(2):159–83.
249. Wilkins JF. Antagonistic coevolution of two imprinted loci with pleiotropic effects. *Evolution* 2009;**64**(1):142–51.
250. Newman JRS, Ghaemmaghami S, Ihmels J, Breslow DK, Noble M, DeRisi JL, et al. Single-cell proteomic analysis of S. cerevisiae reveals the architecture of biological noise. *Nature* 2006;**441**(7095):840–6.
251. Spielman RS, Bastone LA, Burdick JT, Morley M, Ewens WJ, Cheung VG, et al. Common genetic variants account for differences in gene expression among ethnic groups. *Nat Genet* 2007;**39**(2):226–31.

252. Siegal ML, Bergman A. Waddington's canalization revisited: developmental stability and evolution. *Proc Natl Acad Sci USA* 2002;**99**(16):10528–32.
253. Waddington CH. Canalization of development and the inheritance of acquired characters. *Nature* 1942;**150**:563–5.
254. Waddington CH. Canalization of development and genetic assimilation of acquired characters. *Nature* 1959;**183**(4676):1654–5.
255. Wilkins JF. Genomic imprinting and conflict-induced decanalization. *Evolution* 2010;**65**:537–53.

DNA Methylation Changes in Cancer

Minoru Toyota and
Eiichiro Yamamoto

*Department of Biochemistry,
Sapporo Medical University, Chuo-ku,
Sapporo, Japan*

DNA methylation of CpG islands, together with deacetylation of histone and methylation of histone H3 lysines 9 and 27 (K9/K27), can lead to silencing of tumor-suppressor genes. The mechanisms underlying DNA methylation changes in cancer involve alteration of the activity of DNA methyltransferases (DNMTs), inflammation, and viral infection. DNA methylation affects genes involved in cell-cycle checkpoints, apoptosis, angiogenesis, invasion, immune responses, and cellular signaling. Subsets of cancers show DNA methylation of multiple genes, indicating that these tumors have the CpG island methylator phenotype (CIMP). Cancers with CIMP show distinct genetic changes, including microsatellite instability and mutations in the BRAF Ser/Thr kinase gene. Repetitive sequences such as short and long interspersed repeat elements are often hypomethylated in cancer, and are implicated in chromosomal instability. DNA methylation is a reversible phenomenon, and DNMT inhibitors can induce gene expression due to demethylation.

I. Introduction

DNA methylation plays a role in epigenetic gene regulation and has been implicated in genome imprinting, X chromosome inactivation, and silencing of retrotransposon.[1]

Gene promoters often contain CpG-rich sequences (called CpG islands) in their 5′ regions.[2] In normal cells, CpG islands are unmethylated. However, tumors often show aberrant methylation of tumor-suppressor genes. Recent

Progress in Molecular Biology
and Translational Science, Vol. 101
DOI: 10.1016/B978-0-12-387685-0.00014-7

1877-1173/11 $35.00

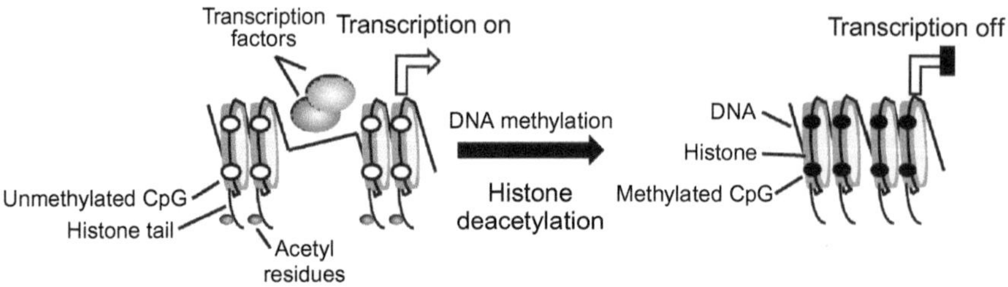

FIG. 1. *Gene silencing by DNA methylation and histone deacetylation*. In normal cells, the 5′ region of the gene is unmethylated, which allows transcription factors to bind promoter region and initiate gene transcription. In cancer cells, the promoter region is methylated, which results in histone deacetylation and gene silencing.

studies have shown that DNA methylation of CpG islands, together with deacetylation of histone and methylation of histone H3K9/K27, leads to gene silencing (Fig. 1).[3] DNA methylation can deactivate about half of tumor-suppressor genes. In this review, we focus on the role of DNA methylation changes in carcinogenesis. We also discuss the molecular mechanisms of DNA methylation changes in cancers.

II. Mechanism Underlying DNA Methylation Changes in Cancer

A. The Role of DNMTs in Cancer

Although the causes of aberrant DNA methylation are largely unknown, several studies indicate that DNA methyltransferases (DNMTs) might play a role in aberrant methylation in cancer. DNA methylation is catalyzed by three DNMTs: DNMT1, DNMT3A, and DNMT3B (see Chapters by Željko M. Svedružić; and Frédéric Chédin). Heterozygotic mice with null mutation of Dnmt1, when treated with 5-aza-dC reduced tumor formation in Apc Min+/− mice.[4] In addition, a Dnmt1 hypomorphic allele (causing partial loss of function) can suppress polyp formation and CpG island methylation.[5] These results indicated that Dnmt1 plays a critical role in murine tumor formation. Expression levels of DNMT1 are significantly elevated in various types of tumors.[6–9] Overexpression of DNMT1 was correlated with the methylation of multiple genes, indicating that DNMT1 may play a role in the *de novo* methylation of genes.[7] Overexpression of DNMT1 was also found in premalignant lesions such as colorectal adenoma, liver cirrhosis, and pancreatic intraepithelial neoplasias.[6,9] Overexpression of DNMT1 may not be a secondary event due to increased cell proliferation, because DNMT1 expression is independent of cell proliferation as determined by the PCNA-labeling index.[10]

The *de novo* DNMTs, DNMT3A and DNMT3B, also play roles in tumorigenesis. Conditional knockout of Dnmt3B in mice decreases the development of macroscopic adenomas.[11] Dnmt3B overexpression in Apc Min+/− mice increases the incidence and size of tumors.[12] DNMT3B expression is upregulated in various types of tumors including those of the colon, breast, and lung.[13–15] The role of DNMT3A in aberrant methylation remains to be determined, but it is likely that DNMT3A also plays a role in methylation changes in cancer. Mice overexpressing Dnmt3a when crossed with Min mice develop adenomatous polyposis. The tumors show a high degree of secreted frizzled-related protein 5 (SFRP5) methylation.[16]

B. Inflammation and Viral Infection

Environmental factors such as inflammation and viral infection may also play roles in methylation changes in cancer. Aberrant DNA methylation has been found in noncancerous colorectal mucosae of patients with inflammatory bowel disease.[17] Infection of *Helicobacter pylori* is associated with increased levels of DNA methylation and tumorigenesis of gastric cancers.[18] Gastric cancer associated with Epstein–Barr virus shows a high degree of DNA methylation.[19] Hepatocellular carcinoma from China and Egypt showed higher frequencies of CpG island methylation than tumors from the United States and Europe, indicating that environmental factors may influence the degree of methylation in hepatocellular carcinomas.[20] The molecular mechanisms underlying induction of aberrant methylation by inflammation remain to be determined. It is plausible that inflammation simply affects cellular proliferation, which increases the chance for epigenetic errors to occur. Alternatively, cellular signaling involved in inflammation may affect epigenetic machinery to alter DNA methylation.

III. The Roles of DNA Methylation in Cancer Signaling Pathways

DNA methylation affects many genes involved in cell-cycle checkpoints, apoptosis, angiogenesis, invasion, and immune response (Fig. 2).[21,22] During the course of genetic screening to identify genes inactivated by DNA methylation, we have found that SFRP1, SFRP2, and SFRP5 are silenced in colorectal cancer.[23] SFRPs are secreted proteins that interact with WNT and its receptor frizzled in order to suppress the WNT signal. Ectopic expression of SFRPs in colorectal cancer cells that have mutations of APC or β-catenin suppress the transcriptional activity of T-cell factor (TCF).

RAS is frequently activated in human tumors, and is involved in cell proliferation, transformation, and invasion.[24] KRAS and BRAF are often activated by mutations.[25] In addition, negative regulators of the Ras signaling

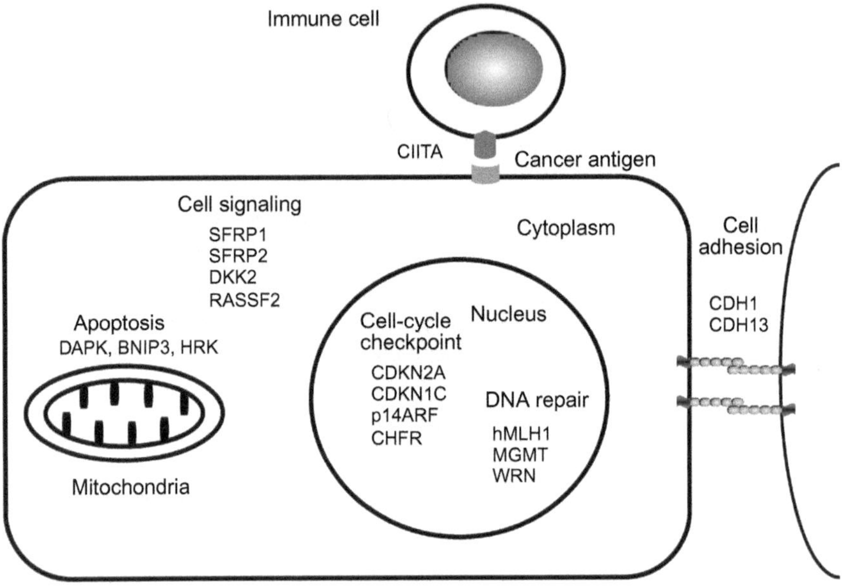

FIG. 2. *Cellular pathways altered by DNA methylation in cancer.* In cancer cells, genes involved in cell-cycle checkpoints, apoptosis, DNA repair, cancer immunity, and cell adhesion are silenced by DNA methylation.

pathway are often silenced by DNA methylation in cancer.[26,27] RASSF1 was first identified as a candidate tumor-suppressor gene on chromosome 3p.[27] One of the 10 RASSF family genes, RASSF1, contains the Ras association domain. RASSF1 is frequently methylated in lung cancer, while RASSF2 is methylated in colorectal, gastric, and oral squamous cell cancers.[26,28,29]

p53 is one of the most frequently mutated tumor-suppressor genes in human neoplasias.[30] In colorectal cancers, about 50% of the cases show mutations of p53, although half of colorectal tumors also show impaired p53 function. Various reports suggest that target genes of p53 are silenced by DNA methylation in cancer. These genes include 14-3-3σ,[31] a cell-cycle checkpoint gene; p14ARF, a gene involved in the stability of p53[32]; and GADD45β, a gene involved in cellular stress.[33]

microRNAs (miRNAs) are short RNAs that regulate the expression of target genes by repressing translation and inhibiting RNA stability. The downregulation of miRNAs is often found in a variety of tumors, indicating that epigenetic mechanisms may be involved in silencing miRNAs. For example, mir-124 was silenced by DNA methylation in bladder cancers.[34] Treatment of cancer cells that lack expression of mir-124 due to DNA methylation results in the downregulation of bcl-6, one of the target genes of mir-124. By screening miRNAs upregulated in DNMT knockout cancer cells, we have found that miR-34b/c, direct transcriptional targets of p53,[35] are frequently silenced in colorectal cancers.[36] Introducing miR-34b/c into cancer cells resulted in decreased expression of CDK6, a cyclin-dependent kinase, and MET, a receptor tyrosine kinase. Epigenetic inactivation of miR-34b/c may then attenuate p53 function, through the dysregulation of the cell cycle and cell growth. Screening for miRNAs downregulated in oral squamous cell carcinoma revealed that mir-137 and mir-193a were frequently silenced by DNA methylation.[37] Because the transcription start sites of pre-miRNA are often unknown, analysis of DNA methylation of miRNA requires detailed analysis of the 5′ends of the genes as well as promoter analysis.

IV. Genetic and Epigenetic Interaction in Cancer

About 10–15% of colorectal cancers show microsatellite instability (MSI). However, mutations of mismatch repair genes are rarely detected in sporadic colorectal cancer with MSI, and another mechanism may be involved in mismatch repair deficiency. In fact, the methylation of hMLH1 turns out to be a major mode of mismatch repair deficiency.[38,39] We have found that subsets of colorectal cancer show the methylation of multiple CpG islands, which we termed the CpG island methylator phenotype (CIMP).[40] The methylation of hMLH1 was tightly associated with CIMP, indicating that a genome-wide methylation defect causes genetic instability. In addition, CIMP-positive tumors show frequent mutations of BRAF.[41] The molecular mechanisms underlying the high concordance between BRAF mutations and CIMP remain to be determined. One plausible explanation is that CIMP may be involved in silencing of the genes in which BRAF has induced cellular senescence (Fig. 3). In fact, IGFBP7 is frequently silenced in tumors with BRAF mutations.[42,43]

Genetic and epigenetic interactions are also found in other types of tumors. A subset of glioma cases show CIMP,[44] indicating that these tumors have a form of CIMP (glioma-CIMP, G-CIMP). G-CIMP cases show a high frequency of mutations of IDH1 and good prognosis.

V. Hypomethylation in Cancer

Cancer often shows genome-wide DNA hypomethylation, though neither the underlying molecular mechanism nor the role of hypomethylation is fully understood. Mice carrying the hypomorphic allele of Dnmt1 show genome-wide hypomethylation.[45] These mice develop lymphoma and exhibit a high frequency of trisomy of chromosome 15, indicating that hypomethylation plays a role in tumorigenesis by inducing genomic instability.

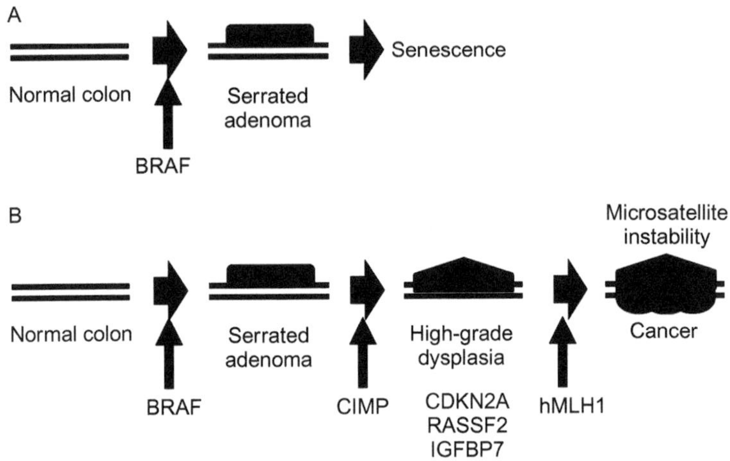

FIG. 3. *CIMP and MSI in colorectal tumorigenesis.* (A) Early colorectal lesions with BRAF mutation alone do not progress to invasive cancer because of the induction of cellular senescence. (B) In colorectal tumors with CIMP, genes involved in cellular senescence are silenced, and tumors progress.

Repetitive sequences such as short and long interspersed repeat elements (SINEs and LINEs) are often hypomethylated in cancer[46,47] (see Chapter by Jon F. Wilkins and Francisco Úbeda). It is generally thought that genome-wide hypomethylation is one of the causes of chromosomal instability in cancer (Fig. 4). Hypomethylation of DNA is found in early lesions including premalignant lesions, such as colorectal adenomas and early breast cancers.[48,49] We have found that LINE1 is hypomethylated in enlarged-fold gastritis, which is a high-risk form of gastric cancer.[50] These results indicated that hypomethylation in noncancerous tissues can be a predictive molecular biomarker for cancer.

Structural changes in these enzymes are often associated with diseases. Mutations of DNMT3B are the cause of immunodeficiency, centromeric region instability, and facial anomalies (ICF) syndrome[51,52] (see Chapter by Jon F. Wilkins and Francisco Úbeda). Patients with ICF syndrome show agammaglobulinemia with B cells, as well as DNA rearrangements of the centromere-adjacent heterochromatic regions of chromosomes 1, 16, and 9 in mitogen-stimulated lymphocytes.[53] Mice carrying mutant Dnmt3b show hypomethylation of the centromeric minor satellite repeat, and can be a model for ICF syndrome.[52] DNMT3A is mutated in a subset of acute myeloid leukemia.[54] Specifically, Arg882His mutation of DNMT3A results in decreased DNA methylation activity. Mutations of DNMT3A may thus be a cause of hypomethylation in tumors.

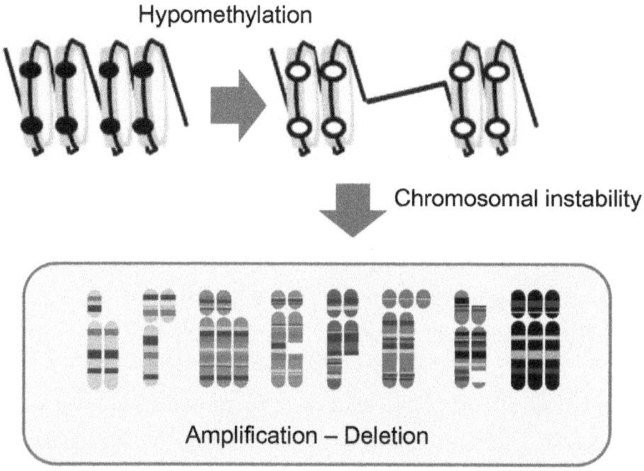

FIG. 4. *The role of hypomethylation of retrotransposons in chromosomal instability.* About 70% of methylated CpG sites in normal cells are located in retrotransposons, such as Alu and LINE1. Demethylation of retrotransposons in cancer causes genome-wide hypomethylation and induces chromosomal instability.

Although hypomethylation of retrotransposons has been associated with chromosomal instability in general, hypomethylation of a specific LINE1 sequence results in activation of cancer-related genes. Specifically, the LINE1 promoter located within the MET proto-oncogene can activate an alternative transcript of MET.[55] In normal cells, LINE1–MET is methylated. In bladder cancer cell lines and primary tumors, LINE1–MET is hypomethylated, which results in gene activation. These results indicated that hypomethylation of retrotransposon is involved not only in chromosomal instability but also in aberrant expression of the gene (see Chapter by Mary G. Goll and Marnie E. Halpern).

VI. Epigenetic Therapy

DNA methylation is a reversible phenomenon, and DNMT inhibitors such as 5-azacytidine or 5′-aza-2′ deoxycytidine can induce gene expression due to demethylation. Recently, 5-azacytidine and 5′-aza-2′ deoxycytidine have been used to treat hematopoietic malignancies such as myelodysplastic syndrome.[56] Histone deacetylase inhibitors, such as suberoylanilide hydroxamic acid (SAHA), have been used to treat cutaneous T-cell lymphoma (CTCL),[57] and combined treatment with a DNMT inhibitor and a histone deacetylase inhibitor has a synergistic antitumor effect in at least some cases.[58–60] Patients with myelodysplastic syndrome and high levels of CpG island methylation have poorer prognoses.[61] Obviously, a genome-wide approach to identify the molecular markers with which to predict responses to DNMT inhibitors is needed.

VII. Conclusions and Future Directions

Although molecular mechanisms of DNA hypermethylation in cancer have been studied, little is known about DNA demethylation. Recently, several genes have been shown to be involved in the active DNA demethylation process, and one of them is often mutated in neoplasia.[62] New technology to detect the active DNA demethylation process may facilitate an understanding of dynamic DNA methylation changes in cancer.

Acknowledgments

We apologize to those of our colleagues whose contributions could not be cited due to space constraints. We thank KN International for editing the chapter. This study was supported by Grants-in-Aid for Scientific Research from the Ministry of Education, Culture, Sports, Science, and Technology of Japan; a Grant-in-Aid for the Third-Term Cancer Control Strategy; and a Grant-in-Aid for Cancer Research from the Ministry of Health, Labor, and Welfare, Japan.

References

1. Reik W, Dean W, Walter J. Epigenetic reprogramming in mammalian development. *Science* 2001;**293**:1089–93.
2. Takai D, Jones PA. Comprehensive analysis of CpG islands in human chromosomes 21 and 22. *Proc Natl Acad Sci USA* 2002;**99**:3740–5.
3. Jones PA, Baylin SB. The epigenomics of cancer. *Cell* 2007;**128**:683–92.
4. Laird PW, Jackson-Grusby L, Fazeli A, Dickinson SL, Jung WE, Li E, et al. Suppression of intestinal neoplasia by DNA hypomethylation. *Cell* 1995;**81**:197–205.
5. Eads CA, Nickel AE, Laird PW. Complete genetic suppression of polyp formation and reduction of CpG-island hypermethylation in Apc(Min/+) Dnmt1-hypomorphic mice. *Cancer Res* 2002;**62**:1296–9.
6. Kanai Y, Ushijima S, Kondo Y, Nakanishi Y, Hirohashi S. DNA methyltransferase expression and DNA methylation of CPG islands and peri-centromeric satellite regions in human colorectal and stomach cancers. *Int J Cancer* 2001;**91**:205–12.
7. Peng DF, Kanai Y, Sawada M, Ushijima S, Hiraoka N, Kitazawa S, et al. DNA methylation of multiple tumor-related genes in association with overexpression of DNA methyltransferase 1 (DNMT1) during multistage carcinogenesis of the pancreas. *Carcinogenesis* 2006;**27**:1160–8.
8. Peng H, Shen N, Qian L, Sun XL, Koduru P, Goodwin LO, et al. Hypermethylation of CpG islands in the mouse asparagine synthetase gene: relationship to asparaginase sensitivity in lymphoma cells. Partial methylation in normal cells. *Br J Cancer* 2001;**85**:930–5.

9. Saito Y, Kanai Y, Sakamoto M, Saito H, Ishii H, Hirohashi S. Expression of mRNA for DNA methyltransferases and methyl-CpG-binding proteins and DNA methylation status on CpG islands and pericentromeric satellite regions during human hepatocarcinogenesis. *Hepatology* 2001;**33**:561–8.
10. Nakagawa T, Kanai Y, Saito Y, Kitamura T, Kakizoe T, Hirohashi S. Increased DNA methyltransferase 1 protein expression in human transitional cell carcinoma of the bladder. *J Urol* 2003;**170**:2463–6.
11. Lin H, Yamada Y, Nguyen S, Linhart H, Jackson-Grusby L, Meissner A, et al. Suppression of intestinal neoplasia by deletion of Dnmt3b. *Mol Cell Biol* 2006;**26**:2976–83.
12. Linhart HG, Lin H, Yamada Y, Moran E, Steine EJ, Gokhale S, et al. Dnmt3b promotes tumorigenesis in vivo by gene-specific de novo methylation and transcriptional silencing. *Genes Dev* 2007;**21**:3110–22.
13. Girault I, Tozlu S, Lidereau R, Bieche I. Expression analysis of DNA methyltransferases 1, 3A, and 3B in sporadic breast carcinomas. *Clin Cancer Res* 2003;**9**:4415–22.
14. Lin RK, Hsu HS, Chang JW, Chen CY, Chen JT, Wang YC. Alteration of DNA methyltransferases contributes to 5′CpG methylation and poor prognosis in lung cancer. *Lung Cancer* 2007;**55**:205–13.
15. Nosho K, Shima K, Irahara N, Kure S, Baba Y, Kirkner GJ, et al. DNMT3B expression might contribute to CpG island methylator phenotype in colorectal cancer. *Clin Cancer Res* 2009;**15**:3663–71.
16. Samuel MS, Suzuki H, Buchert M, Putoczki TL, Tebbutt NC, Lundgren-May T, et al. Elevated Dnmt3a activity promotes polyposis in Apc(Min) mice by relaxing extracellular restraints on Wnt signaling. *Gastroenterology* 2009;**137**:902–13, 913.e901–913.e911.
17. Issa JP, Ahuja N, Toyota M, Bronner MP, Brentnall TA. Accelerated age-related CpG island methylation in ulcerative colitis. *Cancer Res* 2001;**61**:3573–7.
18. Maekita T, Nakazawa K, Mihara M, Nakajima T, Yanaoka K, Iguchi M, et al. High levels of aberrant DNA methylation in Helicobacter pylori-infected gastric mucosae and its possible association with gastric cancer risk. *Clin Cancer Res* 2006;**12**:989–95.
19. Kusano M, Toyota M, Suzuki H, Akino K, Aoki F, Fujita M, et al. Genetic, epigenetic, and clinicopathologic features of gastric carcinomas with the CpG island methylator phenotype and an association with Epstein-Barr virus. *Cancer* 2006;**106**:1467–79.
20. Shen L, Ahuja N, Shen Y, Habib NA, Toyota M, Rashid A, et al. DNA methylation and environmental exposures in human hepatocellular carcinoma. *J Natl Cancer Inst* 2002;**94**:755–61.
21. Esteller M. Epigenetics in cancer. *N Engl J Med* 2008;**358**:1148–59.
22. Toyota M, Suzuki H, Yamashita T, Hirata K, Imai K, Tokino T, et al. Cancer epigenomics: implications of DNA methylation in personalized cancer therapy. *Cancer Sci* 2009;**100**:787–91.
23. Suzuki H, Watkins DN, Jair KW, Schuebel KE, Markowitz SD, Chen WD, et al. Epigenetic inactivation of SFRP genes allows constitutive WNT signaling in colorectal cancer. *Nat Genet* 2004;**36**:417–22.
24. Downward J. Targeting RAS signalling pathways in cancer therapy. *Nat Rev Cancer* 2003;**3**:11–22.
25. Rajagopalan H, Bardelli A, Lengauer C, Kinzler KW, Vogelstein B, Velculescu VE. Tumorigenesis: RAF/RAS oncogenes and mismatch-repair status. *Nature* 2002;**418**:934.
26. Akino K, Toyota M, Suzuki H, Mita H, Sasaki Y, Ohe-Toyota M, et al. The Ras effector RASSF2 is a novel tumor-suppressor gene in human colorectal cancer. *Gastroenterology* 2005;**129**:156–69.
27. Dammann R, Li C, Yoon JH, Chin PL, Bates S, Pfeifer GP. Epigenetic inactivation of a RAS association domain family protein from the lung tumour suppressor locus 3p21.3. *Nat Genet* 2000;**25**:315–9.

28. Imai T, Toyota M, Suzuki H, Akino K, Ogi K, Sogabe Y, et al. Epigenetic inactivation of RASSF2 in oral squamous cell carcinoma. *Cancer Sci* 2008;**99**:958–66.
29. Maruyama R, Akino K, Toyota M, Suzuki H, Imai T, Ohe-Toyota M, et al. Cytoplasmic RASSF2A is a proapoptotic mediator whose expression is epigenetically silenced in gastric cancer. *Carcinogenesis* 2008;**29**:1312–8.
30. Hollstein M, Sidransky D, Vogelstein B, Harris CC. p53 mutations in human cancers. *Science* 1991;**253**:49–53.
31. Suzuki H, Itoh F, Toyota M, Kikuchi T, Kakiuchi H, Imai K. Inactivation of the 14-3-3 sigma gene is associated with 5′ CpG island hypermethylation in human cancers. *Cancer Res* 2000;**60**:4353–7.
32. Esteller M, Tortola S, Toyota M, Capella G, Peinado MA, Baylin SB, et al. Hypermethylation-associated inactivation of p14(ARF) is independent of p16(INK4a) methylation and p53 mutational status. *Cancer Res* 2000;**60**:129–33.
33. Qiu W, Zhou B, Zou H, Liu X, Chu PG, Lopez R, et al. Hypermethylation of growth arrest DNA damage-inducible gene 45 beta promoter in human hepatocellular carcinoma. *Am J Pathol* 2004;**165**:1689–99.
34. Saito Y, Liang G, Egger G, Friedman JM, Chuang JC, Coetzee GA, et al. Specific activation of microRNA-127 with downregulation of the proto-oncogene BCL6 by chromatin-modifying drugs in human cancer cells. *Cancer Cell* 2006;**9**:435–43.
35. He L, He X, Lim LP, de Stanchina E, Xuan Z, Liang Y, et al. A microRNA component of the p53 tumour suppressor network. *Nature* 2007;**447**:1130–4.
36. Toyota M, Suzuki H, Sasaki Y, Maruyama R, Imai K, Shinomura Y, et al. Epigenetic silencing of microRNA-34b/c and B-cell translocation gene 4 is associated with CpG island methylation in colorectal cancer. *Cancer Res* 2008;**68**:4123–32.
37. Kozaki K, Imoto I, Mogi S, Omura K, Inazawa J. Exploration of tumor-suppressive microRNAs silenced by DNA hypermethylation in oral cancer. *Cancer Res* 2008;**68**:2094–105.
38. Cunningham JM, Christensen ER, Tester DJ, Kim CY, Roche PC, Burgart LJ, et al. Hypermethylation of the hMLH1 promoter in colon cancer with microsatellite instability. *Cancer Res* 1998;**58**:3455–60.
39. Herman JG, Umar A, Polyak K, Graff JR, Ahuja N, Issa JP, et al. Incidence and functional consequences of hMLH1 promoter hypermethylation in colorectal carcinoma. *Proc Natl Acad Sci USA* 1998;**95**:6870–5.
40. Toyota M, Ahuja N, Ohe-Toyota M, Herman JG, Baylin SB, Issa JP. CpG island methylator phenotype in colorectal cancer. *Proc Natl Acad Sci USA* 1999;**96**:8681–6.
41. Weisenberger DJ, Siegmund KD, Campan M, Young J, Long TI, Faasse MA, et al. CpG island methylator phenotype underlies sporadic microsatellite instability and is tightly associated with BRAF mutation in colorectal cancer. *Nat Genet* 2006;**38**:787–93.
42. Hinoue T, Weisenberger DJ, Pan F, Campan M, Kim M, Young J, et al. Analysis of the association between CIMP and BRAF in colorectal cancer by DNA methylation profiling. *PLoS One* 2009;**4**:e8357.
43. Suzuki H, Igarashi S, Nojima M, Maruyama R, Yamamoto E, Kai M, et al. IGFBP7 is a p53-responsive gene specifically silenced in colorectal cancer with CpG island methylator phenotype. *Carcinogenesis* 2010;**31**:342–9.
44. Noushmehr H, Weisenberger DJ, Diefes K, Phillips HS, Pujara K, Berman BP, et al. Identification of a CpG island methylator phenotype that defines a distinct subgroup of glioma. *Cancer Cell* 2010;**17**:510–22.
45. Gaudet F, Hodgson JG, Eden A, Jackson-Grusby L, Dausman J, Gray JW, et al. Induction of tumors in mice by genomic hypomethylation. *Science* 2003;**300**:489–92.
46. Ehrlich M. DNA methylation in cancer: too much, but also too little. *Oncogene* 2002;**21**:5400–13.

47. Wilson AS, Power BE, Molloy PL. DNA hypomethylation and human diseases. *Biochim Biophys Acta* 2007;**1775**:138–62.
48. Feinberg AP, Vogelstein B. Hypomethylation distinguishes genes of some human cancers from their normal counterparts. *Nature* 1983;**301**:89–92.
49. Jackson K, Yu MC, Arakawa K, Fiala E, Youn B, Fiegl H, et al. DNA hypomethylation is prevalent even in low-grade breast cancers. *Cancer Biol Ther* 2004;**3**:1225–31.
50. Yamamoto E, Toyota M, Suzuki H, Kondo Y, Sanomura T, Murayama Y, et al. LINE-1 hypomethylation is associated with increased CpG island methylation in Helicobacter pylori-related enlarged-fold gastritis. *Cancer Epidemiol Biomark Prev* 2008;**17**:2555–64.
51. Hansen RS, Wijmenga C, Luo P, Stanek AM, Canfield TK, Weemaes CM, et al. The DNMT3B DNA methyltransferase gene is mutated in the ICF immunodeficiency syndrome. *Proc Natl Acad Sci USA* 1999;**96**:14412–7.
52. Okano M, Bell DW, Haber DA, Li E. DNA methyltransferases Dnmt3a and Dnmt3b are essential for de novo methylation and mammalian development. *Cell* 1999;**99**:247–57.
53. Jeanpierre M, Turleau C, Aurias A, Prieur M, Ledeist F, Fischer A, et al. An embryonic-like methylation pattern of classical satellite DNA is observed in ICF syndrome. *Hum Mol Genet* 1993;**2**:731–5.
54. Yamashita Y, Yuan J, Suetake I, Suzuki H, Ishikawa Y, Choi YL, et al. Array-based genomic resequencing of human leukemia. *Oncogene* 2010;**29**:3723–31.
55. Wolff EM, Byun HM, Han HF, Sharma S, Nichols PW, Siegmund KD, et al. Hypomethylation of a LINE-1 promoter activates an alternate transcript of the MET oncogene in bladders with cancer. *PLoS Genet* 2010;**6**:e1000917.
56. Kantarjian HM, O'Brien S, Huang X, Garcia-Manero G, Ravandi F, Cortes J, et al. Survival advantage with decitabine versus intensive chemotherapy in patients with higher risk myelodysplastic syndrome: comparison with historical experience. *Cancer* 2007;**109**:1133–7.
57. Duvic M, Talpur R, Ni X, Zhang C, Hazarika P, Kelly C, et al. Phase 2 trial of oral vorinostat (suberoylanilide hydroxamic acid, SAHA) for refractory cutaneous T-cell lymphoma (CTCL). *Blood* 2007;**109**:31–9.
58. Chai G, Li L, Zhou W, Wu L, Zhao Y, Wang D, et al. HDAC inhibitors act with 5-aza-2′-deoxycytidine to inhibit cell proliferation by suppressing removal of incorporated abases in lung cancer cells. *PLoS One* 2008;**3**:e2445.
59. Hurtubise A, Momparler RL. Effect of histone deacetylase inhibitor LAQ824 on antineoplastic action of 5-aza-2′-deoxycytidine (decitabine) on human breast carcinoma cells. *Cancer Chemother Pharmacol* 2006;**58**:618–25.
60. Tong WG, Wei Y, Stevenson W, Kuang SQ, Fang Z, Zhang M, et al. Preclinical antileukemia activity of JNJ-26481585, a potent second-generation histone deacetylase inhibitor. *Leuk Res* 2010;**34**:221–8.
61. Shen L, Kantarjian H, Guo Y, Lin E, Shan J, Huang X, et al. DNA methylation predicts survival and response to therapy in patients with myelodysplastic syndromes. *J Clin Oncol* 2010;**28**:605–13.
62. Langemeijer SM, Kuiper RP, Berends M, Knops R, Aslanyan MG, Massop M, et al. Acquired mutations in TET2 are common in myelodysplastic syndromes. *Nat Genet* 2009;**41**:838–42.

Genome-Wide Distribution of DNA Methylation at Single-Nucleotide Resolution

ELEANOR WONG[*,†] AND
CHIA-LIN WEI[*,†,‡]

[*]*Genome Technology and Biology, Genome Institute of Singapore, Singapore*
[†]*Department of Biological Sciences, National University of Singapore, Singapore*
[‡]*Department of Genome Technology, Joint Genome Institute, California, USA*

DNA methylation, a well-known epigenetic modification in mammalian genomes, is important for development and health. Dysregulation of DNA methylation can cause abnormal gene regulation, leading to anomalous development and diseases. Until recently, the ability to understand the functions and dynamics of DNA methylation was limited by the availability of technologies for comprehensively characterizing methylation on a genome-wide scale. Rapid advances in high-throughput approaches (particularly next-generation sequencing), coupled with molecular techniques, have enabled unbiased genome-wide profiling of DNA modifications at single-base resolution and helped to elucidate their impact on gene regulation. Here, we discuss the development of genomic approaches to decipher the global methylome at single-base resolution, the challenges faced, and the emerging new insights.

Progress in Molecular Biology
and Translational Science, Vol. 101
DOI: 10.1016/B978-0-12-387685-0.00015-9

1877-1173/11 $35.00

Our ability to decipher this important epigenetic modification and how it impacts gene expression will provide a framework for understanding numerous disease mechanisms, and suggest means to treat or prevent them in the future.

I. Impact of Single-Nucleotide-Based Detection on DNA Methylome Profiling

DNA methylation is one of the best characterized epigenetic modifications and plays essential roles in regulating mammalian gene expression and development. DNA methylation involves adding a methyl group, through the actions of DNA methyltransferases (DNMTs), to the five carbon of cytosines that primarily reside within CpG dinucleotide context (see Section III). In mammals, CpGs only comprise about 1% of the genome (compared to $1/16 = 6.25\%$ in random sequences) and are nonrandomly distributed.[1] Specifically, CpGs are enriched in regions known as CpG islands (CGIs), commonly located near the 5′ ends of annotated genes.[2] In normal cells, the majority of CpG dinucleotides are methylated, but CGIs at promoters are often protected from DNA methylation.[3,4] The resulting promoter hypomethylation is a landmark of many ubiquitously expressed housekeeping genes essential for normal cell maintenance.

Being a stable and long-term modification, unique DNA methylation patterns have been found in different cell lineages.[5] Further, the status of DNA methylation can be changed rapidly through active demethylation, either genome-wide or gene-specifically, during distinct developmental stages or in response to environmental stimuli.[6,7] The DNA methylome can be dynamically regulated, and its dysregulation is associated with cancers, neurological disorders, and imprinting-related diseases[8] (see Chapters by Jon F. Wilkins and Francisco Úbeda; and Minoru Toyota and Eiichiro Yamamoto). Hence, mapping of DNA methylation changes in different cell types and understanding their impacts in transcription regulation and disease progression have been the center of epigenetic analysis.

Prior to high-resolution DNA methylome profiling, our understanding of how DNA methylation influences other genomic features was restricted to studies of selected loci or specific regulatory regions, that is, promoters, CGIs, or transcription factor (TF)-binding sites.[9,10] Specifically, hypermethylation of promoters for tumor-suppressor genes are common hallmarks of cancers.[5,11–13] TFs such as *POU5f1* and *NANOG* can be selectively silenced by methylation of their promoters during differentiation.[14] DNA methylation also impacts the binding of TFs to their corresponding *cis*-regulatory elements and controls chromatin accessibility.[15] From these limited studies, it was clear that DNA methylation plays critical roles in establishing proper transcription programs leading to cellular differentiation and development.[16,17] However, attempts to expand such knowledge to a high-resolution, genome-wide scale were limited by the lack of robust technology.

Recently, rapid advancement in genome technologies has enabled a variety of approaches to globally interrogate chromatin and epigenetic modifications.[18,19] The development of powerful DNA methylation analyses and generation of comprehensive high-resolution DNA methylomes for complex genomes have created a new paradigm for understanding the dynamics and complexity of DNA methylation in both normal development and disease progression.[20–23] Hence, it is timely to provide an overview for the technical advances and summarize what we have learned from the new approaches and valuable resources.

II. Overview of Molecular Approaches Used for Methylation Studies

For decades, the molecular methods used to understand DNA methylation patterns were mainly focused on a small portion of the genome (such as selected loci, promoters, and CpG-rich regions), due to limitations in the throughput and efficiency of the existing methylation detection technologies. Such technical limitations had prevented the interrogations of noncoding regions that, in recent years, have been suggested to play important regulatory roles.[24] With the revolution in DNA sequencing approaches and advances in molecular biology techniques, many new methods have now been developed for unbiased enrichment and detection of DNA methylation states at single-base-pair resolution, even at genome-wide scale.[25] Generally, these methods can be categorized into two specific areas: enriching/selective labeling of 5-methylcytosine (5mC) and detecting the identity and location of the enriched/labeled 5mC.

A. Methods to Enrich or Label 5mC

Direct detection of 5mC in DNA, through conventional approaches like hybridization or DNA sequencing, has been challenging because methylation of cytosine does not affect its ability to base pair. This is further complicated by the fact that the methyl group on 5mC is not maintained by any *in vitro* DNA amplification such as PCR. Therefore, detecting methylation requires selectively labeling the nucleotides based on their methylation status, as well as the ability to detect the labels at high resolution, preferably at the individual nucleotide level. Toward these goals, methylation-dependent pretreatments of DNA were developed. Initially restricted to localized regions, they have now been expanded to enable genome-wide analysis. These molecular methods can be characterized into three main categories based on (1) enzymatic digestion, (2) differential affinity, and (3) chemical treatment based. As discussed below, the choice of methods plays significant roles in determining the specificity, resolution, and sensitivity of the methylation detection.

1. Enzymatic Digestion Approaches

Some restriction enzymes (REases) used in molecular cloning techniques are sensitive to the presence of methyl groups in their recognition sequences. Thus, the ability of these methylation-sensitive REases to cut DNA can be used to determine the methylation status of their cutting sites.[26] Because most DNA methylation occurs within CpG dinucleotides, the most commonly used REases are the isoschizomers, *Hpa*II and *Msp*I, that both recognize the sequence CCGG. However, only *Hpa*II is blocked by CpG methylation.[27] Variations of the digestion protocols have also evolved to provide higher resolution and sensitivity, including the use of a combination of different REases as well as the methylation-dependent endonuclease, *McrBC,* that specifically cuts methylated DNA[28] (Fig. 1A). When coupled with gene-specific analysis like Southern blot, or PCR across the restriction sites, base-pair resolution DNA methylation patterns can be revealed.[29] REase-based methods can also be extended to study global methylation profiling, through genome-scale approaches such as array hybridization[30,31] or sequencing (Methyl-seq).[32] However, the genome coverage derived from REase-based methods is highly dependent on the choice of REases. In order to increase the coverage, a mix of REases is routinely used; this generally overfragments the genome. Further, the selection of REases becomes a challenge in studies involving *de novo* detection of methylation in the non-CpG context. Therefore, this method is unsuitable for DNA methylation detection in plant genomes where non-CpG methylations are prevalent.

2. Affinity Approaches

Affinity enrichment of 5mC is another powerful approach to study the DNA methylome in complex genomes. This method uses *c*hromatin *i*mmuno*p*recipitation (ChIP) or affinity-based capture to selectively enrich methylated DNA. Among the different affinity-based agents, antibodies specific against 5mC (*me*thylated *D*NA *i*mmuno*p*recipitation; MeDIP)[33] and *m*ethyl-CpG *b*inding *d*omain-based *cap*ture (MBDCap) are the most widely used[34] (Fig. 1B).

In the MBDCap approach, only 5mC in the CpG context are enriched, and different CpG methylation density regions can be selectively collected by using gradient salt fractionation. In contrast, 5mC antibodies enrich for methylcytosines regardless of their nucleotide contexts. Although MeDIP enables the enrichment of 5mC in non-CpG regions, MBDCap gives higher overall enrichment, especially in CpG-dense regions.[35] These methods, though robust and efficient, are biased toward high CpG density, sensitive to copy number variation and, most importantly, unable to provide methylation status at the individual nucleotide level.

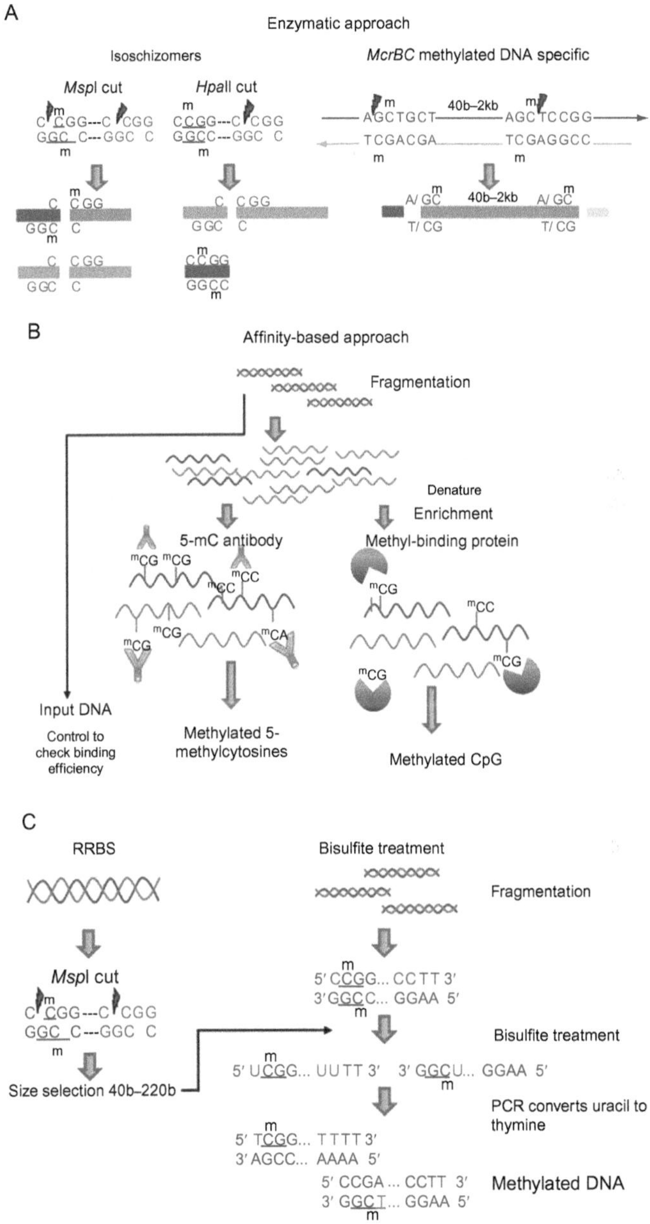

FIG. 1. Overview of methylated DNA enrichment labeling techniques. (A) Restriction enzymes *Msp*I and *Hpa*II both cut a site containing a CpG (CCGG), but of the two, only *Hpa*II is blocked by CpG methylation. These enzymes are commonly used in a process called Methyl-seq,

3. Bisulfite Treatment

The use of sodium bisulfite to detect methylated DNA has opened the door for many different types of methylation analysis.[36,37] Bisulfite conversion chemically deaminates unmethylated cytosine to uracil, while 5mC is resistant to this conversion. Upon PCR amplification, uracil is replaced by thymine while 5mC remains as cytosine and can then be differentiated from the unmethylated cytosine (Fig. 1C). In this method, it is extremely important to achieve near 100% conversion efficiency, as incomplete conversion cannot be distinguished from incomplete methylation in a population of cells. Bisulfite conversion has been crowned as the "gold standard" for methylation detection because it provides an unbiased and robust way to determine the methylation status of each individual cytosine, regardless of its sequence context. In order to reduce sequencing complexity and cost, bisulfite conversion has been combined with REase digestion (*r*educed *r*epresentation of *b*isulfite *s*equencing, RRBS) and, subsequently, was also used in combination with capture hybridization to enrich target-specific regions of the genomes. In RRBS, CpG-dense fragments, enriched by digestion with enzymes like *Msp*I, were further treated with sodium bisulfite to detect the methylated CpG dinucleotides.[10,38] In capture-based techniques, specific regions of interest were further selected from bisulfite-treated DNA through either array-based[21] or padlock solution-based hybridization.[39] Despite the different secondary enrichment methods, the methylation level of individual cytosines from bisulfite-treated DNA can be determined at various resolutions and scales as discussed below.

---

to detect methylated DNA. *McrBC* is an endonuclease that cleaves DNA containing methylcytosine in two half-sites of the form (G/A)^{m}C, where the two half-sites can be 40 bp–2 kb apart. *McrBC*-digested DNA can subsequently be ligated with specific sequencing adapters for high-throughput sequencing or hybridized onto arrays for methylated cytosine detection. (B) Methylated cytosines can be enriched using 5mC-specific antibodies or methyl-CpG DNA-binding protein through the processes of MeDIP and MDB-cap, respectively. In MeDIP, genomic DNA is randomly sheared and denatured, followed by enrichment with anti-5mC antibodies and elution using varying salt concentrations to isolate the methylated cytosines. Similarly, in MDB-cap, the genomic DNA is first fragmented and denatured, followed by the binding of methyl-CpG-binding protein to pull down the mCpG DNA fragments. To assess enrichment efficiency, input DNA without any antibody enrichment serves as control DNA. Next, both the input DNA and enriched methylated DNA can be subjected to various detection techniques for further methylation analysis. (C) Randomly sheared genomic DNA is subjected to bisulfite treatment that converts unmethylated cytosines to uracils, leaving methylated cytosines unchanged. Upon PCR amplification, the uracils are replaced by thymidines, and this allows the differentiation between methylated and unmethylated cytosines. In addition, DNA digested by *Msp*I can be further treated with sodium bisulfite in a method called RRBS for whole-genome methylation profiling at single-base resolution. (See Color Insert.)

B. Technologies for Detecting Labeled 5mC at Single-Nucleotide Resolution

Once the genomic DNA is selectively treated based on the methylation status of each individual cytosine, it can be analyzed using different detection methods to reveal the locations of the 5mC. For locus-specific analysis, the treated DNA can be coupled with PCR amplification across predicted methylation sites. For global scale analysis, array hybridization or sequencing is the method of choice. In recent years, new sequencing technologies were adapted for system-wide epigenome studies,[40] including mapping histone modifications, chromatin interactions, and DNA methylome profiling. With sufficient base coverage, methylation sequencing is able to decipher the whole-genome methylome at single-base resolution even in complex mammalian genomes. Below, we discuss the different detection methods, their associated technical specificities, and issues including coverage, multiplexity, and methylation specific informatic tools.

1. Array-Based Detection

Array hybridization can yield DNA methylation profiles from bisulfite-treated DNA at base-pair resolution. However, to distinguish the unmethylated from methylated cytosines, hybridization of amplified bisulfite-converted DNA to the microarray requires special consideration on the array design as well as the complicated data analysis schemes.[41] Alternatively (and perhaps the most popular option), there is the Illumina Infinium HumanMethylation27 BeadChip method (Illumina, Inc., San Diego, CA, USA). In this approach, two versions of site-specific probes, designed based on whether the locus been methylated or unmethylated, are used to perform multiplexed primer extensions on bisulfite-treated DNA. The fluorescent ddNTPs substrates are incorporated as a single-base extension, and the relative ratio of fluorescent hybridization signals from the methylated (C) and unmethylated (T) probes provide a quantitative methylation measurement for each interrogated CpG locus[42] (Fig. 2A). The current capacity of HumanMethylation27 BeadChips allows parallel profiling of 27,578 CpG sites from 14,495 human Refseq gene promoters. Such coverage has since been extended to more than 450,000 sites in the Infinium HumanMethylation450 BeadArray, which covers CpGs beyond CGIs and promoters, as well as non-CpG sites.

Compared to the sequencing approach, array-based methods offer superior sample multiplex capability but suffer from potential bias resulting from hybridization noise. Although it allows single-base resolution, array-based detection is limited to studies of methylomes from species with commercially available methylation arrays. Further, most of the probe designs fail to cover all of the cytosines in complex genomes. Therefore, this detection approach is not recommended for discovery of novel methylation patterns.

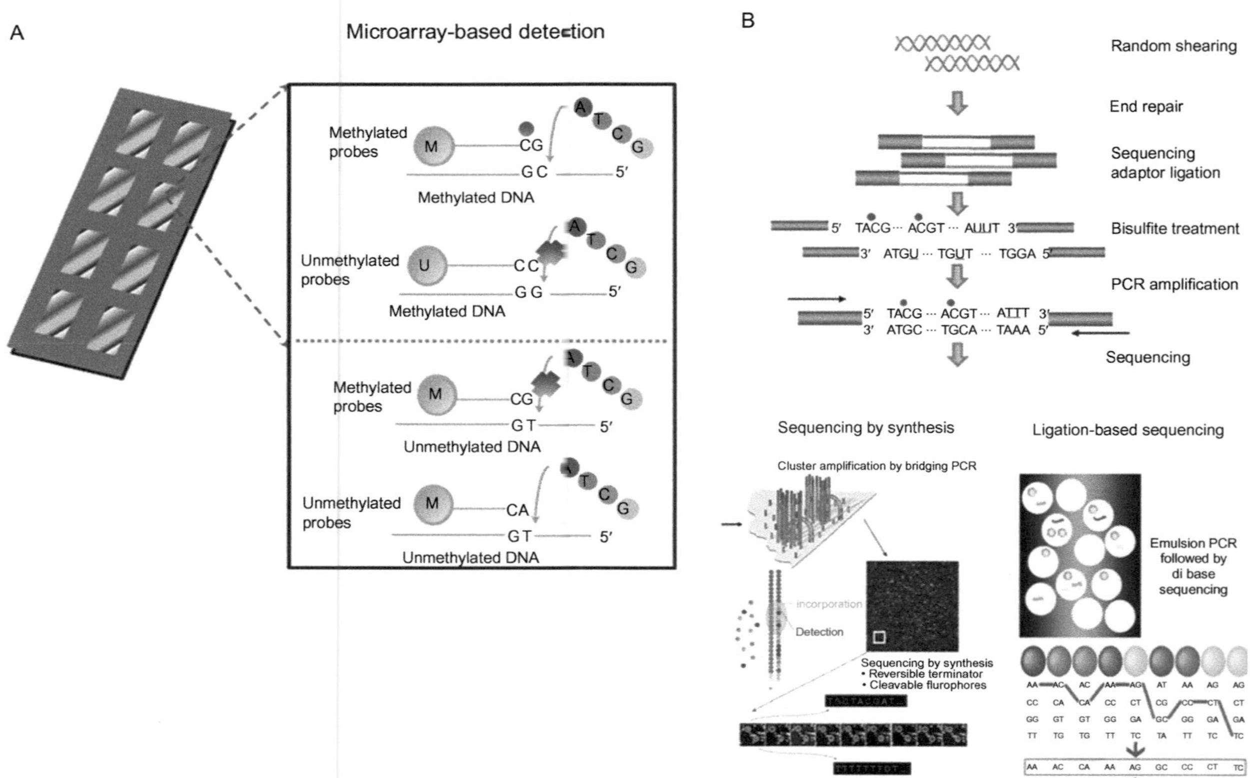

FIG. 2. DNA methylation detection methods. (A) Commercially available microarray chips can be used to detect methylated DNA of specific loci or expanded to whole-genome scale. Multiple bisulfite-treated genomic DNA samples can be hybridized onto the customized arrays, which allows the interrogation of up to 450,000 sites in parallel. A pair of methylated and unmethylated probes for each site was designed to perform primer extensions on

2. Sequencing-Based Detection

With the recent advancement in ultra-high-throughput DNA sequencing, the direct mapping of DNA methylation can be carried out on 2nd-generation sequencing platforms such as Genome Analyzer (Illumina) and SOLiD (Life Technologies, Inc., Carlsbad, CA, USA). This represents a major milestone in epigenetic analysis.[20,22,23,43] Toward this end, sequencing has been coupled with REase-based (Methyl-Seq), affinity-based enrichment (MeDIP and MBDCap), RRBS, and randomly sheared bisulfite-treated DNA (BS-Seq) methods to determine methylation profiles. Among these, direct BS-sequencing is a simple yet powerful method for global profiling of DNA methylomes with high accuracy and reproducibility[44] (Fig. 2B). Initially performed with small genomes such as *Arabidopsis thaliana*,[45,46] BS-Seq has been expanded to handle large complex mammalian genomes such as mouse and human at single-base resolution.[20,22,23]

Ultimately, the accuracy, coverage, and resolution of direct BS-Seq depends on the efficiency of bisulfite conversion, depth of sequencing, and robustness of analysis tools. The challenges in methylation analysis now lie in the postsequencing read mapping, methylation calling, and comparative analysis steps. In mammalian genomes, only a small portion (3–6%) of cytosines is methylated. As a result, bisulfite-treated DNA mostly contains three bases (T, A, G) instead of the usual four. Such reduced sequence complexity causes poor mapability and lowers the specificity of assigning bisulfite sequence reads to their corresponding reference genomes using generic alignment programs. This mapping complexity is further complicated by incomplete bisulfite conversion. Therefore, special computational solutions and bioinformatic tools are required to process bisulfite sequencing data. To overcome the alignment issue, the bisulfite-treated sequence reads and the reference genome need to be converted to a three-base genome by replacing all Cs in the plus strand with Ts and all Gs in the minus strand with As. Thereafter, the mapped sequence tags and

the bisulfate-treated DNA. Fluorescent ddNTPs are added to the arrays and incorporated at single-base extensions. DNA with methyl C will allow the hybridization of the methylated probes and subsequent base extension, while incorporation of ddNTPs at the unmethylated probes will be terminated, and vice versa in the unmethylated locus. The "methylated" and "unmethylated" probes differ in having C or T at a given position. Methylated DNA is identified from the ratio of fluorescent signals recorded when using the unmethylated and methylated probes. (B) Direct bisulfite sequencing (BS-seq) involves the coupling of bisulfite treatment of DNA with 2nd-generation sequencing technologies, to detect methylated DNA at genome-wide scale at single-nucleotide resolution. Genomic DNA is typically sheared into fragments ($\sim$500 bp), end polished, and ligated with specific sequencing adapters. This is followed by the bisulfite treatment and PCR amplification, before subjecting to DNA sequencing by synthesis or ligation-based chemistry to detect the methylcytosines. (See Color Insert.)

reference genome are reverted to their original forms to identify the locations of 5mC. Alignment using seed-and-extension algorithms such as SOAP2[47] and Bowtie[48] can be applied to the converted sequences with minimal mismatches to tolerate sequencing errors and SNPs. To further increase the mapability, longer read-length and paired-end reads information are used, particularly across regions containing repeat elements. With the current modifications, existing alignment software can map 70% of the filtered reads back to the reference genome.

With the SOLiD platform, because of its ligation-based chemistry using di-base-labeled probes, mapping of the bisulfite reads first needs to be performed in color space and then translated back into base space using the same concept.[43] Alternatively, a dynamic and ungapped alignment program (SOCB-s) was used for *A. thaliana* to map color space bisulfite reads directly, without base space conversion.[49] Because of the reduced genome complexity and the amount of sequencing data required for sufficient coverage, high demands of computing resources are the next challenges to achieve exhaustive and accurate mapping processing regardless of the different algorithms used. Hence, further developments in mapping and computational processing are very much needed, in order to disseminate the BS-Seq approach widely to the community.

With BS-Seq as the robust platform for methylome mapping, comparison between different methylomes could soon unveil the dynamics of methylation patterns and complex levels of regulation. However, factors such as genome structure variations could have significant impacts on the relative profiles and differentially methylated regions (DMRs) defined; particularly in regions subject to copy number variation.[35] Thus, proper controls with input should be sequenced in parallel, to correct biases resulting from structural variations. Such information is particularly important for making inferences from comparative analysis of cancer methylomes.

Compared with array-based analysis, sequencing-based methylome assays can be applied to any species for which reference genome sequences are available, and are more flexible for either targeting specific regions or expanding to whole-genome scale. The dynamic range for sequencing detection can be adjusted to increase sensitivity by increasing the sequencing depth, which is specifically useful in resolving hemi- from fully methylated CpGs. The resulting hemimethylation information, through linking with *cis*-strand SNPs, allows identification of *a*llele-*s*pecific DNA *m*ethylation (ASM), which is unlikely to be determined by array hybridization methods.

In summary, different sample enrichment methods and detection approaches have distinct technical specificity and unique advantages (Table I). To date, direct BS-Seq generates the most comprehensive and high-resolution DNA methylomes. Future advances will rely on the development of sophisticated bioinformatics tools, to understand the impact of various methylation

TABLE I

OVERVIEW OF VARIOUS METHYLATION DETECTION AND ENRICHMENT METHODS

Detection methods	Scale	Resolution	Multiplex	Throughput[a]	Enrichment methods	5mC in non-CpG context	Non-5mC methylation[b]	Limitations
PCR amplification	Loci-specific	Quantitative measurement	–	+	RE-based (*Msp*I, *Hpa*II, *McrBC*)	No	No	Suitable for small-scale validation. Unable to pinpoint location of methylcytosines
					Bisulfite-based	Yes	No	
					Affinity-based (MEDIP, MBDcap, 5-hmC)	No	No	
Microarray	Genome-wide. Only for species with commercially available array	Single nucleotide	+++	+++	Restriction enzyme-based (*Msp*I, *Hpa*II, *McrBC*)	No	No	Restricted to mCpG found in RE sites
					Bisulfite-based	Yes	No	Limited by probes design
					Affinity-based (MEDIP, MBDcap)	No	No	CpGs bias
					5-hydroxymethylcytosine	No	No	Dependent on antibody enrichment
Sequencing								
Genome analyzer	Genome-wide. Any species with a complete reference genome	Single nucleotide	++	++++	RE-based (Methyl-seq)	No	No	Restricted to mCpG found in RE sites
SOLiD sequencing					RE + bisulfite (RRBS)	No	No	Restricted to mCpG found in RE sites
					Affinity-based (MeDIP-seq)	No	No	CpGs bias
					Affinity-based (MBDcap-seq)	No	No	
					Bisulfite (BS-seq)	Yes	No	Required high sequencing depth, costly
Nanopore sequencing		Single molecule	++++	+++++	Direct sequencing	Yes	Yes	Require accurate distinction between different types of methylation
Single molecule real-time (SMRT) sequencing		Single molecule	++++	+++++	Direct sequencing	Yes	Yes	

[a]The multiplex capability and throughput of each detection method is rated by the number of "+."

[b]Non-5mC methylation refers to other forms of DNA methylation modification such as 5-hydroxymethylcytosine and N6-methyladenine.

profiles on other genomics/epigenomics regulation and to integrate them with other genome-wide knowledge. Below, we discuss what we have learned so far from systematic and comparative analyses.

III. DNA Methylation Patterns at Single-Nucleotide Resolution

The completion of many different mammalian methylomes at single-nucleotide resolution gives us a clearer view of the distributions of methylation patterns across the entire genome landscape, providing new insights into how this modification influences transcription regulation and yielding a robust foundation for understanding disease progression and developmental regulation.[10,20,22,50] Comparative analyses have revealed the dynamic nature of methylation, pointing toward identification of DMRs as useful for biomarkers and therapeutic targets.

A. Genome-Wide Distribution of Methylation Patterns

The great majority ($\sim$95%) of the human genome is unmethylated.[22,51] Among different cell types surveyed, undifferentiated pluripotent embryonic stem cells had the highest overall methylation level, which decreased with increasing cell differentiation potential.[22] With unbiased whole genome profiling, one of the most interesting findings was the significant level of non-CpG methylation. Uniquely found in pluripotent stem cells, CpA methylation appears to be the most prevalent form of non-CpG methylation,[20,22] suggesting that there may be a functional role for such non-CpG methylation in governing the pluripotency of stem cells. However, the mechanism regulating such asymmetrical methylation is unclear, and its function has yet to be fully explored.

In global analyses, DNA methylation increased in the bodies of actively transcribed genes in plants[52] and mammals,[53] while sharp reduction was found within $\pm$2 kb of the transcription start site (TSS). Zooming in on the transcribed locus, more-abundant methylation was found in exons than in introns. Moreover, sharp exon–intron boundaries are characterized by unique methylation patterns on both plus and minus strands. A distinct spike in methylation level was observed at the 5′exon, which decreased drastically when entering the intron.[22,23] The methylation level was gradually increased across the intron, followed by a sharp increase at the 3′ intron–exon boundary, suggesting that the transition of methylation pattern could be an important signal for regulating mRNA splicing (perhaps by recruiting spliceosomes). Besides DNA methylation, chromatin modifications and nucleosome positioning also influence the regulation of splicing.[54,55] These discoveries yet again indicate that chromatin states, transcription, and DNA methylation could cross-talk and work synergistically to regulate gene expression.

B. Differential Methylation Regions and Tissue-Specific-Methylation

Genome-wide comparative methylation studies not only unveil the complex methylation patterns embedded in each cell type but have also revealed a substantial number of tissue-specific DMRs in mammalian genomes.[3,17] Specifically, tissue-specific methylation was observed among promoters and intragenic CGIs. When associating global methylation patterns, trimethylation of histone H3 lysine 4 (H3K4me3), an epigenetic mark for active promoter, is found to be enriched within unmethylated intragenic CGIs; which suggests that intragenic methylation could regulate alternative promoters to maintain tissue specificity.[56] Beyond tissue-specific intragenic CGIs, genome-wide comparison of high-resolution methylomes has revealed thousands of nonoverlapping DMRs.[20,22,23] As each tissue type has its specific methylation profile, the tissue-specific DMR profile can distinguish among different tissues. Many of the annotated DMRs are associated with genes involved in pluripotency, development, and imprinting. For example, homeobox TFs such as *ALX1* and *CDX1* showed increased promoter and *t*ranscription *t*ermination *s*ite (TTS) methylation.[22] In addition, increased gene-body methylation was also observed in genes for cell adhesion molecules and for G-protein signaling. When measured in cancers, methylation was eliminated within the repetitive sequences and coding regions but increased within the promoters of tumor-suppressor genes. In particular, many miRNAs are embedded within DMRs and were downregulated in cancer cells.[57]

C. Allele-Specific Methylation

The CpG dinucleotide is particularly variable in the genomes of organisms that methylate CpGs. Integration of their methylation status with the sequence variations allows the identification of *a*llele-*s*pecific *m*ethylation (ASM), heterozygous alleles with strand-specific methylation between two haploid genomes,[23] and such feature can only be uncovered through base-pair resolution methylome profiling. Previously known to occur only in imprinting regions, recent studies have suggested that the ASM events are more prevalent than imprinting events, and it is estimated that about 10% of human genes can be regulated through ASM.[58] When combined with sequence-based RNA expression analysis, ASM allows one to distinguish allele-specific expression and understand how differential methylation between two alleles can impact such imbalanced expression.[23]

IV. DNA Methylation, Histone Modifications, and Other Epigenetic Regulation

DNA methylation occurs in a complex chromatin network and is regulated by the intricate interplays between histone modifications, chromatin structures, and noncoding RNA[59] (see Chapters by Xiaodong Cheng and Robert M. Blumenthal; Jafar Sharif and Haruhiko Koseki; Anton Wutz; and Pierre-Antoine Defossez and Irina Stancheva). Emerging from high-resolution DNA methylation profiles is the complex but codependent patterns between methylation and other epigenetic marks.

A. DNA Methylation and Histone Modification

DNA methylation patterns can be directly or indirectly affected by histone modifications or chromatin states. Cross-talk between DNA methylation and histone modification was shown at specific gene loci.[60] For example, promoter DNA hypomethylation at ES cell-specific *Oct4* and *Nanog* genes is closely associated with hyperacetylated histones. DNA methylation can also direct H3K9 methylation through the interactions between DNMTs, H3K9 methyltransferases, and methyl-CpG-binding domain proteins.[61–63] Similarly, H3K27me3 modification is closely associated with underlying DNA methylation through the direct interaction between H3K27 methyltransferase EZH2 and DNMTs.[64] Recent genome-wide DNA methylation studies, particularly in ES cells, indicate that H3K9 and H3K27 methylations direct DNA methylation in a locus-specific manner,[65] potentially through the recruitment of HP1 proteins. In contrast, regions exhibiting active chromatin states, represented by H3K4 di- and trimethylations, are depleted of DNA methylation.[3,10] There are thus mechanistic connections between DNA methylation and chromatin structural modifications (see Chapter by Xiaodong Cheng and Robert M. Blumenthal).

B. DNA Methylation and Noncoding RNAs

Noncoding RNA (ncRNA) provides epigenetic regulation that is important for tumor progression and development. Although the relationship between DNA methylation and ncRNAs is unclear, increasing numbers of ncRNAs, specifically the miRNAs, have been reported to be silenced in cancer cells as a result of cancer-specific DNA hypermethylation.[66] Genome-wide comparative methylome studies in cancers further confirmed that the hypermethylated miRNA loci were found in the intergenic DMRs. Mir-199a-2, a developmentally regulated miRNA implicated in cancer invasiveness, is differentially hypermethylated in invasive cancers,[67] and methylation of mir-137 was identified in early events of colon cancer.[68] Although the mechanism is not fully understood, cancer-specific hypermethylated miRNA genes revealed by the genome-wide DMR analysis could in future be used as potential biomarkers for early cancer detection.

V. Detection of the 6th Base (5-Hydroxymethylcytosine) and Future Perspectives

5mC was long considered to be the only chemical modification of mammalian genomic DNA, until the recent discovery of 5-hydroxymethylcytosine (5hmC). Known now as the 6th base, 5hmC is found predominantly in neuronal and embryonic stem cell chromatin. Its formation is catalyzed by a family of TET proteins.[69,70] Similar to 5mC, 5hmC is resistant to bisulfite treatment. Therefore, 5hmC cannot be differentiated from 5mC by approaches like bisulfite conversion.[71] Because of the lack of tools, the biological function of this special DNA modification remains unclear. Currently, one of the promising methods to profile genome-wide 5hmC locations involves the use of 5hmC-specific glucosyltransferases. The addition of chemically modified glucosyl groups to 5hmC allows biotin affinity purification of the associated chromatin, followed by sequencing analysis.[72] However, the effort to determine its profile at single-base resolution has yet to be successful. Recently, a single-molecule, *r*eal-*t*ime (SMRT) sequencing technology has been applied to directly detect 5mC and 5hmC, as well as other novel types of DNA modifications such as N6-methyladenine.[73] SMRT sequencing determines the nucleotide through direct incorporation of fluorescently labeled nucleotides at each single molecule. By monitoring the changes in DNA polymerase kinetics through fluorescence pulses, the spectra, durations, and intervals reveal DNA templates having different chemical modifications at precise locations. Despite promising results, the SMRT sequencing system still needs to overcome many technical hurdles before it can become a robust platform for a wide spectrum of applications. Nevertheless, the development of many genomic technologies and analysis capabilities to understand the function and dynamics of 5hmC will attract a lot of attention in the foreseeable future.

VI. Concluding Remarks

The maturation of robust technologies and the generation of many whole-genome DNA methylomes at single-base resolution have established the key elements for better understanding this important epigenetic modification. Along with the resolution and whole-genome scale, DNA methylome analysis enables the identification of methylation at non-CpG sites, provides dynamic surveys during cell differentiation, and reveals insights into regulation through cell type-specific DMRs. Further, the methylation patterns beyond gene promoters, particularly within gene body and intergenic regions, may help explain the regulation of alternative promoter usage. Through these global and high-definition approaches, therapeutic and diagnostic tools could result from the identification of disease-associated DMRs. In addition, the cross-talk between DNA methylation and other epigenetic marks underscores the need to integrate high-resolution methylation profiling with other chromatin features.

Bisulfite-based sequencing has revolutionized the way we study DNA methylation. However, even with the drastic reduction of sequencing cost, the generation of a complete mammalian methylome with sufficient depth still is a costly endeavor. Moreover, BS-Seq is unable to uncover novel DNA modifications which are increasingly understood to be biologically important. More robust and distinctive technologies will certainly emerge in the very near future, that should enable the characterization of various DNA modifications in high resolution, and shall rerevolutionize the study of epigenomics.

References

1. Jones PA. The DNA methylation paradox. *Trends Genet* 1999;**15**:34–7.
2. Bird A, Taggart M, Frommer M, Miller OJ, Macleod D. A fraction of the mouse genome that is derived from islands of nonmethylated CpG-rich DNA. *Cell* 1985;**40**:91–9.
3. Weber M, Hellmann I, Stadler MB, Ramos L, Paabo S, Rebhan M, et al. Distribution, silencing potential and evolutionary impact of promoter DNA methylation in the human genome. *Nat Genet* 2007;**39**:457–66.
4. Saxonov S, Berg P, Brutlag DL. A genome wide analysis of CpG dinucleotides in the human genome distinguishes two distinct classes of promoters. *Proc Natl Acad Sci USA* 2006;**103**:1412–7.
5. Esteller M. Epigenetics in cancer. *N Engl J Med* 2008;**358**:1148–59.
6. Bruniquel D, Schwartz RH. Selective, stable demethylation of the interleukin-2 gene enhances transcription by an active process. *Nat Immunol* 2003;**4**:235–40.
7. Monk M, Boubelik M, Lehnert S. Temporal and regional changes in DNA methylation in the embryonic, extraembryonic and germ cell lineages during mouse embryo development. *Development* 1987;**99**:371–82.
8. Pogribny IP, Beland FA. DNA hypomethylation in the origin and pathogenesis of human diseases. *Cell Mol Life Sci* 2009;**66**:2249–61.
9. Straussman R, Nejman D, Roberts D, Steinfeld I, Blum B, Benvenisty N, et al. Developmental programming of CpG island methylation profiles in the human genome. *Nat Struct Mol Biol* 2009;**16**:564–71.
10. Meissner A, Mikkelsen TS, Gu H, Wernig M, Hanna J, Sivachenko A, et al. Genome-scale DNA methylation maps of pluripotent and differentiated cells. *Nature* 2008;**454**:766–70.
11. Esteller M. Cancer epigenomics: DNA methylomes and histone-modification maps. *Nat Rev Genet* 2007;**8**:286–98.
12. Jones PA, Baylin SB. The epigenomics of cancer. *Cell* 2007;**128**:683–92.
13. Feinberg AP, Vogelstein B. Hypomethylation distinguishes genes of some human cancers from their normal counterparts. *Nature* 1983;**301**:89–92.

14. Deb-Rinker P, Ly D, Jezierski A, Sikorska M, Walker PR. Sequential DNA methylation of the Nanog and Oct-4 upstream regions in human NT2 cells during neuronal differentiation. *J Biol Chem* 2005;**280**:6257–60.
15. Bird A. DNA methylation patterns and epigenetic memory. *Genes Dev* 2002;**16**:6–21.
16. Hemberger M, Dean W, Reik W. Epigenetic dynamics of stem cells and cell lineage commitment: digging Waddington's canal. *Nat Rev Mol Cell Biol* 2009;**10**:526–37.
17. Yagi S, Hirabayashi K, Sato S, Li W, Takahashi Y, Hirakawa T, et al. DNA methylation profile of tissue-dependent and differentially methylated regions (T-DMRs) in mouse promoter regions demonstrating tissue-specific gene expression. *Genome Res* 2008;**18**:1969–78.
18. Zhao XD, Han X, Chew JL, Liu J, Chiu KP, Choo A, et al. Whole-genome mapping of histone H3 Lys4 and 27 trimethylations reveals distinct genomic compartments in human embryonic stem cells. *Cell Stem Cell* 2007;**1**:286–98.
19. Fullwood MJ, Liu MH, Pan YF, Liu J, Xu H, Mohamed YB, et al. An oestrogen-receptor-alpha-bound human chromatin interactome. *Nature* 2009;**462**:58–64.
20. Lister R, Pelizzola M, Dowen RH, Hawkins RD, Hon G, Tonti-Filippini J, et al. Human DNA methylomes at base resolution show widespread epigenomic differences. *Nature* 2009;**462**:315–22.
21. Hodges E, Smith AD, Kendall J, Xuan Z, Ravi K, Rooks M, et al. High definition profiling of mammalian DNA methylation by array capture and single molecule bisulfite sequencing. *Genome Res* 2009;**19**:1593–605.
22. Laurent L, Wong E, Li G, Huynh T, Tsirigos A, Ong CT, et al. Dynamic changes in the human methylome during differentiation. *Genome Res* 2010;**3**:320–31.
23. Li Y, Zhu J, Tian G, Li N, Li Q, Ye M, et al. The DNA methylome of human peripheral blood mononuclear cells. *PLoS Biol* 2010;**8**:e1000533.
24. Birney E, Stamatoyannopoulos JA, Dutta A, Guigo R, Gingeras TR, Margulies EH, et al. Identification and analysis of functional elements in 1% of the human genome by the ENCODE pilot project. *Nature* 2007;**447**:799–816.
25. Laird PW. Principles and challenges of genome-wide DNA methylation analysis. *Nat Rev Genet* 2010;**11**:191–203.
26. Bird AP, Southern EM. Use of restriction enzymes to study eukaryotic DNA methylation: I. The methylation pattern in ribosomal DNA from *Xenopus laevis*. *J Mol Biol* 1978;**118**:27–47.
27. McClelland M, Nelson M, Raschke E. Effect of site-specific modification on restriction endonucleases and DNA modification methyltransferases. *Nucleic Acids Res* 1994;**22**:3640–59.
28. Irizarry RA, Ladd-Acosta C, Carvalho B, Wu H, Brandenburg SA, Jeddeloh JA, et al. Comprehensive high-throughput arrays for relative methylation (CHARM). *Genome Res* 2008;**18**:780–90.
29. Van der Ploeg LH, Flavell RA. DNA methylation in the human gamma delta beta-globin locus in erythroid and nonerythroid tissues. *Cell* 1980;**19**:947–58.
30. Omura N, Li CP, Li A, Hong SM, Walter K, Jimeno A, et al. Genome-wide profiling of methylated promoters in pancreatic adenocarcinoma. *Cancer Biol Ther* 2008;**7**:1146–56.
31. Estecio MR, Yan PS, Ibrahim AE, Tellez CS, Shen L, Huang TH, et al. High-throughput methylation profiling by MCA coupled to CpG island microarray. *Genome Res* 2007;**17**:1529–36.
32. Brunner AL, Johnson DS, Kim SW, Valouev A, Reddy TE, Neff NF, et al. Distinct DNA methylation patterns characterize differentiated human embryonic stem cells and developing human fetal liver. *Genome Res* 2009;**19**:1044–56.
33. Weber M, Davies JJ, Wittig D, Oakeley EJ, Haase M, Lam WL, et al. Chromosome-wide and promoter-specific analyses identify sites of differential DNA methylation in normal and transformed human cells. *Nat Genet* 2005;**37**:853–62.

34. Rauch T, Li H, Wu X, Pfeifer GP. MIRA-assisted microarray analysis, a new technology for the determination of DNA methylation patterns, identifies frequent methylation of homeodomain-containing genes in lung cancer cells. *Cancer Res* 2006;**66**:7939–47.
35. Robinson MD, Stirzaker C, Statham AL, Coolen MW, Song JZ, Nair SS, et al. Evaluation of affinity-based genome-wide DNA methylation data: Effects of CpG density, amplification bias, and copy number variation. *Genome Res* 2010;**20**:1719–29.
36. Clark SJ, Harrison J, Paul CL, Frommer M. High sensitivity mapping of methylated cytosines. *Nucleic Acids Res* 1994;**22**:2990–7.
37. Clark SJ, Statham A, Stirzaker C, Molloy PL, Frommer M. DNA methylation: bisulphite modification and analysis. *Nature Protocol* 2006;**1**:2353–64.
38. Gu H, Bock C, Mikkelsen TS, Jager N, Smith ZD, Tomazou E, et al. Genome-scale DNA methylation mapping of clinical samples at single-nucleotide resolution. *Nat Methods* 2010;**7**:133–6.
39. Li JB, Gao Y, Aach J, Zhang K, Kryukov GV, Xie B, et al. Multiplex padlock targeted sequencing reveals human hypermutable CpG variations. *Genome Res* 2009;**19**:1606–15.
40. Zhao X, Ruan Y, Wei CL. Tackling the epigenome in the pluripotent stem cells. *J Genet Genomics* 2008;**35**:403–12.
41. Reinders J, Delucinge Vivier C, Theiler G, Chollet D, Descombes P, Paszkowski J. Genome-wide, high-resolution DNA methylation profiling using bisulfite-mediated cytosine conversion. *Genome Res* 2008;**18**:469–76.
42. Bibikova M, Fan JB. GoldenGate assay for DNA methylation profiling. *Methods Mol Biol* 2009;**507**:149–63.
43. Bormann Chung CA, Boyd VL, McKernan KJ, Fu Y, Monighetti C, Peckham HE, et al. Whole methylome analysis by ultra-deep sequencing using two-base encoding. *PLOS One* 2010;**5**:1–8.
44. Li N, Ye M, Li Y, Yan Z, Butcher LM, Sun J, et al. Whole genome DNA methylation analysis based on high throughput sequencing technology. *Methods* 2010;**52**:203–12.
45. Cokus SJ, Feng S, Zhang X, Chen Z, Merriman B, Haudenschild CD, et al. Shotgun bisulphite sequencing of the Arabidopsis genome reveals DNA methylation patterning. *Nature* 2008;**452**:215–9.
46. Lister R, O'Malley RC, Tonti-Filippini J, Gregory BD, Berry CC, Millar AH, et al. Highly integrated single-base resolution maps of the epigenome in Arabidopsis. *Cell* 2008;**133**:523–36.
47. Li R, Yu C, Li Y, Lam TW, Yiu SM, Kristiansen K, et al. SOAP2: an improved ultrafast tool for short read alignment. *Bioinformatics* 2009;**25**:1966–7.
48. Langmead B, Trapnell C, Pop M, Salzberg SL. Ultrafast and memory-efficient alignment of short DNA sequences to the human genome. *Genome Biol* 2009;**10**:R25.
49. Ondov BD, Cochran C, Landers M, Meredith GD, Dudas M, Bergman HM. An alignment algorithm for bisulfite sequencing using the Applied Biosystems SOLiD System. *Bioinformatics* 2010;**26**:1901–2.
50. Ji H, Ehrlich LI, Seita J, Murakami P, Doi A, Lindau P, et al. Comprehensive methylome map of lineage commitment from haematopoietic progenitors. *Nature* 2010;**467**:338–42.
51. Rollins RA, Haghighi F, Edwards JR, Das R, Zhang MQ, Ju J, et al. Large-scale structure of genomic methylation patterns. *Genome Res* 2006;**16**:157–63.
52. Zhang X, Yazaki J, Sundaresan A, Cokus S, Chan SW, Chen H, et al. Genome-wide high-resolution mapping and functional analysis of DNA methylation in arabidopsis. *Cell* 2006;**126**:1189–201.
53. Ball MP, Li JB, Gao Y, Lee JH, LeProust EM, Park IH, et al. Targeted and genome-scale strategies reveal gene-body methylation signatures in human cells. *Nat Biotechnol* 2009;**27**:361–8.
54. Loomis RJ, Naoe Y, Parker JB, Savic V, Bozovsky MR, Macfarlan T, et al. Chromatin binding of SRp20 and ASF/SF2 and dissociation from mitotic chromosomes is modulated by histone H3 serine 10 phosphorylation. *Mol Cell* 2009;**33**:450–61.

55. Sims 3rd RJ, Millhouse S, Chen CF, Lewis BA, Erdjument-Bromage H, Tempst P, et al. Recognition of trimethylated histone H3 lysine 4 facilitates the recruitment of transcription postinitiation factors and pre-mRNA splicing. *Mol Cell* 2007;**28**:665–76.
56. Maunakea AK, Nagarajan RP, Bilenky M, Ballinger TJ, D'Souza C, Fouse SD, et al. Conserved role of intragenic DNA methylation in regulating alternative promoters. *Nature* 2010;**466**:253–7.
57. Lujambio A, Ropero S, Ballestar E, Fraga MF, Cerrato C, Setien F, et al. Genetic unmasking of an epigenetically silenced microRNA in human cancer cells. *Cancer Res* 2007;**67**:1424–9.
58. Zhang Y, Rohde C, Reinhardt R, Voelcker-Rehage C, Jeltsch A. Non-imprinted allele-specific DNA methylation on human autosomes. *Genome Biol* 2009;**10**:R138.
59. Ikegami K, Ohgane J, Tanaka S, Yagi S, Shiota K. Interplay between DNA methylation, histone modification and chromatin remodeling in stem cells and during development. *Int J Dev Biol* 2009;**53**:203–14.
60. Hattori N, Nishino K, Ko YG, Ohgane J, Tanaka S, Shiota K. Epigenetic control of mouse Oct-4 gene expression in embryonic stem cells and trophoblast stem cells. *J Biol Chem* 2004;**279**:17063–9.
61. Fuks F, Hurd PJ, Deplus R, Kouzarides T. The DNA methyltransferases associate with HP1 and the SUV39H1 histone methyltransferase. *Nucleic Acids Res* 2003;**31**:2305–12.
62. Esteve PO, Chin HG, Smallwood A, Feehery GR, Gangisetty O, Karpf AR, et al. Direct interaction between DNMT1 and G9a coordinates DNA and histone methylation during replication. *Genes Dev* 2006;**20**:3089–103.
63. Fujita N, Watanabe S, Ichimura T, Tsuruzoe S, Shinkai Y, Tachibana M, et al. Methyl-CpG binding domain 1 (MBD1) interacts with the Suv39h1-HP1 heterochromatic complex for DNA methylation-based transcriptional repression. *J Biol Chem* 2003;**278**:24132–8.
64. Vire E, Brenner C, Deplus R, Blanchon L, Fraga M, Didelot C, et al. The Polycomb group protein EZH2 directly controls DNA methylation. *Nature* 2006;**439**:871–4.
65. Ikegami K, Iwatani M, Suzuki M, Tachibana M, Shinkai Y, Tanaka S, et al. Genome-wide and locus-specific DNA hypomethylation in G9a deficient mouse embryonic stem cells. *Genes Cells* 2007;**12**:1–11.
66. Han L, Witmer PD, Casey E, Valle D, Sukumar S. DNA methylation regulates MicroRNA expression. *Cancer Biol Ther* 2007;**6**:1284–8.
67. Migliore C, Petrelli A, Ghiso E, Corso S, Capparuccia L, Eramo A, et al. MicroRNAs impair MET-mediated invasive growth. *Cancer Res* 2008;**68**:10128–36.
68. Balaguer F, Link A, Lozano JJ, Cuatrecasas M, Nagasaka T, Boland CR, et al. Epigenetic silencing of miR-137 is an early event in colorectal carcinogenesis. *Cancer Res* 2010;**70**:6609–18.
69. Kriaucionis S, Heintz N. The nuclear DNA base 5-hydroxymethylcytosine is present in Purkinje neurons and the brain. *Science* 2009;**324**:929–30.
70. Tahiliani M, Koh KP, Shen Y, Pastor WA, Bandukwala H, Brudno Y, et al. Conversion of 5-cytosine to 5-hydroxymethylcytosine in mammalian DNA by MLL partner TET1. *Science* 2009;**324**:930–5.
71. Nestor C, Ruzov A, Meehan R, Dunican D. Enzymatic approaches and bisulfite sequencing cannot distinguish between 5-methylcytosine and 5-hydroxymethylcytosine in DNA. *Biotechniques* 2010;**48**:317–9.
72. Song CX, Szulwach KE, Fu Y, Dai Q, Yi C, Li X, et al. Selective chemical labeling reveals the genome-wide distribution of 5-hydroxymethylcytosine. *Nat Biotechnol* 2011;**29**:68–72.
73. Flusberg BA, Webster DR, Lee JH, Travers KJ, Olivares EC, Clark TA, et al. Direct detection of DNA methylation during single-molecule, real-time sequencing. *Nat Methods* 2010;**7**:461–5.

Index

D

N

O

P

R

S

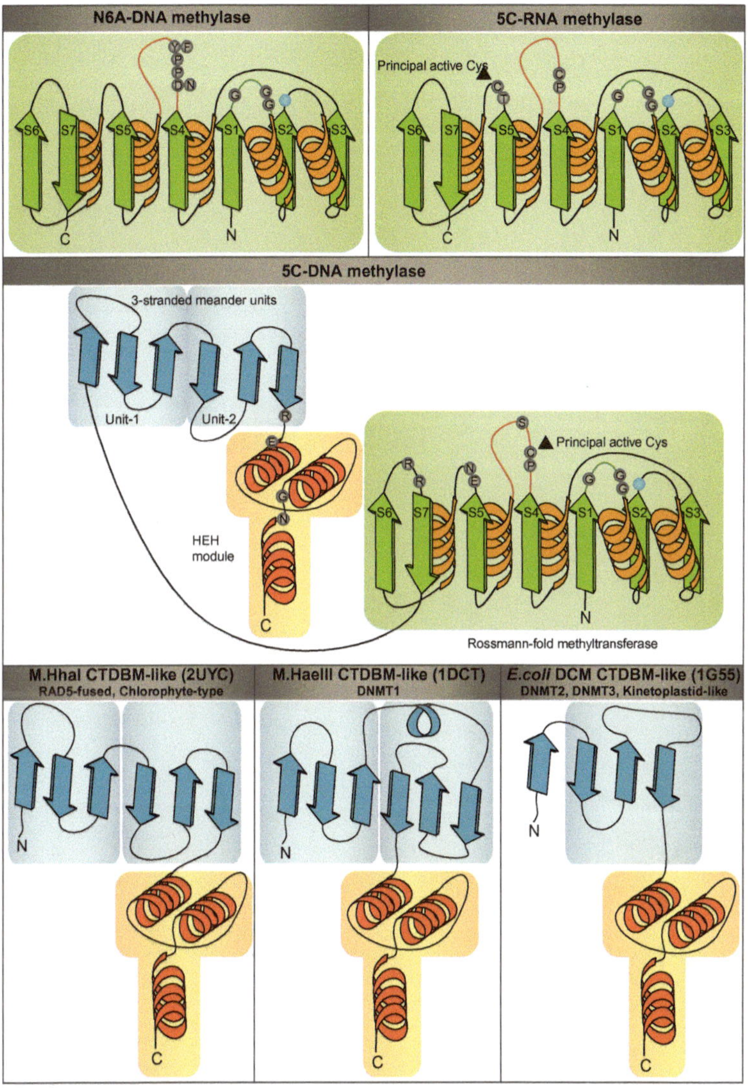

LAKSHMINARAYAN M. IYER *ET AL.*, FIG. 1. Structure and sequence features of DNA and RNA methylases. The methylases, and distinct variants of the DNA 5C-MTase CTDBM, are depicted as cartoon topology diagrams. Strands and helices of the Rossmannoid fold core of the methylases are colored green and orange, whereas those of the CTDBM are colored blue and red, respectively. Strands of the core Rossmannoid fold are labeled S1–S6. Key sequence features described in the text, including those involved in AdoMet binding, catalysis, lineage-specific residues, and residues that are frequently mutated in human DNMT3A in acute myeloid leukemia,[84] are shown in gray circles with the residue abbreviation at the corresponding structural element. The blue circle corresponds to the highly conserved polar position in methylases at the end of strand 2, that H-bonds the AdoMet ribose.

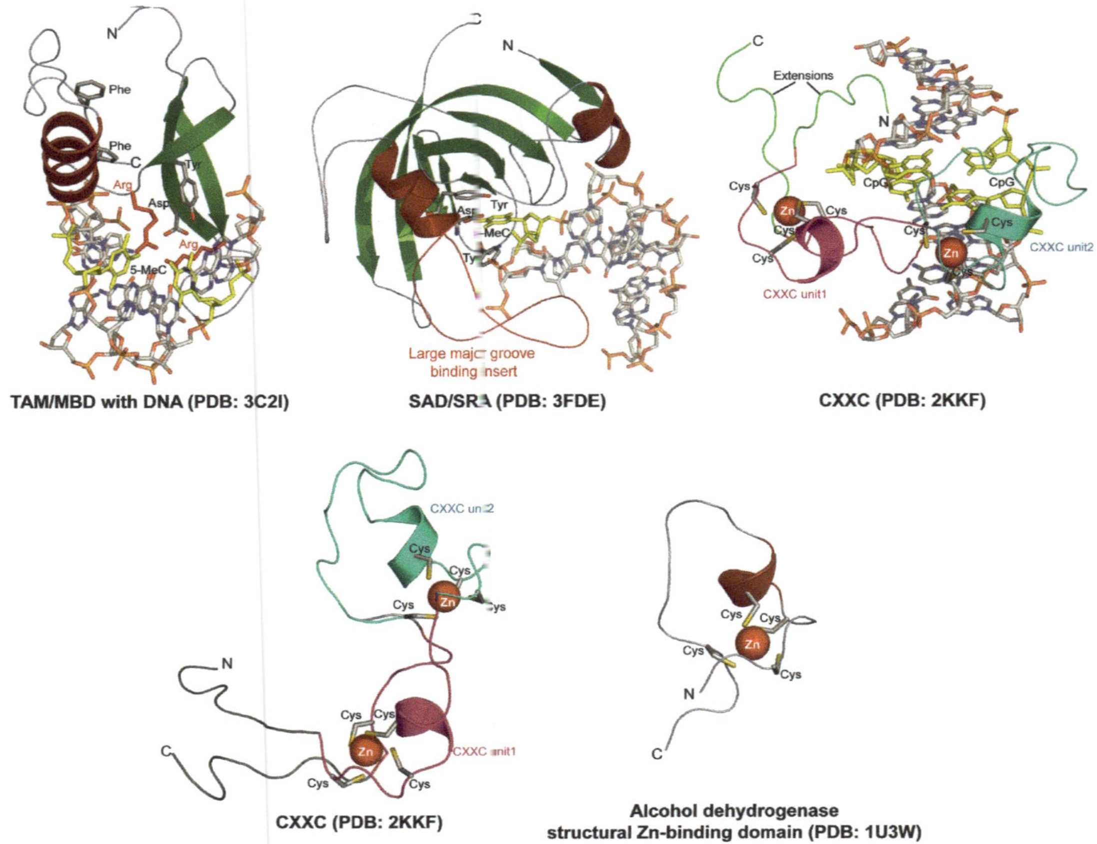

LAKSHMINARAYAN M. IYER *ET AL.*, FIG. 5. DNA methylation-discriminating domains. The top panel illustrates the DNA-recognition mode of the TAM/MBD, SAD/SRA, and CXXC domains. β-Strands are colored green and α-helices brick-red. The two repeat units of the bi-CXXC domain are each shown in magenta and blue, respectively. DNA is shown as a semitransparent stick model with the interacting bases in yellow. Key interacting and zinc-binding residues of the domains are marked. The bottom panel illustrates the duplication in the bi-CXXC domain and its similarity of each unit to the structural zinc-binding domain of medium-chain alcohol dehydrogenases.

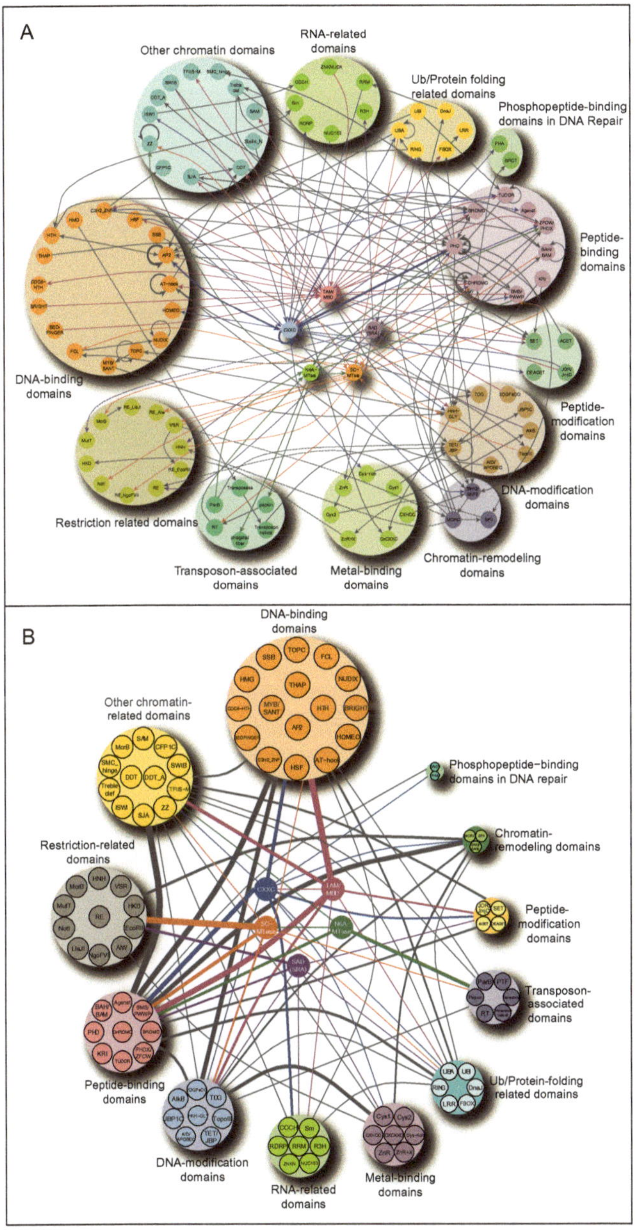

Lakshminarayan M. Iyer *et al.*, Fig. 6. Domain architecture and gene neighborhood network. These are shown as a network graph with nodes representing domains related to DNA methylation, and edges their physical connectivity in a polypeptide or gene neighborhood. The metanetwork is used to highlight the overall trends of associations between different functional types of domains involved in DNA methylation. The arrow heads depict directionality; for domain architectures they point from the N-terminal to the C-terminal domain and for gene neighborhoods from the 5′ gene to the 3′ gene. Gene neighborhood associations are shown as dashed lines. Domains with similar functional roles are in the same color and further grouped into metanodes in the metanetwork. Edges are colored based on the principal domain of an association; 5C MTases: orange, N6A MTase: green, CXXC: blue, TAM/MBD: magenta, and SAD/SRA: purple. Edges not involving these principal domains are colored gray. The edge thickness is proportional to the relative frequency with which linkages between two domains or metanodes reoccur in distinct polypeptides and gene neighborhoods. Conventional abbreviations are used for domain nomenclature. Other domains with nonstandard abbreviations include CFP1C; CFP1 C-terminal domain; ACET, GCN5-like acetyltransferase; AuxHF, a novel version of the chromo-fold predicted to bind methylated histones; Cys1, domain with conserved cysteines associated with fungal TET/JBP-containing transposons; Cys2, a domain with conserved cysteines associated with the AlkB and SAD family of proteins in fungi; Cys-rich, a domain with conserved cysteines inserted in the 2OGFeDO domain of the metazoan TET family; DEACET, RPD3/HDAC-like histone deacetylase; DDT_A, DDT associated domain; RT, reverse transcriptase; RE, restriction endonuclease; and ZnR, zinc ribbon.

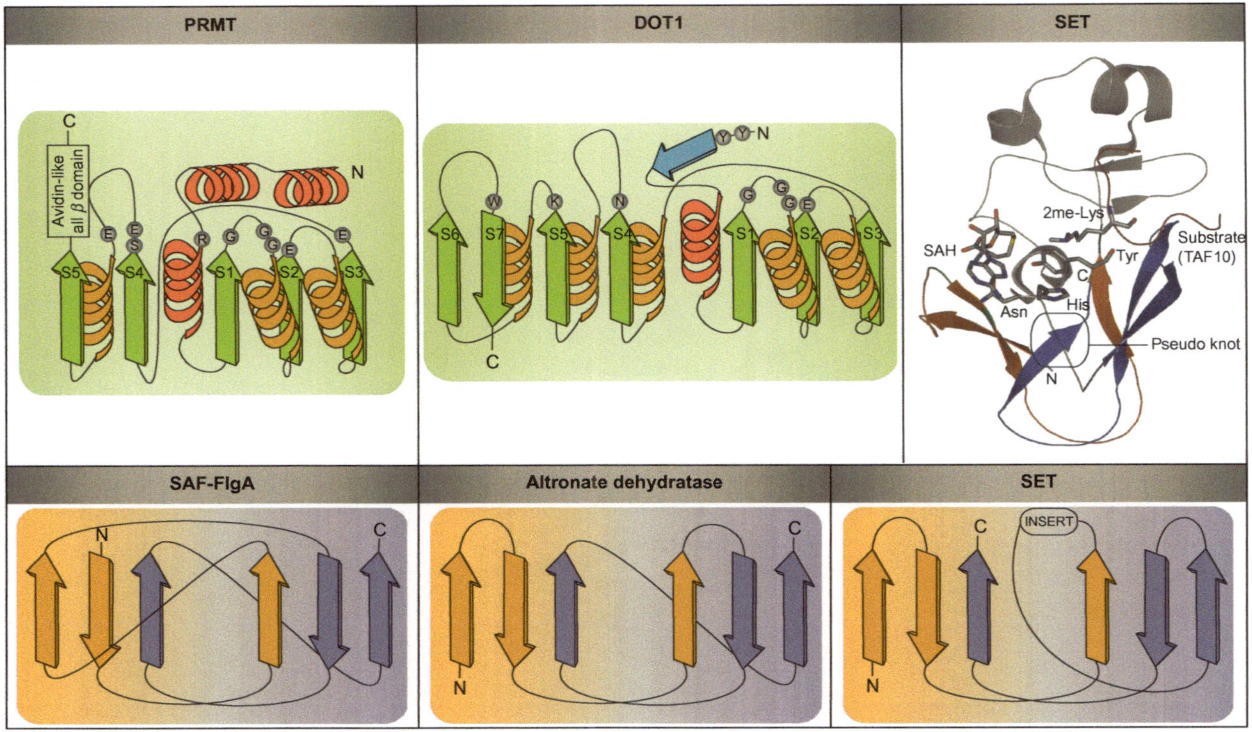

L. ARAVIND *ET AL.*, FIG. 1. Structure and sequence features of protein methyltransferases. The Rossmann fold protein methylases in the PRMT and DOT1 families are depicted as topology diagrams. Strands and helices of the Rossmannoid-fold core of the methylases are colored green and orange. Additional family-specific helices and strands are colored pink and blue, respectively. Strands of the core Rossmannoid fold are labeled S1-S5/S7. Key sequence features described in the text, including those involved in AdoMet-binding and catalysis, are shown in gray circles with the residue abbreviation at the corresponding structural element. The cartoon structure of the SET domain of SETD7 (PDB: 3M59), in complex with its substrate (TAF10) and AdoHcy, illustrates the 3-stranded repeat units within the β-clip fold, the pseudoknot, and key residues involved in AdoMet and substrate-lysine binding, and catalysis. Topology diagrams of the β-clip units of SAF-FlgA, altronate dehydratase, and the SET domain are shown in the bottom panels to illustrate the possible pathway of evolution of a SET-like β-clip unit from a SAF-FlgA-like unit through an altronate dehydratase like intermediate. The N-terminal unit of the β-clip is colored orange and the C-terminal one blue.

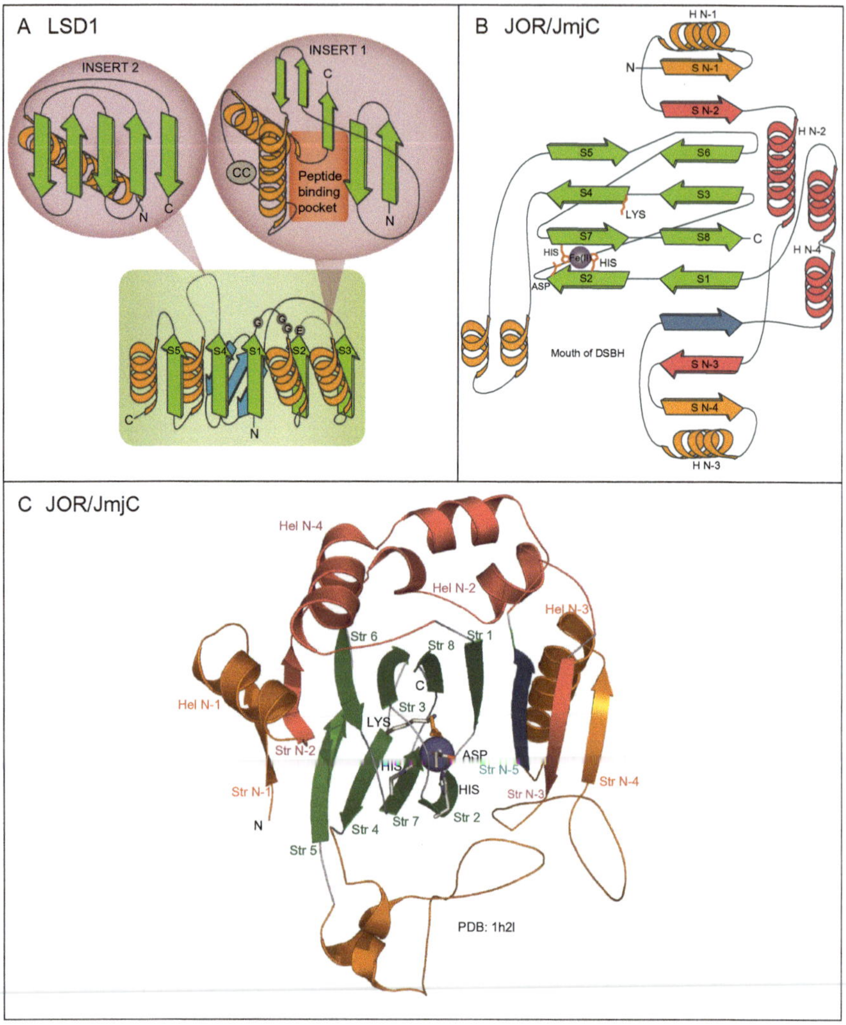

L. Aravind *et al.*, Fig. 4. Structure and sequence features of protein demethylases. (A) Topology diagrams of the LSD1 demethylase. The core secondary structure elements of the Rossmann fold are colored as in Fig. 1. The distinct substrate and flavin-binding domain (insert 1), and the FAD-binding cap domain (insert 2) are shown separately. CC refers to the coiled-coil motif found in insert 1. Shown in (B) and (C) are the topology and cartoon structure diagrams of the JOR/JmjC domain. The coloring scheme illustrates various shared features of the JOR/JmjC superfamily with related DSBH domains. Thus, the conserved strands of the DSBH core are colored green, and the additional strand shared with the metal-binding cupin is colored blue. Secondary structure elements additionally shared with all 2-oxoglutarate Fe^{2+}-dependent dioxygenases are colored pink. Shown in orange are structural elements and inserts that are only present in the JOR/JmjC superfamily. The core of the DSBH unit is labeled strands 1–8. Secondary structure elements N-terminal to the DSBH unit are labeled with a prefix "N" (e.g., Str N-5). The histidine and aspartate residues coordinate a Fe(II), whereas the lysine residue forms a salt bridge with the 2-oxoglutarate.

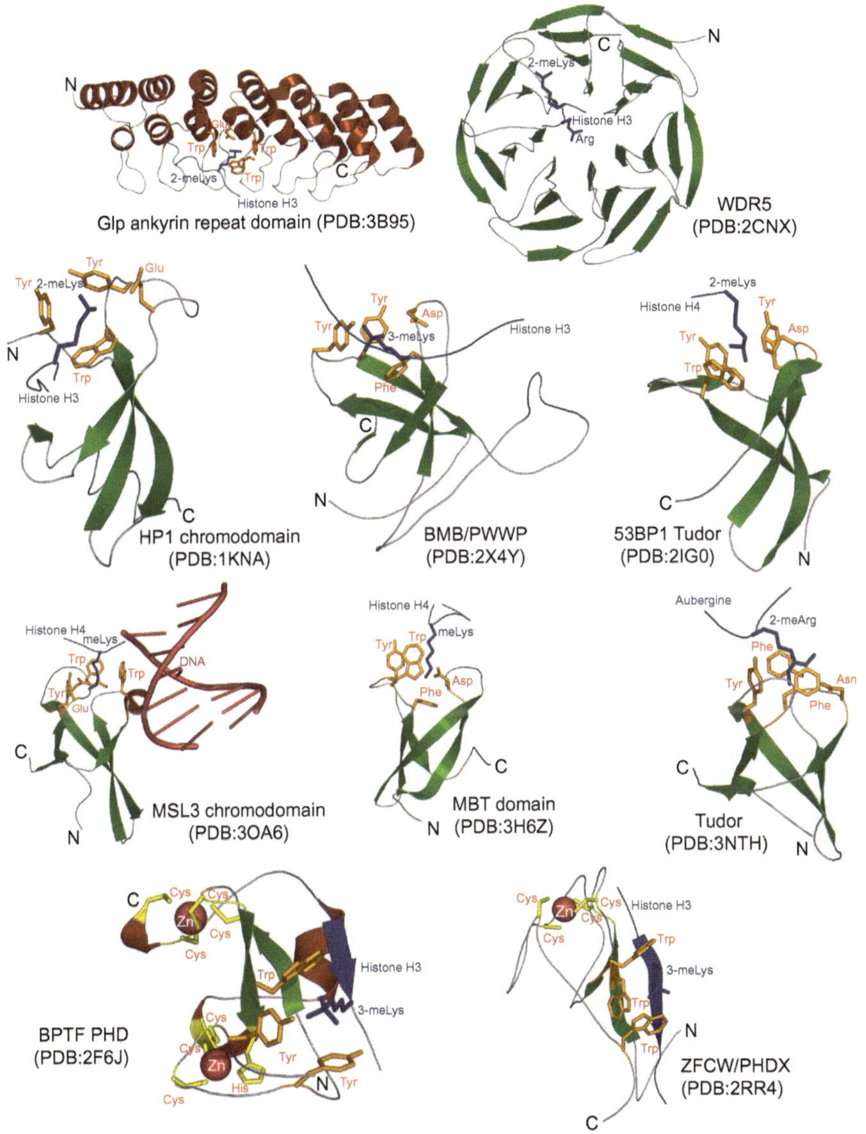

L. ARAVIND *ET AL.*, FIG. 5. Structures of various protein sidechain-methyl binding domains. The structures depict various protein side chain methyl binding domains, their binding modes and substrate contacting residues. Strands are colored green, and helices are shown in brick-red. Substrate polypeptides and amino acid side chains are colored blue and the binding residues are shown as stick models in orange.

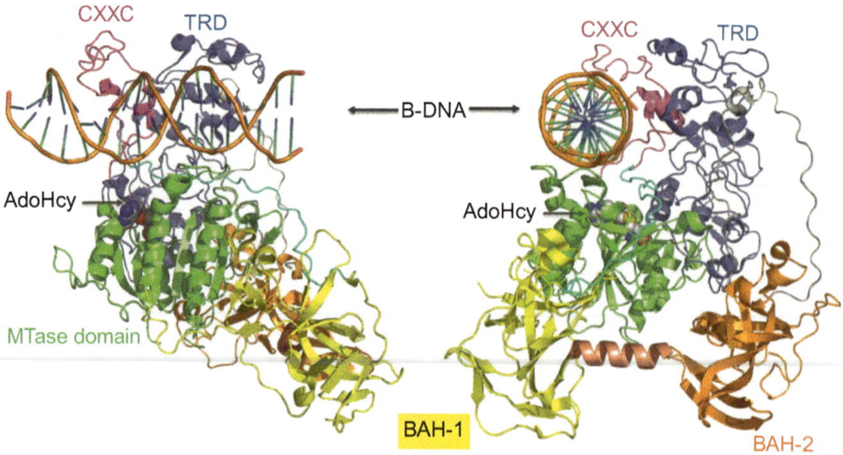

ŽELJKO M. SVEDRUŽIĆ, FIG. 1. Structure of mouse Dnmt1. Two orthogonal views of mouse Dnmt1 (residues 650–1602) in complex with unmethylated DNA substrate and cofactor *S*-adenosyl homocysteine (AdoHcy) (PDB 3PT6).[189] The domain structure from N-to-C terminal are Zn-finger CXXC domain (magenta), BAH-1 (yellow), BAH-2 (orange), methyltransferase (green), and target recognition domain (blue).

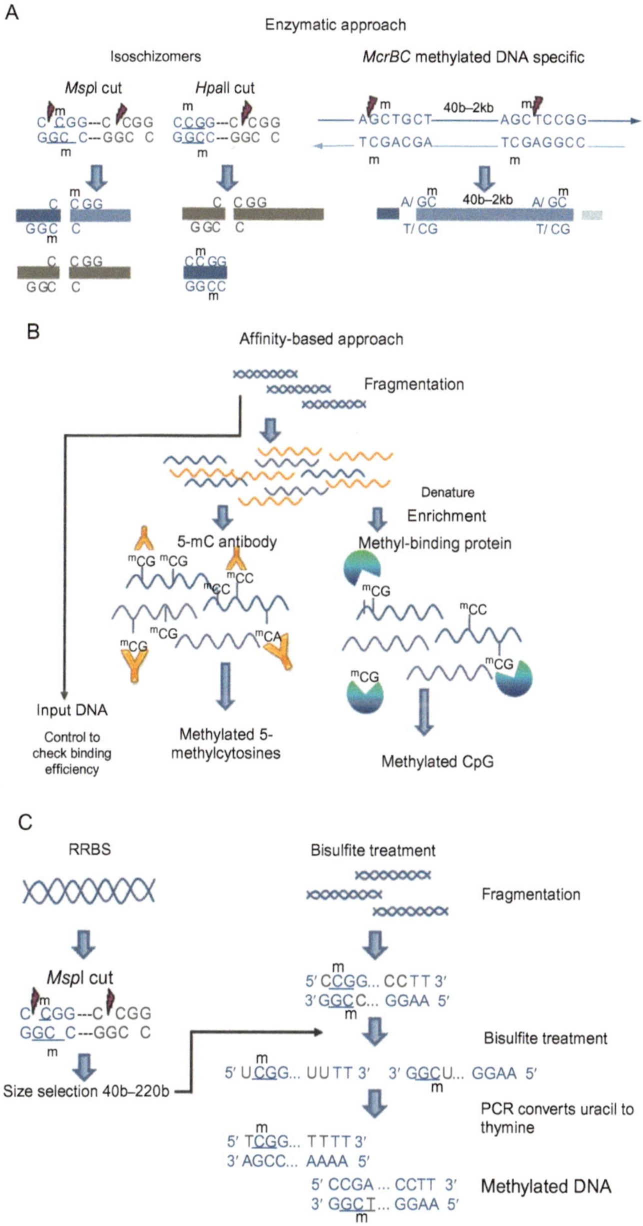

Eleanor Wong and Chia-Lin Wei, Fig. 1. Overview of methylated DNA enrichment labeling techniques. (A) Restriction enzymes *Msp*I and *Hpa*II both cut a site containing a CpG (CCGG), but of the two, only *Hpa*II is blocked by CpG methylation. These enzymes are commonly used in a process called Methyl-seq, to detect methylated DNA. *McrBC* is an endonuclease that cleaves DNA containing methylcytosine in two half-sites of the form (G/A)^{m}C, where the two half-sites can be 40 bp–2 kb apart. *McrBC*-digested DNA can subsequently be ligated with specific sequencing adapters for high-throughput sequencing or hybridized onto arrays for methylated cytosine detection. (B) Methylated cytosines can be enriched using 5mC-specific antibodies or methyl-CpG DNA-binding protein through the processes of MeDIP and MDB-cap, respectively. In MeDIP, genomic DNA is randomly sheared and denatured, followed by enrichment with anti-5mC antibodies and elution using varying salt concentrations to isolate the methylated cytosines. Similarly, in MDB-cap, the genomic DNA is first fragmented and denatured, followed by the binding of methyl-CpG-binding protein to pull down the mCpG DNA fragments. To assess enrichment efficiency, input DNA without any antibody enrichment serves as control DNA. Next, both the input DNA and enriched methylated DNA can be subjected to various detection techniques for further methylation analysis. (C) Randomly sheared genomic DNA is subjected to bisulfite treatment that converts unmethylated cytosines to uracils, leaving methylated cytosines unchanged. Upon PCR amplification, the uracils are replaced by thymidines, and this allows the differentiation between methylated and unmethylated cytosines. In addition, DNA digested by *Msp*I can be further treated with sodium bisulfite in a method called RRBS for whole-genome methylation profiling at single-base resolution.

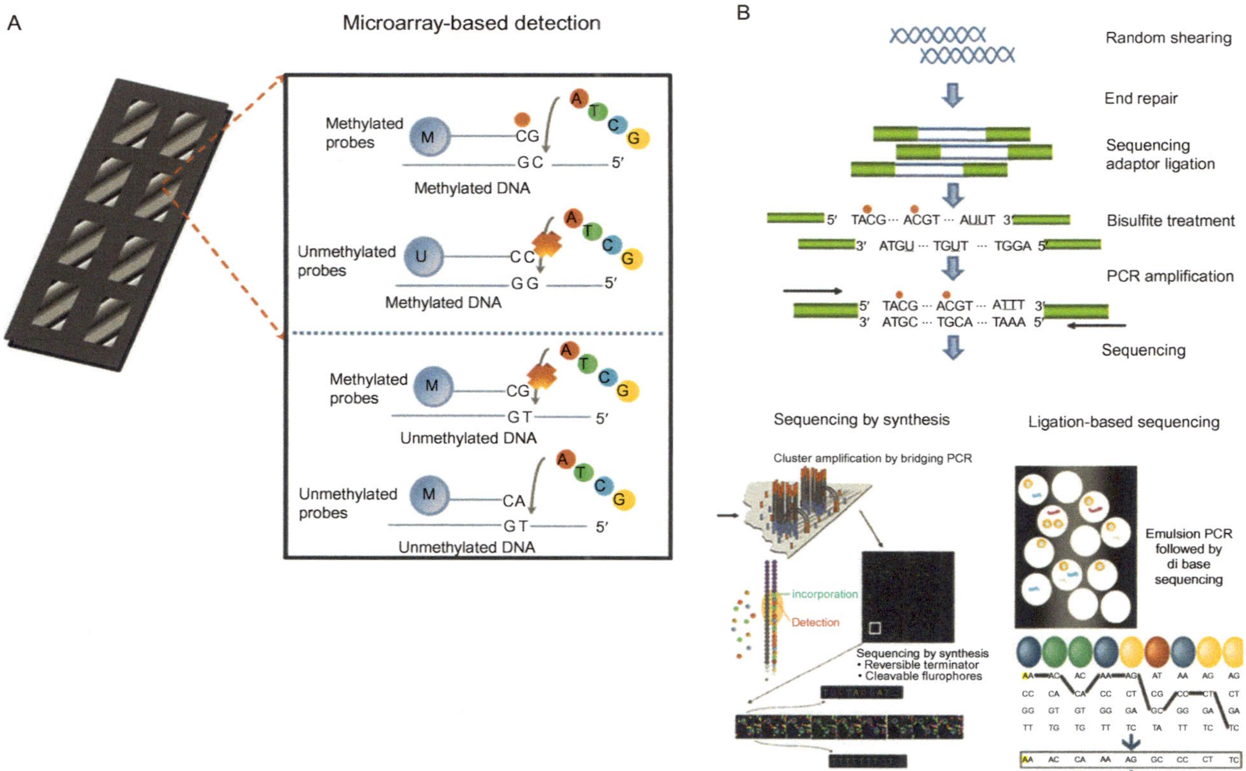

Eleanor Wong and Chia-Lin Wei, Fig. 2. DNA methylation detection methods. (A) Commercially available microarray chips can be used to detect methylated DNA of specific loci or expanded to whole-genome scale. Multiple bisulfite-treated genomic DNA samples can be hybridized onto the customized arrays, which allows the interrogation of up to 450,000 sites in parallel. A pair of methylated and unmethylated probes for each site was designed to perform primer extensions on the bisulfate-treated DNA. Fluorescent ddNTPs are added to the arrays and incorporated at single-base extensions. DNA with methyl C will allow the hybridization of the methylated probes and subsequent base extension, while incorporation of ddNTPs at the unmethylated probes will be terminated, and vice versa in the unmethylated locus. The "methylated" and "unmethylated" probes differ in having C or T at a given position. Methylated DNA is identified from the ratio of fluorescent signals recorded when using the unmethylated and methylated probes. (B) Direct bisulfite sequencing (BS-seq) involves the coupling of bisulfite treatment of DNA with 2nd-generation sequencing technologies, to detect methylated DNA at genome-wide scale at single-nucleotide resolution. Genomic DNA is typically sheared into fragments (~500 bp), end polished, and ligated with specific sequencing adapters. This is followed by the bisulfite treatment and PCR amplification, before subjecting to DNA sequencing by synthesis or ligation-based chemistry to detect the methylcytosines.

FSC
www.fsc.org
MIX
Papier | Fördert gute Waldnutzung
FSC® C083411

Kolibri 360 GmbH
Ferdinand-Jühlke-Straße 7
99095 Erfurt, Deutschland
produktsicherheit@kolibri360.de